# Advances in Pattern Recognition

T0156281

For further volumes:
http://www.springer.com/series/4205

Shigeo Abe

# Support Vector Machines for Pattern Classification

## Second Edition

 Springer

Prof. Dr. Shigeo Abe
Kobe University
Graduate School of Engineering
1-1 Rokkodai-cho
Nada-ku
Kobe
657-8501 Japan
abe@kobe-u.ac.jp

*Series Editor*
Prof. Sameer Singh, PhD
Research School of Informatics
Loughborough University
Loughborough, UK

ISSN 1617-7916
ISBN 978-1-4471-2548-8          e-ISBN 978-1-84996-098-4
DOI 10.1007/978-1-84996-098-4
Springer London Dordrecht Heidelberg New York

British Library Cataloguing in Publication Data
A catalogue record for this book is available from the British Library

Printed on acid-free paper

Springer is part of Springer Science+Business Media (www.springer.com)

# Preface

## Preface to the Second Edition

Since the introduction of support vector machines, we have witnessed the huge development in theory, models, and applications of what is so-called *kernel-based methods*: advancement in generalization theory, kernel classifiers and regressors and their variants, various feature selection and extraction methods, and wide variety of applications such as pattern classification and regressions in biology, medicine, chemistry, as well as computer science.

In *Support Vector Machines for Pattern Classification, Second Edition*, I try to reflect the development of kernel-based methods since 2005. In addition, I have included more intensive performance comparison of classifiers and regressors, added new references, and corrected many errors in the first edition. The major modifications of, and additions to, the first edition are as follows:

Symbols: I have changed the symbols of the mapping function to the feature space from $\mathbf{g}(\mathbf{x})$ to more commonly used $\boldsymbol{\phi}(\mathbf{x})$ and its associated kernel from $H(\mathbf{x}, \mathbf{x}')$ to $K(\mathbf{x}, \mathbf{x}')$.

1.3 Data Sets Used in the Book: I have added publicly available two-class data sets, microarray data sets, multiclass data sets, and regression data sets.

1.4 Classifier Evaluation: Evaluation criteria for classifiers and regressors are discussed.

2.3.2 Kernels: Mahalanobis kernels, graph kernels, etc., are added.

2.3.6 Empirical Feature Space: The high-dimensional feature space is treated implicitly via kernel tricks. This is an advantage and also a disadvantage because we treat the feature space without knowing its structure. The empirical feature space is equivalent to the feature space, in that it gives the same kernel value as that of the feature space. The introduction of the empirical feature space greatly enhances the interpretability and manipulability of the feature space.

2.8.4  Effect of Model Selection by Cross-Validation: In realizing high generalization ability of a support vector machine, selection of kernels and their
parameter values, i.e., model selection, is very important. Here I discuss
how cross-validation, which is one of the most well-used model selection
methods, works to generate a support vector machine with high generalization ability.

3.4  I have deleted the section "Sophisticated Architecture" because it does
not work.

4.3  Sparse Support Vector Machines: Based on the idea of the empirical feature space, sparse support vector machines, which realize smaller numbers
of support vectors than those of support vector machines are discussed.

4.4  Performance Comparison of Different Classifiers: Performance of some
types of support vector machines is compared using benchmark data sets.

4.8  Learning Using Privileged Information: Incorporating prior knowledge
into support vector machines is very useful in improving the generalization
ability. Here, one such approach proposed by Vapnik is explained.

4.9  Semi-supervised Learning: I have explained the difference between semi-
supervised learning and transductive learning.

4.10  Multiple Classifier Systems: Committee machines in the first edition
are renamed and new materials are added.

4.11  Multiple Kernel Learning: A weighted sum of kernels with positive
weights is also a kernel and is called a *multiple kernel*. A learning method
of support vector machines with multiple kernels is discussed.

5.6  Steepest Ascent Methods and Newton's Methods: Steepest ascent methods in the first edition are renamed as Newton's methods and steepest
ascent methods are explained in Section 5.6.1.

5.7  Batch Training by Exact Incremental Training: A batch training method
based on incremental training is added.

5.8  Active Set Training in Primal and Dual: Training methods in the primal
or dual form by variable-size chunking are added.

5.9  Training of Linear Programming Support Vector Machines: Three decomposition techniques for linear programming support vector machines
are discussed.

6  Kernel-Based Methods: Chapter 8 Kernel-Based Methods in the first edition is placed just after Chapter 5 Training methods and kernel discriminant analysis is added.

11.5.3  Active Set Training: Active set training discussed in Section 5.8 is
extended to function approximation.

11.7  Variable Selection: Variable selection for support vector regressors is
added.

# Preface to the First Edition

I was shocked to see a student's report on performance comparisons between support vector machines (SVMs) and fuzzy classifiers that we had developed with our best endeavors. Classification performance of our fuzzy classifiers was comparable, but in most cases inferior, to that of support vector machines. This tendency was especially evident when the numbers of class data were small. I shifted my research efforts from developing fuzzy classifiers with high generalization ability to developing support vector machine-based classifiers.

This book focuses on the application of support vector machines to pattern classification. Specifically, we discuss the properties of support vector machines that are useful for pattern classification applications, several multi-class models, and variants of support vector machines. To clarify their applicability to real-world problems, we compare the performance of most models discussed in the book using real-world benchmark data. Readers interested in the theoretical aspect of support vector machines should refer to books such as [1–4].

Three-layer neural networks are universal classifiers in that they can classify any labeled data correctly if there are no identical data in different classes [5, 6]. In training multilayer neural network classifiers, network weights are usually corrected so that the sum-of-squares error between the network outputs and the desired outputs is minimized. But because the decision boundaries between classes acquired by training are not directly determined, classification performance for the unknown data, i.e., the generalization ability, depends on the training method. And it degrades greatly when the number of training data is small and there is no class overlap.

On the other hand, in training support vector machines the decision boundaries are determined directly from the training data so that the separating margins of decision boundaries are maximized in the high-dimensional space called *feature space*. This learning strategy, based on statistical learning theory developed by Vapnik [1, 2], minimizes the classification errors of the training data and the unknown data.

Therefore, the generalization abilities of support vector machines and other classifiers differ significantly, especially when the number of training data is small. This means that if some mechanism to maximize the margins of decision boundaries is introduced to non-SVM-type classifiers, their performance degradation will be prevented when the class overlap is scarce or nonexistent.[1]

In the original support vector machine, an $n$-class classification problem is converted into $n$ two-class problems, and in the $i$th two-class problem we determine the optimal decision function that separates class $i$ from the remaining classes. In classification, if one of the $n$ decision functions classifies

---

[1] To improve generalization ability of a classifier, a regularization term, which controls the complexity of the classifier, is added to the objective function.

an unknown data sample into a definite class, it is classified into that class. In this formulation, if more than one decision function classify a data sample into definite classes or if no decision functions classify the data sample into a definite class, the data sample is unclassifiable.

Another problem of support vector machines is slow training. Because support vector machines are trained by solving a quadratic programming problem with the number of variables equal to the number of training data, training is slow for a large number of training data.

To resolve unclassifiable regions for multiclass support vector machines we propose fuzzy support vector machines and decision-tree-based support vector machines.

To accelerate training, in this book, we discuss two approaches: selection of important data for training support vector machines before training and training by decomposing the optimization problem into two subproblems.

To improve generalization ability of non-SVM-type classifiers, we introduce the ideas of support vector machines to the classifiers: neural network training incorporating maximizing margins and a kernel version of a fuzzy classifier with ellipsoidal regions [6, pp. 90–93, 119–139].

In Chapter 1, we discuss two types of decision functions: direct decision functions, in which the class boundary is given by the curve where the decision function vanishes, and the indirect decision function, in which the class boundary is given by the curve where two decision functions take on the same value.

In Chapter 2, we discuss the architecture of support vector machines for two-class classification problems. First we explain hard-margin support vector machines, which are used when the classification problem is linearly separable, namely, the training data of two classes are separated by a single hyperplane. Then, introducing slack variables for the training data, we extend hard-margin support vector machines so that they are applicable to inseparable problems. There are two types of support vector machines: L1 soft-margin support vector machines and L2 soft-margin support vector machines. Here, L1 and L2 denote the linear sum and the square sum of the slack variables that are added to the objective function for training. Then we investigate the characteristics of solutions extensively and survey several techniques for estimating the generalization ability of support vector machines.

In Chapter 3, we discuss some methods for multiclass problems: one-against-all support vector machines, in which each class is separated from the remaining classes; pairwise support vector machines, in which one class is separated from another class; the use of error-correcting output codes for resolving unclassifiable regions; and all-at-once support vector machines, in which decision functions for all the classes are determined at once. To resolve unclassifiable regions, in addition to error-correcting codes, we discuss fuzzy support vector machines with membership functions and decision-tree-based support vector machines. To compare several methods for multiclass problems, we show performance evaluation of these methods for the benchmark data sets.

Since support vector machines were proposed, many variants of support vector machines have been developed. In Chapter 4, we discuss some of them: least-squares support vector machines whose training results in solving a set of linear equations, linear programming support vector machines, robust support vector machines, and so on.

In Chapter 5, we discuss some training methods for support vector machines. Because we need to solve a quadratic optimization problem with the number of variables equal to the number of training data, it is impractical to solve a problem with a huge number of training data. For example, for 10,000 training data, 800 MB memory is necessary to store the Hessian matrix in double precision. Therefore, several methods have been developed to speed training. One approach reduces the number of training data by preselecting the training data. The other is to speed training by decomposing the problem into two subproblems and repeatedly solving the one subproblem while fixing the other and exchanging the variables between the two subproblems.

Optimal selection of features is important in realizing high-performance classification systems. Because support vector machines are trained so that the margins are maximized, they are said to be robust for nonoptimal features. In Chapter 7, we discuss several methods for selecting optimal features and show, using some benchmark data sets, that feature selection is important even for support vector machines. Then we discuss feature extraction that transforms input features by linear and nonlinear transformation.

Some classifiers need clustering of training data before training. But support vector machines do not require clustering because mapping into a feature space results in clustering in the input space. In Chapter 8, we discuss how we can realize support vector machine-based clustering.

One of the features of support vector machines is that by mapping the input space into the feature space, nonlinear separation of class data is realized. Thus the conventional linear models become nonlinear if the linear models are formulated in the feature space. They are usually called *kernel-based methods*. In Chapter 6, we discuss typical kernel-based methods: kernel least squares, kernel principal component analysis, and the kernel Mahalanobis distance.

The concept of maximum margins can be used for conventional classifiers to enhance generalization ability. In Chapter 9, we discuss methods for maximizing margins of multilayer neural networks and in Chapter 10 we discuss maximum-margin fuzzy classifiers with ellipsoidal regions and polyhedral regions.

Support vector machines can be applied to function approximation. In Chapter 11, we discuss how to extend support vector machines to function approximation and compare the performance of the support vector machine with that of other function approximators.

# Acknowledgments

I am grateful to those who are involved in the research project, conducted at the Graduate School of Engineering, Kobe University, on neural, fuzzy, and support vector machine-based classifiers and function approximators, for their efforts in developing new methods and programs. Discussions with Dr. Seiichi Ozawa were always helpful. Special thanks are due to them and current graduate and undergraduate students: T. Inoue, K. Sakaguchi, T. Takigawa, F. Takahashi, Y. Hirokawa, T. Nishikawa, K. Kaieda, Y. Koshiba, D. Tsujinishi, Y. Miyamoto, S. Katagiri, T. Yamasaki, T. Kikuchi, K. Morikawa, Y. Kamata, M. Ashihara, Y. Torii, N. Matsui, Y. Hosokawa, T. Nagatani, K. Morikawa, T. Ishii, K. Iwamura, T. Kitamura, S. Muraoka, S. Takeuchi, Y. Tajiri, and R. Yabuwaki; and a then PhD student T. Ban.

I thank A. Ralescu for having used my draft version of the book as a graduate course text and having given me many useful comments. Thanks are also due to V. Kecman, H. Nakayama, S. Miyamoto, J. A. K. Suykens, F. Anouar, G. C. Cawley, H. Motoda, A. Inoue, F. Schwenker, N. Kasabov, B.-L. Lu, and J. T. Dearmon for their valuable discussions and useful comments.

This book includes my students' and my papers published in journals and proceedings of international conferences. I must thank many anonymous reviewers who reviewed our papers for their constructive comments and pointing out errors and missing references in improving the papers and consequently my book.

The Internet was a valuable source of information in writing the book. In preparing the second edition, I extensively used SCOPUS (https://www.scopus.com/home.url) to check the well-cited papers and tried to include those papers that are relevant to my book. Most of the papers listed in References were obtained from the Internet, from either authors' home pages or free downloadable sites such as

ESANN: http://www.dice.ucl.ac.be/esann/proceedings
/electronicproceedings.htm
JMLR: http://jmlr.csail.mit.edu/papers/

CiteSeer: http://citeseer.ist.psu.edu/
NIPS: http://books.nips.cc/

I enjoyed reading papers with innovating ideas. In understanding and including the papers into my book, they were especially helpful if they were well written in that in the abstract and the introduction their proposed methods were explained not by their characteristics but by their approaches and ideas and in the main text the ideas behind the algorithms were explained before their detailed explanation.

Kobe, October 2004, October 2009                         *Shigeo Abe*

# References

1. V. N. Vapnik. *The Nature of Statistical Learning Theory.* Springer-Verlag, New York, 1995.
2. V. N. Vapnik. *Statistical Learning Theory.* John Wiley & Sons, New York, 1998.
3. R. Herbrich. *Learning Kernel Classifiers: Theory and Algorithms.* MIT Press, Cambridge, MA, 2002.
4. B. Schölkopf and A. J. Smola. *Learning with Kernels: Support Vector Machines, Regularization, Optimization, and Beyond.* MIT Press, Cambridge, MA, 2002.
5. S. Young and T. Downs. CARVE—A constructive algorithm for real-valued examples. *IEEE Transactions on Neural Networks*, 9(6):1180–1190, 1998.
6. S. Abe. *Pattern Classification: Neuro-Fuzzy Methods and Their Comparison.* Springer-Verlag, London, UK, 2001.

# Contents

# Symbols

We use lowercase bold letters to denote vectors and uppercase italic letters to denote matrices. The following list shows the symbols used in the book:

| | |
|---|---|
| $\alpha_i$ | Lagrange multiplier for $\mathbf{x}_i$ |
| $\xi_i$ | slack variable associated with $\mathbf{x}_i$ |
| $A^{-1}$ | inverse of matrix $A$ |
| $A^\top$ | transpose of matrix $A$ |
| $B$ | set of bounded support vector indices |
| $b_i$ | bias term of the $i$th hyperplane |
| $C$ | margin parameter |
| $d$ | degree of a polynomial kernel |
| $\phi(\mathbf{x})$ | mapping function from $\mathbf{x}$ to the feature space |
| $\gamma$ | parameter for a radial basis function kernel |
| $K(\mathbf{x}, \mathbf{x}')$ | kernel |
| $l$ | dimension of the feature space |
| $M$ | number of training data |
| $m$ | number of input variables |
| $n$ | number of classes |
| $S$ | set of support vector indices |
| $U$ | set of unbounded support vector indices |
| $\|\mathbf{x}\|$ | Euclidean norm of vector $\mathbf{x}$ |
| $\mathbf{w}_i$ | coefficient vector of the $i$th hyperplane |
| $X_i$ | set for class $i$ training data |
| $|X_i|$ | number of data in the set $X_i$ |
| $\mathbf{x}_i$ | $i$th $m$-dimensional training data |
| $y_i$ | class label 1 or $-1$ for input $\mathbf{x}_i$ for pattern classification and a scalar output for function approximation |

# Chapter 1
# Introduction

Support vector machines and their variants and extensions, often called *kernel-based methods* (or simply kernel methods), have been studied extensively and applied to various pattern classification and function approximation problems. Pattern classification is to classify some object into one of the given categories called *classes*. For a specific pattern classification problem, a classifier, which is computer software, is developed so that objects are classified correctly with reasonably good accuracy. Inputs to the classifier are called *features*, because they are determined so that they represent each class well or so that data belonging to different classes are well separated in the input space.

In general there are two approaches to develop classifiers: a parametric approach [1], in which a priori knowledge of data distributions is assumed, and a nonparametric approach, in which no a priori knowledge is assumed.

Neural networks [2–4], fuzzy systems [5–7], and support vector machines [8, 9] are typical nonparametric classifiers. Through training using input–output pairs, classifiers acquire decision functions that classify an input into one of the given classes.

In this chapter we first classify decision functions for a two-class problem into direct and indirect decision functions. The class boundary given by a direct decision function corresponds to the curve where the function vanishes, while the class boundary given by two indirect decision functions corresponds to the curve where the two functions give the same values. Then we discuss how to define and determine the direct decision functions for multiclass problems and list up benchmark data sets used in the book. Finally we discuss some measures to evaluate performance of classifiers and function approximators for a given data set.

S. Abe, *Support Vector Machines for Pattern Classification*,
Advances in Pattern Recognition, DOI 10.1007/978-1-84996-098-4_1,
© Springer-Verlag London Limited 2010

## 1.1 Decision Functions

### 1.1.1 Decision Functions for Two-Class Problems

Consider classifying an $m$-dimensional vector $\mathbf{x} = (x_1, \ldots, x_m)^\top$ into one of two classes. Suppose that we are given scalar functions $g_1(\mathbf{x})$ and $g_2(\mathbf{x})$ for Classes 1 and 2, respectively, and we classify $\mathbf{x}$ into Class 1 if

$$g_1(\mathbf{x}) > 0, \quad g_2(\mathbf{x}) < 0, \tag{1.1}$$

and into Class 2 if

$$g_1(\mathbf{x}) < 0, \quad g_2(\mathbf{x}) > 0. \tag{1.2}$$

We call these functions *decision functions*. By the preceding decision functions, if for $\mathbf{x}$

$$g_1(\mathbf{x}) \, g_2(\mathbf{x}) > 0 \tag{1.3}$$

is satisfied, $\mathbf{x}$ is not classifiable (see the hatched regions in Fig. 1.1; the arrows show the positive sides of the functions).

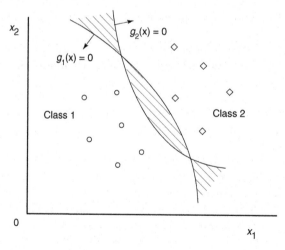

**Fig. 1.1** Decision functions in a two-dimensional space

To resolve unclassifiable regions, we may change (1.1) and (1.2) as follows. We classify $\mathbf{x}$ into Class 1 if

$$g_1(\mathbf{x}) > g_2(\mathbf{x}) \tag{1.4}$$

and into Class 2 if

$$g_1(\mathbf{x}) < g_2(\mathbf{x}). \tag{1.5}$$

In this case, the class boundary is given by (see the dotted curve in Fig. 1.2)

$$g_1(\mathbf{x}) = g_2(\mathbf{x}). \tag{1.6}$$

This means that the class boundary is indirectly obtained by solving (1.6) for $\mathbf{x}$. We call this type of decision function an *indirect decision function*.

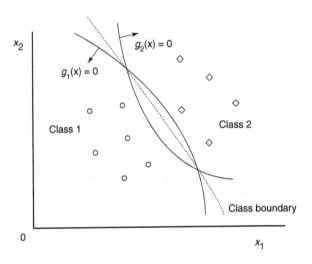

**Fig. 1.2** Class boundary for Fig. 1.1

If we define the decision functions by

$$g_1(\mathbf{x}) = -g_2(\mathbf{x}), \tag{1.7}$$

we classify $\mathbf{x}$ into Class 1 if

$$g_1(\mathbf{x}) > 0 \tag{1.8}$$

and into Class 2 if

$$g_2(\mathbf{x}) > 0. \tag{1.9}$$

Thus the class boundary is given by

$$g_1(\mathbf{x}) = -g_2(\mathbf{x}) = 0. \tag{1.10}$$

Namely, the class boundary corresponds to the curve where the decision function vanishes. We call this type of decision function a *direct decision function*.

If the decision function is linear, namely, $g_1(\mathbf{x})$ is given by

$$g_1(\mathbf{x}) = \mathbf{w}^\top \mathbf{x} + b, \tag{1.11}$$

where $\mathbf{w}$ is an $m$-dimensional vector and $b$ is a bias term, and if one class is on the positive side of the hyperplane, i.e., $g_1(\mathbf{x}) > 0$, and the other class is on the negative side, the given problem is said to be *linearly separable*.

## 1.1.2 Decision Functions for Multiclass Problems

### 1.1.2.1 Indirect Decision Functions

For an $n(> 2)$-class problem, suppose we have indirect decision functions $g_i(\mathbf{x})$ for classes $i$. To avoid unclassifiable regions, we classify $\mathbf{x}$ into class $j$ given by

$$j = \arg \max_{i=1,\ldots,n} g_i(\mathbf{x}), \tag{1.12}$$

where arg returns the subscript with the maximum value of $g_i(\mathbf{x})$. If more than one decision function take the same maximum value for $\mathbf{x}$, namely, $\mathbf{x}$ is on the class boundary, it is not classifiable.

In the following we discuss several methods to obtain the direct decision functions for multiclass problems.

### 1.1.2.2 One-Against-All Formulation

The first approach is to determine the decision functions by the one-against-all formulation [10]. We determine the $i$th decision function $g_i(\mathbf{x})$ $(i = 1,\ldots,n)$, so that when $\mathbf{x}$ belongs to class $i$,

$$g_i(\mathbf{x}) > 0, \tag{1.13}$$

and when $\mathbf{x}$ belongs to one of the remaining classes,

$$g_i(\mathbf{x}) < 0. \tag{1.14}$$

When $\mathbf{x}$ is given, we classify $\mathbf{x}$ into class $i$ if $g_i(\mathbf{x}) > 0$ and $g_j(\mathbf{x}) < 0$ $(j \neq i, j = 1,\ldots,n)$. But by these decision functions, unclassifiable regions exist when more than one decision function are positive or no decision functions are positive, as seen from Fig. 1.3. To resolve these unclassifiable regions we introduce membership functions in Chapter 3.

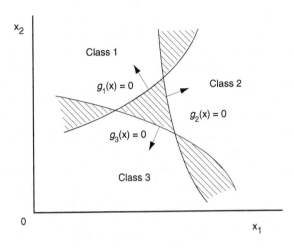

**Fig. 1.3** Class boundaries by one-against-all formulation

### 1.1.2.3 Decision-Tree Formulation

The second approach is based on a decision tree. It is considered to be a variant of one-against-all formulation. We determine the $i$th decision function $g_i(\mathbf{x})\,(i = 1, \ldots, n-1)$, so that when $\mathbf{x}$ belongs to class $i$,

$$g_i(\mathbf{x}) > 0, \tag{1.15}$$

and when $\mathbf{x}$ belongs to one of the classes $\{i+1, \ldots, n\}$,

$$g_i(\mathbf{x}) < 0. \tag{1.16}$$

In classifying $\mathbf{x}$, starting from $g_1(\mathbf{x})$, we find the first positive $g_i(\mathbf{x})$ and classify $\mathbf{x}$ into class $i$. If there is no such $i$ among $g_i(\mathbf{x})\,(i = 1, \ldots, n-1)$, we classify $\mathbf{x}$ into class $n$.

Figure 1.4 shows an example of decision functions for four classes. The decision functions change if we determine decision functions in descending order or in an arbitrary order of class labels. Therefore, in this architecture, we need to determine the decision functions so that classification performance in the upper level of the tree is more accurate than in the lower one. Otherwise, the classification performance may not be good.

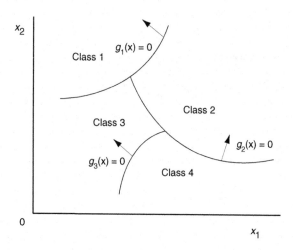

**Fig. 1.4** Decision-tree-based decision functions

### 1.1.2.4 Pairwise Formulation

The third approach is to determine the decision functions by pairwise formulation [11]. For classes $i$ and $j$ we determine the decision function $g_{ij}(\mathbf{x})$ ($i \neq j, i, j = 1, \ldots, n$), so that

$$g_{ij}(\mathbf{x}) > 0 \tag{1.17}$$

when $\mathbf{x}$ belongs to class $i$ and

$$g_{ij}(\mathbf{x}) < 0 \tag{1.18}$$

when $\mathbf{x}$ belongs to class $j$.

In this formulation, $g_{ij}(\mathbf{x}) = -g_{ji}(\mathbf{x})$, and we need to determine $n(n-1)/2$ decision functions. Classification is done by voting, namely, we calculate

$$g_i(\mathbf{x}) = \sum_{j \neq i, j=1}^{n} \text{sign}(g_{ij}(\mathbf{x})), \tag{1.19}$$

where

$$\text{sign}(x) = \begin{cases} 1 & x \geq 0, \\ -1 & x < 0, \end{cases} \tag{1.20}$$

and we classify $\mathbf{x}$ into the class with the maximum $g_i(\mathbf{x})$.[1] By this formulation also, unclassifiable regions exist if $g_i(\mathbf{x})$ take the maximum value for

---

[1] We may define the sign function by

$$\text{sign}(x) = \begin{cases} 1 & x > 0, \\ 0 & x = 0, \\ -1 & x < 0. \end{cases}$$

more than one class (see the hatched region in Fig. 1.5). These can be resolved by decision-tree formulation or by introducing membership functions
as discussed in Chapter 3.

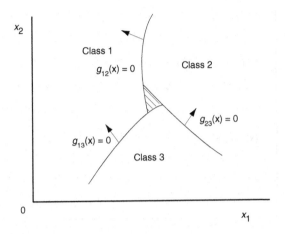

**Fig. 1.5** Class boundaries by pairwise formulation

### 1.1.2.5 Error-Correcting Output Codes

The fourth approach is to use error-correcting codes for encoding outputs [12].
One-against-all formulation is a special case of error-correcting code with no
error-correcting capability, and so is pairwise formulation, as discussed in
Chapter 3, if "don't" care bits are introduced.

### 1.1.2.6 All-at-Once Formulation

The fifth approach is to determine decision functions all at once, namely, we
determine the decision functions $g_i(\mathbf{x})$ by

$$g_i(\mathbf{x}) > g_j(\mathbf{x}) \quad \text{for } j \neq i, \, j = 1, \ldots, n. \tag{1.21}$$

In this formulation we need to determine $n$ decision functions at all once [13,
pp. 174–176], [8, pp. 437–440], [14–20]. This results in solving a problem with
a larger number of variables than the previous methods.

An example of class boundaries is shown in Fig. 1.6. Unlike one-against-all and pairwise formulations, there is no unclassifiable region.

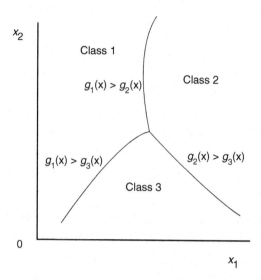

**Fig. 1.6** Class boundaries by all-at-once formulation

## 1.2 Determination of Decision Functions

Determination of decision functions using input–output pairs is called *training*. In training a multilayer neural network for a two-class problem, we can determine a direct decision function if we set one output neuron instead of two. But because for an $n$-class problem we set $n$ output neurons with the $i$th neuron corresponding to the class $i$ decision function, the obtained functions are indirect. Similarly, decision functions for fuzzy classifiers are indirect because membership functions are defined for each class.

Conventional training methods determine the indirect decision functions so that each training input is correctly classified into the class designated by the associated training output. Figure 1.7 shows an example of the decision functions obtained when the training data of two classes do not overlap. Assuming that the circles and squares are training data for Classes 1 and 2, respectively, even if the decision function $g_2(\mathbf{x})$ moves to the right as shown in the dotted curve, the training data are still correctly classified. Thus there are infinite possibilities of the positions of the decision functions that correctly classify the training data. Although the generalization ability is directly affected by the positions, conventional training methods do not consider this.

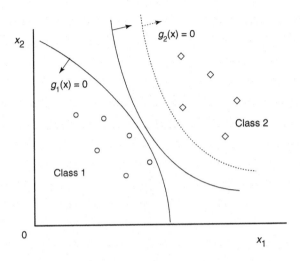

**Fig. 1.7** Class boundary when classes do not overlap

In a support vector machine, the direct decision function that maximizes the generalization ability is determined for a two-class problem. Assuming that the training data of different classes do not overlap, the decision function is determined so that the distance from the training data is maximized. We call this the *optimal decision function*. Because it is difficult to determine a nonlinear decision function, the original input space is mapped into a high-dimensional space called *feature space*. And in the feature space, the optimal decision function, namely, the optimal hyperplane is determined.

Support vector machines outperform conventional classifiers, especially when the number of training data is small and the number of input variables is large. This is because the conventional classifiers do not have the mechanism to maximize the margins of class boundaries. Therefore, if we introduce some mechanism to maximize margins, the generalization ability is improved.

## 1.3 Data Sets Used in the Book

In this book we evaluate methods for pattern classification and function approximation using some benchmark data sets so that advantages and disadvantages of these methods are clarified. In the following we explain these data sets.

Table 1.1 lists the data sets for two-class classification problems [21–23] For each problem the table lists the numbers of inputs, training data, test data, and data sets. Each problem has 100 or 20 training data sets and their corresponding test data sets and is used to compare statistical differences among some classifiers.

**Table 1.1** Benchmark data sets for two-class problems

| Data | Inputs | Training data | Test data | Sets |
|---|---|---|---|---|
| Banana | 2 | 400 | 4,900 | 100 |
| Breast cancer | 9 | 200 | 77 | 100 |
| Diabetes | 8 | 468 | 300 | 100 |
| Flare-solar | 9 | 666 | 400 | 100 |
| German | 20 | 700 | 300 | 100 |
| Heart | 13 | 170 | 100 | 100 |
| Image | 18 | 1,300 | 1,010 | 20 |
| Ringnorm | 20 | 400 | 7,000 | 100 |
| Splice | 60 | 1,000 | 2,175 | 20 |
| Thyroid | 5 | 140 | 75 | 100 |
| Titanic | 3 | 150 | 2,051 | 100 |
| Twonorm | 20 | 400 | 7,000 | 100 |
| Waveform | 21 | 400 | 4,600 | 100 |

Pattern classification technology has been applied to DNA microarray data, which provide expression levels of thousands of genes, to classify cancerous/non-cancerous patients. Microarray data are characterized by a large number of input variables but a small number of training/test data. Thus the classification problems are linearly separable and overfitting occurs quite easily. Therefore, usually, feature selection or extraction is performed to improve generalization ability. Table 1.2 lists the data sets [24] used in this book. For each problem there is one training data set and one test data set.

Table 1.3 shows the data sets for multiclass problems. Each problem has one training data set and the associated test data set.

The Fisher iris data [32, 33] are widely used for evaluating classification performance of classifiers. They consist of 150 data with four features and three classes; there are 50 data per class. We used the first 25 data of each class as the training data and the remaining 25 data of each class as the test data.

The numeral data [34] were collected to identify Japanese license plates of running cars. They include numerals, hiragana, and kanji characters. The original image taken from a TV camera was preprocessed and each numeral was transformed into 12 features, such as the number of holes and the curvature of a numeral at some point.

**Table 1.2** Benchmark data sets for microarray problems

| Data | Inputs | Training data | Test data | Classes |
|------|--------|---------------|-----------|---------|
| Breast cancer (1) [25] | 3,226 | 14 | 8 | 2 |
| Breast cancer (2) [25] | 3,226 | 14 | 8 | 2 |
| Breast cancer (3) [26] | 24,188 | 78 | 19 | 2 |
| Breast cancer (s) [25] | 3,226 | 14 | 8 | 2 |
| Colon cancer [27] | 2,000 | 40 | 20 | 2 |
| Hepatocellular carcinoma [28] | 7,129 | 33 | 27 | 2 |
| High-grade glioma [29] | 12,625 | 21 | 29 | 2 |
| Leukemia [30] | 7,129 | 38 | 34 | 2 |
| Prostate cancer [31] | 12,600 | 102 | 34 | 2 |

The thyroid data [35, 36] include 15 digital features and more than 92% of the data belong to one class. Thus the recognition rate lower than 92% is useless.

The blood cell classification [37] involves classifying optically screened white blood cells into 12 classes using 13 features. This is a very difficult problem; class boundaries for some classes are ambiguous because the classes are defined according to the growth stages of white blood cells.

Hiragana-50 and hiragana-105 data [38, 7] were gathered from Japanese license plates. The original grayscale images of hiragana characters were transformed into (5 × 10)-pixel and (7 × 15)-pixel images, respectively, with the grayscale range being from 0 to 255. Then by performing grayscale shift, position shift, and random noise addition to the images, the training and test data were generated. Then for the hiragana-105 data to reduce the number of input variables, i.e., $7 \times 15 = 105$, the hiragana-13 data [38, 7] were generated by calculating the 13 central moments for the (7 × 15)-pixel images [39, 38].

Satimage data [36] have 36 inputs: 3 × 3 pixels each with four spectral values in a satellite image and are to classify the center pixel into one of the six classes: red soil, cotton crop, grey soil, damp grey soil, soil with vegetation stubble, and very damp grey soil.

USPS data [40] are handwritten numerals in (16 × 16)-pixel grayscale images. They are scanned from envelopes by the United States Postal Services. The MNIST data [41, 42] are handwritten numerals consisting of (28×28)-pixel inputs with 256 grayscale levels; they are often used to compare performance of support vector machines and other classifiers.

Table 1.4 lists the data sets for function approximation used in the book. For all the problems in the table, the number of outputs is 1.

The Mackey–Glass differential equation [43] generates time series data with a chaotic behavior and is given by

$$\frac{dx(t)}{dt} = \frac{0.2\,x(t-\tau)}{1+x^{10}(t-\tau)} - 0.1\,x(t), \tag{1.22}$$

**Table 1.3** Benchmark data specification for multiclass problems

| Data | Inputs | Classes | Training data | Test data |
|------|--------|---------|---------------|-----------|
| Iris | 4 | 3 | 75 | 75 |
| Numeral | 12 | 10 | 810 | 820 |
| Thyroid | 21 | 3 | 3,772 | 3,428 |
| Blood cell | 13 | 12 | 3,097 | 3,100 |
| Hiragana-50 | 50 | 39 | 4,610 | 4,610 |
| Hiragana-105 | 105 | 38 | 8,375 | 8,356 |
| Hiragana-13 | 13 | 38 | 8,375 | 8,356 |
| Satimage | 36 | 6 | 4,435 | 2,000 |
| USPS | 256 | 10 | 7,291 | 2,007 |
| MNIST | 784 | 10 | 60,000 | 10,000 |

where $t$ and $\tau$ denote time and time delay, respectively.

By integrating (1.22), we can obtain the time series data $x(0), x(1), x(2),$ $\ldots, x(t), \ldots$. Using $x$ prior to time $t$, we predict $x$ after time $t$. Setting $\tau = 17$ and using four inputs $x(t - 18), x(t - 12), x(t - 6), x(t)$, we estimate $x(t + 6)$.

The first 500 data from the time series data, $x(118), \ldots, x(1117)$, are used to train function approximators, and the remaining 500 data are used to test performance. This data set is often used as the benchmark data for function approximation and the normalized root-mean-square error (NRMSE), i.e., the root-mean-square error divided by the standard deviation of the time series data is used to measure the performance.

In a water purification plant, to eliminate small particles floating in the water taken from a river, coagulant is added and the water is stirred while these small particles begin sticking to each other. As more particles stick together they form flocs, which fall to the bottom of a holding tank. Potable water is obtained by removing the precipitated flocs and adding chlorine. Careful implementation of the coagulant injection is very important to obtain high-quality water. Usually an operator determines the amount of coagulant needed according to an analysis of the water qualities, observation of floc formation, and prior experience.

To automate this operation, as inputs for water quality, (1) turbidity, (2) temperature, (3) alkalinity, (4) pH, and (5) flow rate were used, and to replace the operator's observation of floc properties by image processing, (1) floc diameter, (2) number of flocs, (3) floc volume, (4) floc density, and (5) illumination intensity were used [44].

The 563 input–output data, which were gathered over a 1-year period, were divided into 478 stationary data and 95 nonstationary data according to whether turbidity values were smaller or larger than a specified value. Then each type of data was further divided into two groups to form a training data set and a test data set; division was done in such a way that both sets had similar distributions in the output space.

The orange juice data are to estimate the level of saccharose of orange juice from its observed near-infrared spectra [45].

The abalone data set predicts the age of abalone from physical measurements [36].

Boston 5 and Boston 14 data sets [46, 47] use the 5th and 14th input variables of the Boston data set as outputs, respectively. The fifth variable is NOX (nitric oxide) concentrations and the 14th variable is the house price in the Boston area. For these data sets training data are only provided.

Techniques for analyzing biological response to chemical structures are called quantitative structure–activity relationships (QSARs). Pyrimidines [36], Triazines [36], and Phenetylamines [48] data sets are well-known QSAR data sets. For these data sets, only the training data are given.

**Table 1.4** Benchmark data specification for function approximation

| Data | Inputs | Training data | Test data |
| --- | --- | --- | --- |
| Mackey–Glass | 4 | 500 | 500 |
| Water purification (stationary) | 10 | 241 | 237 |
| Water purification (nonstationary) | 10 | 45 | 40 |
| Orange juice | 700 | 150 | 68 |
| Abalone | 8 | 4,177 | — |
| Boston 5 | 13 | 506 | — |
| Boston 14 | 13 | 506 | — |
| Pyrimidines | 27 | 74 | — |
| Triazines | 60 | 186 | — |
| Phenetylamines | 628 | 22 | — |

# 1.4 Classifier Evaluation

In developing a classifier for a given problem we repeat determining input variables, namely features, gathering input–output pairs according to the determined features, training the classifier, and evaluating classifier performance. In training the classifier, special care must be taken so that no information on the test data set is used for training the classifier.[2]

Assume that a classifier for an $n$ class problem is tested using $M$ data samples. To evaluate the classifier for a test data set, we generate an $n \times n$ *confusion matrix* $A$, whose element $a_{ij}$ is the number of class $i$ data classified into class $j$. Then the *recognition rate* $R$ or *recognition accuracy* in % is calculated by

---

[2] It is my regret that I could not reevaluate the computer experiments, included in the book, that violate this rule.

$$R = \frac{\sum\limits_{i=1}^{n} a_{ii}}{\sum\limits_{i,j=1}^{n} a_{ij}} \times 100\,(\%). \tag{1.23}$$

Or conversely the *error rate E* is defined by

$$E = \frac{\sum\limits_{i \neq j, i, j=1}^{n} a_{ij}}{\sum\limits_{i,j=1}^{n} a_{ij}} \times 100\,(\%), \tag{1.24}$$

where $R + E = 100\%$, assuming that there are no unclassified data. The recognition rate (error rate) gives the overall performance of a classifier and is used to compare classifiers. To improve reliability in comparing classifiers, we prepare several training data sets and their associated test data sets and check whether there is a statistical difference in the mean recognition rates and their standard deviations of the classifiers.

There may be cases where there are several classification problems with a single training data set and a single test data set each. In such a situation, it is not a good way of simply comparing the average recognition rates of the classifiers. This is because the difference of 1% for a difficult problem is equally treated with that for an easy problem. For the discussions on how to statistically compare classifiers in such a situation, see [49, 50].

In diagnosis problems with negative (normal) and positive (abnormal) classes, data samples for the negative class are easily obtained but those for the positive class are difficult to obtain. In such problems with imbalanced training data, misclassification of positive data into the negative class is fatal compared to misclassification of negative data into the positive class. The confusion matrix for this problem is as shown in Table 1.5, where TP (true positive) is the number of correctly classified positive data, FN (false negative) is the number of misclassified positive data, FP (false positive) is the number of misclassified negative data, and TN (true negative) is the number of correctly classified negative data.

**Table 1.5** Confusion matrix

|                 | Assigned positive | Assigned negative |
|-----------------|-------------------|-------------------|
| Actual positive | TP                | FN                |
| Actual negative | FP                | TN                |

The well-used measures to evaluate classifier performance for diagnosis problems are precision-recall and ROC (receiver operator characteristic) curves. Precision is defined by

$$\text{Precision} = \frac{TP}{TP+FP}.$$

(1.25)

and recall is defined by

$$\text{Recall} = \frac{TP}{TP+FN}.$$

(1.26)

The precision-recall curve is plotted with precision on the $y$-axis and recall on the $x$-axis. A classifier with precision and recall values near 1 is preferable.

The ROC curve is plotted with the true-positive rate defined by

$$\text{True-positive rate} = \frac{TP}{TP+FN}.$$

(1.27)

on the $y$-axis and the false-positive rate defined by

$$\text{False-positive rate} = \frac{FP}{FP+TN}$$

(1.28)

on the $x$-axis. Recall is equivalent to the true-positive rate. The precision-recall and the ROC curves are plotted changing some parameter value of the classifier. The precision-recall curve is better than the ROC curve for heavily unbalanced data. The relations between the two types of curves are discussed in [51].

To see the difference of the measures, we used the thyroid data set shown in Table 1.3. We generated a two-class data set deleting data samples belonging to Class 2. We trained a support vector machine with RBF kernels with $\gamma = 1$ for different values of margin parameter $C$. Tables 1.6 and 1.7 show the confusion matrices for $C = 100$ and 2,000, respectively. And Table 1.8 lists the recognition rate, precision, recall, and the false-positive rate in % for the test data set. Although the recognition rate for $C = 100$ is higher, the recall value for $C = 2,000$ is smaller. But the reverse is true for the precision values, which is less fatal. Therefore, for diagnosis problems, it is better to select classifier with a higher recall value. Because the number of samples for the negative class is extremely large, comparison of the false-positive rate becomes meaningless.

**Table 1.6** Confusion matrix for $C = 100$

|                  | Assigned positive | Assigned negative |
|------------------|-------------------|-------------------|
| Actual positive  | 56                | 17                |
| Actual negative  | 12                | 3,166             |

The often-used performance evaluation measures for regression problems are the mean average error (MAE), the root-mean-square error (RMSE), and the normalized root-mean-square error (NRMSE). Let the input–output pairs be $(\mathbf{x}_i, y_i)$ $(i = 1, \ldots, M)$ and the regression function be $f(\mathbf{x})$. Then the MAE, RMSE, and NRMSE are given, respectively, by

**Table 1.7** Confusion matrix for $C = 2,000$

|  | Assigned positive | Assigned negative |
|---|---|---|
| Actual positive | 63 | 10 |
| Actual negative | 20 | 3,158 |

**Table 1.8** Classification performance of the thyroid data set (in %)

| $C$ | R. rate | Precision | Recall | False-positive rate |
|---|---|---|---|---|
| 100 | **99.11** | 82.35 | 76.71 | 0.38 |
| 2,000 | 99.08 | 75.90 | **86.30** | 0.63 |

$$\text{MAE} = \frac{1}{M} \sum_{i}^{M} |y_i - f(\mathbf{x}_i)|, \tag{1.29}$$

$$\text{RMSE} = \sqrt{\frac{1}{M} \sum_{i}^{M} (y_i - f(\mathbf{x}_i))^2}, \tag{1.30}$$

$$\text{NRMSE} = \frac{1}{\sigma} \sqrt{\frac{1}{M} \sum_{i}^{M} (y_i - f(\mathbf{x}_i))^2}, \tag{1.31}$$

where $\sigma$ is the standard deviation of the observed data samples.

# References

1. K. Fukunaga. *Introduction to Statistical Pattern Recognition, Second Edition.* Academic Press, San Diego, 1990.
2. C. M. Bishop. *Neural Networks for Pattern Recognition.* Oxford University Press, Oxford, 1995.
3. S. Abe. *Neural Networks and Fuzzy Systems: Theory and Applications.* Kluwer Academic Publishers, Norwell, MA, 1997.
4. S. Haykin. *Neural Networks: A Comprehensive Foundation, Second Edition.* Prentice Hall, Upper Saddle River, NJ, 1999.
5. J. C. Bezdek, J. Keller R. Krisnapuram, and N. R. Pal. *Fuzzy Models and Algorithms for Pattern Recognition and Image Processing.* Kluwer Academic Publishers, Norwell, MA, 1999.
6. S. K. Pal and S. Mitra. *Neuro-Fuzzy Pattern Recognition: Methods in Soft Computing.* John Wiley & Sons, New York, 1999.
7. S. Abe. *Pattern Classification: Neuro-Fuzzy Methods and Their Comparison.* Springer-Verlag, London, 2001.
8. V. N. Vapnik. *Statistical Learning Theory.* John Wiley & Sons, New York, 1998.
9. N. Cristianini and J. Shawe-Taylor. *An Introduction to Support Vector Machines and Other Kernel-Based Learning Methods.* Cambridge University Press, Cambridge, 2000.
10. V. N. Vapnik. *The Nature of Statistical Learning Theory.* Springer-Verlag, New York, 1995.

11. U. H.-G. Kreßel. Pairwise classification and support vector machines. In B. Schölkopf, C. J. C. Burges, and A. J. Smola, editors, *Advances in Kernel Methods: Support Vector Learning*, pages 255–268. MIT Press, Cambridge, MA, 1999.

12. T. G. Dietterich and G. Bakiri. Solving multiclass learning problems via error-correcting output codes. *Journal of Artificial Intelligence Research*, 2:263–286, 1995.

13. R. O. Duda and P. E. Hart. *Pattern Classification and Scene Analysis*. John Wiley & Sons, New York, 1973.

14. J. Weston and C. Watkins. Multi-class support vector machines. Technical Report CSD-TR-98-04, Royal Holloway, University of London, London, UK, 1998.

15. J. Weston and C. Watkins. Support vector machines for multi-class pattern recognition. In *Proceedings of the Seventh European Symposium on Artificial Neural Networks (ESANN 1999)*, pages 219–224, Bruges, Belgium, 1999.

16. K. P. Bennett. Combining support vector and mathematical programming methods for classification. In B. Schölkopf, C. J. C. Burges, and A. J. Smola, editors, *Advances in Kernel Methods: Support Vector Learning*, pages 307–326. MIT Press, Cambridge, MA, 1999.

17. E. J. Bredensteiner and K. P. Bennett. Multicategory classification by support vector machines. *Computational Optimization and Applications*, 12(1–3):53–79, 1999.

18. Y. Guermeur, A. Elisseeff, and H. Paugam-Moisy. A new multi-class SVM based on a uniform convergence result. In *Proceedings of the IEEE-INNS-ENNS International Joint Conference on Neural Networks (IJCNN 2000)*, volume 4, pages 183–188, Como, Italy, 2000.

19. C. Angulo, X. Parra, and A. Català. An [*sic*] unified framework for 'all data at once' multi-class support vector machines. In *Proceedings of the Tenth European Symposium on Artificial Neural Networks (ESANN 2002)*, pages 161–166, Bruges, Belgium, 2002.

20. D. Anguita, S. Ridella, and D. Sterpi. A new method for multiclass support vector machines. In *Proceedings of International Joint Conference on Neural Networks (IJCNN 2004)*, volume 1, pages 407–412, Budapest, Hungary, 2004.

21. K.-R. Müller, S. Mika, G. Rätsch, K. Tsuda, and B. Schölkopf. An introduction to kernel-based learning algorithms. *IEEE Transactions on Neural Networks*, 12(2):181–201, 2001.

22. G. Rätsch, T. Onoda, and K.-R. Müller. Soft margins for AdaBoost. *Machine Learning*, 42(3):287–320, 2001.

23. Intelligent Data Analysis Group. http://ida.first.fraunhofer.de/projects/bench/benchmarks.htm.

24. N. Pochet, F. De Smet, J. A. K. Suykens, and B. L. R. De Moor. http://homes.esat.kuleuven.be/npochet/bioinformatics/.

25. I. Hedenfalk, D. Duggan, Y. Chen, M. Radmacher, M. Bittner, R. Simon, P. Meltzer, B. Gusterson, M. Esteller, M. Raffeld, Z. Yakhini, A. Ben-Dor, E. Dougherty, J. Kononen, L. Bubendorf, W. Fehrle, S. Pittaluga, S. Gruvberger, N. Loman, O. Johannsson, H. Olsson, B. Wilfond, G. Sauter, O.-P. Kallioniemi, A. Borg, and J. Trent. Gene-expression profiles in hereditary breast cancer. *The New England Journal of Medicine*, 344(8):539–548, 2001.

26. L. J. van't Veer, H. Dai, M. J. van de Vijver, Y. D. He, A. A. M. Hart, M. Mao, H. L. Peterse, K. van der Kooy, M. J. Marton, A. T. Witteveen, G. J. Schreiber, R. M. Kerkhoven, C. Roberts, P. S. Linsley, R. Bernards, and S. H. Friend. Gene expression profiling predicts clinical outcome of breast cancer. *Nature*, 415:530–536, 2002.

27. U. Alon, N. Barkai, D. A. Notterman, K. Gish, S. Ybarra, D. Mack, and A. J. Levine. Broad patterns of gene expression revealed by clustering analysis of tumor and normal colon tissues probed by oligonucleotide arrays. *Proceedings of the National Academy of Sciences of the United States of America*, 96(12):6745–6750, 1999.

28. N. Iizuka, M. Oka, H. Yamada-Okabe, M. Nishida, Y. Maeda, N. Mori, T. Takao, T. Tamesa, A. Tangoku, H. Tabuchi, K. Hamada, H. Nakayama, H. Ishitsuka, T. Miyamoto, A. Hirabayashi, S. Uchimura, and Y. Hamamoto. Oligonucleotide

microarray for prediction of early intrahepatic recurrence of hepatocellular carcinoma after curative resection. *The Lancet*, 361(9361):923–929, 2003.

29. C. L. Nutt, D. R. Mani, R. A. Betensky, P. Tamayo, J. G. Cairncross, C. Ladd, U. Pohl, C. Hartmann, M. E. McLaughlin, T. T. Batchelor, P. M. Black, A. von Deimling, S. L. Pomeroy, T. R. Golub, and D. N. Louis. Gene expression-based classification of malignant gliomas correlates better with survival than histological classification. *Cancer Research*, 63(7):1602–1607, 2003.

30. T. R. Golub, D. K. Slonim, P. Tamayo, C. Huard, M. Gaasenbeek, J. P. Mesirov, H. Coller, M. L. Loh, J. R. Downing, M. A. Caligiuri, C. D. Bloomfield, and E. S. Lander. Molecular classification of cancer: Class discovery and class prediction by gene expression monitoring. *Science*, 286:531–537, 1999.

31. D. Singh, P. G. Febbo, K. Ross, D. G. Jackson, J. Manola, C. Ladd, P. Tamayo, A. A. Renshaw, A. V. D'Amico, J. P. Richie, E. S. Lander, M. Loda, P. W. Kantoff, T. R. Golub, and W. R. Sellers. Gene expression correlates of clinical prostate cancer behavior. *Cancer Cell*, 1(2):203–209, 2002.

32. R. A. Fisher. The use of multiple measurements in taxonomic problems. *Annals of Eugenics*, 7:179–188, 1936.

33. J. C. Bezdek, J. M. Keller, R. Krishnapuram, L. I. Kuncheva, and N. R. Pal. Will the real iris data please stand up? *IEEE Transactions on Fuzzy Systems*, 7(3):368–369, 1999.

34. H. Takenaga, S. Abe, M. Takato, M. Kayama, T. Kitamura, and Y. Okuyama. Input layer optimization of neural networks by sensitivity analysis and its application to recognition of numerals. *Electrical Engineering in Japan*, 111(4):130–138, 1991.

35. S. M. Weiss and I. Kapouleas. An empirical comparison of pattern recognition, neural nets, and machine learning classification methods. In *Proceedings of the Eleventh International Joint Conference on Artificial Intelligence*, pages 781–787, Detroit, 1989.

36. A. Asuncion and D. J. Newman. UCI machine learning repository, http://www.ics.uci.edu/mlearn/MLRepository.html. 2007.

37. A. Hashizume, J. Motoike, and R. Yabe. Fully automated blood cell differential system and its application. In *Proceedings of the IUPAC Third International Congress on Automation and New Technology in the Clinical Laboratory*, pages 297–302, Kobe, Japan, 1988.

38. M.-S. Lan, H. Takenaga, and S. Abe. Character recognition using fuzzy rules extracted from data. In *Proceedings of the Third IEEE International Conference on Fuzzy Systems*, volume 1, pages 415–420, Orlando, FL, 1994.

39. G. L. Cash and M. Hatamian. Optical character recognition by the method of moments. *Computer Vision, Graphics, and Image Processing*, 39(3):291–310, 1987.

40. USPS Dataset. http://www-i6.informatik.rwth-aachen.de/keysers/usps.html.

41. Y. LeCun, L. Bottou, Y. Bengio, and P. Haffner. Gradient-based learning applied to document recognition. *Proceedings of the IEEE*, 86(11):2278–2324, 1998.

42. Y. LeCun and C. Cortes. The MNIST database of handwritten digits http://yann.lecun.com/exdb/mnist/.

43. R. S. Crowder. Predicting the Mackey-Glass time series with cascade-correlation learning. In D. S. Touretzky, J. L. Elman, T. J. Sejnowski, and G. E. Hinton, editors, *Connectionist Models: Proceedings of the 1990 Summer School*, pages 117–123. Morgan Kaufmann, San Mateo, CA, 1991.

44. K. Baba, I. Enbutu, and M. Yoda. Explicit representation of knowledge acquired from plant historical data using neural network. In *Proceedings of 1990 IJCNN International Joint Conference on Neural Networks*, volume 3, pages 155–160, San Diego, CA, 1990.

45. UCL Machine Learning Group. http://www.ucl.ac.be/mlg/index.php?page=home.

46. D. Harrison and D. L. Rubinfeld. Hedonic prices and the demand for clean air. *Journal of Environmental Economics and Management*, 5(1):81–102, 1978.

47. Delve Datasets. http://www.cs.toronto.edu/delve/data/datasets.html.

48. Milano        Chemometrics        and        QSAR        Research        Group. http://michem.disat.unimib.it/chm/download/download.htm.

49. J. Demšar. Statistical comparisons of classifiers over multiple data sets. *Journal of Machine Learning Research*, 7:1–30, 2006.

50. T. G. Dietterich. Approximate statistical tests for comparing supervised classification learning algorithms. *Neural Computation*, 10(7):1895–1923, 1998.

51. J. Davis and M. Goadrich. The relationship between precision-recall and ROC curves. In *Proceedings of the Twenty-Third International Conference on Machine Learning (ICML 2006)*, pages 233–240, Pittsburgh, PA, 2006.

# Chapter 2
# Two-Class Support Vector Machines

In training a classifier, usually we try to maximize classification performance for the training data. But if the classifier is too fit for the training data, the classification ability for unknown data, i.e., the generalization ability is degraded. This phenomenon is called *overfitting*, namely, there is a trade-off between the generalization ability and fitting to the training data. Various methods have been proposed to prevent overfitting [1, pp. 11–15], [2, pp. 86–91], [3].[1]

For a two-class problem, a support vector machine is trained so that the direct decision function maximizes the generalization ability [4, pp. 127–151], [5, pp. 92–129], namely, the $m$-dimensional input space $\mathbf{x}$ is mapped into the $l$-dimensional ($l \geq m$) feature space $\mathbf{z}$. Then in $\mathbf{z}$, the quadratic programming problem is solved to separate two classes by the optimal separating hyperplane.

In this chapter we discuss support vector machines for two-class problems. First, we discuss hard-margin support vector machines, in which training data are linearly separable in the input space. Then we extend hard-margin support vector machines to the case where training data are not linearly separable and map the input space into the high-dimensional feature space to enhance linear separability in the feature space. The characteristics of support vector machines are then studied theoretically and by computer experiments.

## 2.1 Hard-Margin Support Vector Machines

Let $M$ $m$-dimensional training inputs $\mathbf{x}_i$ $(i = 1, \ldots, M)$ belong to Class 1 or 2 and the associated labels be $y_i = 1$ for Class 1 and $-1$ for Class 2. If these data are linearly separable, we can determine the decision function:

---

[1] One of the main ideas is, like support vector machines, to add a regularization term, which controls the generalization ability, to the objective function.

S. Abe, *Support Vector Machines for Pattern Classification*,
Advances in Pattern Recognition, DOI 10.1007/978-1-84996-098-4_2,
© Springer-Verlag London Limited 2010

$$D(\mathbf{x}) = \mathbf{w}^\top \mathbf{x} + b, \tag{2.1}$$

where $\mathbf{w}$ is an $m$-dimensional vector, $b$ is a bias term, and for $i = 1, \ldots, M$

$$\mathbf{w}^\top \mathbf{x}_i + b \begin{cases} > 0 & \text{for} \quad y_i = 1, \\ < 0 & \text{for} \quad y_i = -1. \end{cases} \tag{2.2}$$

Because the training data are linearly separable, no training data satisfy $\mathbf{w}^\top \mathbf{x} + b = 0$. Thus, to control separability, instead of (2.2), we consider the following inequalities:

$$\mathbf{w}^\top \mathbf{x}_i + b \begin{cases} \geq 1 & \text{for} \quad y_i = 1, \\ \leq -1 & \text{for} \quad y_i = -1. \end{cases} \tag{2.3}$$

Here, 1 and $-1$ on the right-hand sides of the inequalities can be constants $a\,(>0)$ and $-a$, respectively. But by dividing both sides of the inequalities by $a$, (2.3) is obtained. Equation (2.3) is equivalent to

$$y_i \left( \mathbf{w}^\top \mathbf{x}_i + b \right) \geq 1 \quad \text{for } i = 1, \ldots, M. \tag{2.4}$$

The hyperplane

$$D(\mathbf{x}) = \mathbf{w}^\top \mathbf{x} + b = c \quad \text{for } -1 < c < 1 \tag{2.5}$$

forms a separating hyperplane that separates $\mathbf{x}_i$ $(i = 1, \ldots, M)$. When $c = 0$, the separating hyperplane is in the middle of the two hyperplanes with $c = 1$ and $-1$. The distance between the separating hyperplane and the training data sample nearest to the hyperplane is called the *margin*. Assuming that the hyperplanes $D(\mathbf{x}) = 1$ and $-1$ include at least one training data sample, the hyperplane $D(\mathbf{x}) = 0$ has the maximum margin for $-1 < c < 1$. The region $\{\mathbf{x} | -1 \leq D(\mathbf{x}) \leq 1\}$ is the generalization region for the decision function.

Figure 2.1 shows two decision functions that satisfy (2.4). Thus there are an infinite number of decision functions that satisfy (2.4), which are separating hyperplanes. The generalization ability depends on the location of the separating hyperplane, and the hyperplane with the maximum margin is called the *optimal separating hyperplane* (see Fig. 2.1). Assume that no outliers are included in the training data and that unknown test data will obey the same distribution as that of the training data. Then it is intuitively clear that the generalization ability is maximized if the optimal separating hyperplane is selected as the separating hyperplane.

Now consider determining the optimal separating hyperplane. The Euclidean distance from a training data sample $\mathbf{x}$ to the separating hyperplane is given by $|D(\mathbf{x})|/\|\mathbf{w}\|$. This can be shown as follows. Because the vector $\mathbf{w}$ is orthogonal to the separating hyperplane, the line that goes through $\mathbf{x}$ and that is orthogonal to the hyperplane is given by $a\mathbf{w}/\|\mathbf{w}\| + \mathbf{x}$, where $|a|$ is the Euclidean distance from $\mathbf{x}$ to the hyperplane. It crosses the hyperplane

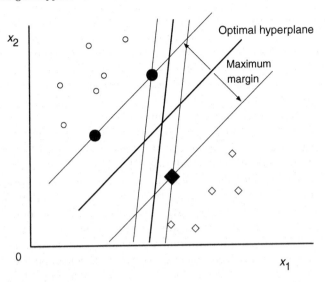

**Fig. 2.1** Optimal separating hyperplane in a two-dimensional space

at the point where

$$D(a\mathbf{w}/\|\mathbf{w}\| + \mathbf{x}) = 0 \tag{2.6}$$

is satisfied. Solving (2.6) for $a$, we obtain $a = -D(\mathbf{x})/\|\mathbf{w}\|$.

Then all the training data must satisfy

$$\frac{y_k D(\mathbf{x}_k)}{\|\mathbf{w}\|} \geq \delta \quad \text{for } k = 1, \dots, M, \tag{2.7}$$

where $\delta$ is the margin.

Now if $(\mathbf{w}, b)$ is a solution, $(a\mathbf{w}, ab)$ is also a solution, where $a$ is a scalar. Thus we impose the following constraint:

$$\delta \|\mathbf{w}\| = 1. \tag{2.8}$$

From (2.7) and (2.8), to find the optimal separating hyperplane, we need to find $\mathbf{w}$ with the minimum Euclidean norm that satisfies (2.4).

Therefore, the optimal separating hyperplane can be obtained by solving the following minimization problem for $\mathbf{w}$ and $b$:

$$\text{minimize} \quad Q(\mathbf{w}, b) = \frac{1}{2} \|\mathbf{w}\|^2 \tag{2.9}$$

$$\text{subject to} \quad y_i \left(\mathbf{w}^\top \mathbf{x}_i + b\right) \geq 1 \quad \text{for } i = 1, \dots, M. \tag{2.10}$$

Here, the square of the Euclidean norm $\|\mathbf{w}\|$ in (2.9) is to make the optimization problem quadratic programming. The assumption of linear separability means that there exist $\mathbf{w}$ and $b$ that satisfy (2.10). We call the solutions that satisfy (2.10) *feasible solutions*. Because the optimization problem has the quadratic objective function with the inequality constraints, even if the solutions are nonunique, the value of the objective function is unique (see Section 2.6.4). Thus nonuniqueness is not a problem for support vector machines. This is one of the advantages of support vector machines over neural networks, which have numerous local minima.

Because we can obtain the same optimal separating hyperplane even if we delete all the data that satisfy the strict inequalities in (2.10), the data that satisfy the equalities are called *support vectors*.[2] In Fig. 2.1, the data corresponding to the filled circles and the filled square are support vectors.

The variables of the convex optimization problem given by (2.9) and (2.10) are $\mathbf{w}$ and $b$. Thus the number of variables is the number of input variables plus 1: $m+1$. When the number of input variables is small, we can solve (2.9) and (2.10) by the quadratic programming technique. But, as will be discussed later, because we map the input space into a high-dimensional feature space, in some cases, with infinite dimensions, we convert (2.9) and (2.10) into the equivalent dual problem whose number of variables is the number of training data.

To do this, we first convert the constrained problem given by (2.9) and (2.10) into the unconstrained problem:

$$Q(\mathbf{w}, b, \boldsymbol{\alpha}) = \frac{1}{2}\mathbf{w}^\top \mathbf{w} - \sum_{i=1}^{M} \alpha_i \left\{ y_i \left( \mathbf{w}^\top \mathbf{x}_i + b \right) - 1 \right\}, \qquad (2.11)$$

where $\boldsymbol{\alpha} = (\alpha_1, \ldots, \alpha_M)^\top$ and $\alpha_i$ are the nonnegative Lagrange multipliers. The optimal solution of (2.11) is given by the saddle point, where (2.11) is minimized with respect to $\mathbf{w}$, maximized with respect to $\alpha_i$ ($\geq 0$), and maximized or minimized with respect to $b$ according to the sign of $\sum_{i=1}^{M} \alpha_i y_i$, and the solution satisfies the following Karush–Kuhn–Tucker (KKT) conditions (see Theorem C.1 on p. 455):

$$\frac{\partial Q(\mathbf{w}, b, \boldsymbol{\alpha})}{\partial \mathbf{w}} = \mathbf{0}, \qquad (2.12)$$

$$\frac{\partial Q(\mathbf{w}, b, \boldsymbol{\alpha})}{\partial b} = 0, \qquad (2.13)$$

$$\alpha_i \left\{ y_i \left( \mathbf{w}^\top \mathbf{x}_i + b \right) - 1 \right\} = 0 \quad \text{for } i = 1, \ldots, M, \qquad (2.14)$$

$$\alpha_i \geq 0 \quad \text{for } i = 1, \ldots, M. \qquad (2.15)$$

---

[2] This definition is imprecise. As shown in Definition 2.1 on p. 72, there are data that satisfy $y_i \left( \mathbf{w}^\top \mathbf{x} + b \right) = 1$ but that can be deleted without changing the optimal separating hyperplane. Support vectors are defined using the solution of the dual problem, as discussed later.

Especially, the relations between the inequality constraints and their associated Lagrange multipliers given by (2.14) are called *KKT complementarity conditions*. In the following, if there is no confusion, the KKT complementarity conditions are called the KKT conditions.

From (2.14), $\alpha_i = 0$, or $\alpha_i \neq 0$ and $y_i(\mathbf{w}^\top \mathbf{x}_i + b) = 1$ must be satisfied. The training data $\mathbf{x}_i$ with $\alpha_i \neq 0$ are called *support vectors*.[3]

Using (2.11), we reduce (2.12) and (2.13), respectively, to

$$\mathbf{w} = \sum_{i=1}^{M} \alpha_i y_i \mathbf{x}_i \tag{2.16}$$

and

$$\sum_{i=1}^{M} \alpha_i y_i = 0. \tag{2.17}$$

Substituting (2.16) and (2.17) into (2.11), we obtain the following dual problem:

$$\text{maximize} \quad Q(\boldsymbol{\alpha}) = \sum_{i=1}^{M} \alpha_i - \frac{1}{2} \sum_{i,j=1}^{M} \alpha_i \alpha_j y_i y_j \mathbf{x}_i^\top \mathbf{x}_j \tag{2.18}$$

$$\text{subject to} \quad \sum_{i=1}^{M} y_i \alpha_i = 0, \qquad \alpha_i \geq 0 \quad \text{for } i = 1, \ldots, M. \tag{2.19}$$

The formulated support vector machine is called the *hard-margin support vector machine*. Because

$$\frac{1}{2} \sum_{i,j=1}^{M} \alpha_i \alpha_j y_i y_j \mathbf{x}_i^\top \mathbf{x}_j = \frac{1}{2} \left( \sum_{i=1}^{M} \alpha_i y_i \mathbf{x}_i \right)^\top \left( \sum_{i=1}^{M} \alpha_i y_i \mathbf{x}_i \right) \geq 0, \tag{2.20}$$

maximizing (2.18) under the constraints (2.19) is a concave quadratic programming problem. If a solution exists, namely, if the classification problem is linearly separable, the global optimal solution $\alpha_i$ $(i = 1, \ldots, M)$ exists. For quadratic programming, the values of the primal and dual objective functions coincide at the optimal solutions if they exist. This is called the *zero duality gap*.

Data that are associated with positive $\alpha_i$ are support vectors for Class 1 or 2. Then from (2.16) the decision function is given by

$$D(\mathbf{x}) = \sum_{i \in S} \alpha_i y_i \mathbf{x}_i^\top \mathbf{x} + b, \tag{2.21}$$

---

[3] In the definition of support vectors, we exclude the data in which both $\alpha_i = 0$ and $y_i(\mathbf{w}^\top \mathbf{x}_i + b) = 1$ hold.

where $S$ is the set of support vector indices, and from the KKT conditions given by (2.14), $b$ is given by

$$b = y_i - \mathbf{w}^\top \mathbf{x}_i \quad \text{for } i \in S. \tag{2.22}$$

From the standpoint of precision of calculations, it is better to take the average among the support vectors as follows[4]:

$$b = \frac{1}{|S|} \sum_{i \in S} (y_i - \mathbf{w}^\top \mathbf{x}_i). \tag{2.23}$$

Then unknown data sample $\mathbf{x}$ is classified into

$$\begin{cases} \text{Class 1} & \text{if } D(\mathbf{x}) > 0, \\ \text{Class 2} & \text{if } D(\mathbf{x}) < 0. \end{cases} \tag{2.24}$$

If $D(\mathbf{x}) = 0$, $\mathbf{x}$ is on the boundary and thus is unclassifiable. When training data are separable, the region $\{\mathbf{x} \,|\, 1 > D(\mathbf{x}) > -1\}$ is a generalization region.

*Example 2.1.* Consider a linearly separable case shown in Fig. 2.2. The inequality constraints given by (2.10) are

$$-w + b \geq 1, \tag{2.25}$$
$$-b \geq 1, \tag{2.26}$$
$$-(w + b) \geq 1. \tag{2.27}$$

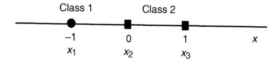

**Fig. 2.2** Linearly separable one-dimensional case

The region of $(w, b)$ that satisfies (2.25), (2.26), and (2.27) is given by the shaded region shown in Fig. 2.3. Thus the solution that minimizes $\|w\|^2$ is given by

$$b = -1, \qquad w = -2. \tag{2.28}$$

Namely, the decision function is given by

---

[4] If we use a training method with fixed-size chunking such as SMO (see Section 5.2), the values of $b$ calculated for $\mathbf{x}_i$ in the working set and those in the fixed set may be different. In such a case it is better to take the average. But if a training method with variable-size chunking is used, in which all the nonzero $\alpha_i$ are in the working set, the average is not necessary.

$$D(x) = -2\,x - 1. \tag{2.29}$$

The class boundary is given by $x = -1/2$. Because the solution is determined by (2.25) and (2.26), $x = 0$ and $-1$ are support vectors.

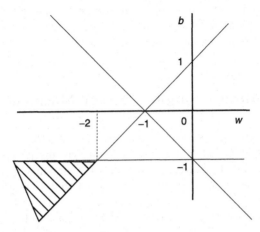

**Fig. 2.3** Region that satisfies constraints

The dual problem is given as follows:

$$\text{maximize} \quad Q(\boldsymbol{\alpha}) = \alpha_1 + \alpha_2 + \alpha_3 - \frac{1}{2}(\alpha_1 + \alpha_3)^2 \tag{2.30}$$

$$\text{subject to} \quad \alpha_1 - \alpha_2 - \alpha_3 = 0, \tag{2.31}$$

$$\alpha_i \geq 0 \quad \text{for } i = 1, 2, 3. \tag{2.32}$$

From (2.31), $\alpha_2 = \alpha_1 - \alpha_3$. Substituting it into (2.30), we obtain

$$Q(\boldsymbol{\alpha}) = 2\,\alpha_1 - \frac{1}{2}(\alpha_1 + \alpha_3)^2, \tag{2.33}$$

$$\alpha_i \geq 0 \quad \text{for } i = 1, 2, 3. \tag{2.34}$$

Because $\alpha_1 \geq 0$ and $\alpha_3 \geq 0$, (2.33) is maximized when $\alpha_3 = 0$. Thus, (2.33) reduces to

$$Q(\boldsymbol{\alpha}) = 2\,\alpha_1 - \frac{1}{2}\alpha_1^2$$

$$= -\frac{1}{2}(\alpha_1 - 2)^2 + 2, \tag{2.35}$$

$$\alpha_1 \geq 0. \tag{2.36}$$

Because (2.35) is maximized for $\alpha_1 = 2$, the optimal solution for (2.30) is

$$\alpha_1 = 2, \quad \alpha_2 = 2, \quad \alpha_3 = 0. \tag{2.37}$$

Therefore, $x = -1$ and $0$ are support vectors and $w = -2$ and $b = -1$, which are the same as the solution obtained by solving the primary problem. In addition, because $Q(w, b) = Q(\alpha) = 2$, the duality gap is zero.

Consider changing the label of $x_3$ into that of the opposite class, i.e., $y_3 = 1$. Then the problem becomes inseparable and (2.27) becomes $w + b \geq 1$. Thus, from Fig 2.3 there is no feasible solution.

## 2.2 L1 Soft-Margin Support Vector Machines

In hard-margin support vector machines, we assumed that the training data are linearly separable. When the data are linearly inseparable, there is no feasible solution, and the hard-margin support vector machine is unsolvable. Here we extend the support vector machine so that it is applicable to an inseparable case.

To allow inseparability, we introduce the nonnegative slack variables $\xi_i$ ($\geq 0$) into (2.4):

$$y_i \left( \mathbf{w}^\top \mathbf{x}_i + b \right) \geq 1 - \xi_i \quad \text{for } i = 1, \ldots, M. \tag{2.38}$$

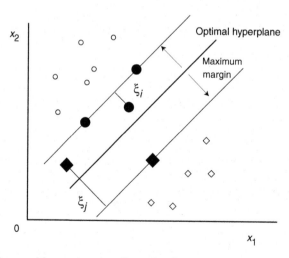

**Fig. 2.4** Inseparable case in a two-dimensional space

By the slack variables $\xi_i$, feasible solutions always exist. For the training data $\mathbf{x}_i$, if $0 < \xi_i < 1$ ($\xi_i$ in Fig. 2.4), the data do not have the maximum margin but are still correctly classified. But if $\xi_i \geq 1$ ($\xi_j$ in Fig. 2.4) the data are misclassified by the optimal hyperplane. To obtain the optimal hyperplane in which the number of training data that do not have the maximum margin is minimum, we need to minimize

$$Q(\mathbf{w}) = \sum_{i=1}^{M} \theta(\xi_i),$$

where

$$\theta(\xi_i) = \begin{cases} 1 & \text{for} \quad \xi_i > 0, \\ 0 & \text{for} \quad \xi_i = 0. \end{cases}$$

But this is a combinatorial optimization problem and difficult to solve. Instead, we consider the following minimization problem:

$$\text{minimize} \quad Q(\mathbf{w}, b, \boldsymbol{\xi}) = \frac{1}{2}\|\mathbf{w}\|^2 + \frac{C}{p}\sum_{i=1}^{M}\xi_i^p \tag{2.39}$$

$$\text{subject to} \quad y_i\left(\mathbf{w}^\top \mathbf{x}_i + b\right) \geq 1 - \xi_i, \quad \xi_i \geq 0 \quad \text{for } i = 1,\dots,M, \tag{2.40}$$

where $\boldsymbol{\xi} = (\xi_1, \dots, \xi_M)^\top$ and $C$ is the margin parameter that determines the trade-off between the maximization of the margin and the minimization of the classification error.[5] We select the value of $p$ as either 1 or 2. We call the obtained hyperplane the *soft-margin hyperplane*. When $p = 1$, we call the support vector machine the *L1 soft-margin support vector machine* or the L1 support vector machine for short (L1 SVM) and when $p = 2$, the L2 soft-margin support vector machine or L2 support vector machine (L2 SVM). In this section, we discuss L1 soft-margin support vector machines.

Similar to the linearly separable case, introducing the nonnegative Lagrange multipliers $\alpha_i$ and $\beta_i$, we obtain

$$Q(\mathbf{w}, b, \boldsymbol{\xi}, \boldsymbol{\alpha}, \boldsymbol{\beta}) = \frac{1}{2}\|\mathbf{w}\|^2 + C\sum_{i=1}^{M}\xi_i$$

$$- \sum_{i=1}^{M}\alpha_i\left(y_i\left(\mathbf{w}^\top\mathbf{x}_i + b\right) - 1 + \xi_i\right) - \sum_{i=1}^{M}\beta_i\,\xi_i, \tag{2.41}$$

where $\boldsymbol{\alpha} = (\alpha_1, \dots, \alpha_M)^\top$ and $\boldsymbol{\beta} = (\beta_1, \dots, \beta_M)^\top$.

For the optimal solution, the following KKT conditions are satisfied (see Theorem C.1 on p. 455):

$$\frac{\partial Q(\mathbf{w}, b, \boldsymbol{\xi}, \boldsymbol{\alpha}, \boldsymbol{\beta})}{\partial \mathbf{w}} = \mathbf{0}, \tag{2.42}$$

---

[5] Orsenigo and Vercellis [6] formulate the discrete support vector machines that maximize the margin and minimize the number of misclassifications. This results in a linear mixed integer programming problem.

$$\frac{\partial Q(\mathbf{w}, b, \boldsymbol{\xi}, \boldsymbol{\alpha}, \boldsymbol{\beta})}{\partial b} = 0, \tag{2.43}$$

$$\frac{\partial Q(\mathbf{w}, b, \boldsymbol{\xi}, \boldsymbol{\alpha}, \boldsymbol{\beta})}{\partial \boldsymbol{\xi}} = \mathbf{0}. \tag{2.44}$$

$$\alpha_i \left( y_i \left( \mathbf{w}^\top \mathbf{x}_i + b \right) - 1 + \xi_i \right) = 0 \quad \text{for } i = 1, \dots, M, \tag{2.45}$$

$$\beta_i \, \xi_i = 0 \quad \text{for } i = 1, \dots, M, \tag{2.46}$$

$$\alpha_i \geq 0, \quad \beta_i \geq 0, \quad \xi_i \geq 0 \quad \text{for } i = 1, \dots, M. \tag{2.47}$$

Using (2.41), we reduce (2.42) to (2.44), respectively, to

$$\mathbf{w} = \sum_{i=1}^{M} \alpha_i \, y_i \, \mathbf{x}_i, \tag{2.48}$$

$$\sum_{i=1}^{M} \alpha_i \, y_i = 0, \tag{2.49}$$

$$\alpha_i + \beta_i = C \quad \text{for } i = 1, \dots, M. \tag{2.50}$$

Thus substituting (2.48), (2.49), and (2.50) into (2.41), we obtain the following dual problem:

$$\text{maximize} \quad Q(\boldsymbol{\alpha}) = \sum_{i=1}^{M} \alpha_i - \frac{1}{2} \sum_{i,j=1}^{M} \alpha_i \, \alpha_j \, y_i \, y_j \, \mathbf{x}_i^\top \mathbf{x}_j \tag{2.51}$$

$$\text{subject to} \quad \sum_{i=1}^{M} y_i \, \alpha_i = 0, \quad C \geq \alpha_i \geq 0 \quad \text{for } i = 1, \dots, M. \tag{2.52}$$

The only difference between L1 soft-margin support vector machines and hard-margin support vector machines is that $\alpha_i$ cannot exceed $C$. The inequality constraints in (2.52) are called *box constraints*.

Especially, (2.45) and (2.46) are called *KKT (complementarity) conditions*. From these and (2.50), there are three cases for $\alpha_i$:

1. $\alpha_i = 0$. Then $\xi_i = 0$. Thus $\mathbf{x}_i$ is correctly classified.
2. $0 < \alpha_i < C$. Then $y_i \left( \mathbf{w}^\top \mathbf{x}_i + b \right) - 1 + \xi_i = 0$ and $\xi_i = 0$. Therefore, $y_i \left( \mathbf{w}^\top \mathbf{x}_i + b \right) = 1$ and $\mathbf{x}_i$ is a support vector. Especially, we call the support vector with $C > \alpha_i > 0$ an *unbounded support vector*.
3. $\alpha_i = C$. Then $y_i \left( \mathbf{w}^\top \mathbf{x}_i + b \right) - 1 + \xi_i = 0$ and $\xi_i \geq 0$. Thus $\mathbf{x}_i$ is a support vector. We call the support vector with $\alpha_i = C$ a *bounded support vector*. If $0 \leq \xi_i < 1$, $\mathbf{x}_i$ is correctly classified, and if $\xi_i \geq 1$, $\mathbf{x}_i$ is misclassified.

The decision function is the same as that of the hard-margin support vector machine and is given by

$$D(\mathbf{x}) = \sum_{i \in S} \alpha_i \, y_i \, \mathbf{x}_i^\top \, \mathbf{x} + b, \tag{2.53}$$

where $S$ is the set of support vector indices. Because $\alpha_i$ are nonzero for the support vectors, the summation in (2.53) is added only for the support vectors. For the unbounded $\alpha_i$,

$$b = y_i - \mathbf{w}^\top \mathbf{x}_i \tag{2.54}$$

is satisfied. To ensure the precision of calculations, we take the average of $b$ that is calculated for unbounded support vectors:

$$b = \frac{1}{|U|} \sum_{i \in U} (y_i - \mathbf{w}^\top \mathbf{x}_i), \tag{2.55}$$

where $U$ is the set of unbounded support vector indices.

Then unknown data sample $\mathbf{x}$ is classified into

$$\begin{cases} \text{Class 1} & \text{if } D(\mathbf{x}) > 0, \\ \text{Class 2} & \text{if } D(\mathbf{x}) < 0. \end{cases} \tag{2.56}$$

If $D(\mathbf{x}) = 0$, $\mathbf{x}$ is on the boundary and thus is unclassifiable. When there are no bounded support vectors, the region $\{\mathbf{x} \,|\, 1 > D(\mathbf{x}) > -1\}$ is a generalization region, which is the same as the hard-margin support vector machine.

## 2.3 Mapping to a High-Dimensional Space

### 2.3.1 Kernel Tricks

In a support vector machine the optimal hyperplane is determined to maximize the generalization ability. But if the training data are not linearly separable, the obtained classifier may not have high generalization ability although the hyperplane is determined optimally. Thus to enhance linear separability, the original input space is mapped into a high-dimensional dot-product space called the *feature space*.

Now using the nonlinear vector function $\boldsymbol{\phi}(\mathbf{x}) = (\phi_1(\mathbf{x}), \dots, \phi_l(\mathbf{x}))^\top$ that maps the $m$-dimensional input vector $\mathbf{x}$ into the $l$-dimensional feature space, the linear decision function in the feature space is given by

$$D(\mathbf{x}) = \mathbf{w}^\top \boldsymbol{\phi}(\mathbf{x}) + b, \tag{2.57}$$

where $\mathbf{w}$ is an $l$-dimensional vector and $b$ is a bias term.

According to the Hilbert–Schmidt theory (see Appendix D), if a symmetric function $K(\mathbf{x}, \mathbf{x}')$ satisfies

$$\sum_{i,j=1}^{M} h_i h_j K(\mathbf{x}_i, \mathbf{x}_j) \geq 0 \qquad (2.58)$$

for all $M$, $\mathbf{x}_i$, and $h_i$, where $M$ takes on a natural number and $h_i$ take on real numbers, there exists a mapping function, $\phi(\mathbf{x})$, that maps $\mathbf{x}$ into the dot-product feature space and $\phi(\mathbf{x})$ satisfies

$$K(\mathbf{x}, \mathbf{x}') = \phi^{\top}(\mathbf{x}) \, \phi(\mathbf{x}'). \qquad (2.59)$$

If (2.59) is satisfied,

$$\sum_{i,j=1}^{M} h_i h_j K(\mathbf{x}_i, \mathbf{x}_j) = \left( \sum_{i=1}^{M} h_i \, \phi^{\top}(\mathbf{x}_i) \right) \left( \sum_{i=1}^{M} h_i \, \phi(\mathbf{x}_i) \right) \geq 0. \qquad (2.60)$$

The condition (2.58) or (2.60) is called *Mercer's condition*, and the function that satisfies (2.58) or (2.60) is called the *positive semidefinite kernel* or the *Mercer kernel*. In the following, if there is no confusion, we simply call it the *kernel*.[6]

The advantage of using kernels is that we need not treat the high-dimensional feature space explicitly. This technique is called *kernel trick*, namely, we use $K(\mathbf{x}, \mathbf{x}')$ in training and classification instead of $\phi(\mathbf{x})$ as shown later. The methods that map the input space into the feature space and avoid explicit treatment of variables in the feature space by kernel tricks are called *kernel methods* or *kernel-based methods*.

Using the kernel, the dual problem in the feature space is given as follows:

$$\text{maximize} \quad Q(\boldsymbol{\alpha}) = \sum_{i=1}^{M} \alpha_i - \frac{1}{2} \sum_{i,j=1}^{M} \alpha_i \, \alpha_j \, y_i \, y_j \, K(\mathbf{x}_i, \mathbf{x}_j) \qquad (2.61)$$

$$\text{subject to} \quad \sum_{i=1}^{M} y_i \, \alpha_i = 0, \quad 0 \leq \alpha_i \leq C \quad \text{for } i = 1, \ldots, M. \qquad (2.62)$$

There are various variants of support vector machines and if the word *regular support vector machine* or *standard support vector machine* is used, it denotes the above L1 soft-margin support vector machine.

Because $K(\mathbf{x}, \mathbf{x}')$ is a positive semidefinite kernel, the optimization problem is a concave quadratic programming problem. And because $\boldsymbol{\alpha} = \mathbf{0}$ is a feasible solution, the problem has the global optimum solution.

The KKT complementarity conditions are given by

---

[6] For the interpretation of indefinite kernels for classification, please see [7].

$$\alpha_i \left( y_i \left( \sum_{j=1}^{M} y_j \, \alpha_j \, K(\mathbf{x}_i, \mathbf{x}_j) + b \right) - 1 + \xi_i \right) = 0$$

$$\text{for} \quad i = 1, \ldots, M, \tag{2.63}$$

$$(C - \alpha_i)\, \xi_i = 0 \quad \text{for} \quad i = 1, \ldots, M, \tag{2.64}$$

$$\alpha_i \geq 0, \quad \xi_i \geq 0 \quad \text{for} \quad i = 1, \ldots, M. \tag{2.65}$$

The decision function is given by

$$D(\mathbf{x}) = \sum_{i \in S} \alpha_i \, y_i \, K(\mathbf{x}_i, \mathbf{x}) + b, \tag{2.66}$$

where $b$ is given by

$$b = y_j - \sum_{i \in S} \alpha_i y_i K(\mathbf{x}_i, \mathbf{x}_j). \tag{2.67}$$

Here, $\mathbf{x}_j$ is an unbounded support vector. To ensure stability of calculations, we take the average:

$$b = \frac{1}{|U|} \sum_{j \in U} \left( y_j - \sum_{i \in S} \alpha_i y_i K(\mathbf{x}_i, \mathbf{x}_j) \right), \tag{2.68}$$

where $U$ is the set of unbounded support vector indices.

Then unknown data are classified using the decision function as follows:

$$\mathbf{x} \in \begin{cases} \text{Class 1} & \text{if} \quad D(\mathbf{x}) > 0, \\ \text{Class 2} & \text{if} \quad D(\mathbf{x}) < 0. \end{cases} \tag{2.69}$$

If $D(\mathbf{x}) = 0$, $\mathbf{x}$ is unclassifiable.

## 2.3.2 Kernels

One of the advantages of support vector machines is that we can improve generalization performance by properly selecting kernels. Thus selection of kernels for specific applications is very important and development of new kernels is one of the ongoing research topics. In the following we discuss some of the kernels that are used in support vector machines. For the properties of kernels, see Appendix D.1. For classification of kernels from a statistical viewpoint, see [8].

## 2.3.2.1 Linear Kernels

If a classification problem is linearly separable in the input space, we need not map the input space into a high-dimensional space. In such a situation we use linear kernels:

$$K(\mathbf{x}, \mathbf{x}') = \mathbf{x}^\top \mathbf{x}'. \tag{2.70}$$

## 2.3.2.2 Polynomial Kernels

The polynomial kernel with degree $d$, where $d$ is a natural number, is given by

$$K(\mathbf{x}, \mathbf{x}') = (\mathbf{x}^\top \mathbf{x}' + 1)^d. \tag{2.71}$$

Here, 1 is added so that cross terms with degrees equal to or less than $d$ are all included.

When $d = 1$, the kernel is the linear kernel plus 1. Thus, by adjusting $b$ in the decision function, it is equivalent to the linear kernel. For $d = 2$ and $m = 2$, the polynomial kernel given by (2.71) becomes

$$\begin{aligned} K(\mathbf{x}, \mathbf{x}') &= 1 + 2x_1 x_1' + 2x_2 x_2' + 2x_1 x_1' x_2 x_2' + x_1^2 x_1'^2 + x_2^2 x_2'^2 \\ &= \boldsymbol{\phi}^\top(\mathbf{x})\, \boldsymbol{\phi}(\mathbf{x}), \end{aligned} \tag{2.72}$$

where $\boldsymbol{\phi}(\mathbf{x}) = (1, \sqrt{2}x_1, \sqrt{2}x_2, \sqrt{2}x_1x_2, x_1^2, x_2^2)^\top$. Thus for $d = 2$ and $m = 2$, polynomial kernels satisfy Mercer's condition. In general, we can prove that polynomial kernels satisfy Mercer's condition (see Appendix D.1).

Instead of (2.71), the following polynomial kernel can be used:

$$K(\mathbf{x}, \mathbf{x}') = (\mathbf{x}^\top \mathbf{x}')^d. \tag{2.73}$$

However, in this book we use (2.71), which is more general.[7]

*Example 2.2.* In Fig. 2.2 let $x_3 (= 1)$ belong to Class 1 so that the problem is inseparable (see Fig. 2.5). Using the polynomial kernel with degree 2, the dual problem is given as follows:

$$\text{maximize} \quad Q(\boldsymbol{\alpha}) = \alpha_1 + \alpha_2 + \alpha_3$$

$$-(2\alpha_1^2 + \frac{1}{2}\alpha_2^2 + 2\alpha_3^2 - \alpha_2(\alpha_1 + \alpha_3)) \tag{2.74}$$

$$\text{subject to} \quad \alpha_1 - \alpha_2 + \alpha_3 = 0, \tag{2.75}$$

$$C \geq \alpha_i \geq 0 \quad \text{for } i = 1, 2, 3. \tag{2.76}$$

From (2.75), $\alpha_2 = \alpha_1 + \alpha_3$. Then substituting it into (2.74), we obtain

---

[7] In Section 2.3.4, we will show that the mapping functions associated with (2.73) are many to one for even $d$, which is unfavorable.

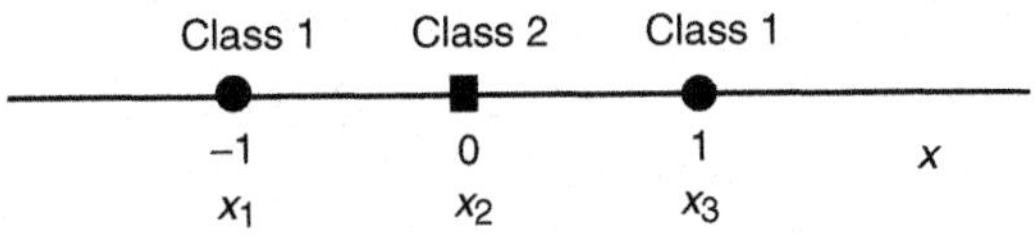

Fig. 2.5 Inseparable one-dimensional case

$$Q(\boldsymbol{\alpha}) = 2\,\alpha_1 + 2\,\alpha_3 - \left(2\,\alpha_1^2 - \frac{1}{2}\,(\alpha_1 + \alpha_3)^2 + 2\,\alpha_3^2\right), \qquad (2.77)$$

$$C \geq \alpha_i \geq 0 \quad \text{for } i = 1, 2, 3. \qquad (2.78)$$

From

$$\frac{\partial Q(\boldsymbol{\alpha})}{\partial \alpha_1} = 2 - 3\,\alpha_1 + \alpha_3 = 0, \qquad (2.79)$$

$$\frac{\partial Q(\boldsymbol{\alpha})}{\partial \alpha_3} = 2 + \alpha_1 - 3\,\alpha_3 = 0, \qquad (2.80)$$

$\alpha_1 = \alpha_3 = 1$. Thus, for $C \geq 2$, the optimal solution is

$$\alpha_1 = 1, \quad \alpha_2 = 2, \quad \alpha_3 = 1. \qquad (2.81)$$

Therefore, for $C \geq 2$, $x = -1$, 0, and 1 are support vectors. From (2.67), $b = -1$. Then the decision function is given by (see Fig. 2.6)

$$\begin{aligned} D(x) &= (x - 1)^2 + (x + 1)^2 - 3 \\ &= 2x^2 - 1. \end{aligned} \qquad (2.82)$$

The decision boundaries are given by $x = \pm\sqrt{2}/2$. Therefore, in the input space the margin for Class 2 is larger than that for Class 1, although they are the same in the feature space.

### 2.3.2.3 Radial Basis Function Kernels

The radial basis function (RBF) kernel is given by

$$K(\mathbf{x}, \mathbf{x}') = \exp(-\gamma \|\mathbf{x} - \mathbf{x}'\|^2), \qquad (2.83)$$

where $\gamma$ is a positive parameter for controlling the radius. Rewriting (2.83),

$$K(\mathbf{x}, \mathbf{x}') = \exp(-\gamma \|\mathbf{x}\|^2) \exp(-\gamma \|\mathbf{x}'\|^2) \exp(2\gamma \mathbf{x}^\top \mathbf{x}'). \qquad (2.84)$$

Because

$$\exp(2\gamma \mathbf{x}^\top \mathbf{x}') = 1 + 2\gamma \mathbf{x}^\top \mathbf{x}' + 2\gamma^2 (\mathbf{x}^\top \mathbf{x}')^2 + \frac{(2\gamma)^3}{3!} (\mathbf{x}^\top \mathbf{x}')^3 + \cdots, \qquad (2.85)$$

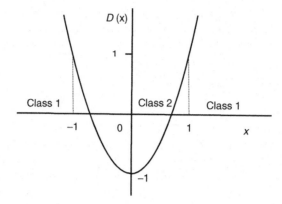

**Fig. 2.6** Decision function for the inseparable one-dimensional case

$\exp(2\gamma\,\mathbf{x}^\top\mathbf{x}')$ is an infinite summation of polynomials. Thus it is a kernel. In addition, $\exp(-\gamma\,\|\mathbf{x}\|^2)$ and $\exp(-\gamma\,\|\mathbf{x}'\|^2)$ are proved to be kernels and the product of kernels is also a kernel (see Appendix D.1). Thus (2.84) is a kernel.

From (2.66), the resulting decision function is given by

$$D(\mathbf{x}) = \sum_{i \in S} \alpha_i\, y_i\, \exp(-\gamma\,\|\mathbf{x}_i - \mathbf{x}\|^2) + b. \tag{2.86}$$

Here, the support vectors are the centers of the radial basis functions.

Because RBF kernels use the Euclidean distance, they are not robust to outliers. To overcome this, Chen [9] proposed the $M$-estimator-based robust kernels, which are robust versions of RBF kernels, introducing the idea of robust statistics.

### 2.3.2.4 Three-Layer Neural Network Kernels

The three-layer neural network kernel is given by

$$K(\mathbf{x}, \mathbf{x}') = \frac{1}{1 + \exp(\nu\,\mathbf{x}^\top\,\mathbf{x}' - a)}, \tag{2.87}$$

where $\nu$ and $a$ are constants. This kernel does not always satisfy Mercer's condition; we need to determine the values of $\nu$ and $a$ so that (2.58) is satisfied [4, p. 141], [5, p. 369].[8]

From (2.66), the resulting decision function is given by

---

[8] In [10], neural network kernels are shown to be indefinite.

$$D(\mathbf{x}) = \sum_{i \in S} \frac{\alpha_i y_i}{1 + \exp(\nu \, \mathbf{x}_i^\top \mathbf{x} - a)} + b. \tag{2.88}$$

The support vectors correspond to the weights between the input and hidden neurons in the three-layer neural network.

Because Mercer's condition is not always satisfied for three-layer neural network kernels, several approaches are made to overcome this problem [11, 12] (see Chapter 9).

### 2.3.2.5 Mahalanobis Kernels

In the RBF kernel, each input variable has equal weight in calculating the kernel value. But usually input variables have different data distributions each other and this may affect the generalization ability of the classifier. One way to avoid this problem is normalizing kernels as will be discussed in Section 2.3.3. Another way is to calculate, for each input variable, the average and the standard deviation of the training data, to subtract the average from the training data, and to divide the resulting value by the standard deviation. A more general method, whitening techniques, will be discussed later. In [13], training of the support vector machines is reformulated so that the radii of the RBF kernel for the input variables are determined during training.

In this section we discuss yet another approach: the Mahalanobis kernel. First we explain the Mahalanobis distance between a data sample and the center vector of a cluster. Let the set of $M$ $m$-dimensional data be $\{\mathbf{x}_1, \dots, \mathbf{x}_M\}$ for the cluster. Then the center vector and the covariance matrix of the data are given, respectively, by

$$\mathbf{c} = \frac{1}{M} \sum_{i=1}^{M} \mathbf{x}_i, \tag{2.89}$$

$$Q = \frac{1}{M} \sum_{i=1}^{M} (\mathbf{x}_i - \mathbf{c})(\mathbf{x}_i - \mathbf{c})^\top. \tag{2.90}$$

The Mahalanobis distance of $\mathbf{x}$ is given by

$$d(\mathbf{x}) = \sqrt{(\mathbf{x} - \mathbf{c})^\top Q^{-1} (\mathbf{x} - \mathbf{c})}. \tag{2.91}$$

Because the Mahalanobis distance is normalized by the covariance matrix, it is linear translation invariant [14]. This is especially important because we need not worry about the scales of input variables.

Another interesting characteristic is that the average of the square of Mahalanobis distances for the training data is $m$ [14]:

$$\frac{1}{M}\sum_{i=1}^{M}(\mathbf{x}_i - \mathbf{c})^{\top} Q^{-1}(\mathbf{x}_i - \mathbf{c}) = m. \qquad (2.92)$$

Based on the definition of the Mahalanobis distance, we define the Mahalanobis kernel [15, 16] by

$$K(\mathbf{x}, \mathbf{x}') = \exp\left(-(\mathbf{x} - \mathbf{x}')^{\top} A (\mathbf{x} - \mathbf{x}')\right), \qquad (2.93)$$

where $A$ is a positive definite matrix. Here, the Mahalanobis distance is calculated between $\mathbf{x}$ and $\mathbf{x}'$, not between $\mathbf{x}$ and $\mathbf{c}$.

The Mahalanobis kernel is an extension of the RBF kernel, namely, by setting

$$A = \gamma I, \qquad (2.94)$$

where $I$ is the $m \times m$ unit matrix, we obtain the RBF kernel:

$$K(\mathbf{x}, \mathbf{x}') = \exp(-\gamma\|\mathbf{x} - \mathbf{x}'\|^2). \qquad (2.95)$$

For a two-class problem, the Mahalanobis kernel is used for the data belonging to either of the two classes. Assuming that $X = \{\mathbf{x}_1, \ldots, \mathbf{x}_M\}$ is the set of all the training data, we calculate the center and the covariance matrix by (2.89) and (2.90), respectively.

Then we approximate the Mahalanobis kernel by

$$K(\mathbf{x}, \mathbf{x}') = \exp\left(-\frac{\delta}{m}(\mathbf{x} - \mathbf{x}')^{\top} Q^{-1}(\mathbf{x} - \mathbf{x}')\right), \qquad (2.96)$$

where $\delta\, (> 0)$ is the scaling factor to control the Mahalanobis distance.

From (2.92), by dividing the square of the Mahalanobis distance by $m$, the average of the square of the Mahalanobis distance is normalized to 1 irrespective of the number of input variables. If the support vectors are near $\mathbf{c}$, we can expect the effect of normalization, and this may enable to limit the search of the optimal $\delta$ value in a small range.

If we use the full covariance matrix, it will be time-consuming for a large number of input variables. Thus we consider two cases: Mahalanobis kernels with diagonal covariance matrices and Mahalanobis kernels with full covariance matrices. We call the former *diagonal Mahalanobis kernels* and the latter *non-diagonal Mahalanobis kernels*.

Mahalanobis kernels are closely related to whitening techniques. The whitening technique transforms the correlated input vector into the uncorrelated vector with each component having a unit variance.

For $m$-dimensional input $\mathbf{x}$, we calculate the center vector $\mathbf{c}$ and the covariance matrix $Q$ given, respectively, by (2.89) and (2.90). Then we calculate the eigenvalues and eigenvectors of $Q$:

$$Q S = S \Lambda, \qquad (2.97)$$

where $S$ is the $m \times m$ matrix with the $i$th column corresponding to the $i$th eigenvector of $Q$, and $\Lambda$ is the diagonal matrix with the $i$th diagonal element corresponding to the $i$th eigenvalue. Then we define vector $\mathbf{y}$ as follows:

$$\mathbf{y} = \Lambda^{-1/2} S^\top (\mathbf{x} - \mathbf{c}). \tag{2.98}$$

It is easy to show that $\mathbf{y}$ has a zero mean and a diagonal covariance matrix with all diagonal elements being 1. In addition, because

$$\begin{aligned} \mathbf{y}^\top \mathbf{y} &= (\mathbf{x} - \mathbf{c})^\top S\Lambda^{-1} S^\top (\mathbf{x} - \mathbf{c}) \\ &= (\mathbf{x} - \mathbf{c})^\top Q^{-1}(\mathbf{x} - \mathbf{c}), \end{aligned} \tag{2.99}$$

$\mathbf{y}^\top \mathbf{y}$ is the square of the Mahalanobis distance. Thus, a non-diagonal Mahalanobis kernel behaves similarly to an RBF kernel with the whitening input.

### 2.3.2.6 Kernels for Odd or Even Functions

If we know some prior knowledge about a classification or regression problem, incorporating it into the development process will lead to improving the generalization ability of the classifier or regressor. One piece of such prior knowledge for regression is that the decision function is odd or even [17, 18]. Let the decision function be

$$D(\mathbf{x}) = \mathbf{w}^\top \phi(\mathbf{x}) + b. \tag{2.100}$$

Then $D(\mathbf{x})$ is even or odd if

$$\mathbf{w}^\top \phi(\mathbf{x}) = a\,\mathbf{w}^\top \phi(-\mathbf{x}), \tag{2.101}$$

where $a = 1$ for even and $a = -1$ for odd.

Assume that the kernel that we are going to use has the following properties:

$$K(\mathbf{x}, -\mathbf{x}') = K(-\mathbf{x}, \mathbf{x}'), \qquad K(\mathbf{x}, \mathbf{x}') = K(-\mathbf{x}, -\mathbf{x}'). \tag{2.102}$$

Here, all the preceding kernels satisfy the above equations. Then training of a support vector machine with the following constraints

$$\mathbf{w}^\top \phi(\mathbf{x}_i) = a\,\mathbf{w}^\top \phi(-\mathbf{x}_i) \quad \text{for } i = 1, \ldots, M \tag{2.103}$$

results in training the support vector machine with the following equivalent kernel:

$$K_{\text{eq}}(\mathbf{x}, \mathbf{x}') = \frac{1}{2}\left(K(\mathbf{x}, \mathbf{x}') + a\,K(\mathbf{x}, -\mathbf{x}')\right). \tag{2.104}$$

### 2.3.2.7 Other Kernels

In most applications, RBF kernels are widely used but kernels suited for specific applications are developed such as spline functions [19–21] for regression. Howley and Madden proposed [22] tree-structured kernels, e.g.,

$$K(\mathbf{x}, \mathbf{x}') = \exp(((\mathbf{x}+\mathbf{x})-\mathbf{x}')^{\top}((\mathbf{x}'+\mathbf{x}')-\mathbf{x}) \times 58.35/(22.15 \times 27.01)), \quad (2.105)$$

where the argument of the exponential function is expressed in a tree structure and is determined by genetic programming.

In addition, many kernels have been developed for specific applications, where inputs have variable lengths such as image processing, text classification, and speech recognition. Because support vector machines are based on fixed-length inputs, we need to extract fixed-length features or extend kernels that handle variable-length inputs. In the following, we discuss some of their kernels.

Image Kernels

Because correlation among image pixels is localized, global kernels such as polynomial kernels are inadequate for image classification. One way to overcome this problem is to devise new features. For instance, local histograms of edge directions called *histograms of oriented gradients* [23] are known to be good features for object detection. Another approach is to develop local kernels for image processing [24–28]. Barla et al. [26] discussed two image kernels: histogram intersection kernels, which measure the similarity of two color images, and Hausdorff kernels, which measure the similarity of two grayscale images. Hotta [27] used the summation of RBF kernels for robust occluded face recognition. Grauman and Darrel [28] proposed pyramid match kernels which partially match multiresolution histograms obtained from sets of unordered local features.

String Kernels

In text classification, documents are classified into one of several topics. The common approach uses, as features, a histogram of words called a *bag of words* [29]. And there are much work on developing kernels that treat strings called *string kernels* [30–35]. Lodhi et al. [30, 31] used string kernels that give the similarity of a common substring in two documents. Let the substring length be 2 and the words be *cat* and *bat*. Then we obtain five substrings: c-a, c-t, a-t, b-a, and b-t, which constitute variables in the feature space. The mapping function is defined so that the substring that is nearer in a document (in this case a word) has a higher score, as shown in Table 2.1, where $0 < \lambda < 1$. For example, c-a in *cat*, which is a contiguous substring, has a higher score than c-t in *cat*. The resulting kernel is calculated as follows:

$$K(cat, bat) = \lambda^4$$
$$K(cat, cat) = K(bat, bat) = 2\lambda^4 + \lambda^6.$$

The whole document is mapped into one feature space, ignoring punctuation and retaining spaces. For a long substring length, evaluation of kernels is sped up by dynamic programming.

**Table 2.1** Mapping function for *cat* and *bat*

|  | c-a | c-t | a-t | b-a | b-t |
|---|---|---|---|---|---|
| $\phi(cat)$ | $\lambda^2$ | $\lambda^3$ | $\lambda^2$ | 0 | 0 |
| $\phi(bat)$ | 0 | 0 | $\lambda^2$ | $\lambda^2$ | $\lambda^3$ |

Leslie et al. [33, 34] developed mismatch kernels, which belong to a class of string kernels, for protein classification. The mapping function of the $(k, m)$-mismatch kernel maps the space of all finite strings with an alphabet of size $l$ to an $l^k$-dimensional space, i.e., all the combinations of $k$-length strings. For the input string of length $k$, the value of a variable in the feature space is 1 if the input string differs from the variable string at at most $k$ mismatches and 0 otherwise. For an arbitrary input, the value of a variable is a mismatch count of all $k$-substrings included in the input. Then the kernel value is calculated by the dot-product of the variables in the feature space.

Graph Kernels

Graphs, which consist of nodes and edges that connect nodes, are used to represent symbolic data structures. There are two types of graph kernels: those that define similarities of two nodes in a graph [36–39] and those that define similarities between graphs [40–43]. In the following, we discuss these two types of graph kernels.

We begin with necessary definitions to discuss graph kernels defined in a graph. Consider an undirected graph with $k$ nodes. If nodes $i$ and $j$ are connected with an edge, a positive weight $w_{ij}$ is assigned, $w_{ij} = 0$ otherwise. If we do not consider weights, $w_{ij} = 1$ if nodes $i$ and $j$ are connected, and 0 otherwise. Then the adjacency matrix $A$ is $A = \{a_{ij}\} = \{w_{ij}\}$. The Laplacian matrix $L$ is given by $L = D - A$, where $D$ is a diagonal matrix and the diagonal elements $d_{ii}$ are given by

$$d_{ii} = \sum_{j=1}^{k} a_{ij} \qquad \text{for} \quad i = 1, \ldots, k. \tag{2.106}$$

The Laplacian matrix is symmetric and positive semidefinite.

Graph kernels are defined in the form of kernel matrices. The diffusion kernel matrix $K_D$ [36, 37] is defined by

$$K_D = \exp(-\beta L), \tag{2.107}$$

where $\beta$ is a positive parameter, $\exp(-\beta L)$ is a $k \times k$ matrix and is called a *matrix exponential function* defined by

$$\exp(-\beta L) = I - \beta L + \frac{1}{2!}(\beta L)^2 - \frac{1}{3!}(\beta L)^3 + \cdots. \tag{2.108}$$

The $(i, j)$ element in $L^j$ reflects the relation between nodes $i$ and $j$ affected by paths, consisting of $j$ edges, that connect these nodes. Therefore, kernel $K_D(i, j)$, which is the $(i, j)$ element of $K_D$, reflects the aggregate of relations between nodes $i$ and $j$ affected by all the paths.

The regularized Laplacian kernel matrix [37, 38] is defined by

$$K_{RL} = \sum_{k=0}^{\infty} (-\beta L_\gamma)^k = (I + \beta L_\gamma)^{-1}, \tag{2.109}$$

where $L_\gamma = \gamma D - A$ and $\beta$ and $\gamma$ are positive parameters and $0 \leq \gamma \leq 1$. The second term in the above equation to converge, the product of $\beta$ and the maximum eigenvalue of $L_\gamma$ needs to be less than 1. The kernel without the $\gamma$ parameter is proposed in [37] and is extended in [38].

Graph kernels are used for analyzing databases such as in the computer science [38, 39], chemical, and biological fields.

Kernels for defining similarities between graphs are also used in the computer science, chemical [44], and biological fields [45]. There are several approaches to define kernels between graphs [40–43]. Here we discuss graph kernels [42] based on dissimilarity measures [46]. We define the graph edit distance $d(g, g')$ between graphs $g$ and $g'$ as the minimum distance that is needed to transform one graph into another by insertion, deletion, and/or substitution of nodes and/or edges. There may be several transformations to reach from one graph to another. To select the minimum distance, we assign a cost for each edit operation and the minimum edit distance is defined by the minimum sum cost of the edit operations.

From given training graphs, we select $m$ prototypes $\{g_1, \cdots, g_m\}$. In [42], five prototype selection methods are shown. Here, we introduce one of them called *targetsphere prototype selector*. First, from the training graphs we select the center graph $g_c$, in which the maximum distance to the remaining graphs is minimum. Then we find the graph $g_f$ with the maximum distance from the center graph. These two graphs are selected as prototypes. Then as additional prototypes we select $m - 2$ graphs with the distances to the center graph nearest to $k \, d(g_c, g_f)/(m - 1)$ $(k = 1, \ldots, m - 2)$.

Then we define the mapping function $\phi(g)$ by

$$\phi(g) = (d(g, g_1), \ldots, d(g, g_m))^\top. \tag{2.110}$$

Using (2.110), graph kernels can be defined as follows:

$$\text{linear kernels:} \quad K(g, g') = \phi^\top(g) \, \phi(g') \tag{2.111}$$
$$\text{polynomial kernels:} \, K(g, g') = (\phi^\top(g) \, \phi(g') + 1)^d \tag{2.112}$$
$$\text{RBF kernels:} \quad K(g, g') = \exp(-\gamma \|\phi(g) - \phi(g')\|^2) \tag{2.113}$$

where $d$ is the degree of the polynomial and $\gamma$ is a positive parameter.

Kernels for Speech Recognition

In speech recognition, sequences of data with different lengths need to be matched [47–49]. Shimodaira et al. [47] proposed dynamic time-alignment kernels. Let two sequences of vectors be $X = (\mathbf{x}_1, \ldots, \mathbf{x}_m)$ and $Y = (\mathbf{y}_1, \ldots, \mathbf{y}_n)$, where $m$ and $n$ are, in general, different. The dynamic time-alignment kernel is defined by

$$K_s(X, V) = \max_{\phi, \theta} \frac{1}{M_{\phi\theta}} \sum_{k=1}^{L} m(k) \, \phi^\top(\mathbf{x}_{\phi(k)}) \, \phi(\mathbf{x}_{\theta(k)})$$

$$= \max_{\phi, \theta} \frac{1}{M_{\phi\theta}} \sum_{k=1}^{L} m(k) \, K(\mathbf{x}_{\phi(k)}, \mathbf{x}_{\theta(k)}), \tag{2.114}$$

where $L$ is the normalized length, $m(k)$ are weights, $M_{\phi\theta}$ is a normalizing factor, $\phi(\cdot)$ and $\theta(\cdot)$ are time-alignment functions that align two sequences by dynamic programming so that the two become similar.

## 2.3.3 Normalizing Kernels

Each input variable has a different physical meaning and thus has a different range. But if we use the original input range, and if we do not use Mahalanobis kernels, the result may not be good because support vector machines are not invariant for the linear transformation of inputs as will be discussed in Section 2.9. Thus to make each input variable to work equally in classification, we normalizes the input variables either by scaling the ranges of input variables into $[0, 1]$ or $[-1, 1]$ or by whitening the input variables.

However, even if we normalize the input variables, if the number of input variables is very large, the value of a kernel becomes so small or large that training of support vector machines becomes difficult. To overcome this, it is advisable to normalize kernels.

For a polynomial kernel with degree $d$, the maximum value is $(m+1)^d$ for the input range of $[0,1]$, where $m$ is the number of input variables. Thus we use the following normalized polynomial:

$$K(\mathbf{x}, \mathbf{x}') = \frac{(\mathbf{x}^\top \mathbf{x}' + 1)^d}{(m+1)^d} \qquad (2.115)$$

for large $m$. In the computer experiments in this book, we used (2.115) for $m$ larger than or equal to 100.

For an RBF kernel, the maximum value of $\|\mathbf{x} - \mathbf{x}'\|^2$ is $m$ for the input range of $[0,1]$. Thus we use the following normalized RBF kernel:

$$K(\mathbf{x}, \mathbf{x}') = \exp\left(-\frac{\gamma}{m}\|\mathbf{x} - \mathbf{x}'\|^2\right). \qquad (2.116)$$

The use of (2.116) is favorable for choosing a proper value of $\gamma$ for problems with different numbers of input variables.

### 2.3.4 Properties of Mapping Functions Associated with Kernels

#### 2.3.4.1 Preserving Neighborhood Relations

Is the neighborhood of $\mathbf{x}$ mapped into the neighborhood of $\phi(\mathbf{x})$? This is proved to be true for RBF kernels [50], namely, for $\mathbf{x}$, $\mathbf{x}_1$, and $\mathbf{x}_2$ that satisfy

$$\|\mathbf{x} - \mathbf{x}_1\| < \|\mathbf{x} - \mathbf{x}_2\|, \qquad (2.117)$$

the following inequality holds:

$$\|\phi(\mathbf{x}) - \phi(\mathbf{x}_1)\| < \|\phi(\mathbf{x}) - \phi(\mathbf{x}_2)\|. \qquad (2.118)$$

This can be proved as follows: Equation (2.118) is equivalent to

$$K(\mathbf{x}, \mathbf{x}) + K(\mathbf{x}_1, \mathbf{x}_1) - 2K(\mathbf{x}, \mathbf{x}_1) < K(\mathbf{x}, \mathbf{x}) + K(\mathbf{x}_2, \mathbf{x}_2) - 2K(\mathbf{x}, \mathbf{x}_2). \qquad (2.119)$$

For RBF kernels, $K(\mathbf{x}, \mathbf{x}) = 1$ holds. Thus, (2.119) reduces to

$$K(\mathbf{x}, \mathbf{x}_1) > K(\mathbf{x}, \mathbf{x}_2), \qquad (2.120)$$

which means

$$\exp(-\gamma\|\mathbf{x} - \mathbf{x}_1\|^2) > \exp(-\gamma\|\mathbf{x} - \mathbf{x}_2\|^2). \tag{2.121}$$

From (2.117), (2.121) is satisfied. Therefore, (2.118) is satisfied.

Likewise, for Mahalanobis kernels, instead of (2.117), if we define the neighborhood in the input space by

$$(\mathbf{x} - \mathbf{x}_1)^\top Q^{-1}(\mathbf{x} - \mathbf{x}_1) < (\mathbf{x} - \mathbf{x}_2)^\top Q^{-1}(\mathbf{x} - \mathbf{x}_2), \tag{2.122}$$

(2.118) is satisfied.

For polynomial kernels given by (2.71), the neighborhood relation given by (2.117) and (2.118) is satisfied if $\|\mathbf{x}\| = \|\mathbf{x}_1\| = \|\mathbf{x}_2\|$, namely, if data are normalized with equal norms.

This is an interesting property; if we want to know the neighborhood relations between mapped training data, instead of calculating the Euclidian distances in the feature space, we need only to calculate the Euclidian or Mahalanobis distances in the input space. For example, $k$-nearest neighbors in the feature space are equivalent to those in the input space.

### 2.3.4.2 One-to-One Mapping

If the mapping function associated with a nonlinear kernel is many-to-one mapping, namely, more than one point in the input space are mapped to a single point in the feature space, a classification problem may become inseparable in the feature space even if the problem is separable in the input space. In the following, we discuss that this does not happen for polynomial kernels with constant terms and RBF kernels. But for polynomial kernels without constant terms, many-to-one mapping may occur.

The mapping function $\phi(\mathbf{x})$ associated with the polynomial kernel given by (2.71) is

$$\phi(\mathbf{x}) = (1, \sqrt{d}\,x_1, \ldots, \sqrt{d}\,x_m, \ldots, x_1^d, \ldots, x_m^d)^\top. \tag{2.123}$$

Thus, if $\phi(\mathbf{x}) = \phi(\mathbf{x}')$, $\mathbf{x} = \mathbf{x}'$. Therefore, mapping is one to one.

But for $K(\mathbf{x}, \mathbf{x}') = (\mathbf{x}^\top \mathbf{x}')^d$ with even $d$, the mapping is many to one. For example, for $d = 2$,

$$\phi(\mathbf{x}) = (x_1^2, \ldots, x_m^2, \sqrt{2}\,x_1\,x_2, \ldots, \sqrt{2}\,x_{m-1}\,x_m)^\top. \tag{2.124}$$

Thus, if $\phi(\mathbf{x}) = \phi(\mathbf{x}')$,

$$x_1^2 = x_1'^2,$$
$$\cdots$$
$$x_m^2 = x_m'^2,$$
$$x_1\,x_2 = x_1'\,x_2',$$
$$\cdots$$

$$x_{m-1}\, x_m \;=\; x'_{m-1}\, x'_m.$$

This set of simultaneous equations is satisfied if $\mathbf{x} = \mathbf{x}'$ or $\mathbf{x} = -\mathbf{x}'$. For the input range of $[0,1]$, the mapping is one to one, because we can exclude $\mathbf{x} = -\mathbf{x}'$. But for the range of $[-1,1]$, the mapping is two to one. A similar discussion holds for even $d$. Therefore, we should avoid using the scale $[-1,1]$ for even $d$.

For the RBF kernel, if $\phi(\mathbf{x}) = \phi(\mathbf{x}')$,

$$
\begin{aligned}
K(\mathbf{x}, \mathbf{x}') = \phi^\top(\mathbf{x})\, \phi(\mathbf{x}') &= \exp\left(-\gamma \, \|\mathbf{x} - \mathbf{x}'\|^2\right) \\
&= \phi^\top(\mathbf{x})\, \phi(\mathbf{x}) \\
&= K(\mathbf{x}, \mathbf{x}) \\
&= 1.
\end{aligned}
$$

Therefore, $\mathbf{x} = \mathbf{x}'$. Thus, the mapping is one to one. This is also true for Mahalanobis kernels.

### 2.3.4.3 Nonconvexity of Mapped Regions

We clarify that the mapped region is not convex, namely, the preimage of a linear sum of the mapped input data does not exist.

Let $G$ be the mapped region of the $m$-dimensional input space $R^m$ by $\phi(\cdot)$:

$$G = \{\phi(\mathbf{x}) \,|\, \mathbf{x} \in R^m\}. \tag{2.125}$$

Then for polynomial, RBF, and Mahalanobis kernels, $G$ is not convex, namely, for $\mathbf{z}$ and $\mathbf{z}' \in G$, there exists $\alpha$ $(1 > \alpha > 0)$ such that $\alpha \mathbf{z} + (1 - \alpha)\mathbf{z}'$ is not included in $G$. This can be shown as follows.

For a polynomial kernel, it is sufficient to show nonconvexity for the axis $x_i^d$. Because $\alpha x_i^d + (1 - \alpha) x_i'^d$ $(1 > \alpha > 0)$ is a line segment that connects $x_i^d$ and $x_i'^d$, there is no $x$ that satisfies $x^d = \alpha x_i^d + (1 - \alpha) x_i'^d$ (see Fig. 2.7). Thus, $G$ is not convex.

For an RBF kernel, because $K(\mathbf{x}, \mathbf{x}) = \phi^\top(\mathbf{x})\, \phi(\mathbf{x}) = 1$, $\phi(\mathbf{x})$ is on the surface of the hypersphere with the radius of 1 and the center being at the origin. Thus $\alpha \mathbf{z} + (1 - \alpha)\mathbf{z}'$ for $1 > \alpha > 0$ is not in $G$. Similarly, for a Mahalanobis kernel, the above relation holds.

Now, for $\mathbf{z} \notin G$, there is no $\mathbf{x}$ that satisfies $\phi(\mathbf{x}) = \mathbf{z}$, namely, the preimage of $\mathbf{z}$ exists if and only if $\mathbf{z} \in G$. And, in general for $\mathbf{z}_1, \ldots, \mathbf{z}_k \in G$, the preimage of a linear sum of $\mathbf{z}_1, \ldots, \mathbf{z}_k$ does not exist.

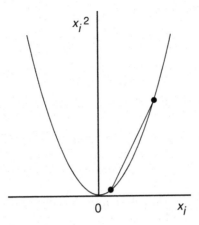

**Fig. 2.7** Nonconvexity of the mapped region. For $d = 2$, the line segment in the $x_i^2$ axis is outside of the mapped region

## 2.3.5 Implicit Bias Terms

If a kernel includes a constant term, instead of (2.57) we can use the decision function without an explicit bias term:

$$D(\mathbf{x}) = \mathbf{w}^\top \boldsymbol{\phi}(\mathbf{x}). \tag{2.126}$$

In this case, the dual optimization problem becomes as follows:

$$\text{maximize} \quad Q(\boldsymbol{\alpha}) = \sum_{i=1}^{M} \alpha_i - \frac{1}{2} \sum_{i,j=1}^{M} \alpha_i \, \alpha_j \, y_i \, y_j \, K(\mathbf{x}_i, \mathbf{x}_j) \tag{2.127}$$

$$\text{subject to} \quad C \geq \alpha_i \geq 0 \quad \text{for } i = 1, \dots, M. \tag{2.128}$$

The decision function is given by

$$D(\mathbf{x}) = \sum_{i \in S} \alpha_i \, y_i \, K(\mathbf{x}_i, \mathbf{x}). \tag{2.129}$$

Elimination of the bias term results in elimination of the equality constraint in the dual problem. Thus the problem is more easily solved. Polynomial, RBF, and Mahalanobis kernels include bias terms. And even if a kernel does not include a constant term, adding 1, we obtain the kernel with a constant term. For $K(\mathbf{x}, \mathbf{x}') = \boldsymbol{\phi}^\top(\mathbf{x}) \, \boldsymbol{\phi}(\mathbf{x})$, the kernel

$$K'(\mathbf{x}, \mathbf{x}') = K(\mathbf{x}, \mathbf{x}') + 1$$
$$= (\boldsymbol{\phi}^\top(\mathbf{x}), 1) \, (\boldsymbol{\phi}^\top(\mathbf{x}), 1)^\top \tag{2.130}$$

has a constant term. For example, for the linear kernel $K(\mathbf{x}, \mathbf{x}') = \mathbf{x}^\top \mathbf{x}$, $K'(\mathbf{x}, \mathbf{x}') = \mathbf{x}^\top \mathbf{x}' + 1$ has a constant term.

Unfortunately, the solutions of both formulations are, in general, different even for the linear kernels [51, p. 22] as Example 2.3 shows. According to the computer experiment by Huang and Kecman [52], the solution of the support vector machine without bias terms has a lager number of support vectors. Therefore, to suppress the increase, they developed a model that includes Mangasarian and Musicant's model as a special case.

*Example 2.3.* Suppose that there are only two data: $\mathbf{x}_1$ belonging to Class 1 and $\mathbf{x}_2$ belonging to Class 2. Let $K'(\mathbf{x}, \mathbf{x}') = K(\mathbf{x}, \mathbf{x}') + 1$ and assume that $K(\mathbf{x}_i, \mathbf{x}_j) \geq 0$ for $i, j = 1, 2$.[9]

We study the relation of the solutions with and without bias terms. Assume that $\mathbf{x}_1 \neq \mathbf{x}_2$. Then the problem is linearly separable. So we consider the hard-margin support vector machine. The dual problem with the explicit bias term is as follows:

$$\text{maximize} \quad Q(\boldsymbol{\alpha}) = \alpha_1 + \alpha_2 - \frac{1}{2}\left(\alpha_1^2 K_{11} + \alpha_2^2 K_{22} - 2\alpha_1 \alpha_2 K_{12}\right) \quad (2.131)$$

$$\text{subject to} \quad \alpha_1 - \alpha_2 = 0, \quad \alpha_1 \geq 0, \quad \alpha_2 \geq 0. \quad (2.132)$$

Here, for example, $K_{12} = K(\mathbf{x}_1, \mathbf{x}_2)$.

Substituting the first equation in (2.132) into (2.131), we obtain the following solution:

$$\alpha_1 = \alpha_2 = \frac{2}{K}, \quad (2.133)$$

$$b = \frac{1}{K}\left(K_{22} - K_{11}\right), \quad (2.134)$$

where $K = K_{11} + K_{22} - 2K_{12} > 0$.

Thus the decision function is given by

$$D(\mathbf{x}) = \frac{2}{K}\left(K(\mathbf{x}_1, \mathbf{x}) - K(\mathbf{x}_2, \mathbf{x})\right) + \frac{1}{K}\left(K_{22} - K_{11}\right). \quad (2.135)$$

From (2.135), $D(\mathbf{x}_1) = 1$ and $D(\mathbf{x}_2) = -1$, which satisfy the KKT conditions. For $K(\mathbf{x}, \mathbf{x}') = \mathbf{x}^\top \mathbf{x}'$, (2.135) becomes

$$D(\mathbf{x}) = \frac{1}{\|\mathbf{x}_1 - \mathbf{x}_2\|^2}\left(2\left(\mathbf{x}_1 - \mathbf{x}_2\right)^\top \mathbf{x} + \mathbf{x}_2^\top \mathbf{x}_2 - \mathbf{x}_1^\top \mathbf{x}_1\right). \quad (2.136)$$

Thus the decision function is orthogonal to $(\mathbf{x}_1 - \mathbf{x}_2)$ and the decision boundary passes through the middle point of $\mathbf{x}_1$ and $\mathbf{x}_2$.

Now consider the special case where the input variable has one dimension. For $K(x, x') = (x\,x')^d$, the decision function is

---

[9] This assumption is satisfied when the input variables are nonnegative and $K(\mathbf{x}, \mathbf{x}') = (\mathbf{x}^\top \mathbf{x})^d$.

$$D(\mathbf{x}) = \frac{2x^d - x_1^d - x_2^d}{x_1^d - x_2^d}. \tag{2.137}$$

Thus the decision function vanishes when

$$x = \sqrt[d]{\frac{x_1^d + x_2^d}{2}}. \tag{2.138}$$

This means that for $d = 1$, the decision boundary passes through the middle point of $x_1$ and $x_2$, but for $d$ larger than 1, it passes through the middle point of $\phi(x_1) = x_1^d$ and $\phi(x_2) = x_2^d$, which is different from $(x_1 + x_2)/2$.

The dual problem with the implicit bias term is as follows:

maximize $\quad Q(\boldsymbol{\alpha}) = \alpha_1 + \alpha_2 - \dfrac{1}{2}(\alpha_1^2 K_{11}' + \alpha_2^2 K_{22}' - 2\alpha_1\alpha_2 K_{12}') \tag{2.139}$

subject to $\quad \alpha_1 \geq 0, \alpha_2 \geq 0. \tag{2.140}$

Partially differentiating (2.139) with respect to $\alpha_1$ and $\alpha_2$ and equating them to 0, we obtain the following solution:

$$\alpha_1 = \frac{K_{12}' + K_{22}'}{K_{11}' K_{22}' - (K_{12}')^2}, \tag{2.141}$$

$$\alpha_2 = \frac{K_{11}' + K_{12}'}{K_{11}' K_{22}' - (K_{12}')^2}. \tag{2.142}$$

According to the assumption, $\alpha_1$ and $\alpha_2$ are nonnegative.

Then the decision function is given by

$$\begin{aligned}
D(\mathbf{x}) &= \frac{1}{K_{11}' K_{22}' - (K_{12}')^2} \\
&\quad \times ((K_{12}' + K_{22}')\, K'(\mathbf{x}_1, \mathbf{x}) - (K_{11}' + K_{12}')\, K'(\mathbf{x}_2, \mathbf{x})) \\
&= \frac{1}{(K_{11} + 1)(K_{22} + 1) - (K_{12} + 1)^2} \\
&\quad \times ((K_{12} + K_{22} + 2)\, K(\mathbf{x}_1, \mathbf{x}) - (K_{11} + K_{12} + 2)\, K(\mathbf{x}_2, \mathbf{x}) \\
&\quad + K_{22} - K_{11}). \tag{2.143}
\end{aligned}$$

Thus, $D(\mathbf{x}_1) = 1$ and $D(\mathbf{x}_2) = -1$, which satisfy the KKT conditions. For $K'(\mathbf{x}, \mathbf{x}') = \mathbf{x}^\top \mathbf{x}' + 1$, (2.143) becomes

$$\begin{aligned}
D(\mathbf{x}) &= \frac{1}{(\mathbf{x}_1^\top \mathbf{x}_1 + 1)(\mathbf{x}_2^\top \mathbf{x}_2 + 1) - (\mathbf{x}_1^\top \mathbf{x}_2 + 1)^2} \\
&\quad \times ((((\mathbf{x}_1 + \mathbf{x}_2)^\top \mathbf{x}_2 + 2)\, \mathbf{x}_1^\top - (\mathbf{x}_1^\top(\mathbf{x}_1 + \mathbf{x}_2) + 2)\, \mathbf{x}_2^\top)\, \mathbf{x} \\
&\quad + \mathbf{x}_2^\top \mathbf{x}_2 - \mathbf{x}_1^\top \mathbf{x}_1). \tag{2.144}
\end{aligned}$$

Because $D(\mathbf{x}_1) = 1$ and $D(\mathbf{x}_2) = -1$ and (2.144) is linear, the decision boundary passes through the middle point of $\mathbf{x}_1$ and $\mathbf{x}_2$. But because in general (2.144) is not orthogonal to $(\mathbf{x}_1 - \mathbf{x}_2)$, (2.137) and (2.144) are different.

Now consider the special case where the two decision functions are equal. Let the dimension of the input variable be 1 and $K'(x, x') = (x\,x')^d + 1$. Then the decision function (2.143) becomes

$$D(\mathbf{x}) = \frac{2x^d - x_2^d - x_1^d}{x_1^d - x_2^d},\qquad (2.145)$$

which is equivalent to (2.137).

## 2.3.6 Empirical Feature Space

One of the advantages of support vector machines is that using kernels we can realize nonlinear separation without treating the variables in the feature space. However, because the feature space is implicit and for RBF kernels the dimension is infinite, it is difficult to analyze the mapped data in the feature space. To solve this problem Xiong et al. [53] proposed the empirical feature space whose dimension is the number of training data at most and whose kernel matrix is equivalent to that for the feature space. In this section we discuss the empirical feature space based on [53, 54]. First, we define the empirical feature space spanned by the training data and prove that the kernels associated with the feature space and with the empirical feature space give the same value if one of the argument of the kernels is mapped in the empirical feature space by the mapping function associated with the feature space.

### 2.3.6.1 Definition of Empirical Feature Space

Let the kernel be $K(\mathbf{x}, \mathbf{x}') = \boldsymbol{\phi}^\top(\mathbf{x})\,\boldsymbol{\phi}(\mathbf{x})$, where $\boldsymbol{\phi}(\mathbf{x})$ is the mapping function that maps $m$-dimensional vector $\mathbf{x}$ into the $l$-dimensional space. For the $M$ $m$-dimensional data $\mathbf{x}_i$, the $M \times M$ kernel matrix $K = \{K_{ij}\}$ $(i, j = 1, \ldots M)$, where $K_{ij} = K(\mathbf{x}_i, \mathbf{x}_j)$, is symmetric and positive semidefinite. Let the rank of $K$ be $N\,(\leq M)$. Then $K$ is expressed by

$$K = U\,S\,U^\top,\qquad (2.146)$$

where the column vectors of $U$ are eigenvectors of $K$ and $U\,U^\top = U^\top U = I_{M \times M}$, $I_{M \times M}$ is the $M \times M$ unit matrix, and $S$ is given by

$$S = \begin{pmatrix} \sigma_1 & & 0 & \\ & \ddots & & 0_{N \times (M-N)} \\ 0 & & \sigma_N & \\ \hline 0_{(M-N) \times N} & & 0_{(M-N) \times (M-N)} \end{pmatrix}. \tag{2.147}$$

Here, $\sigma_i (> 0)$ are eigenvalues of $K$, whose eigenvectors correspond to the $i$th columns of $U$, and for instance $0_{(M-N) \times N}$ is the $(M - N) \times N$ zero matrix.

Defining the first $N$ vectors of $U$ as the $M \times N$ matrix $P$ and

$$\Lambda = \begin{pmatrix} \sigma_1 & & 0 \\ & \ddots & \\ 0 & & \sigma_N \end{pmatrix}, \tag{2.148}$$

we can rewrite (2.146) as follows:

$$K = P \Lambda P^{\top}, \tag{2.149}$$

where $P^{\top} P = I_{N \times N}$ but $P P^{\top} \neq I_{M \times M}$.

We must notice that if $N < M$, the determinant of $K$ vanishes. Thus, from $K(\mathbf{x}, \mathbf{x}') = \boldsymbol{\phi}^{\top}(\mathbf{x}) \boldsymbol{\phi}(\mathbf{x})$, the following equation holds:

$$\sum_{i=1}^{M} a_i \boldsymbol{\phi}^{\top}(\mathbf{x}_i) = 0, \tag{2.150}$$

where $a_i (i = 1, \ldots, M)$ are constant and some of them are nonzero, namely, if $N < M$, the mapped training data $\boldsymbol{\phi}(\mathbf{x}_i)$ are linearly dependent. And if $N = M$, they are linearly independent and there are no non-zero $a_i$ that satisfy (2.150).

Now we define the mapping function that maps the $m$-dimensional vector $\mathbf{x}$ into the $N$-dimensional space called *empirical feature space* [53]:

$$\mathbf{h}(\mathbf{x}) = \Lambda^{-1/2} P^{\top} (K(\mathbf{x}_1, \mathbf{x}), \ldots, K(\mathbf{x}_M, \mathbf{x}))^{\top}. \tag{2.151}$$

We define the kernel associated with the empirical feature space by

$$K_{\mathrm{e}}(\mathbf{x}, \mathbf{x}') = \mathbf{h}^{\top}(\mathbf{x}) \mathbf{h}(\mathbf{x}'). \tag{2.152}$$

Clearly, the dimension of the empirical feature space is $N$, namely, the empirical feature space is spanned by the linearly independent mapped training data.

Now we prove that the kernel for the empirical feature space is equivalent to the kernel for the feature space if they are evaluated using the training data, namely [53],

$$K_e(\mathbf{x}_i, \mathbf{x}_j) = K(\mathbf{x}_i, \mathbf{x}_j) \quad \text{for } i, j = 1, \ldots, M. \tag{2.153}$$

From (2.151),

$$\begin{aligned}
K_e(\mathbf{x}_i, \mathbf{x}_j) &= \mathbf{h}^\top(\mathbf{x}_i)\, \mathbf{h}(\mathbf{x}_j) \\
&= (K(\mathbf{x}_1, \mathbf{x}_i), \ldots, K(\mathbf{x}_M, \mathbf{x}_i)) P \Lambda^{-1} P^\top \\
&\quad \times (K(\mathbf{x}_1, \mathbf{x}_j), \ldots, K(\mathbf{x}_M, \mathbf{x}_j))^\top.
\end{aligned} \tag{2.154}$$

From (2.149),

$$(K(\mathbf{x}_1, \mathbf{x}_i), \ldots, K(\mathbf{x}_M, \mathbf{x}_i))^\top = P \Lambda \mathbf{q}_i, \tag{2.155}$$

where $\mathbf{q}_i$ is the $i$th column vector of $P^\top$. Substituting (2.155) into (2.154), we obtain

$$\begin{aligned}
K_e(\mathbf{x}_i, \mathbf{x}_j) &= \mathbf{q}_i^\top \Lambda P^\top P \Lambda^{-1} P^\top P \Lambda \mathbf{q}_j \\
&= \mathbf{q}_i^\top \Lambda \mathbf{q}_j \\
&= K(\mathbf{x}_i, \mathbf{x}_j).
\end{aligned} \tag{2.156}$$

The relation given by (2.153) is very important because a problem expressed using kernels can be interpreted, without introducing any approximation, as the problem defined in the associated empirical feature space. The dimension of the feature space is sometimes very high, e.g., that of the feature space induced by RBF kernels is infinite. But the dimension of the empirical feature space is the number of training data at most. Thus, instead of analyzing the feature space, we only need to analyze the associated empirical feature space.

We further prove that the kernel values associated with the feature space and the empirical feature space are the same if one of the argument of the kernels is mapped into the empirical feature space by the mapping function associated with the feature space, namely, the following theorem holds:

**Theorem 2.1.** *For* $\mathbf{x}$ *and* $\mathbf{x}'$, *where* $\phi(\mathbf{x}') = \sum_{i=1}^{M} a_i \phi(\mathbf{x}_i)$, $a_i$ $(i = 1, \ldots, M)$ *are constants, and some* $a_i$ *are nonzero, the following relation is satisfied:*

$$K_e(\mathbf{x}, \mathbf{x}') = K(\mathbf{x}, \mathbf{x}'). \tag{2.157}$$

*Proof.* First we assume that $a_i = 0$ except for $a_j = 1$. Then $\mathbf{x}' = \mathbf{x}_j$. Let $K'$

be the kernel matrix in which $\mathbf{x}$ is added as an input to the kernel, namely,

$$K' = \begin{pmatrix}
K(\mathbf{x}_1, \mathbf{x}_1) & \cdots & K(\mathbf{x}_1, \mathbf{x}_M) & K(\mathbf{x}_1, \mathbf{x}) \\
\vdots & \ddots & \vdots & \vdots \\
K(\mathbf{x}_M, \mathbf{x}_1) & \cdots & K(\mathbf{x}_M, \mathbf{x}_M) & K(\mathbf{x}_M, \mathbf{x}) \\
K(\mathbf{x}, \mathbf{x}_1) & \cdots & K(\mathbf{x}, \mathbf{x}_M) & K(\mathbf{x}, \mathbf{x})
\end{pmatrix}. \tag{2.158}$$

Now, we consider the following two cases.

1. Assume that the rank of $K'$ is the same with that of $K$. Then, the determinant of $K'$ vanishes. Expanding the determinant of $K'$ in the $(M+1)$th row, we obtain

$$\phi(\mathbf{x}) = \sum_{i=1}^{M} b_i \, \phi(\mathbf{x}_i), \qquad (2.159)$$

where $b_i$ $(i = 1, \dots, M)$ are constants and all of them are not zero. Then

$$K(\mathbf{x}, \mathbf{x}_j) = \phi^{\top}(\mathbf{x}) \, \phi(\mathbf{x}_j) = \sum_{i=1}^{M} b_i \, K(\mathbf{x}_i, \mathbf{x}_j). \qquad (2.160)$$

From (2.151), (2.155), and (2.159),

$$K_{\mathrm{e}}(\mathbf{x}, \mathbf{x}_j) = \sum_{i=1}^{M} b_i \, \mathbf{q}_i^{\top} \, \varLambda \, \mathbf{q}_j = \sum_{i=1}^{M} b_i \, K(\mathbf{x}_i, \mathbf{x}_j). \qquad (2.161)$$

Thus, (2.157) holds.

2. Assume that the rank of $K'$ is larger than that of $K$ by 1. Then

$$\phi(\mathbf{x}) = \sum_{i=1}^{M} b_i \, \phi(\mathbf{x}_i) + \mathbf{y}, \qquad (2.162)$$

where $b_i$ $(i = 1, \dots, M)$ are constants, some of $b_i$ are not zero, and $\mathbf{y}$ is the non-zero $l$-dimensional vector that satisfies $\phi^{\top}(\mathbf{x}_i) \, \mathbf{y} = 0$ for $i = 1, \dots, M$. Then because $\phi^{\top}(\mathbf{x}_i) \, \mathbf{y} = 0$, (2.160) and (2.161) are satisfied. Thus (2.157) holds.

Now because (2.157) holds for $\mathbf{x}' = \mathbf{x}_j$, it is easy to show that (2.157) holds for the general case. ∎

**Corollary 2.1.** *For* $\mathbf{x}$ *and* $\mathbf{x}'$, *let*

$$\phi(\mathbf{x}) = \sum_{i=1}^{M} a_i \phi(\mathbf{x}_i) + \mathbf{y}, \qquad (2.163)$$

$$\phi(\mathbf{x}') = \sum_{i=1}^{M} a'_i \phi(\mathbf{x}_i) + \mathbf{y}' \qquad (2.164)$$

*where* $\mathbf{y}$ *and* $\mathbf{y}'$ *are the non-zero* $l$-*dimensional vectors that satisfy* $\phi^{\top}(\mathbf{x}_i) \, \mathbf{y} = 0$ *and* $\phi^{\top}(\mathbf{x}_i) \, \mathbf{y}' = 0$ *for* $i = 1, \dots, M$, *respectively. Then the following relation is not always satisfied:*

$$K(\mathbf{x}, \mathbf{x}') = K_{\mathrm{e}}(\mathbf{x}, \mathbf{x}'). \qquad (2.165)$$

*Proof.* Because $K(\mathbf{x}, \mathbf{x}') = \boldsymbol{\phi}^\top(\mathbf{x})\, \boldsymbol{\phi}(\mathbf{x}')$,

$$K(\mathbf{x}, \mathbf{x}') = \sum_{i,j=1}^{M} a_i\, a_j'\, K(\mathbf{x}_i, \mathbf{x}_j) + \mathbf{y}^\top \mathbf{y}'. \tag{2.166}$$

But

$$K_{\mathrm{e}}(\mathbf{x}, \mathbf{x}') = \sum_{i,j=1}^{M} a_i\, a_j'\, K(\mathbf{x}_i, \mathbf{x}_j). \tag{2.167}$$

Then if $\mathbf{y}^\top \mathbf{y}' \neq 0$, (2.165) is not satisfied. ∎

*Example 2.4.* Consider one-dimensional input $x$, a polynomial kernel with degree 3: $K(x, x') = (1 + xx')^3$, and two training data $x_1 = 1$ and $x_2 = -1$. Because

$$K = \begin{pmatrix} 8 & 0 \\ 0 & 8 \end{pmatrix}, \qquad h(x) = \begin{pmatrix} \dfrac{(1+x)^3}{2\sqrt{2}} \\[2mm] \dfrac{(1-x)^3}{2\sqrt{2}} \end{pmatrix}, \tag{2.168}$$

$$K_{\mathrm{e}}(x, x') = \frac{1}{8}(1+x)^3(1+x')^3 + \frac{1}{8}(1-x)^3(1-x')^3. \tag{2.169}$$

Therefore,

$$\begin{aligned}
K(1,1) &= K_{\mathrm{e}}(1,1) = 8, \\
K(-1,-1) &= K_{\mathrm{e}}(-1,-1) = 8, \\
K(1,-1) &= K_{\mathrm{e}}(1,-1) = 0, \\
K(1,2) &= K_{\mathrm{e}}(1,2) = 27, \\
K(2,2) &= 125 \neq K_{\mathrm{e}}(2,2) = \frac{365}{4}.
\end{aligned}$$

### 2.3.6.2 Separability in the Feature Space

For a hyperplane obtained by a kernel-based method in the feature space there is an equivalent hyperplane in the empirical feature space. Therefore, linear separability of the mapped training data in the feature space is equivalent to that in the empirical feature space.

For a given kernel and training data, let the rank of the kernel matrix, $N$, be $N \geq M - 1$, namely $N = M - 1$ or $M$. Because the dimension of the empirical feature space is $N$, the training data with arbitrary assigned class labels of 1 or $-1$ are linearly separable by a hyperplane in the empirical feature space. (For details see the discussions on VC dimension in Section 2.8.2.2.)

But if $N < M - 1$, the linear separability of the mapped training data in the feature space is not guaranteed. Thus, we obtain the following theorem.

**Theorem 2.2.** *If the rank of the kernel matrix satisfies $N \geq M - 1$, the training data are linearly separable in the feature space.*

We must notice that the above theorem is a sufficient condition that guarantees linear separability for arbitrary assigned class labels.

Consider an RBF kernel: $\exp(-\gamma\|\mathbf{x}-\mathbf{x}'\|^2)$, where $\gamma$ is a positive parameter and let $\mathbf{x}_i \neq \mathbf{x}_j$ for $i \neq j$ and $i, j = 1, \ldots, M$. As $\gamma$ approaches 0, the elements of the kernel matrix approach 1. In the extreme case, where all the elements of the kernel matrix are 1, its rank is 1. On the other hand, as $\gamma$ approaches $\infty$, the values of the off-diagonal elements approach zero, while those of the diagonal elements approach 1. Thus, for sufficiently large $\gamma$, the rank of the kernel matrix is $M$. Therefore, for a large value of $\gamma$, the training data are linearly separable in the feature space. And for a small value of $\gamma$, the training data may not be linearly separable.

### 2.3.6.3 Training and Testing in the Empirical Feature Space

In kernels-based methods, pattern classification or function approximation problems are formulated using kernels and no explicit treatment of variables in the feature space is included. Thus, in training a kernel-based machine, kernels defined in the feature space and the empirical feature space are equivalent. Thus the following theorem holds:

**Theorem 2.3.** *Training a kernel-based method defined in the feature space is equivalent to training the associated kernel-based method defined in the empirical feature space.*

This theorem means that even if the dimension of the feature space is infinite, we can train the kernel-based method in the empirical feature space, whose dimension is the maximum number of the independent mapped training data. Thus, we can solve the primal problem of the kernel-based method as well as the dual problem.

For the trained kernel-based method, the data sample to be classified may be mapped into the complementary subspace of the empirical feature space. In such a case also, because of Theorem 2.1, we can classify the data sample in the empirical feature space:

**Theorem 2.4.** *Testing a kernel-based method defined in the feature space is equivalent to testing the associated kernel-based method defined in the empirical feature space.*

## 2.4 L2 Soft-Margin Support Vector Machines

Instead of the linear sum of the slack variables $\xi_i$ in the objective function, the L2 soft-margin support vector machine (L2 SVM) uses the square sum of the slack variables, namely, training is done by solving the following optimization problem:

$$\text{minimize} \quad Q(\mathbf{w}, b, \boldsymbol{\xi}) = \frac{1}{2} \mathbf{w}^\top \mathbf{w} + \frac{C}{2} \sum_{i=1}^{M} \xi_i^2 \tag{2.170}$$

$$\text{subject to} \quad y_i \left( \mathbf{w}^\top \boldsymbol{\phi}(\mathbf{x}_i) + b \right) \geq 1 - \xi_i \quad \text{for } i = 1, \dots, M. \tag{2.171}$$

Here, $\mathbf{w}$ is the $l$-dimensional vector, $b$ is the bias term, $\boldsymbol{\phi}(\mathbf{x})$ is the mapping function that maps the $m$-dimensional vector $\mathbf{x}$ into the $l$-dimensional feature space, $\xi_i$ is the slack variable for $\mathbf{x}_i$, and $C$ is the margin parameter.

Introducing the Lagrange multipliers $\alpha_i \, (\geq 0)$, we obtain

$$Q(\mathbf{w}, b, \boldsymbol{\alpha}, \boldsymbol{\xi}) = \frac{1}{2} \mathbf{w}^\top \mathbf{w} + \frac{C}{2} \sum_{i=1}^{M} \xi_i^2$$
$$- \sum_{i=1}^{M} \alpha_i \left( y_i \left( \mathbf{w}^\top \boldsymbol{\phi}(\mathbf{x}_i) + b \right) - 1 + \xi_i \right). \tag{2.172}$$

Here, we do not need to introduce the Lagrange multipliers associated with $\boldsymbol{\xi}$. As is shown immediately, $C\xi_i = \alpha_i$ is satisfied for the optimal solution. Hence $\xi_i$ is nonnegative, so long as $\alpha_i$ is nonnegative.

For the optimal solution the following KKT conditions are satisfied:

$$\frac{\partial Q(\mathbf{w}, b, \boldsymbol{\alpha}, \boldsymbol{\xi})}{\partial \mathbf{w}} = \mathbf{w} - \sum_{i=1}^{M} y_i \, \alpha_i \, \boldsymbol{\phi}(\mathbf{x}_i) = \mathbf{0}, \tag{2.173}$$

$$\frac{\partial Q(\mathbf{w}, b, \boldsymbol{\alpha}, \boldsymbol{\xi})}{\partial \xi_i} = C \, \xi_i - \alpha_i = 0, \tag{2.174}$$

$$\frac{\partial Q(\mathbf{w}, b, \boldsymbol{\alpha}, \boldsymbol{\xi})}{\partial b} = \sum_{i=1}^{M} y_i \, \alpha_i = 0, \tag{2.175}$$

$$\alpha_i \left( y_i \left( \mathbf{w}^\top \boldsymbol{\phi}(\mathbf{x}_i) + b \right) - 1 + \xi_i \right) = 0 \quad \text{for } i = 1, \dots, M. \tag{2.176}$$

We call $\mathbf{x}_i$ with positive $\alpha_i$ *support vectors*. For L1 support vector machines, the absolute values of the decision function evaluated at unbounded support vectors is 1. But because of (2.174), for L2 support vector machines they are smaller than 1, namely, the margins of support vectors are always smaller than 1.

Equation (2.176) gives the KKT complementarity conditions; from (2.173), (2.174), and (2.176), the optimal solution must satisfy either $\alpha_i = 0$ or

$$y_i \left( \sum_{j=1}^{M} \alpha_j\, y_j \left( K(\mathbf{x}_j, \mathbf{x}_i) + \frac{\delta_{ij}}{C} \right) + b \right) - 1 = 0, \qquad (2.177)$$

where $K(\mathbf{x}, \mathbf{x}') = \phi^\top(\mathbf{x})\, \phi(\mathbf{x})$ and $\delta_{ij}$ is Kronecker's delta function, in which $\delta_{ij} = 1$ for $i = j$ and 0 otherwise. Thus the value of the bias term $b$ is calculated for $\alpha_i > 0$:

$$b = y_j - \sum_{i \in S} \alpha_i\, y_i \left( K(\mathbf{x}_i, \mathbf{x}_j) + \frac{\delta_{ij}}{C} \right), \qquad (2.178)$$

where $\mathbf{x}_j$ is a support vector and $S$ is a set of support vector indices. To ensure stability of calculations we take the average:

$$b = \frac{1}{|S|} \sum_{j \in S} \left( y_j - \sum_{i \in S} \alpha_i\, y_i \left( K(\mathbf{x}_i, \mathbf{x}_j) + \frac{\delta_{ij}}{C} \right) \right). \qquad (2.179)$$

Equations (2.178) and (2.179) are different from those of the L1 support vector machine in that $\delta_{ij}/C$ are added to the diagonal elements. But the decision function is the same:

$$D(\mathbf{x}) = \sum_{i \in S} \alpha_i\, y_i\, K(\mathbf{x}_i, \mathbf{x}) + b. \qquad (2.180)$$

Substituting (2.173), (2.174), and (2.175) into (2.172), we obtain the dual objective function:

$$Q(\boldsymbol{\alpha}) = \sum_{i=1}^{M} \alpha_i - \frac{1}{2} \sum_{i,j=1}^{M} y_i\, y_j\, \alpha_i\, \alpha_j \left( K(\mathbf{x}_i, \mathbf{x}_j) + \frac{\delta_{ij}}{C} \right). \qquad (2.181)$$

Thus the following dual problem is obtained:

$$\text{maximize} \quad Q(\boldsymbol{\alpha}) = \sum_{i=1}^{M} \alpha_i - \frac{1}{2} \sum_{i,j=1}^{M} y_i\, y_j\, \alpha_i\, \alpha_j \left( K(\mathbf{x}_i, \mathbf{x}_j) + \frac{\delta_{ij}}{C} \right) \quad (2.182)$$

$$\text{subject to} \quad \sum_{i=1}^{M} y_i\, \alpha_i = 0, \qquad \alpha_i \geq 0 \quad \text{for } i = 1, \dots, M. \qquad (2.183)$$

This is similar to a hard-margin support vector machine. The difference is the addition of $\delta_{ij}/C$ in (2.182). Therefore, for the L1 support vector machine, if we replace $K(\mathbf{x}_j, \mathbf{x}_i)$ with $K(\mathbf{x}_j, \mathbf{x}_i) + \delta_{ij}/C$ and remove the upper bound given by $C$ for $\alpha_i$, we obtain the L2 support vector machine. But we must notice that when we calculate the decision function in (2.180) we must not add $\delta_{ij}/C$.

Because $1/C$ is added to the diagonal elements of the matrix $K = \{K(\mathbf{x}_i, \mathbf{x}_j)\}$ called the *kernel matrix*, the resulting matrix becomes positive definite. Thus the associated optimization problem is more computationally stable than that of the L1 support vector machine.

L2 soft-margin support vector machines look similar to hard-margin support vector machines. Actually, letting

$$\tilde{\mathbf{w}} = \begin{pmatrix} \mathbf{w} \\ \sqrt{\dfrac{C}{2}}\,\boldsymbol{\xi} \end{pmatrix}, \quad \tilde{b} = b, \quad \tilde{\boldsymbol{\phi}}(\mathbf{x}_i) = \begin{pmatrix} \phi(\mathbf{x}_i) \\ \sqrt{\dfrac{2}{C}}\,y_i\,\mathbf{e}_i \end{pmatrix}, \quad (2.184)$$

where $\mathbf{e}_i$ is the $M$-dimensional vector with the $i$th element being 1 and the remaining elements 0, training of the L2 support vector machine given by (2.170) and (2.171) is converted into the following problem:

$$\text{minimize} \quad \frac{1}{2}\tilde{\mathbf{w}}^\top \tilde{\mathbf{w}} \quad\quad\quad\quad\quad\quad (2.185)$$

$$\text{subject to} \quad y_i\left(\tilde{\mathbf{w}}^\top \tilde{\boldsymbol{\phi}}(\mathbf{x}_i) + \tilde{b}\right) \geq 1 \quad \text{for } i = 1, \ldots, M. \quad (2.186)$$

Therefore, the L2 support vector machine is equivalent to the hard-margin support vector machine with the augmented feature space. Because the L2 support vector machine always has a solution because of the slack variables, the equivalent hard-margin support vector machine also has a solution. But this only means that the solution is nonoverlapping in the augmented feature space. Therefore, there may be cases where the solution is overlapped in the original feature space, and thus the recognition rate of the training data for the L2 support vector machine is not 100%.

## 2.5 Advantages and Disadvantages

Based on the support vector machines explained so far, we discuss their advantages over multilayer neural networks, which are representative nonparametric classifiers. Then we discuss disadvantages of support vector machines and some ways to solve these problems.

### 2.5.1 Advantages

The advantages of support vector machines over multilayer neural network classifiers are as follows.

1. **Maximization of generalization ability.** In training a multilayer neural network classifier, the sum-of-squares error between outputs and desired

training outputs is minimized. Thus, the class boundaries change as the initial weights change. So does the generalization ability. Thus, especially when training data are scarce and linearly separable, the generalization ability deteriorates considerably. But because a support vector machine is trained to maximize the margin, the generalization ability does not deteriorate very much, even under such a condition [14].

2. **No local minima.** A multilayer neural network classifier is known to have numerous local minima, and there have been extensive discussions on how to avoid a local minimum in training. But because a support vector machine is formulated as a quadratic programming problem, there is a global optimum solution.

3. **Wide range of applications.** Generalization ability of a multiplayer neural network classifier is controlled by changing the number of hidden layers or the number of hidden units. But usually sigmoid or RBF functions are used as output functions of neurons. In support vector machines, generalization ability is controlled by changing a kernel, its parameter, and the margin parameter. Especially, in addition to sigmoid and RBF kernels by developing a kernel specific to a given application, we can improve generalization ability.

4. **Robustness to outliers.** Multilayer neural network classifiers are vulnerable to outliers because they use the sum-of-squares errors. Thus to prevent the effect of outliers, outliers need to be eliminated before training, or some mechanism for suppressing outliers needs to be incorporated in training. In support vector machines the margin parameter $C$ controls the misclassification error. If a large value is set to $C$, misclassification is suppressed, and if a small value is set, training data that are away from the gathered data are allowed to be misclassified. Thus by properly setting a value to $C$, we can suppress outliers.

## 2.5.2 Disadvantages

The disadvantages of support vector machines explained so far are as follows.

1. **Extension to multiclass problems.** Unlike multilayer neural network classifiers, support vector machines use direct decision functions. Thus an extension to multiclass problems is not straightforward, and there are several formulations. One of the purposes of this book is to clarify relations between these formulations (see Chapter 3).

2. **Long training time.** Because training of a support vector machine is done by solving the associated dual problem, the number of variables is equal to the number of training data. Thus for a large number of training data, solving the dual problem becomes difficult from both the memory size and the training time (see Chapter 5 for training speedup).

3. **Selection of parameters.** In training a support vector machine, we need to select an appropriate kernel and its parameters, and then we need to set the value to the margin parameter $C$. To select the optimal parameters to a given problem is called *model selection*. This is the same situation as that of neural network classifiers, namely, we need to set the number of hidden units, initial values of weights, and so on. In support vector machines, model selection is done by estimating the generalization ability through repeatedly training support vector machines. But because this is time consuming, several indices for estimating the generalization ability have been proposed (see Section 2.8).

## 2.6 Characteristics of Solutions

Here we analyze characteristics of solutions for the L1 and L2 soft-margin support vector machines. For the L1 soft-margin support vector machine we find $\alpha_i$ $(i = 1, \ldots, M)$ that satisfy

$$\text{maximize} \quad Q(\boldsymbol{\alpha}) = \sum_{i=1}^{M} \alpha_i - \frac{1}{2} \sum_{i,j=1}^{M} \alpha_i \, \alpha_j \, y_i \, y_j \, K(\mathbf{x}_i, \mathbf{x}_j) \qquad (2.187)$$

$$\text{subject to} \quad \sum_{i=1}^{M} y_i \, \alpha_i = 0, \qquad C \geq \alpha_i \geq 0 \quad \text{for } i = 1, \ldots, M. \quad (2.188)$$

For the L2 soft-margin support vector machine we find $\alpha_i$ $(i = 1, \ldots, M)$ that satisfy

$$\text{maximize} \quad Q(\boldsymbol{\alpha}) = \sum_{i=1}^{M} \alpha_i - \frac{1}{2} \sum_{i,j=1}^{M} \alpha_i \, \alpha_j \, y_i \, y_j \left( K(\mathbf{x}_i, \mathbf{x}_j) + \frac{\delta_{ij}}{C} \right) \quad (2.189)$$

$$\text{subject to} \quad \sum_{i=1}^{M} y_i \, \alpha_i = 0, \qquad \alpha_i \geq 0 \quad \text{for } i = 1, \ldots, M. \qquad (2.190)$$

### 2.6.1 Hessian Matrix

Rewriting (2.187) for the L1 support vector machine using the mapping function $\phi(\mathbf{x})$, we have

$$Q(\boldsymbol{\alpha}) = \sum_{i=1}^{M} \alpha_i - \frac{1}{2} \left( \sum_{i=1}^{M} \alpha_i \, y_i \, \phi(\mathbf{x}_i) \right)^{\top} \sum_{i=1}^{M} \alpha_i \, y_i \, \phi(\mathbf{x}_i). \qquad (2.191)$$

Solving the first equation in (2.188) for $\alpha_s$ ($s \in \{1, \ldots, M\}$),

$$\alpha_s = -y_s \sum_{\substack{i \neq s, \\ i=1}}^{M} y_i \, \alpha_i. \tag{2.192}$$

Substituting (2.192) into (2.191), we obtain

$$Q(\alpha') = \sum_{\substack{i \neq s, \\ i=1}}^{M} (1 - y_s \, y_i) \, \alpha_i - \frac{1}{2} \left( \sum_{\substack{i \neq s, \\ i=1}}^{M} \alpha_i \, y_i \, (\phi(\mathbf{x}_i) - \phi(\mathbf{x}_s)) \right)^{\top}$$

$$\times \sum_{\substack{i \neq s, \\ i=1}}^{M} \alpha_i \, y_i \, (\phi(\mathbf{x}_i) - \phi(\mathbf{x}_s)), \tag{2.193}$$

where $\alpha'$ is obtained by deleting $\alpha_s$ from $\alpha$. Thus the Hessian matrix of $-Q(\alpha')$, $K_{\mathrm{L1}}$,[10] which is an $(M-1) \times (M-1)$ matrix, is given by

$$\begin{aligned} K_{\mathrm{L1}} &= -\frac{\partial^2 Q(\alpha')}{\partial \alpha'^2} \\ &= \left( \cdots y_i \, (\phi(\mathbf{x}_i) - \phi(\mathbf{x}_s)) \cdots \right)^{\top} \left( \cdots y_j \, (\phi(\mathbf{x}_j) - \phi(\mathbf{x}_s)) \cdots \right). \end{aligned} \tag{2.194}$$

Because $K_{\mathrm{L1}}$ is expressed by the product of the transpose of a matrix and the matrix, $K_{\mathrm{L1}}$ is positive semidefinite. Let $N_\phi$ be the maximum number of linearly independent vectors among $\{\phi(\mathbf{x}_i) - \phi(\mathbf{x}_s)|\, i \in \{i, \ldots, M, i \neq s\}\}$. Then the rank of $K_{\mathrm{L1}}$ is $N_\phi$ [14, pp. 311–312]. Because $N_\phi$ does not exceed the dimension of the feature space $l$,

$$N_\phi \leq l \tag{2.195}$$

is satisfied. Therefore, if $M > (l+1)$, $K_{\mathrm{L1}}$ is positive semidefinite. For the linear kernel, $l = m$, where $m$ is the number of input variables, for the polynomial kernel with degree $d$, $l = {}_{m+d}C_d$ [55, pp. 38–41], and for the RBF kernel, $l = \infty$.

The Hessian matrix $K_{\mathrm{L2}}$ in which one variable is eliminated, for the L2 support vector machine, is expressed by

$$K_{\mathrm{L2}} = K_{\mathrm{L1}} + \left\{ \frac{y_i \, y_j + \delta_{ij}}{C} \right\}. \tag{2.196}$$

The matrix $K_{\mathrm{L1}}$ is positive semidefinite, and the matrix $\{\delta_{ij}/C\}$ is positive definite. Because

---

10 Conditionally positive semidefiniteness, which is positive semidefiniteness under an equality constraint discussed in Appendix D.1, is equivalent to positive semidefiniteness of $K_{\mathrm{L1}}$.

$$\{y_i y_j\} = (y_1 \cdots y_M)^\top (y_1 \cdots y_M), \tag{2.197}$$

the matrix $\{y_i y_j / C\}$ is positive semidefinite. Here, the sum of positive definite and positive semidefinite matrices is positive definite. Therefore, unlike $K_{L1}$, $K_{L2}$ is positive definite irrespective of the dimension of the feature space.

## 2.6.2 Dependence of Solutions on C

According to the analysis of Pontil and Verri [56], we discuss the dependence of the solution of the L1 support vector machine on the margin parameter $C$.

Let the sets of support vector indices be

$$U = \{i \,|\, 0 < \alpha_i < C\}, \tag{2.198}$$
$$B = \{i \,|\, \alpha_i = C\}, \tag{2.199}$$
$$S = U \cup B. \tag{2.200}$$

From the KKT complementarity conditions given by (2.63), (2.64), and (2.65), for $i \in U$,

$$y_i \left( \sum_{j \in S} \alpha_j \, y_j \, K(\mathbf{x}_j, \mathbf{x}_i) + b \right) = 1. \tag{2.201}$$

Thus, from (2.201) for a fixed $s \in U$, $b$ is given by

$$b = y_s - \sum_{j \in S} \alpha_j \, y_j \, K(\mathbf{x}_j, \mathbf{x}_s). \tag{2.202}$$

Solving the equality constraint (2.188) for $\alpha_s$ $(s \in U)$, we obtain

$$\alpha_s = - \sum_{\substack{j \neq s, \\ j \in S}} y_j \, y_s \, \alpha_j. \tag{2.203}$$

Substituting (2.202) and (2.203) into (2.201), we obtain

$$\sum_{\substack{j \neq s, \\ j \in S}} \alpha_j K_{ji} = 1 - y_i \, y_s, \tag{2.204}$$

where

$$K_{ji} = y_j \, y_i \, (K(\mathbf{x}_j, \mathbf{x}_i) - K(\mathbf{x}_s, \mathbf{x}_i) - K(\mathbf{x}_j, \mathbf{x}_s) + K(\mathbf{x}_s, \mathbf{x}_s)). \tag{2.205}$$

Separating the unbounded and bounded $\alpha_i$ in (2.204), we obtain

$$\sum_{\substack{j \neq s, \\ j \in U}} \alpha_j \, K_{ji} + C \sum_{j \in B} K_{ji} = 1 - y_i \, y_s. \tag{2.206}$$

In a matrix form, (2.206) becomes

$$K_{U'} \, \boldsymbol{\alpha}_{U'} + C \, K_{U'B} \mathbf{1}_{U'} = \mathbf{1}_{U'} - \mathbf{y}_{U'}, \tag{2.207}$$

where $U' = U - \{s\}$, $K_{U'} = \{K_{ij} \,|\, i, j \in U'\}$, $\boldsymbol{\alpha}_{U'} = (\cdots \alpha_i \cdots)^\top$ $(i \in U')$, $K_{U'B} = \{K_{ij} \,|\, i \in U', j \in B\}$, $\mathbf{y}_{U'} = (\cdots y_s \, y_i \cdots)^\top$ $(i \in U')$, and $\mathbf{1}_U$ is a $|U'|$-dimensional vector with all elements equal to 1.

From (2.59), $K_{ij}$ is expressed by

$$K_{ij} = y_i \, y_j \, (\boldsymbol{\phi}(\mathbf{x}_i) - \boldsymbol{\phi}(\mathbf{x}_s))^\top (\boldsymbol{\phi}(\mathbf{x}_j) - \boldsymbol{\phi}(\mathbf{x}_s)). \tag{2.208}$$

Thus $K_{U'}$ is a symmetric, positive semidefinite matrix. As will be stated in Theorem 2.10 on p. 73, $K_{U'}$ is not always positive definite. But here we assume that it *is* positive definite. Then, from (2.207) and (2.203), we obtain the following theorem.

**Theorem 2.5.** *If the Hessian matrix $K_{U'}$ is positive definite, the unbounded $\alpha_i$ are given by*

$$\boldsymbol{\alpha}_{U'} = K_{U'}^{-1} \, (\mathbf{1}_{U'} - \mathbf{y}_{U'}) - C \, K_{U'}^{-1} \, K_{U'B} \, \mathbf{1}_B, \tag{2.209}$$

$$\alpha_s = -\mathbf{y}_{U'}^\top \, \boldsymbol{\alpha}_{U'} - C \, \mathbf{y}_B^\top \, \mathbf{1}_B, \tag{2.210}$$

*where $\mathbf{y}_B = (\cdots y_s \, y_i \cdots)^\top$ $(i \in B)$ and $\mathbf{1}_B$ is a $|B|$-dimensional vector with all elements equal to 1.*

Therefore, if $B = \phi$, namely, $0 < \alpha_i < C$ for all support vectors, $\boldsymbol{\alpha}_{U'} = K_{U'}^{-1}(\mathbf{1}_{U'} - \mathbf{y}_{U'})$ and $\alpha_s = -\mathbf{y}_{U'}^\top \, \boldsymbol{\alpha}_{U'}$.

**Theorem 2.6.** *The coefficient vector of the separating hyperplane, $\mathbf{w}$, in the feature space is given by*

$$\mathbf{w} = \mathbf{w}_1 + C \, \mathbf{w}_2, \tag{2.211}$$

*where $\mathbf{w}_1$ and $\mathbf{w}_2$ are given by (2.215) and (2.216), and*

$$\mathbf{w}_1^\top \mathbf{w}_2 = 0. \tag{2.212}$$

*Proof.* Because

$$\mathbf{w} = \sum_{i \in S} \alpha_i \, y_i \, \boldsymbol{\phi}(\mathbf{x}_i)$$

$$= \sum_{i \in U'} \alpha_i \, y_i \, \boldsymbol{\phi}(\mathbf{x}_i) + \alpha_s \, y_s \, \boldsymbol{\phi}(\mathbf{x}_s) + C \sum_{i \in B} y_i \, \boldsymbol{\phi}(\mathbf{x}_i)$$

$$= \sum_{i \in U'} \alpha_i \, y_i \, (\boldsymbol{\phi}(\mathbf{x}_i) - \boldsymbol{\phi}(\mathbf{x}_s)) + C \sum_{i \in B} y_i \, (\boldsymbol{\phi}(\mathbf{x}_i) - \boldsymbol{\phi}(\mathbf{x}_s)) \quad \text{(from (2.203))}$$

$$= \sum_{i \in U'} r_i \, y_i \left(\phi(\mathbf{x}_i) - \phi(\mathbf{x}_s)\right) + C \sum_{i \in U'} t_i \, y_i \left(\phi(\mathbf{x}_i) - \phi(\mathbf{x}_s)\right)$$
$$+ C \sum_{i \in B} y_i \left(\phi(\mathbf{x}_i) - \phi(\mathbf{x}_s)\right),$$

where from (2.209)

$$\mathbf{r} = (\cdots r_i \cdots)^\top = K_{U'}^{-1}(\mathbf{1}_{U'} - \mathbf{y}_{U'}), \tag{2.213}$$
$$\mathbf{t} = (\cdots t_i \cdots)^\top = -K_{U'}^{-1} K_{U'B} \, \mathbf{1}_B, \tag{2.214}$$

we obtain

$$\mathbf{w}_1 = \sum_{i \in U'} r_i \, y_i \left(\phi(\mathbf{x}_i) - \phi(\mathbf{x}_s)\right), \tag{2.215}$$

$$\mathbf{w}_2 = \sum_{i \in U'} t_i \, y_i \left(\phi(\mathbf{x}_i) - \phi(\mathbf{x}_s)\right) + \sum_{i \in B} y_i \left(\phi(\mathbf{x}_i) - \phi(\mathbf{x}_s)\right). \tag{2.216}$$

Then

$$\mathbf{w}_1^\top \mathbf{w}_2 = \sum_{i,j \in U'} r_i \, K_{ij} \, t_j + \sum_{i \in U', j \in B} r_i \, K_{ij}$$
$$= \mathbf{r}^\top K_{U'} \, \mathbf{t} + \mathbf{r}^\top K_{U'B} \, \mathbf{1}_B$$
$$= 0 \quad \text{(from (2.214))}. \ \blacksquare \tag{2.217}$$

Now consider changing the margin parameter $C$. Let $[C_k, C_{k+1})$ be the interval of $C$, in which the set of support vectors does not change. Here, we consider that if unbounded support vectors change to bounded support vectors, the set is changed. We add suffix $k$ to the sets of support vectors for the interval $[C_k, C_{k+1})$, namely, $S_k$, $U_k$, and $B_k$. Then, from (2.209) and (2.211), for $C$ in $[C_k, C_{k+1})$,

$$\alpha_i = a_i + C \, b_i \quad \text{for } i \in U_k, \tag{2.218}$$
$$\alpha_i = C \quad \text{for } i \in B_k, \tag{2.219}$$

where $a_i$ and $b_i$ are constant. Notice that so long as $U_k = U_{k'}$ and $B_k = B_{k'}$, $a_i$ and $b_i$ are the same for the intervals $[C_k, C_{k+1})$ and $[C_{k'}, C_{k'+1})$.

Because the number of training data is finite, the combinations of sets of support vectors are finite. But if some sets of support vectors appear infinitely as we increase $C$ to infinity, the intervals $[C_k, C_{k+1})$ may be infinite. The following theorem shows that this does not happen.

**Theorem 2.7.** *For $C$ in $[0, \infty)$, there are finite points $C_i \, (i = 1, \ldots, \max)$ where the set of support vectors changes. And for any $C$ in $[C_i, C_{i+1})$ or $[C_{\max}, \infty)$ the set of support vectors does not change.*

*Proof.* First we show that as $C$ approaches infinity, the weight vector $\mathbf{w}_2$ in (2.211) approaches $\mathbf{0}$. From (2.211), the quadratic term in the dual objective function of an L1 support vector machine is expressed by

$$\frac{1}{2}\mathbf{w}^{\top}\mathbf{w} = \frac{1}{2}(\mathbf{w}_1 + C\mathbf{w}_2)^{\top}(\mathbf{w}_1 + C\mathbf{w}_2)$$

$$= \frac{1}{2}\mathbf{w}_1^{\top}\mathbf{w}_1 + C\mathbf{w}_1^{\top}\mathbf{w}_2 + \frac{1}{2}C^2\mathbf{w}_2^{\top}\mathbf{w}_2, \qquad (2.220)$$

where $\mathbf{w}$ is the weight vector for the optimal solution with $C$.

Therefore, if $\mathbf{w}_2$ is not a zero vector, the value of (2.220) goes to infinity as $C$ approaches infinity. Thus, the dual objective function becomes negative for a large value of $C$. But because $\alpha_i = 0 \, (i = 1, \ldots, M)$ is a feasible solution with the objective function being 0, $\mathbf{w}$ cannot be optimal. Therefore, there exists $C_0$ where for $C$ larger than $C_0$, $\mathbf{w} = \mathbf{w}_1$.

Assume that for $C$ larger than $C_0$, two sets of support vectors $S_k$ and $S_{k+1}$ alternately appear infinitely. For $C > C_0$, $\mathbf{w} = \mathbf{w}_1$. Thus, from (2.218) and (2.219), the dual objective function $Q(C)$ for $C$ in $[C_k, C_{k+1})$ is given by

$$Q(C) = \sum_{i \in U_k} a_i + C\left(\sum_{i \in U_k} b_i + |B_k|\right) - \frac{1}{2}\mathbf{w}_1^{\top}\mathbf{w}_1, \qquad (2.221)$$

where $\mathbf{w}_1$ is a constant vector in $[C_k, C_{k+1})$.

For the infinite intervals in $[C_0, \infty)$, where the sets of support vectors are the same as $S_k$, the objective function is given by (2.221).

According to the assumption, the sets of support vectors $S_k$ and $S_{k+1}$ alternate, namely,

$$S_k = S_{k+2} = S_{k+4} = \cdots, \qquad (2.222)$$

$$S_{k+1} = S_{k+3} = S_{k+5} = \cdots. \qquad (2.223)$$

Thus, the objective functions (2.221) for $S_k$ and $S_{k+1}$ cross at $C = C_{k+1}$, but because they are monotonic functions, they do not cross at $C$ larger than $C_{k+1}$ (see Fig. 2.8). Because the optimization problem is continuous, namely, the objective function is continuous for the change of $C$, discontinuity of the objective function does not occur. Thus, the assumption that the two sets of support vectors alternate infinitely does not happen. Therefore the number of intervals $[C_k, C_{k+1})$ is finite and the interval ends with $[C_{\max}, \infty)$, where "max" is the maximum number of intervals. Thus, the theorem holds. ∎

For the L2 support vector machine, the following theorem, similar to Theorem 2.7, holds.

**Theorem 2.8.** *For an L2 support vector machine, there are finite points $C_i' \, (i = 1, \ldots, \max')$ for $C \in [0, \infty)$ where the set of support vectors changes. And for any $C$ in $(C_i', C_{i+1}')$ or $(C_{\max'}', \infty)$, the set of support vectors is the same.*

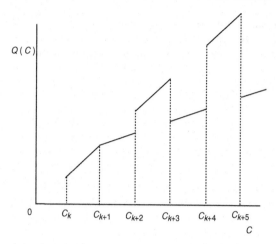

**Fig. 2.8** Counterexample of infinite intervals of $C$

*Proof.* Similar to (2.209) and (2.210), the support vectors for an L2 support vector machine are given by

$$\boldsymbol{\alpha}_{S'} = K_{S'}^{-1}(\mathbf{1}_{S'} - \mathbf{y}_{S'}), \qquad (2.224)$$

$$\alpha_s = -\mathbf{y}_{S'}^\top \boldsymbol{\alpha}_{S'}, \qquad (2.225)$$

where $S' = S - \{s\}$, $K_{S'} = \{K'_{ij} \,|\, i, j \in S'\}$, $K'_{ij} = K_{ij} + \delta_{ij}/C + y_i y_j/C$, $K_{ij}$ is given by (2.205), $\boldsymbol{\alpha}_{S'} = (\cdots \alpha_i \cdots)^\top \, (i \in S')$, $\mathbf{y}_{S'} = (\cdots y_s y_i \cdots)^\top \, (i \in S')$, and $\mathbf{1}_{S'}$ is an $|S'|$-dimensional vector with all elements equal to 1.

Similar to (2.193), the objective function for the L2 support vector machine, in a matrix form, is given by

$$Q(C) = (\mathbf{1}_{S'} - \mathbf{y}_{S'})^\top \boldsymbol{\alpha}_{S'}^\top - \frac{1}{2} \boldsymbol{\alpha}_{S'}^\top K_{S'} \boldsymbol{\alpha}_{S'}. \qquad (2.226)$$

Substituting (2.224) into (2.226) gives

$$Q(C) = \frac{1}{2} (\mathbf{1}_{S'} - \mathbf{y}_{S'})^\top K_{S'}^{-1} (\mathbf{1}_{S'} - \mathbf{y}_{S'}). \qquad (2.227)$$

Because the matrix $K_{S'}$ is positive definite, $K_{S'}^{-1}$ is also positive definite. By increasing the value of $C$, the eigenvalues of $K_{S'}$ decrease, thus those of $K_{S'}^{-1}$ increase. Therefore, $Q(C)$ monotonically increases and saturates as $C$ approaches infinity.

Suppose that there are infinite intervals of $[C'_k, C'_{k+1})$ and that the sets of support vectors $S'_k$ and $S'_{k+1}$ alternate infinitely. Then the objective functions (2.227) for $S'_k$ and $S'_{k+1}$ cross at $C = C'_{k+1}$ and they do not cross at $C$ larger than $C'_{k+1}$. But because the objective function is continuous for the change of $C$, discontinuity of the objective function does not occur. Thus, the assumption that the two sets of support vectors alternate infinitely does not happen. Therefore the number of intervals $[C'_k, C'_{k+1})$ is finite and the interval ends with $[C'_{\max'}, \infty)$, where "max'" is the maximum number of intervals. $\blacksquare$

## 2.6.3 Equivalence of L1 and L2 Support Vector Machines

In this section, we clarify the condition in which L1 and L2 support vector machines are equivalent.

Setting $C = \infty$ in L1 and L2 support vector machines, we obtain the hard-margin support vector machines by solving the following optimization problem:

$$\text{maximize} \quad Q(\boldsymbol{\alpha}) = \sum_{i=1}^{M} \alpha_i - \frac{1}{2} \sum_{i,j=1}^{M} \alpha_i \, \alpha_j \, y_i \, y_j \, K(\mathbf{x}_i, \mathbf{x}_j) \quad (2.228)$$

$$\text{subject to} \quad \sum_{i=1}^{M} y_i \, \alpha_i = 0, \qquad \alpha_i \geq 0. \quad (2.229)$$

In other words, as $C$ approaches infinity, the solution of the L1 support vector machine approaches that of the associated L2 support vector machine. In the following, we discuss this in more detail.

For the L2 support vector machine, the weight vector is not expressed by (2.211), but as $C$ approaches infinity, from our previous discussions, the weight vector converges to $\mathbf{w}_1$, namely, the following theorem holds.

**Theorem 2.9.** *For $C$ in $[\max(C_{\max}, C'_{\max'}), \infty)$, the sets of support vectors $S_{\max}$ and $S'_{\max'}$ are the same, and for L1 and L2 support vector machines, the weight vectors in the feature space converge to vector $\mathbf{w}_1$ as $C$ approaches infinity.*

In the following, we discuss the equivalence for the separable and inseparable classification problems.

If the problem is separable, the L1 support vector machine has a finite optimum solution:

$$0 \leq \alpha_i < \infty \quad \text{for } i = 1, \dots, M. \quad (2.230)$$

Let

$$C_{\max} = \max_{i=1,\dots,M} \alpha_i. \quad (2.231)$$

Then, for the L1 support vector machine, the solution is the same for any $C \in [C_{\max}, \infty)$, namely, for any $C \in [C_{\max}, \infty)$, the support vectors are unbounded and do not change. Thus the resulting optimal hyperplane does not change.

Now assume that the given problem is inseparable in the feature space. Then the solution for (2.228) and (2.229) diverges without bounds. Let $C_{max}$ and $C'_{max'}$ be as defined before. Then for any $C \in [\max(C_{max}, C'_{max}), \infty)$, the sets of support vectors for the L1 and L2 support vector machines are the same. Then although $\alpha_i$ goes to infinity as $C$ approaches infinity, from Theorem 2.9, the weight vector converges to a constant vector.

*Example 2.5.* Consider an inseparable case with a linear kernel, shown in Fig. 2.9. The dual problem of the L1 support vector machine is given by

$$\text{maximize} \quad Q(\boldsymbol{\alpha}) = \alpha_1 + \alpha_2 + \alpha_3 + \alpha_4 - \frac{1}{2}\left(-0.2\,\alpha_2 + 0.4\,\alpha_3 - \alpha_4\right)^2 \quad (2.232)$$

$$\text{subject to} \quad \alpha_1 + \alpha_3 = \alpha_2 + \alpha_4, \quad C \ge \alpha_i \ge 0, \quad i = 1, \ldots, 4. \quad (2.233)$$

**Fig. 2.9** Inseparable one-dimensional case

From Fig. 2.9, we can assume that $x_2$ and $x_3$ are bounded support vectors, namely, $\alpha_2 = \alpha_3 = C$. Thus from (2.233), $\alpha_1 = \alpha_4$. Therefore, (2.232) reduces to

$$Q(\boldsymbol{\alpha}) = 2\,\alpha_1 + 2\,C - \frac{1}{2}\left(\alpha_1 - 0.2\,C\right)^2. \quad (2.234)$$

Equation (2.234) reaches the maximum value of $2.4\,C + 2$ when $\alpha_1 = 0.2\,C + 2$. Thus, the optimal solution is given by

$$\alpha_1 = \alpha_4 = 0.2\,C + 2, \quad \alpha_2 = \alpha_3 = C. \quad (2.235)$$

Therefore, the solution diverges indefinitely as $C$ approaches infinity. Substituting (2.235) into the quadratic term of (2.232), we obtain 4, namely, the quadratic term is constant for $C \ge 2.5$. If this does not hold, an unbounded solution is not obtained.

Using (2.235), the decision function is given by

$$D(x) = -2\,x + 1. \quad (2.236)$$

Thus the class boundary is $x = 0.5$ for $C \ge 2.5$.

*Example 2.6.* To evaluate the equivalence of L1 and L2 support vector machines, we used the iris data listed in Table 1.3. We selected the training data and test data for Classes 2 and 3, which are not linearly separable for the linear kernels. For $d = 3$, the data are linearly separable, and for the L1 support vector machine, $C_{\max} = 4290.39$. Thus, for $C \geq C_{\max}$, the L1 support vector machine gives the same solution. Table 2.2 lists the support vectors and their $\alpha_i$.

**Table 2.2** Support vectors for the iris data with $d = 3$

| $i$ | Class | $\alpha_i$ |
|---|---|---|
| 7 | 2 | 4,290.39 |
| 21 | 2 | 4.28 |
| 24 | 2 | 2,011.89 |
| 26 | 3 | 763.04 |
| 30 | 3 | 16.74 |
| 32 | 3 | 4,123.90 |
| 37 | 3 | 240.29 |
| 38 | 3 | 1,162.58 |

Table 2.3 shows the number of support vectors and $\alpha_7$ of the L2 support vector machine for the change of $C$. Except for $C = 10^3$ with the number of support vectors of 11, the support vectors are the same as those of the L1 support vector machine. For $C = 10^{10}$, the value of $\alpha_7$ coincides with that for the L1 support vector machine for the first six digits.

**Table 2.3** L2 support vectors for the change of $C$ $(d = 3)$

| $C$ | SVs | $\alpha_7$ |
|---|---|---|
| $10^3$ | 11 | 1,015.69 |
| $10^4$ | 8 | 3,215.73 |
| $10^5$ | 8 | 4,151.95 |
| $10^6$ | 8 | 4,276.14 |
| $10^7$ | 8 | 4,288.96 |
| $10^8$ | 8 | 4,290.24 |
| $10^9$ | 8 | 4,290.38 |
| $10^{10}$ | 8 | 4,290.39 |

Next, we examined the nonseparable case, i.e., $d = 1$. Table 2.4 shows the numbers of support vectors and coefficients of the optimal hyperplanes for the L1 and L2 support vector machines against $C$. For the L1 support vector machine, the numerals in parentheses in the "SVs" column show the bounded support vectors. From the table, there are five unbounded L1 support vectors, which corresponds to the maximum number of support vectors, i.e., $m+1 = 5$.

For the L1 support vector machine, the coefficient $w_1$ reached almost constant at $C = 10^4$.

In theory, for $C = \infty$ the optimal hyperplane of the L2 support vector machine converges to that of the L1 support vector machine. But from Table 2.4, the difference between the two is still large for $C = 10^{10}$. Although the optimal hyperplane of the L2 support vector machine changes as $C$ is increased, the change is small and convergence to that of the L1 support vector machine is very slow compared to the case where the problem is separable in the feature space.

**Table 2.4** Coefficients of the optimal hyperplane for L1 and L2 support vector machines ($d = 1$)

| C | L1 SVM | | L2 SVM | |
|---|---|---|---|---|
| | SVs | $w_1$ | SVs | $w_1$ |
| $10^2$ | 11 (6) | 1.07 | 15 | 1.58 |
| $10^3$ | 8 (3) | 4.18 | 10 | 2.75 |
| $10^4$ | 7 (2) | 11.49 | 10 | 3.45 |
| $10^5$ | 7 (2) | 11.50 | 10 | 3.53 |
| $10^6$ | 7 (2) | 11.50 | 10 | 3.54 |
| $10^{10}$ | 7 (2) | 11.49 | 10 | 3.54 |

## 2.6.4 Nonunique Solutions

Because support vector machines are formulated as quadratic optimization problems, there is a global maximum (minimum) of the objective function. This is one of the advantages over multilayer neural networks with numerous local minima. Although the objective function has the global maximum (minimum), there may be cases where solutions are not unique [57]. This is not a serious problem because the values of the objective function are the same. In the following, we discuss nonuniqueness of the solution.

If a convex (concave) function gives a minimum (maximum) at a point, not in an interval, the function is called *strictly convex (concave)*. In general, if the objective function of a quadratic programming problem constrained in a convex set is strictly convex (concave) or the associated Hessian matrix is positive (negative) definite, the solution is unique. And if the objective function is convex (concave), there may be cases where the solution is nonunique (see Fig. 2.10). Convexity (concavity) of objective functions for different support vector machine architectures is summarized in Table 2.5. The symbols in parentheses show the variables.

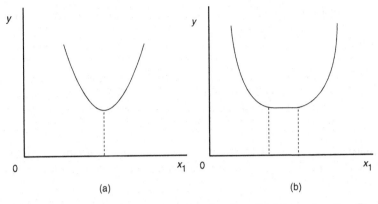

**Fig. 2.10** Convex functions: (**a**) strictly convex function. (**b**) Convex function. Reprinted from [58, p. 92, ©IEEE 2002]

**Table 2.5** Convexity (Concavity) of objective functions. Adapted from [58, p. 92, ©IEEE 2002]

|  | Hard margin | L1 soft margin | L2 soft margin |
|---|---|---|---|
| Primal | Strictly convex $(\mathbf{w}, b)$ | Convex $(\mathbf{w}, b, \boldsymbol{\xi})$ | Strictly convex $(\mathbf{w}, b, \boldsymbol{\xi})$ |
| Dual | Concave $(\boldsymbol{\alpha})$ | Concave $(\boldsymbol{\alpha})$ | Strictly concave $(\boldsymbol{\alpha})$ |

We must notice that because $b$ is not included in the dual problem, even if the solution of the dual problem is unique, the solution of the primal problem may not be unique.

Assume that the hard-margin support vector machine has a solution, i.e., the given problem is separable in the feature space. Then because the objective function of the primal problem is $\|\mathbf{w}\|^2/2$, which is strictly convex, the primal problem has a unique solution for $\mathbf{w}$ and $b$. But because the Hessian matrix of the dual problem is positive semidefinite, the solution for $\alpha_i$ may be nonunique. As discussed previously, the hard-margin support vector machine for a separable problem is equivalent to the L1 soft-margin support vector machine without bounded support vectors.

The objective function of the primal problem for the L2 soft-margin support vector machine is strictly convex. Thus, $\mathbf{w}$ and $b$ are uniquely determined if we solve the primal problem. In addition, because the Hessian matrix of the dual objective function is positive definite, $\alpha_i$ are uniquely determined. And because of the uniqueness of the primal problem, $b$ is determined uniquely using the KKT conditions.

Because the L1 soft-margin support vector machine includes the linear sum of $\xi_i$, the primal objective function is convex. Likewise, the Hessian matrix of the dual objective function is positive semidefinite. Thus the primal and dual solutions may be nonunique [58].

Before we discuss nonuniqueness of the solution, we clarify some properties of support vectors.

The KKT complementarity conditions for the L1 support vector machine are given by (2.63), (2.64), and (2.65). Thus, in some situation, $\alpha_i = 0$ and $y_i\left(\mathbf{w}^\top\boldsymbol{\phi}(\mathbf{x}_i) + b\right) = 1$ are satisfied simultaneously. In this case, $\mathbf{x}_i$ is not a support vector.

**Definition 2.1.** For the L1 support vector machine, we call the data that satisfy $y_i\left(\mathbf{w}^\top\boldsymbol{\phi}(\mathbf{x}_i) + b\right) = 1$ and that are not support vectors *boundary vectors*.

*Example 2.7.* Consider the two-dimensional case shown in Fig. 2.11, in which $\mathbf{x}_1$ belongs to Class 1, $\mathbf{x}_2$ and $\mathbf{x}_3$ belong to Class 2, and $\mathbf{x}_1 - \mathbf{x}_2$ and $\mathbf{x}_3 - \mathbf{x}_2$ are orthogonal. The dual problem with linear kernels is given as follows:

$$
\begin{aligned}
\text{maximize} \quad Q(\boldsymbol{\alpha}) &= \alpha_1 + \alpha_2 + \alpha_3 - \frac{1}{2}\left(\alpha_1\,\mathbf{x}_1 - \alpha_2\,\mathbf{x}_2 - \alpha_3\,\mathbf{x}_3\right)^\top \\
&\quad \times \left(\alpha_1\,\mathbf{x}_1 - \alpha_2\,\mathbf{x}_2 - \alpha_3\,\mathbf{x}_3\right) \quad (2.237)
\end{aligned}
$$

$$
\text{subject to} \quad \alpha_1 - \alpha_2 - \alpha_3 = 0, \quad C \geq \alpha_i \geq 0, \quad i = 1, 2, 3. \quad (2.238)
$$

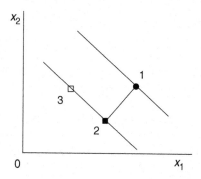

**Fig. 2.11** An example of a nonsupport vector. Reprinted from [58, p. 93, ©IEEE 2002]

Substituting $\alpha_3 = \alpha_1 - \alpha_2$ and $\alpha_2 = a\,\alpha_1\,(a \geq 0)$ into (2.237), we obtain

$$
\begin{aligned}
Q(\boldsymbol{\alpha}) &= 2\,\alpha_1 - \frac{1}{2}\,\alpha_1^2 \\
&\quad \times (\mathbf{x}_1 - \mathbf{x}_3 - a\,(\mathbf{x}_2 - \mathbf{x}_3))^\top(\mathbf{x}_1 - \mathbf{x}_3 - a\,(\mathbf{x}_2 - \mathbf{x}_3)). \quad (2.239)
\end{aligned}
$$

Defining

$$
d^2(a) = (\mathbf{x}_1 - \mathbf{x}_3 - a(\mathbf{x}_2 - \mathbf{x}_3))^\top(\mathbf{x}_1 - \mathbf{x}_3 - a(\mathbf{x}_2 - \mathbf{x}_3)), \quad (2.240)
$$

(2.239) becomes

$$Q(\boldsymbol{\alpha}) = 2\alpha_1 - \frac{1}{2}\alpha_1^2 d^2(a).$$

(2.241)

When

$$C \geq \frac{2}{d^2(a)},$$

(2.242)

$Q(\boldsymbol{\alpha})$ is maximized at $\alpha_1 = 2/d^2(a)$ and takes the maximum

$$Q\left(\frac{2}{d^2(a)}\right) = \frac{2}{d^2(a)}.$$

(2.243)

Because $\mathbf{x}_1 - \mathbf{x}_2$ and $\mathbf{x}_3 - \mathbf{x}_2$ are orthogonal, $d(a)$ is minimized at $a = 1$. Thus $Q(2/d^2(a))$ is maximized at $a = 1$, namely, $\alpha_1 = \alpha_2 = 2/d^2(1)$ and $\alpha_3 = 0$. Because $y_3\,(\mathbf{w}^\top\mathbf{x}_3 + b) = 1$, $\mathbf{x}_3$ is a boundary vector.

**Definition 2.2.** For the L1 support vector machine, a set of support vectors is *irreducible* if deletion of all the boundary vectors and any support vector results in the change of the optimal hyperplane. It is *reducible* if the optimal hyperplane does not change for deletion of all the boundary vectors and some support vectors.

Deletion of nonsupport vectors from the training data set does not change the solution. In [59], after training, an irreducible set is obtained by deleting linearly dependent support vectors in the feature space.

In the following theorem, the Hessian matrix associated with a set of support vectors means that the Hessian matrix is calculated for the support vectors, not for the entire training data.

**Theorem 2.10.** *For the L1 support vector machine, let all the support vectors be unbounded. Then the Hessian matrix associated with an irreducible set of support vectors is positive definite and the Hessian matrix associated with a reducible set of support vectors is positive semidefinite.*

*Proof.* Let the set of support vectors be irreducible. Then, because deletion of any support vector results in the change of the optimal hyperplane, any $\phi(\mathbf{x}_i) - \phi(\mathbf{x}_s)$ cannot be expressed by the remaining $\phi(\mathbf{x}_j) - \phi(\mathbf{x}_s)$. Thus the associated Hessian matrix is positive definite. If the set of support vectors is reducible, deletion of some support vector, e.g., $\mathbf{x}_i$ does not result in the change of the optimal hyperplane. This means that $\phi(\mathbf{x}_i) - \phi(\mathbf{x}_s)$ is expressed by the linear sum of the remaining $\phi(\mathbf{x}_j) - \phi(\mathbf{x}_s)$. Thus the associated Hessian matrix is positive semidefinite. ∎

**Theorem 2.11.** *For the L1 support vector machine, let the dimension of the feature space be finite. Then the number of unbounded support vectors in* *the irreducible set cannot exceed the dimension of the feature space plus 1. For the infinite feature space, the maximum number of unbounded support vectors is the number of training data. For the L2 support vector machine, the maximum number of support vectors is the number of training data.*

*Proof.* It is clear from Theorem 2.10 that the theorem holds for the L1 support vector machine. For the L2 support vector machine, because the Hessian matrix associated with the training data set is positive definite, the theorem holds. ∎

**Theorem 2.12.** *For the L1 support vector machine, if there is only one irreducible set of support vectors and the support vectors are all unbounded, the solution is unique.*

*Proof.* Delete the nonsupport vectors from the training data. Then because the set of support vectors is irreducible, the associated Hessian matrix is positive definite. Thus, the solution is unique for the irreducible set. Because there is only one irreducible set, the solution is unique for the given problem. ∎

*Example 2.8.* Consider the two-dimensional case shown in Fig. 2.12, in which $\mathbf{x}_1$ and $\mathbf{x}_2$ belong to Class 1 and $\mathbf{x}_3$ and $\mathbf{x}_4$ belong to Class 2. Because $\mathbf{x}_1$, $\mathbf{x}_2$, $\mathbf{x}_3$, and $\mathbf{x}_4$ form a square, $\{\mathbf{x}_1, \mathbf{x}_3\}$ and $\{\mathbf{x}_2, \mathbf{x}_4\}$ are irreducible sets of support vectors for the linear kernel.

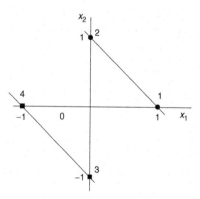

**Fig. 2.12** Nonunique solutions. Reprinted from [58, p. 95, ©IEEE 2002]

Training of the L1 support vector machine with linear kernels is given by

$$\text{maximize} \quad Q(\boldsymbol{\alpha}) = \alpha_1 + \alpha_2 + \alpha_3 + \alpha_4$$
$$-\frac{1}{2}\left((\alpha_1 + \alpha_4)^2 + (\alpha_2 + \alpha_3)^2\right) \tag{2.244}$$

$$\text{subject to} \quad \alpha_1 + \alpha_2 = \alpha_3 + \alpha_4, \quad C \geq \alpha_i \geq 0, \quad i = 1, \ldots, 4. \tag{2.245}$$

For $C \geq 1$, $(\alpha_1, \alpha_2, \alpha_3, \alpha_4) = (1, 0, 1, 0)$ and $(0, 1, 0, 1)$ are two solutions. Thus,

$$(\alpha_1, \alpha_2, \alpha_3, \alpha_4) = (\beta, 1 - \beta, \beta, 1 - \beta), \tag{2.246}$$

where $0 \leq \beta \leq 1$, is also a solution. Then $(\alpha_1, \alpha_2, \alpha_3, \alpha_4) = (0.5, 0.5, 0.5, 0.5)$ is a solution.

For the L2 support vector machine, the objective function becomes

$$Q(\boldsymbol{\alpha}) = \alpha_1 + \alpha_2 + \alpha_3 + \alpha_4$$
$$- \frac{1}{2} \left( (\alpha_1 + \alpha_4)^2 + (\alpha_2 + \alpha_3)^2 + \frac{\alpha_1^2 + \alpha_2^2 + \alpha_3^2 + \alpha_4^2}{C} \right). \tag{2.247}$$

Then for $\alpha_i = 1/(2 + 1/C)$ $(i = 1, \ldots, 4)$, (2.247) becomes

$$Q(\boldsymbol{\alpha}) = \frac{1}{1 + \dfrac{1}{2C}}. \tag{2.248}$$

For $\alpha_1 = \alpha_3 = 1/(1 + 1/C)$ and $\alpha_2 = \alpha_4 = 0$, (2.247) becomes

$$Q(\boldsymbol{\alpha}) = \frac{1}{1 + \dfrac{1}{C}}. \tag{2.249}$$

Thus, for $C > 0$, $Q(\boldsymbol{\alpha})$ given by (2.249) is smaller than that given by (2.248). Therefore, $\alpha_1 = \alpha_3 = 1/(1 + 1/C)$ and $\alpha_2 = \alpha_4 = 0$ or $\alpha_2 = \alpha_4 = 1/(1 + 1/C)$ and $\alpha_1 = \alpha_3 = 0$ are not optimal, but $\alpha_i = 1/(2 + 1/C)$ $(i = 1, \ldots, 4)$ are.

In general, the number of support vectors for an L2 support vector machine is equal to or greater than that for an L1 support vector machine. And for sufficiently large $C$, the sets of support vectors are the same as shown in Theorem 2.9. In the original formulation of support vector machines, there is no mechanism for reducing the number of support vectors. Drezet and Harrison [60] reformulated support vector machines so that the number of support vectors is reduced. More detailed discussions on sparse support vector machines, please see Section 4.3.

Example 2.8 shows nonuniqueness of the dual problem, but the primal problem is unique because there are unbounded support vectors. Nonunique solutions occur when there are no unbounded support vectors. Burges and Crisp [57] derived conditions in which the dual problem is unique but the primal solution is nonunique. In the following, we discuss nonuniqueness of primal problems. First consider an example discussed in [57].

*Example 2.9.* Consider a one-dimensional example shown in Fig. 2.13.
The dual problem for linear kernels is given as follows:

$$\text{maximize } Q(\boldsymbol{\alpha}) = \alpha_1 + \alpha_2 - \frac{1}{2} (\alpha_1^2 + \alpha_2^2 + 2\,\alpha_1\,\alpha_2) \tag{2.250}$$

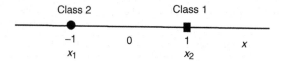

**Fig. 2.13** Nonunique solutions

$$\text{subject to} \quad \alpha_1 - \alpha_2 = 0, \tag{2.251}$$
$$C \geq \alpha_i \geq 0 \quad \text{for } i = 1, 2. \tag{2.252}$$

From (2.251), $\alpha_2 = \alpha_1$. Then substituting it into (2.250), we obtain

$$Q(\boldsymbol{\alpha}) = 2\alpha_1 - 2\alpha_1^2$$
$$= -2\left(\alpha_1 - \frac{1}{2}\right)^2 + \frac{1}{2}, \tag{2.253}$$
$$C \geq \alpha_i \geq 0 \quad \text{for } i = 1, 2. \tag{2.254}$$

For $C \geq 1/2$, (2.253) is maximized when $\alpha_1 = 1/2$. Thus from (2.48), $w = 1$, and from (2.54), $b = 0$. For $C < 1/2$, the optimal solution is given by

$$\alpha_1 = \alpha_2 = C. \tag{2.255}$$

Therefore, $w = 2C$. But because both $\alpha_1$ and $\alpha_2$ are bounded, from (2.45)

$$\xi_1 = 1 + b - 2C \geq 0, \tag{2.256}$$
$$\xi_2 = 1 - b - 2C \geq 0. \tag{2.257}$$

Thus, because $\xi_1 + \xi_2 = 2 - 4C$, the primal objective function does not change if $\xi_1$ and $\xi_2$ change so long as $\xi_1 + \xi_2 = 2 - 4C$. Therefore

$$-1 + 2C \leq b \leq 1 - 2C. \tag{2.258}$$

Because $\alpha_1 = \alpha_2 = C$ and $b$ is not included in the dual problem, the dual problem is unique, but the primal problem is not, namely, the primal objective function $w^2/2 + C(\xi_1 + \xi_2)$ is maximized for $w = 2C$ and $(\xi_1, \xi_2)$ that satisfy (2.256) and (2.257).

For the L2 soft-margin support vector machine, the dual problem is given by

$$\text{maximize} \quad Q(\boldsymbol{\alpha}) = \alpha_1 + \alpha_2 - \frac{1}{2}\left(\left(1 + \frac{1}{C}\right)(\alpha_1^2 + \alpha_2^2) + 2\alpha_1\alpha_2\right) \tag{2.259}$$
$$\text{subject to} \quad \alpha_1 - \alpha_2 = 0, \tag{2.260}$$
$$\alpha_i \geq 0 \quad \text{for } i = 1, 2. \tag{2.261}$$

Substituting $\alpha_2 = \alpha_1$ into (2.259), we obtain

$$Q(\boldsymbol{\alpha}) = 2\alpha_1 - \left(2 + \frac{1}{C}\right)\alpha_1^2 \tag{2.262}$$

$$\text{subject to} \quad \alpha_1 \geq 0. \tag{2.263}$$

Thus, (2.262) is maximized when

$$\alpha_1 = \alpha_2 = \frac{C}{2C+1}, \tag{2.264}$$

and from (2.177), $b = 0$. Therefore, the problem is uniquely solved.

Burges and Crisp [57] derived general conditions in which primal problems have nonunique solutions. Here, we discuss a simplified version.

**Theorem 2.13.** *If the support vectors are all bounded, $b$ is not uniquely determined and the numbers of support vectors belonging to Classes 1 and 2 are the same.*

*Proof.* The KKT conditions are given by

$$(C - \alpha_i)\,\xi_i = 0, \tag{2.265}$$

$$\alpha_i\left(y_i\left(\mathbf{w}^\top \boldsymbol{\phi}(\mathbf{x}_i) + b\right) - 1 + \xi_i\right) = 0. \tag{2.266}$$

If $\alpha_i \neq 0$ and $\alpha_i \neq C$, $b$ is uniquely determined by (2.266). Thus, if $\alpha_i \neq 0$, then $\alpha_i = C$. Because $\sum_{i \in S} y_i \alpha_i = 0$, the numbers of support vectors for Classes 1 and 2 need to be the same for $\alpha_i = C\ (i \in S)$.

For $\alpha_i = C$,

$$\xi_i = 1 - y_i\left(\mathbf{w}^\top \boldsymbol{\phi}(\mathbf{x}_i) + b\right) \geq 0. \tag{2.267}$$

Because the primal objective function concerning $\mathbf{w}$ is strictly positive definite, $\mathbf{w}$ is uniquely determined. In addition, because the numbers of support vectors for Classes 1 and 2 are the same,

$$\sum_{i \in S} \xi_i = |S| - \sum_{i \in S} y_i\,\mathbf{w}^\top \boldsymbol{\phi}(\mathbf{x}_i), \tag{2.268}$$

which is constant irrespective of the values of $\xi_i$. This means that the primal objective function is constant for different values of $\xi_i$. Therefore, $b$ is not unique, and from (2.267),

$$\min_{\substack{i \in S, \\ y_i = 1}} \left(1 - \mathbf{w}^\top \boldsymbol{\phi}(\mathbf{x}_i)\right) \geq b \geq \max_{\substack{i \in S, \\ y_i = -1}} \left(-1 - \mathbf{w}^\top \boldsymbol{\phi}(\mathbf{x}_i)\right). \quad\blacksquare \tag{2.269}$$

## 2.6.5 Reducing the Number of Support Vectors

Because the solution of a support vector machine is expressed by support vectors, which are a small portion of the training data, the solution is sometimes called *sparse*. However, if the number of training data is very large or the classification problem is very complicated, support vectors will increase and thus classification will be slowed down. There are several approaches to overcome this problem [61–64]. We leave detailed discussions on realizing sparse support vector machines to Section 4.3.

Here we first show the geometrical interpretation of hard-margin support vector machines [65], and then we consider the possibility of reducing the number of support vectors [66].

The solution of a hard-margin support vector machine can be geometrically interpreted [65]. Let $U_1$ and $U_2$ be the convex hulls for Class 1 and Class 2 training data, respectively. Because the solution exists, $U_1$ and $U_2$ do not overlap. Let $\mathbf{u}_1^*$ and $\mathbf{u}_2^*$ give the minimum distance between $U_1$ and $U_2$:

$$\min_{\mathbf{u}_1 \in U_1, \mathbf{u}_2 \in U_2} \|\mathbf{u}_1 - \mathbf{u}_2\|. \tag{2.270}$$

Then the optimal hyperplane passes through the middle point of $\mathbf{u}_1^*$ and $\mathbf{u}_2^*$ and perpendicular to $\mathbf{u}_1^* - \mathbf{u}_2^*$. Thus, $\mathbf{w}$ and $b$ are given by

$$\mathbf{w} = \frac{2(\mathbf{u}_1^* - \mathbf{u}_2^*)}{\|\mathbf{u}_1^* - \mathbf{u}_2^*\|^2}, \tag{2.271}$$

$$b = -\frac{\|\mathbf{u}_1^*\|^2 - \|\mathbf{u}_2^*\|^2}{\|\mathbf{u}_1^* - \mathbf{u}_2^*\|^2}. \tag{2.272}$$

It is noted that $\mathbf{u}_1^*$ and $\mathbf{u}_2^*$ are not necessarily training data. In Fig. 2.14a, one is a training data sample but the other is not; in Fig. 2.14b, both are the training data. Thus in a special case, the optimum separating hyperplane is expressed by two support vectors. This interpretation holds for L1 and L2 support vector machines so long as the problem is linearly separable in the feature space.

From this discussion, we can show the following theorem.

**Theorem 2.14.** *An optimal hyperplane, which is expressed by a set of support vectors and a bias term, can be expressed by one unbounded support vector, the associated vector in the feature space, and the bias term.*

*Proof.* Let the index set of support vectors for a given problem be $S$. Then the weight vector is given by

$$\mathbf{w} = \sum_{i \in S} y_i \, \alpha_i \, \boldsymbol{\phi}(\mathbf{x}_i). \tag{2.273}$$

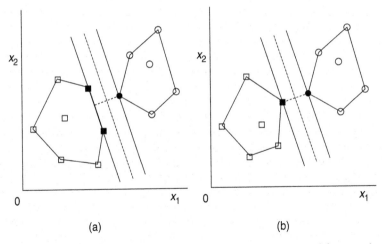

**Fig. 2.14** Geometrical interpretation of linearly separable solution: (**a**) more than two support vectors. (**b**) Two support vectors

Select one unbounded support vector and let this be $\mathbf{x}^+$, which belongs to Class 1. Define $\mathbf{z}^-$ in the feature space by

$$\mathbf{z}^- = \phi(\mathbf{x}^+) - 2\delta^2\,\mathbf{w}, \tag{2.274}$$

where $\delta$ is the margin of the separating hyperplane. Because $\mathbf{w}$ is orthogonal to the hyperplane and $2\delta$ is the distance between the two hyperplanes on which the unbounded support vectors reside, $\mathbf{z}^-$ satisfies $D(\mathbf{z}^-) = -1$. For the two vectors $\mathbf{x}^+$ and $\mathbf{z}^-$, the optimal separating hyperplane goes through $(\phi(\mathbf{x}^+) + \mathbf{z}^-)/2$ and is orthogonal to $(\phi(\mathbf{x}^+) - \mathbf{z}^-)/2$. Thus it is the same as the original separating hyperplane. ∎

If $\mathbf{x}^-$ that satisfies $\mathbf{z}^- = \phi(\mathbf{x}^-)$, i.e., the preimage of $\mathbf{z}^-$, exists, the support vectors can be reduced to two. Or if $\mathbf{v}$ that satisfies $\phi(\mathbf{v}) = \mathbf{w}$ exists, we can evaluate the decision function by $D(\mathbf{x}) = K(\mathbf{x}, \mathbf{v}) + b$, which will result in a considerable speedup.

In [10, pp. 544–545], a simple calculation method of the preimage is proposed if it exists and if $K(\mathbf{x}, \mathbf{x}') = f(\mathbf{x}^\top \mathbf{x}')$, where $f(\cdot)$ is some scalar function. Let $\{\mathbf{e}_1, \ldots, \mathbf{e}_m\}$ be the basis of the input space. Then

$$
\begin{aligned}
K(\mathbf{v}, \mathbf{e}_j) &= f(v_j) \\
&= \mathbf{w}^\top \phi(\mathbf{e}_j) \\
&= \sum_{i \in S} \alpha_i\, y_i\, f(x_{ij}),
\end{aligned}
\tag{2.275}
$$

where $x_{ij}$ is the $j$th element of $\mathbf{x}_i$. For the polynomial kernel, $f(v_j) = (v_j+1)^d$. Thus, if $d$ is odd, the inverse exists and

$$v_j = f^{-1}\left(\sum_{i \in S} \alpha_i\, y_i\, f(x_{ij})\right). \tag{2.276}$$

This seems to be correct. Indeed, with linear kernels, $\phi^{-1}(\mathbf{z}^-) = \mathbf{z}^-$, or $\phi^{-1}(\mathbf{w}) = \mathbf{w}$. But if an $m$-dimensional vector $\mathbf{x}$ is mapped into an $l$-dimensional space ($l > m$), the inverse of $\mathbf{w}$ does not exist. Consider the case where $K(\mathbf{x}, \mathbf{x}') = \mathbf{x}^\top \mathbf{x}' + 1$ and $\phi(\mathbf{v}) = a_1\phi(\mathbf{x}_1) + a_2\phi(\mathbf{x}_2)$. Thus, $\phi(\mathbf{x}) = (1, \mathbf{x}^\top)^\top$ and $l = m + 1$. Then the following $(m + 1)$ equations must be satisfied for $m$ variables:

$$1 = a_1 + a_2, \tag{2.277}$$
$$v_1 = a_1\, x_{11} + a_2\, x_{21}, \tag{2.278}$$
$$\dots$$
$$v_m = a_1\, x_{1m} + a_2\, x_{2m}. \tag{2.279}$$

This set of simultaneous equations is solved only when (2.277) is satisfied.

For the polynomial kernel with degree 2 with a one-dimensional input $x$, $K(x, x') = (1 + x\,x')^2$ and $\phi(x)$ is given by

$$\phi(x) = (1, \sqrt{2}\,x, x^2)^\top. \tag{2.280}$$

Thus, the following equations must be satisfied:

$$1 = a_1 + a_2, \tag{2.281}$$
$$v = a_1\, x_1 + a_2\, x_2, \tag{2.282}$$
$$v^2 = a_1\, x_1^2 + a_2\, x_2^2, \tag{2.283}$$

which is, in general, unsolvable.

In general, a set of $l$ equations must be satisfied for $m$ variables. Thus, if $l \neq m$, the inverse does not exist. This is caused by the fact that the region of $\phi(\mathbf{x})$ is nonconvex, as discussed in Section 2.3.4.

This is discouraging because Theorem 2.14 is useful only when $d = 1$, which is trivial.

To evaluate the speedup by reducing the number of support vectors to two, we used the data sets listed in Table 1.3. Table 2.6 shows the results for $d = 1$. The columns "SVs," "Original," "Reduced," and "Speedup" show the number of support vectors, the classification time of the training and test data using the support vectors, the classification time using the reduced support vectors, and the speedup by reduction, respectively. From the table, it is seen, except for the thyroid data, that the speedup is roughly the number of support vectors divided by two.

**Table 2.6** Classification speedup by reducing the number of support vectors to two $(d = 1)$

| Data | SVs | Original (s) | Reduced (s) | Speedup |
|------|-----|--------------|-------------|---------|
| Blood cell | 14 | 8 | 1 | 8.0 |
| Thyroid | 80 | 33 | 10 | 3.3 |
| Hiragana-50 | 11 | 263 | 49 | 5.4 |
| Hiragana-13 | 7 | 113 | 31 | 3.6 |
| Hiragana-105 | 13 | 1,029 | 158 | 6.5 |

## 2.6.6 Degenerate Solutions

Rifkin et al. [67] discussed degenerate solutions in which $\mathbf{w} = \mathbf{0}$ for L1 support vector machines. Fernández [68] derived similar results for L1 support vector machines, although he did not refer to degeneracy. Degeneracy occurs also for L2 support vector machines. In the following, we discuss degenerate solutions following the proof in [68].

**Theorem 2.15.** *Let* $C = K C_0$, *where* $K$ *and* $C_0$ *are positive parameters and* $\boldsymbol{\alpha}^*$ *be the solution of the L1 support vector machine with* $K = 1$. *Define*

$$\mathbf{w}(\boldsymbol{\alpha}) = \sum_{i=1}^{M} \alpha_i \, y_i \, \boldsymbol{\phi}(\mathbf{x}_i). \tag{2.284}$$

*Then the necessary and sufficient condition for*

$$\mathbf{w}(\boldsymbol{\alpha}^*) = \mathbf{0} \tag{2.285}$$

*is that* $K \, \boldsymbol{\alpha}^*$ *is also a solution for any* $K \, (> 1)$.

*Proof.* We prove the theorem for L2 support vector machines. The proof for L1 support vector machines is obtained by deleting $\boldsymbol{\alpha}^\top \boldsymbol{\alpha}/(2C)$ in the following proof.
**Necessary condition.** Let $\boldsymbol{\alpha}'$ be the optimal solution for $C = K C_0$ $(K > 1)$ and $\mathbf{w}(\boldsymbol{\alpha}') \neq \mathbf{0}$. Define $\boldsymbol{\alpha}' = K \boldsymbol{\alpha}''$. Then, because $\boldsymbol{\alpha}''$ satisfies the equality constraint, it is a nonoptimal solution for $C = C_0$. Then for $C = C_0$,

$$Q(\boldsymbol{\alpha}^*) = \sum_{i=1}^{M} \alpha_i^* - \frac{\boldsymbol{\alpha}^{*\top} \boldsymbol{\alpha}^*}{2 C_0}$$

$$\geq Q(\boldsymbol{\alpha}'') = \sum_{i=1}^{M} \alpha_i'' - \frac{1}{2} \mathbf{w}(\boldsymbol{\alpha}'')^\top \mathbf{w}(\boldsymbol{\alpha}'') - \frac{\boldsymbol{\alpha}''^\top \boldsymbol{\alpha}''}{2 C_0}. \tag{2.286}$$

For $C = K C_0$ $(K > 1)$,

$$Q(K\boldsymbol{\alpha}^*) = K \sum_{i=1}^{M} \alpha_i^* - \frac{K\boldsymbol{\alpha}^{*\top}\boldsymbol{\alpha}^*}{2\,C_0} \leq Q(K\boldsymbol{\alpha}'')$$

$$= K \sum_{i=1}^{M} \alpha_i'' - \frac{K^2}{2} \mathbf{w}(\boldsymbol{\alpha}'')^\top \mathbf{w}(\boldsymbol{\alpha}'') - \frac{K\boldsymbol{\alpha}''^\top\boldsymbol{\alpha}''}{2\,C_0}. \qquad (2.287)$$

Multiplying all the terms in (2.286) by $K$ and comparing it with (2.287), we see the contradiction. Thus, $K\boldsymbol{\alpha}^*$ is the optimal solution for $K > 1$.

**Sufficient condition.**  Suppose $K\boldsymbol{\alpha}^*$ is the optimal solution for any $C = K\,C_0\,(\geq 1)$. Thus for any $C = K\,C_0\,(\geq 1)$,

$$Q(K\boldsymbol{\alpha}^*) = \sum_{i=1}^{M} K\,\alpha_i^* - \frac{1}{2} K^2\,\mathbf{w}^\top(\boldsymbol{\alpha}^*)\,\mathbf{w}(\boldsymbol{\alpha}^*) - \frac{K\boldsymbol{\alpha}^{*\top}\boldsymbol{\alpha}^*}{2\,C_0}$$

$$\geq Q(\boldsymbol{\alpha}^*) = \sum_{i=1}^{M} \alpha_i^* - \frac{1}{2}\mathbf{w}^\top(\boldsymbol{\alpha}^*)\,\mathbf{w}(\boldsymbol{\alpha}^*) - \frac{\boldsymbol{\alpha}^{*\top}\boldsymbol{\alpha}^*}{2\,C_0}. \qquad (2.288)$$

Rewriting (2.288), we have

$$\sum_{i=1}^{M} \alpha_i^* \geq \frac{K+1}{2}\,\mathbf{w}^\top(\boldsymbol{\alpha}^*)\,\mathbf{w}(\boldsymbol{\alpha}^*) + \frac{\boldsymbol{\alpha}^{*\top}\boldsymbol{\alpha}^*}{2\,C_0}. \qquad (2.289)$$

Because (2.289) is satisfied for any large $K$, $\mathbf{w}(\boldsymbol{\alpha}^*) = \mathbf{0}$ must be satisfied. Otherwise $K\boldsymbol{\alpha}^*$ cannot be the optimal solution. ∎

*Example 2.10.* Now reconsider the case shown in Fig. 2.5 on p. 35. Here, we use the linear kernel. The inequality constraints given by (2.40) are

$$-w + b \geq 1 - \xi_1, \qquad (2.290)$$

$$-b \geq 1 - \xi_2, \qquad (2.291)$$

$$w + b \geq 1 - \xi_3. \qquad (2.292)$$

The dual problem for the L1 support vector machine is given by

$$\text{maximize} \quad Q(\boldsymbol{\alpha}) = \alpha_1 + \alpha_2 + \alpha_3 - \frac{1}{2}(-\alpha_1 + \alpha_3)^2 \qquad (2.293)$$

$$\text{subject to} \quad \alpha_1 - \alpha_2 + \alpha_3 = 0, \qquad (2.294)$$

$$C \geq \alpha_i \geq 0 \quad \text{for } i = 1, 2, 3. \qquad (2.295)$$

From (2.294), $\alpha_2 = \alpha_1 + \alpha_3$. Then substituting it into (2.293), we obtain

$$Q(\boldsymbol{\alpha}) = 2\,\alpha_1 + 2\,\alpha_3 - \frac{1}{2}(-\alpha_1 + \alpha_3)^2, \qquad (2.296)$$

$$C \geq \alpha_i \geq 0 \quad \text{for } i = 1, 2, 3. \qquad (2.297)$$

Because $Q(\boldsymbol{\alpha})$ is symmetric for $\alpha_1$ and $\alpha_3$, it is maximized when $\alpha_1 = \alpha_3$. Thus the optimal solution is given by

$$\alpha_1 = \frac{C}{2}, \quad \alpha_2 = C, \quad \alpha_3 = \frac{C}{2}. \tag{2.298}$$

Therefore, $x = -1$, $0$, and $1$ are support vectors; and $w = 0$ and $b = 1$. Because the two unbounded support vectors, $\alpha_1$ and $\alpha_3$, belong to the same class, this is an abnormal solution; all the data are classified into Class 1, irrespective of the input.

The dual objective function for the L2 support vector machine is given by

$$Q(\boldsymbol{\alpha}) = \alpha_1 + \alpha_2 + \alpha_3 - \frac{1}{2}(-\alpha_1 + \alpha_3)^2 - \frac{\alpha_1^2 + \alpha_2^2 + \alpha_3^2}{2C}. \tag{2.299}$$

The objective function is maximized when

$$\alpha_1 = \alpha_3 = \frac{2C}{3}, \quad \alpha_2 = \frac{4C}{3}. \tag{2.300}$$

Thus, $w = 0$ and $b = 1$. Therefore, any data sample is classified into Class 1.

## 2.6.7 Duplicate Copies of Data

Nonunique solutions occur when duplicate copies of a data sample with the same label are included in the training data. If $\mathbf{x}_i = \mathbf{x}_j$, a solution $\boldsymbol{\alpha}^*$ satisfies $\alpha_i^* = \alpha_j^*$. This can be shown as follows. Suppose $\alpha_i^* \neq \alpha_j^*$. Then because of the symmetry of the variables, $(\alpha_i, \alpha_j) = (\alpha_j^*, \alpha_i^*)$ is also a solution. Because for quadratic programming problems the nonunique solutions cannot be isolated [57], the nonunique solution satisfies

$$\alpha_i = \beta\,\alpha_i^* + (1 - \beta)\,\alpha_j^*, \tag{2.301}$$
$$\alpha_j = (1 - \beta)\,\alpha_i^* + \beta\,\alpha_j^*, \tag{2.302}$$

where $0 \leq \beta \leq 1$. Then setting $\beta = 1/2$,

$$\alpha_i = \alpha_j = \frac{\alpha_i^* + \alpha_j^*}{2}. \tag{2.303}$$

Let $\{\mathbf{x}_1, \ldots, \mathbf{x}_M\}$ be a set of training data with each data sample different from the others. Assume that for each $\mathbf{x}_i$ we add $\zeta_i - 1$ copies of $\mathbf{x}_i$ to the set. Then using the preceding result, the same $\zeta_i$ data share the same Lagrange multiplier. Thus the training of the L1 support vector machine for the set is given by [69]

$$\text{maximize} \quad Q(\boldsymbol{\alpha}) = \sum_{i=1}^{M} \zeta_i \, \alpha_i - \frac{1}{2} \sum_{i,j=1}^{M} \zeta_i \, \alpha_i \, \zeta_j \, \alpha_j \, y_i \, y_j \, K(\mathbf{x}_i, \mathbf{x}_j) \quad (2.304)$$

$$\text{subject to} \quad \sum_{i=1}^{M} y_i \, \zeta_i \, \alpha_i = 0, \qquad 0 \le \alpha_i \le C \quad \text{for } i = 1, \ldots, M. \quad (2.305)$$

Changing the variables by $\alpha_i' = \zeta_i \, \alpha_i$, the optimization problem given by (2.304) and (2.305) becomes

$$\text{maximize} \quad Q(\boldsymbol{\alpha}') = \sum_{i=1}^{M} \alpha_i' - \frac{1}{2} \sum_{i,j=1}^{M} \alpha_i' \, \alpha_j' \, y_i \, y_j \, K(\mathbf{x}_i, \mathbf{x}_j) \quad (2.306)$$

$$\text{subject to} \quad \sum_{i=1}^{M} y_i \, \alpha_i' = 0, \qquad 0 \le \alpha_i' \le \zeta_i \, C \quad \text{for } i = 1, \ldots, M. \quad (2.307)$$

Thus, if the solution does not have bounded support vectors, the solution is the same as that without copies.[11]

The L2 support vector machine for the set of training data is trained by

$$\text{maximize} \quad Q(\boldsymbol{\alpha}) = \sum_{i=1}^{M} \zeta_i \, \alpha_i$$

$$-\frac{1}{2} \sum_{i,j=1}^{M} \zeta_i \, \alpha_i \, \zeta_j \, \alpha_j \, y_i \, y_j \left( K(\mathbf{x}_i, \mathbf{x}_j) + \frac{\delta_{ij}}{\zeta_i \, C} \right) \quad (2.308)$$

$$\text{subject to} \quad \sum_{i=1}^{M} y_i \, \zeta_i \, \alpha_i = 0, \qquad \alpha_i \ge 0 \quad \text{for } i = 1, \ldots, M. \quad (2.309)$$

Changing the variables by $\alpha_i' = \zeta_i \, \alpha_i$, the optimization problem given by (2.308) and (2.309) becomes

$$\text{maximize} \quad Q(\boldsymbol{\alpha}') = \sum_{i=1}^{M} \alpha_i' - \frac{1}{2} \sum_{i,j=1}^{M} \alpha_i' \, \alpha_j' \, y_i \, y_j \left( K(\mathbf{x}_i, \mathbf{x}_j) + \frac{\delta_{ij}}{\zeta_i \, C} \right) (2.310)$$

$$\text{subject to} \quad \sum_{i=1}^{M} y_i \, \alpha_i' = 0, \qquad \alpha_i' \ge 0 \quad \text{for } i = 1, \ldots, M. \quad (2.311)$$

Thus, an addition of copied data affects the solution of the L2 support vector machine.

To estimate the generalization ability of classifiers using a small set of data, resampling with replacement is often used. To speed up evaluation,

---

[11] The formulation given by (2.306) and (2.307) is the same as that of the fuzzy support vector machine discussed in [70].

it is advisable to use (2.306) and (2.307), or (2.310) and (2.311) instead of copying data.

## 2.6.8 Imbalanced Data

In medical diagnosis, usually, misclassification of abnormal data into the normal class is more fatal than misclassification of normal data into the abnormal class. To control misclassification, Veropoulos, Campbell, and Cristianini [71] proposed preparing different margin parameters $C^+$ and $C^-$ for the normal $(+1)$ and abnormal $(-1)$ classes, respectively, and setting $C^- > C^+$. Lee et al. [21] applied this method to diagnosis problems with imbalanced data. By setting the ratio of $C^+$ and $C^-$ about the ratio of the number of data for Classes $-1$ and 1, better classification performance was obtained than by the regular support vector machine. Xu and Chan [72] determined the ratios of a one-against-all support vector machine by genetic algorithms. Wu and Chang [73] adjusted the class boundary, modifying the kernel matrix according to the imbalanced data distribution.

For imbalanced data, resampling, which allows multiple selection of the same data sample, is usually used. In that situation, we had better use the formulation of support vector machines discussed in Section 2.6.7.

In some classification problems, a priori class probabilities are given, and they are different from the ratios of class data to the total data. To compensate for this disparity, Cawley and Talbot [69] proposed the formulation discussed in Section 2.6.7, namely, for the class $j$ data sample $\mathbf{x}_i$ we set

$$\zeta_i = \frac{p_j^o}{p_j^t}, \tag{2.312}$$

where $p_j^o$ is the a priori probability for class $j$ and $p_j^t$ is given by the number of class $j$ data divided by the number of total training data.

In [74], the bias term is adjusted for the unbalanced data from the link between the least-squares support vector machines and the kernel discriminant analysis.

## 2.6.9 Classification for the Blood Cell Data

In this section we show the effect of parameters on classification using the blood cell data for Classes 2 and 3 in Table 1.3. Except for the Class 2 training data set, which includes 399 data, training and test data sets include 400 data each. The blood cell data are very difficult to classify, and Classes 2 and 3 are the most difficult.

We investigated the effect of kernel parameters and the margin parameter $C$ on the recognition rates and the number of support vectors using L1 and L2 support vector machines.

Figure 2.15 shows the recognition rates for the change of the polynomial degree fixing $C = 5,000$. For $d = 1$ (i.e., linear kernels), the recognition rates of the training data are about 90% for L1 and L2 support vector machines and for $d$ larger than 2, the recognition rates are 100%. Thus the training data are not linearly separable. For the change of $d$, the recognition rates of the test data do not change very much for L1 and L2 support vector machines. This means that overfitting does not occur in this case.

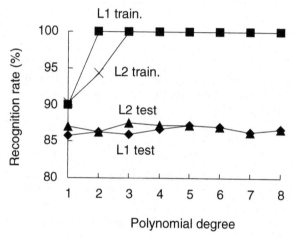

**Fig. 2.15** Recognition rates of the blood cell data with polynomial kernels

Figure 2.16 shows the number of support vectors for the change of the polynomial degree under the same conditions as those in Fig. 2.15. As seen from Fig. 2.16, because the blood cell data are not linearly separable, especially for the L2 support vector machine, a large number of support vectors are necessary for $d = 1$. For the polynomial degree lower than 4, the number of support vectors for the L2 support vector machine is larger than that for the L1 support vector machine, but for degrees higher than 3, they are almost the same. From Figs. 2.15 and 2.16, for $d$ larger than 3, behaviors of the L1 and L2 support vector machines are almost the same.

Figure 2.17 shows the recognition rates for the RBF kernel with different values of $\gamma$ fixing $C = 5,000$. The recognition rates of the L1 and L2 support vector machines are almost the same, and for $\gamma$ lager than 1, the recognition rates of the training data are 100%. For the change of $\gamma$, the recognition rates of the test data do not change very much for L1 and L2 support vector machines.

Figure 2.18 shows the number of support vectors for the RBF kernel with different values of $\gamma$ under the same conditions as those in Fig. 2.17. For $\gamma$ smaller than 1, the number of support vectors for the L2 support vector machine is larger than for the L1 support vector machine, but for $\gamma$ larger than or equal to 1, they are almost the same. For $\gamma = 100$, the numbers of support vectors are 396. As $\gamma$ becomes smaller, the radius of RBF becomes smaller. Thus, a larger number of support vectors are necessary to classify the training data correctly. Similar to the case with the polynomial kernels, the behaviors of the L1 and L2 support vector machines are almost the same for $\gamma$ larger than 1.

Figure 2.19 shows the recognition rates when $C$ is changed with the polynomial degree of 4. As the value of $C$ is increased, the weight for the sum (square sum) of the slack variables is increased. Thus the recognition rate of the training data is improved, and the recognition rate reaches 100% for $C = 1,000$. On the other hand, the recognition rate of the test data gradually decreases. Thus there is a tendency of overfitting. Similar to the previous results, the recognition rates of L1 and L2 support vector machines are almost the same.

Figure 2.20 shows the numbers of support vectors for the change of $C$. As the value of $C$ increases, the weight of the (square) sum of slack variables is increased. Thus the number of support vectors decreases. At $C = 1,000$, the number of support vectors for the L1 and L2 support vector machines is 93 each.

To investigate the robustness of support vector machines, as outliers, we added 10 data belonging to classes other than 2 and 3 to Class 2 training data. Figure 2.21 shows the recognition rate of the test data against the margin parameter $C$ when outliers were added and not added. We used the

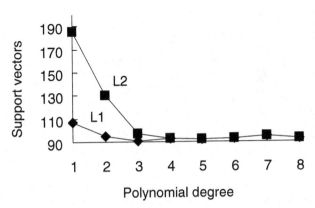

**Fig. 2.16** Number of support vectors for the blood cell data with polynomial kernels

L2 support vector machine. For $C = 0.01$, the recognition rate with outliers is much worse than without outliers. But with $C$ larger than or equal to 0.01, the difference is not so significant. Thus if an appropriate value is set to $C$, the support vector machine is robust against outliers.

From the computer experiment for the blood cell data, the following tendencies are seen.

1. The recognition rate of the training data increased as the polynomial degree, $\gamma$ in RBF kernels, or the value of $C$ is increased. But the recognition rate of the test data does not change very much.
2. In theory, for $C = \infty$, L1 and L2 support vector machines are equivalent. But by computer simulations, this condition may be loosened; for appropriately large $d$ or $\gamma$ and $C$, the recognition rates of the training and test data and the number of support vectors are almost the same for the L1 and L2 support vector machines.
3. For an appropriately chosen value of $C$, the effect of outliers on the recognition rate is not significant.

## 2.7 Class Boundaries for Different Kernels

In this section we discuss how class boundaries change as kernels are changed for a two-class problem. For the linear kernel, the class boundary is a hyperplane. For the polynomial kernel with degree 2, the class boundary is a hyperplane, a hyperellipsoid (see Fig. 2.22), or a hyperparabola (see Fig. 2.23).

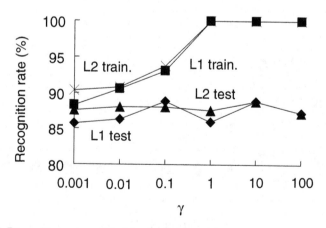

**Fig. 2.17** Recognition rates of the blood cell data with RBF kernels

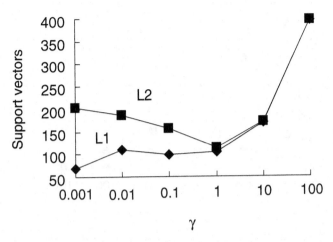

**Fig. 2.18** Number of support vectors for the blood cell data with RBF kernels

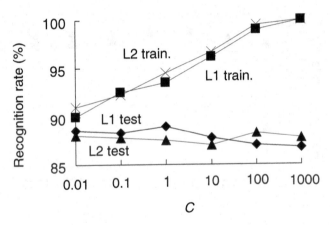

**Fig. 2.19** Recognition rates of the blood cell data with the polynomial kernel $(d = 4)$ for the change of $C$

This class boundary is equivalent to that of the fuzzy classifier with ellipsoidal regions.

To clarify how class regions are clustered for a given kernel, we restrict our discussion to one input variable. For a polynomial kernel with degree 3,

$$K(x, x') = (x\,x' + 1)^3$$
$$= 1 + 3\,x'\,x + 3\,{x'}^2\,x^2 + {x'}^3\,x^3$$

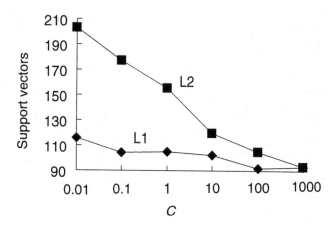

**Fig. 2.20** Number of support vectors for the blood cell data with polynomial kernels ($d = 4$) for the change of $C$

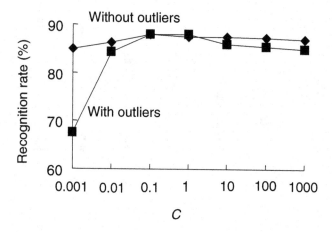

**Fig. 2.21** Recognition rates against the margin parameter $C$ for the inclusion of outliers

$$= (1, \sqrt{3}\,x, \sqrt{3}\,x^2, x^3)\,(1, \sqrt{3}\,x', \sqrt{3}\,x'^2, x'^3)^{\top}. \qquad (2.313)$$

Thus the function that maps $x$ into the feature space $\mathbf{z}$ is given by $\phi(x) = (1, \sqrt{3}\,x, \sqrt{3}\,x^2, x^3)^{\top}$. Therefore, the hyperplane in the feature space is given by

$$w_0 + \sqrt{3}\,w_1\,x + \sqrt{3}\,w_2\,x^2 + w_3\,x^3 = 0. \qquad (2.314)$$

Because the maximum number of solutions of (2.314) for $x$ is 3, the input regions of both Classes 1 and 2 are divided into two subregions in maximum

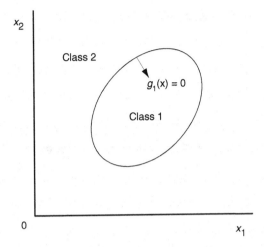

**Fig. 2.22** An ellipsoidal boundary by a polynomial kernel with the degree of two

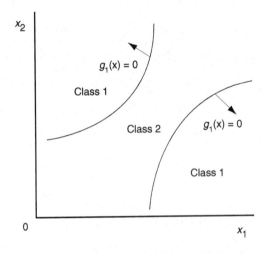

**Fig. 2.23** Parabolic boundaries by a polynomial kernel with the degree of two

(see Fig. 2.24). In general, for a polynomial kernel with degree $n$, $n + 1$ subregions are generated in maximum. If $n$ is even, the input regions of both Classes 1 and 2 are divided into $(n + 2)/2$ and $n/2$ subregions in maximum. And if $n$ is odd, those of Classes 1 and 2 are divided into $(n+1)/2$ subregions in maximum.

For an RBF kernel, the induced feature space has infinite dimensions. Therefore, the hyperplane in the feature space can divide the input space into any number of subregions. This is also true for a three-layer neural

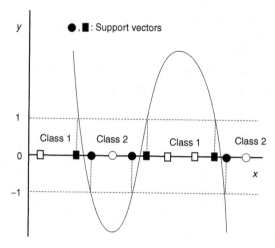

**Fig. 2.24** Class regions by a polynomial kernel with the degree of three. Regions for Classes 1 and 2 are divided into two

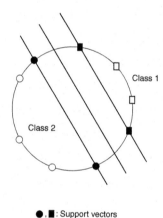

$\bullet$, $\blacksquare$ : Support vectors

**Fig. 2.25** Class regions by an RBF kernel in the feature space

network kernel. Thus we need not worry about clustering the training data before training.

In the feature space induced by an RBF kernel, data are on the surface of the unit hypersphere centered at the origin because $\phi^\top(\mathbf{x})\,\phi(\mathbf{x}) = \exp(-\gamma\|\mathbf{x}-\mathbf{x}\|^2) = 1$ [75]. Thus, the optimal separating hyperplane $D(\mathbf{x}) = 0$ and hyperplanes $D(\mathbf{x}) = \pm 1$ must intersect with the hypersphere. Figure 2.25 shows an illustration of class regions for a hard-margin support vector machine. In this way training data that are not linearly separable in the input space are separated in the feature space.

## 2.8 Developing Classifiers

In this section, we discuss how to develop classifiers with high generalization ability. Here we do not consider optimizing the input features. Thus our aim is to develop the classifier that realizes the best generalization ability for the given input–output training pairs. Here we call the classifier with the best generalization ability as the *optimal classifier*.

### 2.8.1 Model Selection

In training a support vector machine we need to select a kernel and determine values for the kernel parameter and the margin parameter. Thus to develop the optimal classifier for a given kernel, we need to determine the optimal kernel parameter and the margin parameter values. Determining the optimal classifier is called *model selection*.

The model selection is usually done by estimating the generalization abilities for the grid points in the kernel-parameter-and-$C$ plane, and selecting the classifier that realizes the highest generalization ability.

The most reliable but time-consuming method of estimating the generalization ability is cross-validation based on repetitive training of support vector machines. Thus to shorten model selection time, several measures for estimating the generalization ability have been proposed.

### 2.8.2 Estimating Generalization Errors

*Cross-validation* is a widely used technique to estimate the generalization error of a classifier. Instead of using cross-validation, many measures to estimate the generalization error of support vector machines, which are based on statistical learning theory, are proposed. In the following, we briefly summarize these measures.

#### 2.8.2.1 Cross-Validation

Cross-validation is used to measure the generalization error of classifiers for a limited number of gathered data. In cross-validation, the $M$ given data are divided into two data sets, $S_i^{\mathrm{tr}}$ ($i = 1, \ldots, {}_M C_l$), which includes $l$ training data, and $S_i^{\mathrm{ts}}$, which includes $M - l$ test data. Then for the training data set $S_i^{\mathrm{tr}}$ the classifier is trained and tested for the test data set $S_i^{\mathrm{ts}}$. This is repeated for all the combinations (${}_M C_l$) of the partitioned training and test data sets, and the total recognition rate for all the test data sets is calculated as the estimation of the classification performance. But because this is a time-consuming task, we usually use $k$-fold cross-validation.

In $k$-fold cross-validation, training data are randomly divided into approximately equal-sized $k$ subsets, and a classifier is trained using $k - 1$ subsets and tested using the remaining subset. Training is repeated $k$ times, and the total recognition rate for all the $k$ subsets that are not included in the training data is calculated. A leave-one-out method (LOO) is a special case of cross-validation ($l = M - 1$) and $k$-fold cross-validation. The leave-one-out error is known to be an unbiased estimate of the test error [76, p. 265].

For classifiers other than support vector machines, LOO is a time-consuming task when the number of training data is large. But for support vector machines, once we have trained a support vector machine, we need to apply LOO only to support vectors. This is because even if we delete the training data other than support vectors, these data are correctly classified. Cauwenberghs and Poggio's decremental training [77] further speeds up LOO by deleting one support vector from the trained support vector machine.

Saadi et al. [78] discussed acceleration of the leave-one-out procedure of least-squares support vector machines (see Section 4.1). For leaving the $j$th data sample out, we strike out the $j$th row and $j$th column of the coefficient matrix in the set of linear equations that determines $\boldsymbol{\alpha}$ and $b$. Thus in solving the set of equations by the Gauss–Jordan elimination, we can avoid calculation of the first to $(j - 2)$nd elimination by caching these results obtained for the $(j - 1)$th data sample. Ying and Keong [79] trained a least-squares support vector machine using all the training data and then used the matrix inversion lemma to speed up the leave-one-out procedure.

### 2.8.2.2 Error Bound by VC Dimension

The VC (Vapnik–Chervonenkis) dimension is the theoretical basis of support vector machines and is defined as the maximum number of samples that can be separated into any combination of two sets by the set of functions. Because the set of $m$-dimensional hyperplanes can separate at most $m + 1$ samples, the VC dimension of the set is $m + 1$ (see Fig. 2.26).

According to Vapnik's theory [80, pp. 1–15], the generalization error of a support vector machine is bounded with the probability of at least $1 - \eta$ by

$$R(\mathbf{w}, b) \leq R_{\mathrm{emp}}(\mathbf{w}, b) + \phi, \tag{2.315}$$

where $R_{\mathrm{emp}}(\mathbf{w}, b)$ is the empirical risk (classification error) for the $M$ training data and $\phi$ is the confidence interval (classification error) for the unknown data:

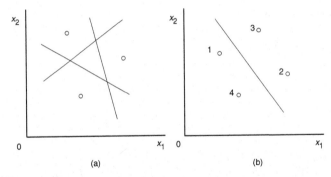

**Fig. 2.26** VC dimension of a set of lines: (**a**) any three data can be separated into any combination of two sets of data by a single line. (**b**) Sets of data {1, 4} and {2, 4} can be separated into the two sets by a line but sets of data {1, 2} and {3, 4} cannot. Thus the VC dimension of the set of lines is three

$$\phi = \sqrt{\frac{h\left[\ln\left(\frac{2M}{h}\right) + 1\right] - \ln\left(\frac{\eta}{4}\right)}{M}}. \tag{2.316}$$

Here $h$ is the VC dimension of a set of hyperplanes, and for the hard-margin optimal separating hyperplane, the VC dimension of the hyperplanes with margin $\|\mathbf{w}\|^{-1}$, $h$, is bounded by [4, 81, 76]

$$h \leq \min(D^2\|\mathbf{w}\|^2, l) + 1. \tag{2.317}$$

Here $D$ is the diameter of the smallest hypersphere that includes all the training data. We can determine $R$ by the method discussed in Section 8.1.

When the training data are linearly separable in the feature space, the empirical risk $R_{\text{emp}}(\mathbf{w}, b)$ is zero.[12] Thus if we minimize $\phi$, we can maximize the generalization ability. From (2.316), $\phi$ is the monotonic function of the VC dimension $h$. Thus $\phi$ is minimized by minimizing $h$. From (2.317), this is realized by maximizing the margin $\|\mathbf{w}\|^{-1}$. When the training data are not linearly separable in the feature space, $R_{\text{emp}}(\mathbf{w}, b)$ is not zero. In this case the generalization ability is determined by the trade-off between the empirical risk and the confidence interval via the margin parameter $C$.

### 2.8.2.3 LOO Error Rate Estimators

Several measures for estimating the LOO error rate have been proposed. By taking the average of the LOO error rates for possible combinations of

---

[12] Here, to simplify discussions, we exclude the case where $R_{\text{emp}}(\mathbf{w}, b)$ is not zero for a small value of $C$.

training data, we obtain the error bound for the test data. The error rate estimators discussed in the following need to train the support vector machine using the training data once.

1. Vapnik [4]

   The LOO error rate is bounded by

   $$\varepsilon_{\text{loo}} \leq \frac{|S|}{M}, \tag{2.318}$$

   where $\varepsilon_{\text{loo}}$ is the LOO error rate for the given training data, $|S|$ is the number of support vectors, and $M$ is the number of training data.

   Assume that we delete a data sample, which is not a support vector, from the training data set. Then the support vector machine trained using the reduced training data set correctly classifies the deleted data sample. But if a support vector is deleted, the classification result cannot be estimated. Thus assuming that all the support vectors are misclassified, (2.318) is obtained.

2. Joachims $\xi\alpha$ estimators [82, 29]

   The LOO error rate is bounded by

   $$\varepsilon_{\text{loo}} \leq \frac{|\{i \,|\, 2\,\alpha_i\,R_\Delta^2 + \xi_i \geq 1\}|}{M}, \tag{2.319}$$

   where $R_\Delta^2$ is the upper bound that satisfies

   $$c \leq K(\mathbf{x}, \mathbf{x}') \leq c + R_\Delta^2 \tag{2.320}$$

   for all $\mathbf{x}$ and $\mathbf{x}'$. Because the inequality on the right-hand side of (2.319) may not hold for small $\alpha_i$, (2.319) is an improved version of (2.318).

3. Vapnik and Chapelle [76, 83]

   The LOO error rate is bounded by

   $$\varepsilon_{\text{loo}} \leq \frac{S_m \max(D, 1/\sqrt{C}) \sum\limits_{i \in U} \alpha_i + |B|}{M}, \tag{2.321}$$

   where $D$ is the diameter of the smallest hypersphere that includes the training data, $U$ is the set of unbounded support vector indices, $B$ is the set of bounded support vector indices, and $S_m$ is the span of support vectors and is defined as follows:

   $$S_m = \max_p S_p,$$
   $$S_p^2 = \min_{\mathbf{x} \in \Lambda_p} (\mathbf{x}_p - \mathbf{x})^2,$$

$$\Lambda_p = \left\{ \sum_{i \in S, i \neq p} \lambda_i \mathbf{x}_i \,\Big|\, \sum_{i \in S, i \neq p} \lambda_i = 1, \alpha_i + y_i\, y_p\, \alpha_p\, \lambda_i \geq 0 \quad \text{for } \forall i \neq p \right\}.$$

Here $S_p$ is the distance between support vector $\mathbf{x}_p$ and the set $\Lambda_p$, and $S$ is the set of support vector indices. The bounded support vectors are assumed to be misclassified when deleted by the LOO procedure.

To optimize parameters of support vector machines, the error bound need not be accurate so long as it shows the correct tendency for the parameter change. Duan et al. [84] evaluated the fivefold cross-validation, the LOO bounds by (2.319), $\phi$ in (2.315) with $h$ evaluated by (2.317), and (2.321) for three benchmark data sets with 400–1,300 training data. The RBF kernels were used, and $C$ and $\gamma$ were tuned using the preceding four measures for the training data. Then the best recognition rates of the test data sets were compared with the recognition rates obtained by model selection. The fivefold cross-validation gave the best results, and the bound by (2.319) gave the second best. Because $S_m$ in (2.321) was replaced by $D_U$, which is the diameter of the smallest hypersphere that includes the unbounded support vectors, the results were not good.

Anguita et al. [85] evaluated the bound given by (2.317) for the six data sets with 80–958 data by the first-order accurate bootstrap [86]. For a fixed $\gamma$ for the RBF kernels, 1,000 training data sets were generated by drawing from the original training data set with replacement. Then for each data set, the support vector machine was trained and the average and standard deviation of the right-hand side of (2.317) were evaluated. Although the bound given by (2.317) was loose, it had the same tendency with the recognition rates evaluated for the data not included in the generated training data.

## 2.8.3 Sophistication of Model Selection

Because model selection is time consuming, various methods have been developed to ease model selection [87–106]. Cristianini and Campbell [87] showed that the bound on the generalization error is smooth in the kernel parameter, namely, when the margin is optimal, small variations in the kernel parameter result in small variations in the margin. Then the model selection for the RBF kernel is done as follows. Train the support vector machine with a small value of $\sigma$, evaluate the error bound, increment the value of $\sigma$, and repeat the procedure until the optimal parameter is obtained. Another approach interprets support vector machines from the standpoint of Gaussian processes [91, 101].

Friedrichs and Igel [97] used evolution strategies to further tune the parameters obtained by grid search. They showed, by computer experiments, that using the generalized RBF kernel (Mahalanobis kernel):

$$K(\mathbf{x}, \mathbf{x}') = \exp\left(-(\mathbf{x} - \mathbf{x}')^\top A (\mathbf{x} - \mathbf{x}')\right), \qquad (2.322)$$

where $A$ is a positive definite matrix, the generalization ability was improved with a smaller number of support vectors.

Lebrun et al. [98] used the vector quantization technique to replace training data with a smaller number of prototypes and then to speed up model selection.

Hastie et al. [99] proposed calculating the entire solution path for the change of the margin parameter, solving a set of linear equations, instead of a quadratic programming problem. This method is extended to support vector regressors [103] and one-class support vector machines [104].

Keerthi and Lin [96] analyzed the solution behavior of a support vector machine with RBF kernels for the changes of the margin and kernel parameter values and derived the heuristic relation, $\log \sigma^2 = \log C - \log \tilde{C}$, that realizes the good generalization ability, where $2\sigma^2 = 1/\gamma$ and $\tilde{C}$ is the best $C$ for the linear support vector machine, which is the limit of the support vector machine with RBF kernels as $\sigma$ approaches infinity. After searching the best $\tilde{C}$, the optimal parameter values are determined by line search according to the relation.

Another approach uses a criterion such as kernel discriminant analysis (KDA) [92, 105] and the inter-class distance [102], which is a simplified version of KDA. In [102], the inter-class distances are evaluated for the discrete points in a range of the kernel parameter value and the kernel parameter value that maximizes the inter-class distance is selected. In [106] the kernel parameter value is calculated by the second-order method for RBF kernels.

## 2.8.4 Effect of Model Selection by Cross-Validation

In this section, we show how cross-validation is effective for model selection using the two-class data sets shown in Table 1.1. We normalized the input range into $[0, 1]$. First we compare performance of L1 and L2 support vector machines for RBF kernels. Then for L1 support vector machines, we evaluate model selection by cross-validation using polynomial, RBF, and Mahalanobis kernels.

For RBF kernels we selected the value of $\gamma$ from $\{0.1, 0.5, 1, 5, 10, 15\}$. And we selected the value of the margin parameter $C$ from $\{1, 10, 50, 100, 500, 1,000, 2,000\}$. For each value of the kernel parameter, we performed fivefold cross-validation for the first five training data sets and calculated the average recognition rate for the validation data sets. Then we selected the value of margin parameter that realized the best average recognition rate and selected the median value of the margin parameter $C$ among five $C$ values determined by fivefold cross-validation. If different values of $C$ gave the same average recognition rate, we broke the tie by taking the largest $C$ value to obtain the simpler architecture (the smaller number of support vectors).

Table 2.7 lists the results for RBF kernels. The column "$\gamma$, $C$" lists the selected values for $\gamma$ and $C$, the column "Error" lists the average classification error in % and its standard deviation of the test data sets, the column "SVs" lists the average number of support vectors and its standard deviation.

From the table, the selected parameter values for L1 and L2 support vector machines are similar for each data set and the classification errors are also similar. We statistically analyzed the classification error by the $t$ test with the significance level of 0.05. If the results are statistically different, the better one is shown in boldface. There is not much difference in the generalization ability between L1 and L2 support vector machines. As previously discussed, the number of support vectors for L2 support vector machines is larger.

**Table 2.7** Comparison of L1 and L2 support vector machines with RBF kernels

| Data | L1 SVM | | | L2 SVM | | |
|---|---|---|---|---|---|---|
| | $\gamma$, $C$ | Error | SVs | $\gamma$, $C$ | Error | SVs |
| Banana | 15, 100 | 10.5±0.46 | 101±10 | 15, 500 | 10.6±0.53 | 157±9.6 |
| B. cancer | 1, 10 | 25.6±4.4 | 114±5.5 | 0.5, 10 | 25.9±4.5 | 188±4.3 |
| Diabetes | 0.5, 50 | 23.3±1.8 | 250±7.9 | 10, 1 | 23.3±1.8 | 408±8.0 |
| Flare-solar | 1, 10 | **32.5±1.7** | 448±13 | 0.5, 10 | 33.4±1.7 | 598±9.5 |
| German | 0.5, 50 | 23.5±2.2 | 388±11 | 0.1, 1,000 | 23.3±2.2 | 598±12 |
| Heart | 0.1, 50 | 16.1±3.1 | 74±5.8 | 0.1, 10 | 16.1±3.3 | 148±5.1 |
| Image | 15, 500 | 2.80±0.44 | 146±2.1 | 10, 2,000 | 2.76±0.37 | 186±9.9 |
| Ringnorm | 15, 1 | 2.64±0.35 | 131±5.6 | 15, 10 | **2.27±0.30** | 111±7.4 |
| Splice | 10, 2,000 | 10.8±0.71 | 740±13 | 10, 50 | 10.8±0.71 | 763±13 |
| Thyroid | 5, 1,000 | 4.05±2.3 | 13±1.9 | 10, 500 | 3.82±2.2 | 15±2.2 |
| Titanic | 5, 2,000 | **22.3±0.98** | 66±9.8 | 0.5, 10 | 22.7±**0.76** | 149±2.7 |
| Twonorm | 0.5, 1 | 2.13±0.65 | 255±8.0 | 0.1, 10 | **1.93±0.60** | 315±7.7 |
| Waveform | 15, 1 | 9.88±0.47 | 149±8.7 | 10, 1 | 9.79±**0.35** | 216±9.2 |

Then we performed cross-validation for L1 support vector machines using polynomial and Mahalanobis kernels. For polynomial kernels we selected the degree $d$ from $\{1, 2, 3, 4, 5, 6, 7, 8\}$, and for Mahalanobis kernels the value of $\delta$ from $\{0.1, 0.5, 1, 1.5, 2.0\}$. Table 2.8 shows the results of L1 support vector machines with polynomial and Mahalanobis kernels. For polynomial kernels, linear kernels were selected six times. In most cases, the average error rates for polynomial and Mahalanobis kernels are comparable, but in some cases, polynomial kernels performed worse.

If we select kernels from among RBF, Mahalanobis, and polynomial kernels, we select the kernel that realizes the best recognition rate for the

**Table 2.8** Comparison of L1 support vector machines with polynomial and Mahalanobis kernels

| Data | Polynomial | | | Mahalanobis | | |
|---|---|---|---|---|---|---|
| | $d, C$ | Error | SVs | $\delta, C$ | Error | SVs |
| Banana | 8, 2,000 | 10.7±0.50 | 99±11 | 1, 50 | 10.5±0.44 | 99±9.0 |
| B. cancer | 2, 1 | 26.1±4.7 | 108±5.8 | 1, 1 | 25.9±4.2 | 132±5.8 |
| Diabetes | 1, 500 | 23.5±1.7 | 241±8.3 | 0.1, 1 | 23.2±1.6 | 281±7.0 |
| Flare-solar | 1, 1 | 32.4±1.8 | 445±15 | 0.1, 1 | 32.5±1.6 | 504±10 |
| German | 1, 50 | 24.0±2.2 | 373±12 | 0.5, 1 | 23.8±2.2 | 423±9.6 |
| Heart | 1, 1 | 16.2±3.1 | 71±6.0 | 0.1, 10 | 16.5±3.1 | 73±6.0 |
| Image | 3, 50 | 3.27±0.40 | 152±7.6 | 0.5, 1,000 | 3.03±0.57 | 152±8.1 |
| Ringnorm | 2, 10 | 6.40±0.82 | 134±7.0 | 2, 2,000 | 1.67±0.12 | 152±11 |
| Splice | 2, 2,000 | 12.6±0.76 | 440±19 | 1.5, 2,000 | 10.8±0.75 | 784±9.4 |
| Thyroid | 5, 100 | 4.21±2.2 | 13±1.7 | 0.1, 1,000 | 3.99±2.3 | 16±2.2 |
| Titanic | 1, 50 | 23.0±1.6 | 66±10 | 0.5, 2,000 | 22.4±1.0 | 66±9.8 |
| Twonorm | 1, 1 | 2.68±0.20 | 75±4.9 | 2, 1 | 2.68±0.15 | 238±6.9 |
| Waveform | 2, 1 | 10.6±0.51 | 106±10 | 1.5, 1 | 9.99±0.43 | 187±8.6 |

validation data set. To check if this strategy worked well for the two-class problems, in the "Selected" columns of Table 2.9 we list the kernels of the L1 support vector machine in the ascending order of the average error rate for the validation data sets and in the "Actual" columns the ascending order of the error rates for the test data sets. In the table, the boldface letters means that the associated average recognition rate is more than 1% better than that with the kernel next to it on the right.

**Table 2.9** Model selection performance

| Data | Selected | | | Actual | | |
|---|---|---|---|---|---|---|
| | 1 | 2 | 3 | 1 | 2 | 3 |
| Banana | Maha | RBF | Poly | Maha | RBF | Poly |
| B. cancer | RBF | Maha | Poly | RBF | Maha | Poly |
| Diabetes | RBF | Maha | Poly | Maha | RBF | Poly |
| Flare-solar | RBF | Poly | Maha | Poly | Maha | RBF |
| German | RBF | Poly | Maha | RBF | Maha | Poly |
| Heart | RBF | Maha | Poly | RBF | Poly | Maha |
| Image | RBF | Maha | Poly | RBF | Maha | Poly |
| Ringnorm | **Maha** | **RBF** | Poly | Maha | **RBF** | Poly |
| Splice | RBF | Maha | Poly | RBF | Maha | Poly |
| Thyroid | RBF | Maha | Poly | Maha | RBF | Poly |
| Titanic | Maha | RBF | Poly | RBF | Maha | Poly |
| Twonorm | RBF | Maha | Poly | RBF | Poly | Maha |
| Waveform | RBF | Maha, Poly | | RBF | Maha | Poly |

From the table, the orders are the same for 6 times out of 13 and the number of cases where the kernel in the first place in the "Selected" columns appear in the second place at the lowest in "Actual" columns is 12. Thus, we can consider the cross-validation worked well in this case. Among the three kernels, RBF kernels showed the best recognition rates for the test data sets eight times. On the contrary Mahalanobis kernels four times, and polynomial kernels once. Thus, we can conclude that in most cases the best choice is RBF kernels and polynomial kernels are the last choice. And there is not much difference between RBF and Mahalanobis kernels and Mahalanobis kernels may show good generalization ability in some cases such as for the ringnorm problem.

Here, we must bear in mind that the recognition performance depends heavily on the parameter ranges adopted. It may also depend on the stopping conditions of the training method, normalization of the input range, or the tie breaking strategy in cross-validation. Comparing the results of L1 SVM with RBF kernels in Table 2.7 with those under different evaluation conditions listed in Table 10.5, the average error rates listed in Table 2.7 are comparable or better except for the ringnorm problem. The error rate of the ringnorm problem in Table 10.5 is comparable with that of the Mahalanobis kernels in Table 2.8. The reason is that under the normalization of the input range into $[0, 1]$, the parameter range of $\gamma$ is small for this problem. Table 2.10 shows the results when the $\gamma$ values of 50 and 100 were added in cross-validation. From the table, the average error rate of the L1 support vector machine is comparable with that shown in Table 10.5. For the L2 support vector machine two values of $C$ were obtained according to whether we chose smallest or largest value of $C$ for the tie break. In this case, the average error rate with $C = 1$ is better than with $C = 2,000$. If we do not normalize the input range of the ringnorm problem, the average error rate of the test data is $1.59 \pm 0.10$ for $\gamma = 0.1$ as will be shown in Table 4.16 on p. 213. Therefore, for the ringnorm problem, normalization of the input range led to the wider search range of parameter values.

**Table 2.10** Results of the ringnorm problem for the extended range of $\gamma$

| SVM | $\gamma, C$ | Error | SVs |
|-----|-------------|-------|-----|
| L1 | 100, 1 | $1.69 \pm 0.12$ | $169 \pm 15$ |
| L2 | 100, 1 | $1.66 \pm 0.12$ | $273 \pm 18$ |
| L2 | 100, 2,000 | $1.74 \pm 0.13$ | $168 \pm 15$ |

## 2.9 Invariance for Linear Transformation

Because fuzzy classifiers with ellipsoidal regions [14, pp. 208–209] are based on the Mahalanobis distance, they are invariant for the linear transformation of input variables; specifically translation, scaling, and rotation-invariant. But most classifiers, such as multilayer neural networks, are not. Thus, to avoid the influence of variables with large input ranges on the generalization ability, we usually scale the ranges of input variables into $[0, 1]$ or $[-1, 1]$.

In [10, pp. 333–358] and [107], invariance of the kernel method, in which a small variance of the input does not affect the classification results, is discussed.

In this section, we discuss invariance of support vector machines for linear transformation of input variables [108]. Here, we consider translation, scaling, and rotation. Then, we clarify the relationships between the input ranges $[0, 1]$ and $[-1, 1]$.

The Euclidean distance is used to calculate the margins, and it is rotation- and translation-invariant but not scale-invariant. Therefore, support vector machines with linear kernels are rotation- and translation-invariant but not scale-invariant. In general, the Euclidean distance is not scale-invariant, but if all the input variables are scaled with the same factor, the Euclidean distance changes with that factor. Therefore here we consider the following transformation:

$$\mathbf{z} = s\, A\, \mathbf{x} + \mathbf{c}, \tag{2.323}$$

where $s\,(> 0)$ is a scaling factor, $A$ is an orthogonal matrix and satisfies $A^\top A = I$, and $\mathbf{c}$ is a constant vector.

Now the RBF kernel $K(\mathbf{z}, \mathbf{z}')$ is given by

$$\begin{aligned}
K(\mathbf{z}, \mathbf{z}') &= \exp(-\gamma' \, \| s\, A\, \mathbf{x} + \mathbf{c} - s\, A\, \mathbf{x}' - \mathbf{c} \|^2) \\
&= \exp(-\gamma' \, \| s\, A\, (\mathbf{x} - \mathbf{x}') \|^2) \\
&= \exp(-\gamma' \, s^2 \, \| \mathbf{x} - \mathbf{x}' \|^2).
\end{aligned} \tag{2.324}$$

Therefore, RBF kernels are translation- and rotation-invariant. For $s \neq 1$, if

$$\gamma' s^2 = \gamma, \tag{2.325}$$

$K(\mathbf{z}, \mathbf{z}') = K(\mathbf{x}, \mathbf{x}')$. Thus, if (2.325) is satisfied, the optimal solutions for a training data set and the data set transformed by (2.323) are the same.

The neural network kernel $K(\mathbf{z}, \mathbf{z}')$ is given by

$$K(\mathbf{z}, \mathbf{z}') = \frac{1}{1 + \exp(\nu' \, (s\, A\, \mathbf{x} + \mathbf{c})^\top (s\, A\, \mathbf{x}' + \mathbf{c}) - a)}. \tag{2.326}$$

If $\mathbf{c} \neq 0$, (2.326) is not invariant. Setting $\mathbf{c} = 0$, (2.326) becomes

$$K(\mathbf{z}, \mathbf{z}') = \frac{1}{1 + \exp(\nu' s^2 \mathbf{x}^\top \mathbf{x}' - a)}. \tag{2.327}$$

Therefore, neural network kernels are rotation-invariant. If

$$\nu' s^2 = \nu \tag{2.328}$$

is satisfied, $K(\mathbf{z}, \mathbf{z}') = K(\mathbf{x}, \mathbf{x}')$. Thus, if (2.328) is satisfied, the optimal solutions for a training data set and the data set transformed by (2.323) with $\mathbf{c} = \mathbf{0}$ are the same.

For the linear kernel, $K(\mathbf{z}, \mathbf{z}')$ is given by

$$\begin{aligned}
K(\mathbf{z}, \mathbf{z}') &= (s\, A\mathbf{x} + \mathbf{c})^\top (s\, A\mathbf{x}' + \mathbf{c}) \\
&= s^2 \mathbf{x}^\top \mathbf{x}' + s\, \mathbf{c}^\top A\mathbf{x}' + s\, \mathbf{x}^\top A^\top \mathbf{c} + \mathbf{c}^\top \mathbf{c}.
\end{aligned} \tag{2.329}$$

Training of the L1 support vector machine with the data set transformed by (2.323) is as follows:

$$\text{maximize} \quad Q(\boldsymbol{\alpha}') = \sum_{i=1}^{M} \alpha'_i - \frac{1}{2} \sum_{i,j=1}^{M} \alpha'_i \alpha'_j y_i y_j$$
$$\times (s^2 \mathbf{x}_i^\top \mathbf{x}_j + s\, \mathbf{c}^\top A\mathbf{x}_j + s\, \mathbf{x}_i^\top A^\top \mathbf{c} + \mathbf{c}^\top \mathbf{c}) \tag{2.330}$$

$$\text{subject to} \quad \sum_{i=1}^{M} y_i \alpha'_i = 0, \quad 0 \le \alpha'_i \le C' \quad \text{for } i = 1, \dots, M. \tag{2.331}$$

Using (2.331), the objective function in (2.330) becomes

$$\begin{aligned}
Q(\boldsymbol{\alpha}') &= \sum_{i=1}^{M} \alpha'_i - \frac{1}{2} \sum_{i,j=1}^{M} \alpha'_i \alpha'_j y_i y_j s^2 \mathbf{x}_i^\top \mathbf{x}_j \\
&= s^{-2} \left( \sum_{i=1}^{M} s^2 \alpha'_i - \frac{1}{2} \sum_{i,j=1}^{M} s^2 \alpha'_i s^2 \alpha'_j y_i y_j \mathbf{x}_i^\top \mathbf{x}_j \right).
\end{aligned} \tag{2.332}$$

Thus, setting $\alpha_i = s^2 \alpha'_i$, the inequality constraint in (2.331) becomes

$$0 \le \alpha_i \le s^2 C' \quad \text{for } i = 1, \dots, M. \tag{2.333}$$

Therefore, the optimal solutions of the L1 support vector machine with the linear kernel for a training data set and the data set transformed by (2.323) are the same when

$$C = s^2 C'. \tag{2.334}$$

This also holds for L2 support vector machines.

For the polynomial kernel, $K(\mathbf{z}, \mathbf{z}')$ is given by

$$K(\mathbf{z}, \mathbf{z}') = \left((s\,A\mathbf{x} + \mathbf{c})^{\top}(s\,A\mathbf{x}' + \mathbf{c}) + 1\right)^{d}$$
$$= \left(s^2\,\mathbf{x}^{\top}\mathbf{x}' + s\,\mathbf{c}^{\top}A\mathbf{x}' + s\,\mathbf{x}^{\top}A^{\top}\mathbf{c} + \mathbf{c}^{\top}\mathbf{c} + 1\right)^{d}. \quad (2.335)$$

Therefore, polynomial kernels are neither scale- nor translation-invariant but for $\mathbf{c} = \mathbf{0}$ rotation-invariant. This is also true for L2 support vector machines using polynomial kernels. Assuming that (2.335) is approximated by the term with the highest degree of $s$ (at least $s > 1$ is necessary):

$$K(\mathbf{z}, \mathbf{z}') = s^{2d}\,(\mathbf{x}^{\top}\mathbf{x}')^{d}, \quad (2.336)$$

similar to the discussions for the linear kernel, the support vector machines with a data set and the data set transformed by (2.323) perform similarly when

$$C = s^{2d}\,C'. \quad (2.337)$$

In training support vector machines, we normalize the range of input variables into $[0, 1]$ or $[-1, 1]$, without knowing their difference. Using our previous discussions, however, we can clarify relations of the solutions. Because the transformation from $[0, 1]$ to $[-1, 1]$ is given by

$$\mathbf{z} = 2\,\mathbf{x} - \mathbf{1}, \quad (2.338)$$

it is a combination of translation and scaling. Thus according to the previous discussions, we can obtain the parameter values that give the same or roughly the same results for the two input ranges. Table 2.11 summarizes this result.

**Table 2.11** Parameters that give the same or roughly the same solutions

| Kernel | $[0, 1]$ | $[-1, 1]$ |
|---|---|---|
| Linear | $4\,C$ | $C$ |
| Polynomial | $\approx 4^d\,C$ | $C$ |
| RBF | $4\,\gamma$ | $\gamma$ |
| NN | $\approx 4\,\nu$ | $\nu$ |

To see the validity of Table 2.11, especially for the polynomial kernels, we conducted the computer experiment using the blood cell data and the thyroid data sets listed in Table 1.3. For both data sets, we selected data for Classes 2 and 3. The numbers of training and test data are listed in Table 2.12. We trained the L1 support vector machine for the blood cell data and the L2 support vector machine for the thyroid data. For the input range of $[-1, 1]$, we set $C = 5,000$ and for $[0, 1]$, we set it appropriately according to Table 2.11. For the polynomial kernels, we changed $C$ for $[0, 1]$ from $4 \times 5,000 = 20,000$ to $4^d \times 5,000$.

**Table 2.12** Training and test data for the blood cell and thyroid data

| Data | Training data | | Test data | |
|------|---------|---------|---------|---------|
| | Class 2 | Class 3 | Class 2 | Class 3 |
| Blood cell | 399 | 400 | 400 | 400 |
| Thyroid | 191 | 3,488 | 177 | 3,178 |

**Table 2.13** Solutions of the L1 support vector machine for the blood cell data

| Kernel | Range | PARM | Test rate (%) | Train. rate (%) | SVs | $Q(\alpha)$ |
|--------|-------|------|-----------|------------|-----|--------|
| Linear | $[0,1]$ | $C20,000$ | 87.00 | 90.23 | 103 (89) | 1,875,192 |
| | $[-1,1]$ | $C5,000$ | 87.00 | 90.23 | 103 (89) | 1,875,192/4 |
| d2 | $[0,1]$ | $C5,000$ | 88.50 | 92.23 | 101 (52) | 331,639 |
| | $[0,1]$ | $C20,000$ | 86.25 | 94.24 | 103 (51) | 1,191,424 |
| | $[0,1]$ | $C80,000$ | 86.75 | 95.99 | 96 (34) | 4,060,006 |
| | $[-1,1]$ | $C5,000$ | 86.75 | 95.49 | 99 (35) | 4,137,900/16 |
| d3 | $[0,1]$ | $C5,000$ | 88.25 | 96.24 | 99 (31) | 237,554 |
| | $[0,1]$ | $C20,000$ | 86.00 | 97.49 | 98 (19) | 672,345 |
| | $[0,1]$ | $C80,000$ | 85.75 | 99.00 | 97 (4) | 1,424,663 |
| | $[0,1]$ | $C320,000$ | 86.50 | 100 | 93 | 1,839,633 |
| | $[-1,1]$ | $C5,000$ | 86.00 | 100 | 90 (1) | 2,847,139/64 |
| RBF | $[0,1]$ | $\gamma4$ | 89.00 | 92.48 | 99 (58) | 358,168 |
| | $[-1,1]$ | $\gamma1$ | 89.00 | 92.48 | 99 (58) | 358,168 |

Table 2.13 lists the recognition rates of the blood cell test and training data, the number of support vectors, and the value of $Q(\alpha)$ for the L1 support vector machine. The numerals in parentheses show the numbers of bounded support vectors. For the linear kernel, as the theory tells us, the solution with the ranges of $[0,1]$ and $C = 20,000$ and that with $[-1,1]$ and $C = 5,000$ are the same. For the RBF kernels also, the solution with $[0,1]$ and $\gamma = 4$ and that with $[-1,1]$ and $\gamma = 1$ are the same.

For the polynomial kernel with $d = 2$, the solution with $[0,1]$ and $C = 4^d \times 5,000 = 80,000$ and that with $[-1,1]$ and $C = 5,000$ are similar. The value of $Q(\alpha)$ with $C = 80,000$ is near the value of $Q(\alpha) \times 16$ with $C = 5,000$. Similar results hold for $d = 3$, although the difference between the values of $Q(\alpha)$ are widened compared to that for $d = 2$.

Table 2.14 lists the recognition rates of the thyroid test and training data, the number of support vectors, and the value of $Q(\alpha)$ for the L2 support vector machine.

For the linear and RBF kernels, the solutions with the range of $[0,1]$ and the associated solutions are the same.

For the polynomial kernels, the solution with $[0, 1]$ and $C = 5,000$ and that with $[-1, 1]$ and $4^d \times 5,000$ are similar.

**Table 2.14** Solutions of the L2 support vector machine for the thyroid data

| Kernel | Range | PARM | Test rate (%) | Train. rate (%) | SVs | $Q(\alpha)$ |
|--------|-------|------|---------------|------------------|-----|-------------|
| Linear | $[0, 1]$ | $C20,000$ | 97.50 | 98.34 | 474 | 2,096,156 |
|        | $[-1, 1]$ | $C5,000$ | 97.50 | 98.34 | 474 | 2,096,156/4 |
| $d2$ | $[0, 1]$ | $C5,000$ | 98.12 | 99.18 | 275 | 298,379 |
|      | $[0, 1]$ | $C20,000$ | 98.18 | 99.29 | 216 | 974,821 |
|      | $[0, 1]$ | $C80,000$ | 98.21 | 99.37 | 191 | 3,360,907 |
|      | $[-1, 1]$ | $C5,000$ | 97.85 | 99.37 | 201 | 3,357,314/16 |
| $d3$ | $[0, 1]$ | $C5,000$ | 97.97 | 99.40 | 217 | 206,335 |
|      | $[0, 1]$ | $C20,000$ | 98.15 | 99.57 | 168 | 642,435 |
|      | $[0, 1]$ | $C80,000$ | 98.06 | 99.76 | 131 | 1,993,154 |
|      | $[0, 1]$ | $C320,000$ | 97.94 | 99.86 | 106 | 5,626,809 |
|      | $[-1, 1]$ | $C5,000$ | 97.65 | 99.92 | 125 | 4,691,633/64 |
| $\gamma$ | $[0, 1]$ | $\gamma4$ | 97.91 | 97.35 | 237 | 3,816,254 |
|          | $[-1, 1]$ | $\gamma1$ | 97.91 | 97.35 | 237 | 3,816,254 |

Theoretical analysis and the computer experiments showed that the input ranges of $[0, 1]$ and $[-1, 1]$ are interchangeable for polynomial kernels with constant terms and RBF kernels, namely, we can use either range. But for polynomial kernels without constant terms, it is unfavorable to use $[-1, 1]$ as discussed in Section 2.3.4.

# References

1. C. M. Bishop. *Neural Networks for Pattern Recognition.* Oxford University Press, Oxford, 1995.
2. S. Abe. *Neural Networks and Fuzzy Systems: Theory and Applications.* Kluwer Academic Publishers, Norwell, MA, 1997.
3. T. Evgeniou, M. Pontil, and T. Poggio. Regularization networks and support vector machines. *Advances in Computational Mathematics*, 13(1):1–50, 2000.
4. V. N. Vapnik. *The Nature of Statistical Learning Theory.* Springer-Verlag, New York, 1995.
5. V. Cherkassky and F. Mulier. *Learning from Data: Concepts, Theory, and Methods.* John Wiley & Sons, New York, 1998.
6. C. Orsenigo and C. Vercellis. Discrete support vector decision trees via tabu search. *Computational Statistics & Data Analysis*, 47(2):311–322, 2004.
7. B. Haasdonk. Feature space interpretation of SVMs with indefinite kernels. *IEEE Transactions on Pattern Analysis and Machine Intelligence*, 27(4):482–492, 2005.

8. M. G. Genton. Classes of kernels for machine learning: A statistics perspective. *Journal of Machine Learning Research*, 2:299–312, 2001.

9. J.-H. Chen. M-estimator based robust kernels for support vector machines. In *Proceedings of the Seventeenth International Conference on Pattern Recognition (ICPR 2004)*, volume 1, pages 168–171, Cambridge, 2004.

10. B. Schölkopf and A. J. Smola. *Learning with Kernels: Support Vector Machines, Regularization, Optimization, and Beyond*. MIT Press, Cambridge, MA, 2002.

11. T. Nishikawa and S. Abe. Maximizing margins of multilayer neural networks. In *Proceedings of the Ninth International Conference on Neural Information Processing (ICONIP '02)*, volume 1, pages 322–326, Singapore, 2002.

12. J. A. K. Suykens and J. Vandewalle. Training multilayer perceptron classifiers based on a modified support vector method. *IEEE Transactions on Neural Networks*, 10(4):907–911, 1999.

13. Y. Grandvalet and S. Canu. Adaptive scaling for feature selection in SVMs. In S. Becker, S. Thrun, and K. Obermayer, editors, *Advances in Neural Information Processing Systems 15*, pages 569–576. MIT Press, Cambridge, MA, 2003.

14. S. Abe. *Pattern Classification: Neuro-Fuzzy Methods and Their Comparison*. Springer-Verlag, London, 2001.

15. S. Abe. Training of support vector machines with Mahalanobis kernels. In W. Duch, J. Kacprzyk, E. Oja, and S. Zadrożny, editors, *Artificial Neural Networks: Formal Models and Their Applications (ICANN 2005)—Proceedings of Fifteenth International Conference, Part II, Warsaw, Poland*, pages 571–576. Springer-Verlag, Berlin, Germany, 2005.

16. Y. Kamada and S. Abe. Support vector regression using Mahalanobis kernels. In F. Schwenker and S. Marinai, editors, *Artificial Neural Networks in Pattern Recognition: Proceedings of Second IAPR Workshop, ANNPR 2006, Ulm, Germany*, pages 144–152. Springer-Verlag, Berlin, Germany, 2006.

17. S. Chen, A. Wolfgang, C. J. Harris, and L. Hanzo. Symmetric kernel detector for multiple-antenna aided beamforming systems. In *Proceedings of the 2007 International Joint Conference on Neural Networks (IJCNN 2007)*, pages 2486–2491, Orlando, FL, 2007.

18. M. Espinoza, J. A. K. Suykens, and B. De Moor. Imposing symmetry in least squares support vector machines regression. In *Proceedings of the Forty-Fifth IEEE Conference on Decision and Control and European Control Conference 2005 (CDC-ECC'05)*, pages 5716–5721, Orlando, FL, 2005.

19. V. N. Vapnik. *Statistical Learning Theory*. John Wiley & Sons, New York, 1998.

20. S. R. Gunn and M. Brown. SUPANOVA: A sparse, transparent modelling approach. In *Neural Networks for Signal Processing IX—Proceedings of the 1999 IEEE Signal Processing Society Workshop*, pages 21–30, 1999.

21. K. K. Lee, S. R. Gunn, C. J. Harris, and P. A. S. Reed. Classification of imbalanced data with transparent kernels. In *Proceedings of International Joint Conference on Neural Networks (IJCNN '01)*, volume 4, pages 2410–2415, Washington, DC, 2001.

22. T. Howley and M. G. Madden. An evolutionary approach to automatic kernel construction. In S. Kollias, A. Stafylopatis, W. Duch, and E. Oja, editors, *Artificial Neural Networks (ICANN 2006)—Proceedings of the Sixteenth International Conference, Athens, Greece, Part II*, pages 417–426. Springer-Verlag, Berlin, Germany, 2006.

23. N. Dalal and B. Triggs. Histograms of oriented gradients for human detection. In *Proceedings of the 2005 IEEE Computer Society Conference on Computer Vision and Pattern Recognition (CVPR 2005)*, volume 1, pages 886–893, San Diego, CA, 2005.

24. B. Schölkopf, P. Simard, A. Smola, and V. Vapnik. Prior knowledge in support vector kernels. In M. I. Jordan, M. J. Kearns, and S. A. Solla, editors, *Advances in Neural Information Processing Systems 10*, pages 640–646. MIT Press, Cambridge, MA, 1998.

25. V. L. Brailovsky, O. Barzilay, and R. Shahave. On global, local, mixed and neighborhood kernels for support vector machines. *Pattern Recognition Letters*, 20(11–13):1183–1190, 1999.

26. A. Barla, E. Franceschi, F. Odone, and A. Verri. Image kernels. In S.-W. Lee and A. Verri, editors, *Pattern Recognition with Support Vector Machines: Proceedings of First International Workshop, SVM 2002, Niagara Falls, Canada*, pages 83–96. Springer-Verlag, Berlin, Germany, 2002.

27. K. Hotta. Support vector machine with local summation kernel for robust face recognition. In *Proceedings of the Seventeenth International Conference on Pattern Recognition (ICPR 2004)*, volume 3, pages 482–485, Cambridge, UK, 2004.

28. K. Grauman and T. Darrell. The pyramid match kernel: Efficient learning with sets of features. *Journal of Machine Learning Research*, 8:725–760, 2007.

29. T. Joachims. *Learning to Classify Text Using Support Vector Machines: Methods, Theory and Algorithms*. Kluwer Academic Publishers, Norwell, MA, 2002.

30. H. Lodhi, J. Shawe-Taylor, N. Cristianini, and C. Watkins. Text classification using string kernels. In T. K. Leen, T. G. Dietterich, and V. Tresp, editors, *Advances in Neural Information Processing Systems 13*, pages 563–569, 2001.

31. H. Lodhi, C. Saunders, J. Shawe-Taylor, N. Cristianini, and C. Watkins. Text classification using string kernels. *Journal of Machine Learning Research*, 2:419–444, 2002.

32. C. Leslie, E. Eskin, and W. S. Noble. The spectrum kernel: A string kernel for SVM protein classification. In *Proceedings of the Pacific Symposium on Biocomputing*, pages 564–575, 2002.

33. C. Leslie, E. Eskin, J. Weston, and W. S. Noble. Mismatch string kernels for SVM protein classification. In S. Becker, S. Thrun, and K. Obermayer, editors, *Advances in Neural Information Processing Systems 15*, pages 1441–1448. MIT Press, Cambridge, MA, 2003.

34. C. S. Leslie, E. Eskin, A. Cohen, J. Weston, and W. S. Noble. Mismatch string kernels for discriminative protein classification. *Bioinformatics*, 20(4):467–476, 2004.

35. H. Saigo, J.-P. Vert, N. Ueda, and T. Akutsu. Protein homology detection using string alignment kernels. *Bioinformatics*, 20(11):1682–1689, 2004.

36. R. I. Kondor and J. Lafferty. Diffusion kernels on graphs and other discrete structures. In C. Sammut and A. Günther and Hoffmann, editors, *Machine Learning, Proceedings of the Nineteenth International Conference (ICML 2002), Sydney, Australia*, pages 315–322. Morgan Kaufmann Publishers, July 2002.

37. A. J. Smola and R. Kondor. Kernels and regularization on graphs. In Schölkopf and M. K. Warmuth, editors, *Learning Theory and Kernel Machines: Proceedings of Sixteenth Annual Conference on Learning Theory and Seventh Kernel Workshop, COLT/Kernel 2003, Washington, DC*, pages 144–158. Springer-Verlag, Berlin, Germany, 2003.

38. T. Ito, M. Shimbo, T. Kudo, and Y. Matsumoto. Application of kernels to link analysis. In *KDD-2005: Proceedings of the Eleventh ACM SIGKDD International Conference on Knowledge Discovery and Data Mining*, pages 586–592, Chicago, IL, August 2005.

39. F. Fouss, L. Yen, A. Pirotte, and M. Saerens. An experimental investigation of graph kernels on a collaborative recommendation task. In *Proceedings of the Sixth IEEE International Conference on Data Mining (ICDM 2006)*, pages 863–868, Hong Kong, China, 2006.

40. T. Gärtner, P. Flach, and S. Wrobel. On graph kernels: Hardness results and efficient alternatives. In Schölkopf and M. K. Warmuth, editors, *Learning Theory and Kernel Machines: Proceedings of Sixteenth Annual Conference on Computational Learning Theory and Seventh Kernel Workshop, COLT/Kernel 2003, Washington, DC*, pages 129–143. Springer-Verlag, Berlin, Germany, 2003.

41. H. Kashima, K. Tsuda, and A. Inokuchi. Kernels for graphs. In B. Schölkopf, K. Tsuda, and J.-P. Vert, editors, *Kernel Methods in Computational Biology*, pages 155–170. MIT Press, Cambridge, MA, 2004.

42. K. Riesen, M. Neuhaus, and H. Bunke. Graph embedding in vector spaces by means of prototype selection. In F. Escolano and M. Vento, editors, *Graph-Based Representations in Pattern Recognition: Proceedings of Sixth IAPR-TC-15 International Workshop, GbRPR 2007, Alicante, Spain*, pages 383–393. Springer-Verlag, Berlin, Germany, June 2007.

43. K. Riesen and H. Bunke. Kernel k-means clustering applied to vector space embedding of graphs. In L. Prevost, S. Marinai, and F. Schwenker, editors, *Artificial Neural Networks in Pattern Recognition: Proceedings of Third IAPR Workshop, ANNPR 2008, Paris, France*, pages 24–35. Springer-Verlag, Berlin, Germany, 2008.

44. L. Ralaivola, S. J. Swamidass, H. Saigo, and P. Baldi. Graph kernels for chemical informatics. *Neural Networks*, 18(8):1093–1110, 2005.

45. K. M. Borgwardt, C. S. Ong, S. Schönauer, S. V. N. Vishwanathan, A. J. Smola, and H.-P. Kriegel. Protein function prediction via graph kernels. *Bioinformatics*, 21(Suppl. 1):i47–i56, 2005.

46. E. Pękalska and R. P. W. Duin. *The Dissimilarity Representation for Pattern Recognition: Foundations and Applications*. World Scientific Publishing, Singapore, 2005.

47. H. Shimodaira, K. Noma, M. Nakai, and S. Sagayama. Dynamic time-alignment kernel in support vector machine. In T. G. Dietterich, S. Becker, and Z. Ghahramani, editors, *Advances in Neural Information Processing Systems 14*, volume 2, pages 921–928, MIT Press, Cambridge, MA, 2002.

48. N. Smith and M. Gales. Speech recognition using SVMs. In T. G. Dietterich, S. Becker, and Z. Ghahramani, editors, *Advances in Neural Information Processing Systems 14*, volume 2, pages 1197–1204, MIT Press, Cambridge, MA, 2002.

49. C. Cortes, P. Haffner, and M. Mohri. Rational kernels. In S. Becker, S. Thrun, and K. Obermayer, editors, *Advances in Neural Information Processing Systems 15*, pages 617–624. MIT Press, Cambridge, MA, 2003.

50. H. Shin and S. Cho. Invariance of neighborhood relation under input space to feature space mapping. *Pattern Recognition Letters*, 26(6):707–718, 2005.

51. S. R. Gunn. Support vector machines for classification and regression. Technical Report ISIS-1-98, School of Electronics and Computer Science, University of Southampton, 1998.

52. T. M. Huang and V. Kecman. Bias term b in SVMs again. In *Proceedings of the Twelfth European Symposium on Artificial Neural Networks (ESANN 2004)*, pages 441–448, Bruges, Belgium, 2004.

53. H. Xiong, M. N. S. Swamy, and M. O. Ahmad. Optimizing the kernel in the empirical feature space. *IEEE Transactions on Neural Networks*, 16(2):460–474, 2005.

54. S. Abe. Sparse least squares support vector training in the reduced empirical feature space. *Pattern Analysis and Applications*, 10(3):203–214, 2007.

55. R. Herbrich. *Learning Kernel Classifiers: Theory and Algorithms*. MIT Press, Cambridge, MA, 2002.

56. M. Pontil and A. Verri. Properties of support vector machines. *Neural Computation*, 10(4):955–974, 1998.

57. C. J. C. Burges and D. J. Crisp. Uniqueness of the SVM solution. In S. A. Solla, T. K. Leen, and K.-R. Müller, editors, *Advances in Neural Information Processing Systems 12*, pages 223–229. MIT Press, Cambridge, MA, 2000.

58. S. Abe. Analysis of support vector machines. In H. Bourlard, T. Adali, S. Bengio, J. Larsen, and S. Douglas, editors, *Neural Networks for Signal Processing XII— Proceedings of the 2002 IEEE Signal Processing Society Workshop*, pages 89–98, 2002.

59. T. Downs, K. E. Gates, and A. Masters. Exact simplification of support vector solutions. *Journal of Machine Learning Research*, 2:293–297, 2001.

60. P. M. L. Drezet and R. F. Harrison. A new method for sparsity control in support vector classification and regression. *Pattern Recognition*, 34(1):111–125, 2001.

61. C. J. C. Burges. Simplified support vector decision rules. In L. Saitta, editor, *Machine Learning, Proceedings of the Thirteenth International Conference (ICML '96), Bari, Italy*, pages 71–77. Morgan Kaufmann, San Francisco, 1996.

62. D. Mattera, F. Palmieri, and S. Haykin. Simple and robust methods for support vector expansions. *IEEE Transactions on Neural Networks*, 10(5):1038–1047, 1999.

63. S. Chen, S. R. Gunn, and C. J. Harris. The relevance vector machine technique for channel equalization application. *IEEE Transactions on Neural Networks*, 12(6):1529–1532, 2001.

64. S. Chen, S. R. Gunn, and C. J. Harris. Errata to "The relevance vector machine technique for channel equalization application." *IEEE Transactions on Neural Networks*, 13(4):1024, 2002.

65. S. S. Keerthi, S. K. Shevade, C. Bhattacharyya, and K. R. K. Murthy. A fast iterative nearest point algorithm for support vector machine classifier design. *IEEE Transactions on Neural Networks*, 11(1):124–136, 2000.

66. D. Tsujinishi, Y. Koshiba, and S. Abe. Why pairwise is better than one-against-all or all-at-once. In *Proceedings of International Joint Conference on Neural Networks (IJCNN 2004)*, volume 1, pages 693–698, Budapest, Hungary, 2004.

67. R. M. Rifkin, M. Pontil, and A. Verri. A note on support vector machine degeneracy. In O. Watanabe and T. Yokomori, editors, *Algorithmic Learning Theory: Proceedings of the Tenth International Conference on Algorithmic Learning Theory (ALT '99), Tokyo, Japan*, pages 252–263. Springer-Verlag, Berlin, Germany, 1999.

68. R. Fernández. Behavior of the weights of a support vector machine as a function of the regularization parameter C. In *Proceedings of the Eighth International Conference on Artificial Neural Networks (ICANN '98)*, volume 2, pages 917–922, Skövde, Sweden, 1998.

69. G. C. Cawley and N. L. C. Talbot. Manipulation of prior probabilities in support vector classification. In *Proceedings of International Joint Conference on Neural Networks (IJCNN '01)*, volume 4, pages 2433–2438, Washington, DC, 2001.

70. C.-F. Lin and S.-D. Wang. Fuzzy support vector machines. *IEEE Transactions on Neural Networks*, 13(2):464–471, 2002.

71. K. Veropoulos, C. Campbell, and N. Cristianini. Controlling the sensitivity of support vector machines. In *Proceedings of the Sixteenth International Joint Conference on Artificial Intelligence (IJCAI-99), Workshop ML3*, pages 55–60, Stockholm, Sweden, 1999.

72. P. Xu and A. K. Chan. Support vector machines for multi-class signal classification with unbalanced samples. In *Proceedings of International Joint Conference on Neural Networks (IJCNN 2003)*, volume 2, pages 1116–1119, Portland, OR, 2003.

73. G. Wu and E. Y. Chang. KBA: Kernel boundary alignment considering imbalanced data distribution. *IEEE Transactions on Knowledge and Data Engineering*, 17(6):786–795, 2005.

74. T. Van Gestel, J. A. K. Suykens, J. De Brabanter, B. De Moor, and J. Vandewalle. Least squares support vector machine regression for discriminant analysis. In *Proceedings of International Joint Conference on Neural Networks (IJCNN '01)*, volume 4, pages 2445–2450, Washington, DC, 2001.

75. C. Yuan and D. Casasent. Support vector machines for class representation and discrimination. In *Proceedings of International Joint Conference on Neural Networks (IJCNN 2003)*, volume 2, pages 1611–1616, Portland, OR, 2003.

76. V. Vapnik and O. Chapelle. Bounds on error expectation for SVM. In A. J. Smola, P. L. Bartlett, B. Schölkopf, and D. Schuurmans, editors, *Advances in Large Margin Classifiers*, pages 261–280. MIT Press, Cambridge, MA, 2000.

77. G. Cauwenberghs and T. Poggio. Incremental and decremental support vector machine learning. In T. K. Leen, T. G. Dietterich, and V. Tresp, editors, *Advances in*

*Neural Information Processing Systems 13*, pages 409–415. MIT Press, Cambridge, MA, 2001.

78. K. Saadi, G. C. Cawley, and N. L. C. Talbot. Fast exact leave-one-out cross-validation of least-squares support vector machines. In *Proceedings of the Tenth European Symposium on Artificial Neural Networks (ESANN 2002)*, pages 149–154, Bruges, Belgium, 2002.

79. Z. Ying and K. C. Keong. Fast leave-one-out evaluation and improvement on inference for LS-SVMs. In *Proceedings of the Seventeenth International Conference on Pattern Recognition (ICPR 2004)*, volume 3, pages 494–497, Cambridge, UK, 2004.

80. B. Schölkopf, C. J. C. Burges, and A. J. Smola, editors. *Advances in Kernel Methods: Support Vector Learning*. MIT Press, Cambridge, MA, 1999.

81. C. J. C. Burges. A tutorial on support vector machines for pattern recognition. *Data Mining and Knowledge Discovery*, 2(2):121–167, 1998.

82. T. Joachims. Estimating the generalization performance of an SVM efficiently. In *Proceedings of the Seventeenth International Conference on Machine Learning (ICML-2000)*, pages 431–438, Stanford, CA, 2000.

83. O. Chapelle and V. Vapnik. Model selection for support vector machines. In S. A. Solla, T. K. Leen, and K.-R. Müller, editors, *Advances in Neural Information Processing Systems 12*, pages 230–236. MIT Press, Cambridge, MA, 2000.

84. K. Duan, S. S. Keerthi, and A. N. Poo. An empirical evaluation of simple performance measures for tuning SVM hyperparameters. In *Proceedings of the Eighth International Conference on Neural Information Processing (ICONIP-2001)*, Paper ID# 159, Shanghai, China, 2001.

85. D. Anguita, A. Boni, and S. Ridella. Evaluating the generalization ability of support vector machines through the bootstrap. *Neural Processing Letters*, 11(1):51–58, 2000.

86. B. Efron and R. J. Tibshirani. *An Introduction to the Bootstrap*. Chapman & Hall/CRC Press, Boca Raton, FL, 1993.

87. N. Cristianini and C. Campbell. Dynamically adapting kernels in support vector machines. In M. S. Kearns, S. A. Solla, and D. A. Cohn, editors, *Advances in Neural Information Processing Systems 11*, pages 204–210. MIT Press, Cambridge, MA, 1999.

88. B. Schölkopf, J. Shawe-Taylor, A. J. Smola, and R. C. Williamson. Kernel-dependent support vector error bounds. In *Proceedings of the Ninth International Conference on Artificial Neural Networks (ICANN '99)*, volume 1, pages 103–108, Edinburgh, UK, 1999.

89. M. Seeger. Bayesian model selection for support vector machines, Gaussian processes and other kernel classifiers. In S. A. Solla, T. K. Leen, and K.-R. Müller, editors, *Advances in Neural Information Processing Systems 12*, pages 603–609. MIT Press, Cambridge, MA, 2000.

90. J. T.-Y. Kwok. The evidence framework applied to support vector machines. *IEEE Transactions on Neural Networks*, 11(5):1162–1173, 2000.

91. P. Sollich. Bayesian methods for support vector machines: Evidence and predictive class probabilities. *Machine Learning*, 46(1–3):21–52, 2002.

92. L. Wang and K. L. Chan. Learning kernel parameters by using class separability measure. In *Sixth Kernel Machines Workshop, In conjunction with Neural Information Processing Systems (NIPS)*, 2002.

93. O. Chapelle, V. Vapnik, O. Bousquet, and S. Mukherjee. Choosing multiple parameters for support vector machines. *Machine Learning*, 46(1–3):131–159, 2002.

94. C. S. Ong and A. J. Smola. Machine learning using hyperkernels. In T. Fawcett and N. Mishra, editors, *Machine Learning, Proceedings of the Twentieth International Conference (ICML 2003), Washington, DC*, pages 568–575. AAAI Press, Menlo Park, CA, 2003.

95. C. S. Ong, A. J. Smola, and R. C. Williamson. Hyperkernels. In S. Thrun S. Becker and K. Obermayer, editors, *Advances in Neural Information Processing Systems 15*, pages 495–502. MIT Press, Cambridge, MA, 2003.

96. S. S. Keerthi and C.-J. Lin. Asymptotic behaviors of support vector machines with Gaussian kernel. *Neural Computation*, 15(7):1667–1689, 2003.

97. F. Friedrichs and C. Igel. Evolutionary tuning of multiple SVM parameters. In *Proceedings of the Twelfth European Symposium on Artificial Neural Networks (ESANN 2004)*, pages 519–524, Bruges, Belgium, 2004.

98. G. Lebrun, C. Charrier, and H. Cardot. SVM training time reduction using vector quantization. In *Proceedings of the Seventeenth International Conference on Pattern Recognition (ICPR 2004)*, volume 1, pages 160–163, Cambridge, UK, 2004.

99. T. Hastie, S. Rosset, R. Tibshirani, and J. Zhu. The entire regularization path for the support vector machine. *Journal of Machine Learning Research*, 5:1391–1415, 2004.

100. C. S. Ong, A. J. Smola, and R. C. Williamson. Learning the kernel with hyperkernels. *Journal of Machine Learning Research*, 6:1043–1071, 2005.

101. C. Gold, A. Holub, and P. Sollich. Bayesian approach to feature selection and parameter tuning for support vector machine classifiers. *Neural Networks*, 18(5–6):693–701, 2005.

102. K.-P. Wu and S.-D. Wang. Choosing the kernel parameters of support vector machines according to the inter-cluster distance. In *Proceedings of the 2006 International Joint Conference on Neural Networks (IJCNN 2006)*, pages 2184–2190, Vancouver, Canada, 2006.

103. G. Gasso, K. Zapien, and S. Canu. Computing and stopping the solution paths for ν-SVR. In *Proceedings of the Fifteenth European Symposium on Artificial Neural Networks (ESANN 2007)*, pages 253–258, 2007.

104. A. Rakotomamonjy and M. Davy. One-class SVM regularization path and comparison with alpha seeding. In *Proceedings of the Fifteenth European Symposium on Artificial Neural Networks (ESANN 2007)*, pages 271–276, 2007.

105. M. M. Beigi and A. Zell. A novel kernel-based method for local pattern extraction in random process signals. In *Proceedings of the Fifteenth European Symposium on Artificial Neural Networks (ESANN 2007)*, pages 265–270, 2007.

106. S. Abe. Optimizing kernel parameters by second-order methods. In *Proceedings of the Fifteenth European Symposium on Artificial Neural Networks (ESANN 2007)*, pages 259–264, 2007.

107. C. J. C. Burges. Geometry and invariance in kernel based methods. In B. Schölkopf, C. J. C. Burges, and A. J. Smola, editors, *Advances in Kernel Methods: Support Vector Learning*, pages 89–116. MIT Press, Cambridge, MA, 1999.

108. S. Abe. On invariance of support vector machines. *Presented at the Fourth International Conference on Intelligent Data Engineering and Learning (IDEAL 2003)*, but not included in the proceedings (http://www2.kobe-u.ac.jp/abe/pdf/ideal2003.pdf), 2003.

# Chapter 3
# Multiclass Support Vector Machines

As discussed in Chapter 2, support vector machines are formulated for two-class problems. But because support vector machines employ direct decision functions, an extension to multiclass problems is not straightforward. There are roughly four types of support vector machines that handle multiclass problems:

1. one-against-all support vector machines,
2. pairwise support vector machines,
3. error-correcting output code (ECOC) support vector machines, and
4. all-at-once support vector machines.

According to Vapnik's formulation [1], in one-against-all support vector machines, an $n$-class problem is converted into $n$ two-class problems and for the $i$th two-class problem, class $i$ is separated from the remaining classes. But by this formulation unclassifiable regions exist if we use the discrete decision functions.

To solve this problem, in pairwise support vector machines, Kreßel [2] converts the $n$-class problem into $n(n-1)/2$ two-class problems, which cover all pairs of classes. But by this method unclassifiable regions also exist.

We can resolve unclassifiable regions by introducing membership functions [3, 4], decision trees [5–10], or error-correcting output codes [11] or by determining the decision functions all at once [12–15].

Especially for one-against-all support vector machines, if we use continuous decision functions instead of discrete decision functions, unclassifiable regions are resolved.

In the preceding methods the code words are discrete and fixed before training. But there are some approaches to optimize codes using continuous code words [16, 17].

In the following, we discuss the four types of support vector machines and their variants that resolve unclassifiable regions, and we clarify their relationships, advantages, and disadvantages through theoretical analysis and computer experiments. Specifically, we prove that one-against-all support vector

S. Abe, *Support Vector Machines for Pattern Classification*,
Advances in Pattern Recognition, DOI 10.1007/978-1-84996-098-4_3,
© Springer-Verlag London Limited 2010

machines with continuous decision functions are equivalent to one-against-all fuzzy support vector machines [18]. We show that generalization ability of decision-tree-based support vector machines [5–7] depends on their structure and discuss how to optimize their structure [8–10]. And we clarify the relationship between fuzzy support vector machines and ECOC support vector machines.

Discussions are based on L1 support vector machines but an extension to L2 support vector machines is straightforward. And we assume that if a multiclass classifier consists of several two-class classifiers, each classifier is trained with the same kernel parameter and margin parameter values. Otherwise, we need to adjust outputs [19]. For instance, if the kernel parameter values are different, each two-class classifier is defined in a different feature space and comparison of the classifier outputs is meaningless.

## 3.1 One-Against-All Support Vector Machines

In this section, first we discuss a one-against-all support vector machine with discrete decision functions and its problem that unclassifiable regions exist. Then we discuss three methods to solve the problem: one-against-all support vector machines with continuous decision functions, fuzzy support vector machines, and decision-tree-based support vector machines. We show that one-against-all support vector machines with continuous decision functions and fuzzy support vector machines are equivalent.

### 3.1.1 Conventional Support Vector Machines

Consider an $n$-class problem. For a one-against-all support vector machine, we determine $n$ direct decision functions that separate one class from the remaining classes. Let the $i$th decision function, with the maximum margin that separates class $i$ from the remaining classes, be

$$D_i(\mathbf{x}) = \mathbf{w}_i^\top \phi(\mathbf{x}) + b_i, \tag{3.1}$$

where $\mathbf{w}_i$ is the $l$-dimensional vector, $\phi(\mathbf{x})$ is the mapping function that maps $\mathbf{x}$ into the $l$-dimensional feature space, and $b_i$ is the bias term.

The hyperplane $D_i(\mathbf{x}) = 0$ forms the optimal separating hyperplane, and if the classification problem is separable, the training data belonging to class $i$ satisfy $D_i(\mathbf{x}) \geq 1$ and those belonging to the remaining classes satisfy $D_i(\mathbf{x}) \leq -1$. Especially, support vectors satisfy $|D_i(\mathbf{x})| = 1$. If the problem is inseparable, unbounded support vectors satisfy $|D_i(\mathbf{x})| = 1$ and bounded support vectors belonging to class $i$ satisfy $D_i(\mathbf{x}) \leq 1$ and those belonging to a class other than class $i$, $D_i(\mathbf{x}) \geq -1$. (For L2 support vector machines support vectors belonging to class $i$ satisfy $D_i(\mathbf{x}) < 1$ and those belonging to a class other than class $i$, $D_i(\mathbf{x}) > -1$.)

In classification, if for the input vector $\mathbf{x}$

$$D_i(\mathbf{x}) > 0 \qquad\qquad (3.2)$$

is satisfied for one $i$, $\mathbf{x}$ is classified into class $i$. Because only the sign of the decision function is used, the decision is discrete.

If (3.2) is satisfied for plural $i$'s or if there is no $i$ that satisfies (3.2), $\mathbf{x}$ is unclassifiable. Consider the three-class problem with two-dimensional input as shown in Fig. 3.1, where the arrows show the positive sides of the hyperplanes. For data sample 1, $\mathbf{x}_1$, the three decision functions are

$$D_1(\mathbf{x}_1) > 0, \quad D_2(\mathbf{x}_1) > 0, \quad D_3(\mathbf{x}_1) < 0.$$

Because $\mathbf{x}_1$ belongs to both Classes 1 and 2, $\mathbf{x}_1$ is unclassifiable. Likewise, for data sample 2, $\mathbf{x}_2$, the three decision functions are

$$D_1(\mathbf{x}_2) < 0, \quad D_2(\mathbf{x}_2) < 0, \quad D_3(\mathbf{x}_2) < 0.$$

Thus, $\mathbf{x}_2$ is unclassifiable.

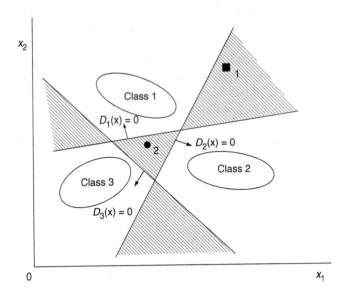

**Fig. 3.1** Unclassifiable regions by the one-against-all formulation

To avoid this, instead of discrete decision functions, continuous decision functions are proposed for classification, namely, data sample $\mathbf{x}$ is classified into the class

$$\arg \max_{i=1,\ldots,n} D_i(\mathbf{x}). \tag{3.3}$$

Then data sample 1 in Fig. 3.1 is classified into Class 1 because $D_1(\mathbf{x}_1)$ is the maximum among the three. Likewise, Data sample 2 is classified into Class 2.

In Section 3.1.3, the meaning of (3.3) is explained from the membership functions defined in the directions orthogonal to the optimal hyperplanes.

## 3.1.2 Fuzzy Support Vector Machines

In this section, we introduce membership functions into one-against-all support vector machines to resolve unclassifiable regions, while realizing the same classification results for the data that are classified by conventional one-against-all support vector machines. We introduce two operators: minimum and average operators to define membership functions for classes.

### 3.1.2.1 One-Dimensional Membership Functions

In a conventional set theory, for a given set, an element belongs either to the set or not. In fuzzy logic, the degree of membership of an element to a fuzzy set is defined by a membership function associated with the set. The range of the membership function is $[0, 1]$ and the degree of membership of 1 means that the element belongs to the fuzzy set 100% and that of 0 means that the element does not belong to the fuzzy set at all. And the degree of membership between 0 and 1 shows the degree that the element belongs to the fuzzy set.

For class $i$ we define one-dimensional membership functions $m_{ij}(\mathbf{x})$ in the directions orthogonal to the optimal separating hyperplanes $D_j(\mathbf{x}) = 0$ as follows:

1. For $i = j$

$$m_{ii}(\mathbf{x}) = \begin{cases} 1 & \text{for} \quad D_i(\mathbf{x}) \geq 1, \\ D_i(\mathbf{x}) & \text{otherwise.} \end{cases} \tag{3.4}$$

2. For $i \neq j$

$$m_{ij}(\mathbf{x}) = \begin{cases} 1 & \text{for} \quad D_j(\mathbf{x}) \leq -1, \\ -D_j(\mathbf{x}) & \text{otherwise.} \end{cases} \tag{3.5}$$

Figure 3.2 shows the membership function $m_{11}(\mathbf{x})$ for the two-dimensional input space. Because for a separable classification problem only the class $i$ training data exist when $D_i(\mathbf{x}) \geq 1$, we assume that the degree of class $i$

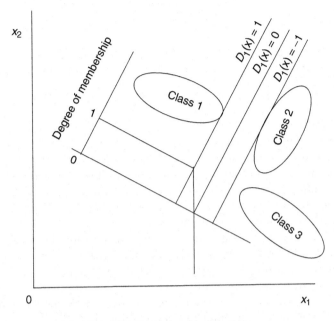

**Fig. 3.2** Definition of one-dimensional membership function

membership is 1 for $D_i(\mathbf{x}) \geq 1$ and $D_i(\mathbf{x})$ otherwise. We assume that the same is true even if inseparable. Here, we allow the negative degree of membership so that any data not on the boundary can be classified.

For $i \neq j$, class $i$ is on the negative side of $D_j(\mathbf{x}) = 0$. In this case, support vectors may not include class $i$ data (as is the case with Class 3 in Fig. 3.2), but when $D_j(\mathbf{x}) \leq -1$, we assume that the degree of class $i$ degree of membership is 1, otherwise $-D_j(\mathbf{x})$.

### 3.1.2.2 Membership Functions for Classes

We define the class $i$ membership function of $\mathbf{x}$ by the minimum operation for $m_{ij}(\mathbf{x})$ $(j = 1, \ldots, n)$:

$$m_i(\mathbf{x}) = \min_{j=1,\ldots,n} m_{ij}(\mathbf{x}), \qquad (3.6)$$

or the average operation:

$$m_i(\mathbf{x}) = \frac{1}{n} \sum_{j=1,\ldots,n} m_{ij}(\mathbf{x}). \qquad (3.7)$$

The data sample $\mathbf{x}$ is classified into the class

$$\arg \max_{i=1,\dots,n} m_i(\mathbf{x}). \tag{3.8}$$

Now consider the difference of membership functions given by (3.6) and (3.7). By the definition of $m_{ij}(\mathbf{x})$ given by (3.4) and (3.5), for $\mathbf{x} \in R_i$ where

$$R_i = \{\mathbf{x} \mid D_i(\mathbf{x}) > 1, D_j(\mathbf{x}) < -1, \ j \neq i, \ j = 1,\dots,n\}, \tag{3.9}$$

$m_i(\mathbf{x}) = 1$ for both (3.6) and (3.7). Because $m_i(\mathbf{x}) = 1$ is satisfied for only one $i$, $\mathbf{x} \in R_i$ is classified into class $i$. Thus, both membership functions give the same classification result for the data in $R_i$. Therefore, the difference of the membership functions occurs for $m_i(\mathbf{x}) < 1$. It is shown in Section 3.1.3 that the class boundaries by the support vector machine with the minimum or average operator are the same as those by the one-against-all support vector machine with continuous decision functions.

Figure 3.3 shows the membership functions $m_1(\mathbf{x})$ for the minimum and average operators for two decision functions. For the minimum operator, a contour line, which has the same degree of membership, lies in parallel to the surface of $R_1$. The membership function with the average operator has a similar shape to that with the minimum operator for the region, where the degree of one of the two one-dimensional functions is 1.

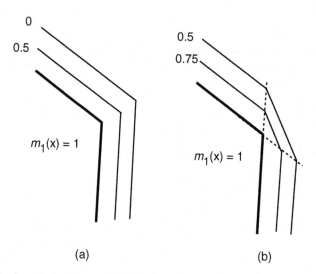

**Fig. 3.3** Membership functions: **(a)** Minimum operator. **(b)** Average operator

According to the formulation, the unclassifiable regions shown in Fig. 3.1 are resolved as shown in Fig. 3.4. This gives the similar class boundaries proposed by Bennett [20].

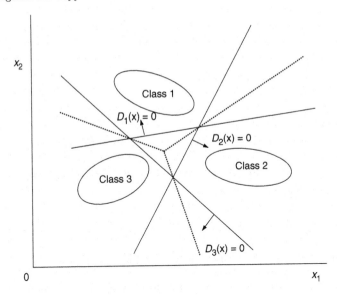

**Fig. 3.4** Extended generalization regions

Because the decision boundary between classes $i$ and $j$ is given by $m_i(\mathbf{x}) = m_j(\mathbf{x})$, the decision boundary changes as the output of the decision functions is normalized. Mayoraz and Alpaydin [21] discussed three ways to normalize the outputs. By this normalization, however, the classification results change only for the data in the unclassifiable regions caused by discrete decision functions.

### 3.1.3 Equivalence of Fuzzy Support Vector Machines and Support Vector Machines with Continuous Decision Functions

Here, we show that one-against-all support vector machines with continuous decision functions and one-against-all fuzzy support vector machines with minimum or average operators are equivalent in that they give the same classification result for the same input [22].

Let $m_i^{\mathrm{m}}(\mathbf{x})$ and $m_i^{\mathrm{a}}(\mathbf{x})$ be the membership functions for class $i$ using the minimum and average operators, respectively.

Then (3.6) and (3.7) are rewritten as follows:

$$m_i^{\mathrm{m}}(\mathbf{x}) = \min\left(\min(1, D_i(\mathbf{x})), \min_{\substack{k \neq i, \\ k=1,\ldots,n}} \min(1, -D_k(\mathbf{x}))\right), \quad (3.10)$$

$$m_i^{\mathrm{a}}(\mathbf{x}) = \frac{1}{n}\left(\min(1, D_i(\mathbf{x})) + \sum_{k=1, k \neq i}^{n} \min(1, -D_k(\mathbf{x}))\right). \qquad (3.11)$$

Thus $m_i^{\mathrm{a}}(\mathbf{x}) - m_j^{\mathrm{a}}(\mathbf{x})$ is given by

$$
\begin{aligned}
m_i^{\mathrm{a}}(\mathbf{x}) - m_j^{\mathrm{a}}(\mathbf{x}) = \frac{1}{n}(&\min(1, D_i(\mathbf{x})) + \min(1, -D_j(\mathbf{x})) \\
&- \min(1, D_j(\mathbf{x})) - \min(1, -D_i(\mathbf{x}))).
\end{aligned}
\qquad (3.12)
$$

Now we prove the equivalence classifying the cases into three:

1. $D_i(\mathbf{x}) > 0$, $D_j(\mathbf{x}) \leq 0$ $(j = 1, \ldots, n, j \neq i)$

By the support vector machine with continuous decision functions, input $\mathbf{x}$ is classified into class $i$.

From (3.10) and the conditions on the signs of $D_k$ $(k = 1, \ldots, n)$,

$$m_i^{\mathrm{m}}(\mathbf{x}) \geq 0, \quad m_j^{\mathrm{m}}(\mathbf{x}) < 0. \qquad (3.13)$$

Thus by the fuzzy support vector machine with minimum operators, input $\mathbf{x}$ is classified into class $i$.

From (3.12),

$$
\begin{aligned}
m_i^{\mathrm{a}}(\mathbf{x}) - m_j^{\mathrm{a}}(\mathbf{x}) = \frac{1}{n}\,(&\min(1, D_i(\mathbf{x})) + \min(1, -D_j(\mathbf{x})) \\
&- D_j(\mathbf{x}) + D_i(\mathbf{x})) > 0.
\end{aligned}
\qquad (3.14)
$$

Thus by the fuzzy support vector machine with average operators, input $\mathbf{x}$ is classified into class $i$.

2. $0 > D_i(\mathbf{x}) > D_j(\mathbf{x})$ $(j = 1, \ldots, n, j \neq i)$

By the support vector machine with continuous decision functions, $\mathbf{x}$ is classified into class $i$.

From (3.10) and the conditions on the signs of $D_k(\mathbf{x})$ $(k = 1, \ldots, n)$,

$$m_i^{\mathrm{m}}(\mathbf{x}) > m_j^{\mathrm{m}}(\mathbf{x}). \qquad (3.15)$$

Thus input $\mathbf{x}$ is classified into class $i$ by the fuzzy support vector machine with minimum operators.

From (3.12),

$$
\begin{aligned}
m_i^{\mathrm{a}}(\mathbf{x}) - m_j^{\mathrm{a}}(\mathbf{x}) = \frac{1}{n}\,(&D_i(\mathbf{x}) - D_j(\mathbf{x}) \\
&- \min(1, -D_i(\mathbf{x})) + \min(1, -D_j(\mathbf{x}))) > 0.
\end{aligned}
\qquad (3.16)
$$

Thus input $\mathbf{x}$ is classified into class $i$ by the fuzzy support vector machine with average operators.

3. $D_i(\mathbf{x}) > D_j(\mathbf{x}) > 0 > D_k(\mathbf{x})$, where $j \in N_1$, $k \in N_2$, $N_1 \cap N_2 = \phi$, $(N_1 \cup N_2) \cap \{i\} = \phi$, $N_1 \cup N_2 \cup \{i\} = \{1, \ldots, n\}$

Input $\mathbf{x}$ is classified into class $i$ by the support vector machine with continuous decision functions.

From (3.10),

$$m_i^{\mathrm{m}}(\mathbf{x}) = \min_{j \in N_1} -D_j(\mathbf{x}), \tag{3.17}$$

$$m_j^{\mathrm{m}}(\mathbf{x}) = -D_i(\mathbf{x}) \quad \text{for } j \in N_1, \tag{3.18}$$

$$m_k^{\mathrm{m}}(\mathbf{x}) = \min(-D_i(\mathbf{x}), D_k(\mathbf{x})) \quad \text{for } k \in N_2. \tag{3.19}$$

Thus,

$$m_i^{\mathrm{m}}(\mathbf{x}) > m_j^{\mathrm{m}}(\mathbf{x}) \quad \text{for } j \in N_1 \cup N_2. \tag{3.20}$$

Therefore, $\mathbf{x}$ is classified into class $i$ by the fuzzy support vector machine with minimum operators.

From

$$m_i^{\mathrm{a}}(\mathbf{x}) - m_j^{\mathrm{a}}(\mathbf{x}) = \frac{1}{n} (\min(1, D_i(\mathbf{x})) - D_j(\mathbf{x})$$
$$- \min(1, D_j(\mathbf{x})) + D_i(\mathbf{x})) > 0 \quad \text{for } j \in N_1 \tag{3.21}$$

and from (3.14),

$$m_i^{\mathrm{a}}(\mathbf{x}) > m_j^{\mathrm{a}}(\mathbf{x}) \quad \text{for } j \in N_1 \cup N_2. \tag{3.22}$$

Thus, input $\mathbf{x}$ is classified into class $i$ by the fuzzy support vector machine with average operators.

Therefore, one-against-all support vector machines with continuous decision functions and one-against-all fuzzy support vector machines with minimum or average operators are equivalent. This gives an interpretation of the classification strategy of one-against-all support vector machines with continuous decision functions from membership functions. If negative degrees of membership are not favorable, we can replace them with the equivalent errors in error-correcting output codes as discussed in Section 3.3.3.

## 3.1.4 Decision-Tree-Based Support Vector Machines

To resolve unclassifiable regions in one-against-all support vector machines, in this section we discuss decision-tree-based support vector machines [8], namely, we train $n - 1$ support vector machines; the $i$th ($i = 1, \ldots, n - 1$) support vector machine is trained so that it separates class $i$ data from data belonging to one of classes $i + 1, i + 2, \ldots, n$. After training, classification is performed from the first to the $(n - 1)$th support vector machines. If the $i$th support vector machine classifies a data sample into class $i$, classification terminates. Otherwise, classification is performed until the data sample is classified into the definite class.

Figure 3.5 shows an example of class boundaries for four classes, when linear kernels are used. As seen from the figure, the classes with smaller class numbers have larger class regions. Thus the processing order affects the generalization ability. In some applications, the structure of a decision tree is determined by the relationships of inclusion among classes [23], but in most cases we need to determine the structure. In a usual decision tree, each node separates one set of classes from another set of classes. And to divide the set of classes into two, in [24, 25] the $k$-means clustering algorithm is used. With $k = 2$, the data in the set are clustered into two clusters. And if the data in one class are clustered into the two clusters, the class data are considered to reside in the cluster with the larger number of data.

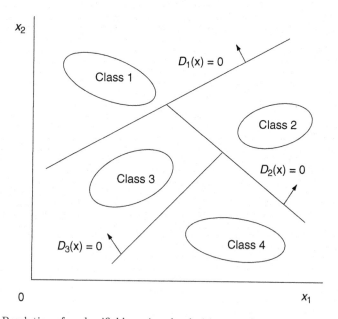

**Fig. 3.5** Resolution of unclassifiable regions by decision-tree formulation

In the following, we discuss determining the structure of decision trees using distance measures.

### 3.1.4.1 Architecture of Decision Trees

Because the more the data are misclassified at the upper node of the decision tree, the worse the classification performance becomes, the classes that are easily separated need to be separated at the upper node of the decision tree. To determine the decision tree, we use the fact that the neighborhood relations of data in the input space are kept in the feature space. We use four types of decision trees as follows:

1. **Type 1 decision tree.** At each node one class is separated from the remaining classes using the Euclidean distance as a separability measure.
2. **Type 2 decision tree.** At each node some classes are separated from the remaining classes using the Euclidean distance as a separability measure.
3. **Type 3 decision tree.** At each node one class is separated from the remaining classes using classification errors by the Mahalanobis distance as a separability measure.
4. **Type 4 decision tree.** At each node some classes are separated from the remaining classes using classification errors by the Mahalanobis distance as a separability measure.

In the following, we discuss these algorithms in detail for an $n$-class problem.

### 3.1.4.2 Type 1 Decision Tree

In this method, we calculate the Euclidean distances between the class centers and recursively separate the farthest class from the remaining classes.

1. Calculate the class centers $c_i$ $(i = 1, \ldots, n)$ by

$$c_i = \frac{1}{|X_i|} \sum_{x \in X_i} x \tag{3.23}$$

and the distance between class $i$ and class $j$, $d_{ij}$ $(i, j = 1, \ldots, n)$, by

$$d_{ij} (= d_{ji}) = \|c_i - c_j\|. \tag{3.24}$$

Here $X_i$ is a set of training data included in class $i$ and $|X_i|$ is the number of elements in $X_i$.

2. Find the smallest value of $d_{ij}$ for class $i$:

$$l_i = \min_{j \neq i} d_{ij} \tag{3.25}$$

and regard the class that has the largest $l_i$ as the farthest class and calculate the optimal hyperplane that separates this class, namely, separate class $k$ from the others. Here $k = \arg\max\limits_i l_i$. If plural $k$'s exist for these classes, compare the next smallest distance $l_i'$ and $k = \arg\max\limits_i l_i'$.

3. Delete class $k$ and repeat Step 2 until all the classes are separated.

### 3.1.4.3 Type 2 Decision Tree

Using the distances between class centers, repeat merging the two nearest classes until two clusters are obtained and separate the clusters by the optimal hyperplane. For the classes in the generated clusters, we repeat the above procedure until each class is separated.

1. Calculate the class centers by (3.23) and the distances between class $i$ and class $j$, $d_{ij}(i, j = 1, \ldots, n)$, by (3.24). Initially, we assume that all the classes belong to different clusters.
2. For all the class pairs in different clusters, calculate the smallest value of distances by (3.25) and merge the associated two clusters into one.
3. Repeat Step 2 until the number of clusters becomes two.
4. Calculate the optimal hyperplane that separates the clusters generated in Step 3.
5. If the separated cluster in Step 4 has more than one class, regard the classes as belonging to different clusters and go to Step 2. If no cluster has more than one class, terminate the algorithm.

### 3.1.4.4 Type 3 Decision Tree

First, we classify the training data using the Mahalanobis distance and determine the optimal hyperplane that separates the class with the smallest misclassifications from the remaining classes. For the remaining classes we repeat the above procedure until all the classes are separated.

1. For each class, calculate the covariance matrix $Q_i$ $(i = 1, \ldots, n)$ by

$$Q_i = \frac{1}{|X_i|} \sum_{\mathbf{x} \in X_i} (\mathbf{x} - \mathbf{c}_i)(\mathbf{x} - \mathbf{c}_i)^\top, \tag{3.26}$$

where $\mathbf{c}_i$ is the center vector of class $i$ given by (3.23). Calculate the Euclidean distance between class centers using (3.24). For all the data, calculate the Mahalanobis distance $d_i(\mathbf{x})$

$$d_i{}^2(\mathbf{x}) = (\mathbf{x} - \mathbf{c}_i)^\top Q_i{}^{-1}(\mathbf{x} - \mathbf{c}_i) \qquad \text{for} \quad i = 1, \ldots, n \tag{3.27}$$

and classify them to the nearest classes.

2. Calculate the number of misclassified data for class $i$ by

$$\sum_{\substack{j \neq i, \\ j=1,\ldots,n}} (e_{ij} + e_{ji}),$$

where $e_{ij}$ is the number of class $i$ data misclassified into class $j$ and separate the class that has the smallest value from the others. If plural classes have the same value, separate the class with the farthest Euclidean distance among these classes.

3. Deleting the separated class, repeat Step 2 until all the classes are separated.

### 3.1.4.5 Type 4 Decision Tree

In this method, we first repeat merging the two most misclassified classes by the Mahalanobis distance until two clusters are generated and separate the two clusters by the optimal hyperplane. Considering each class in the generated cluster as a cluster we repeat merging them into two clusters and generating the optimal hyperplane until all the classes are separated.

1. Initially, we assume that all classes belong to different clusters. Calculate $Q_i$ $(i = 1, \ldots, n)$ and $e_{ij}$ $(i, j = 1, \ldots, n)$.
2. For all the class pairs in different clusters, find the largest value of $e_{ij}$ and merge the associated clusters into one.
3. Repeat Step 2 until the number of clusters becomes two.
4. Calculate a hyperplane that separates these two clusters.
5. If the separated cluster in Step 4 has more than one class, regard the classes as belonging to different clusters and go to Step 2. If no cluster has more than one class, terminate the algorithm.

### 3.1.4.6 Performance Evaluation

Because there is not much difference of generalization abilities among Type 1 to Type 4 decision trees, in the following we show only the results for Type 1 decision trees.

We compared the recognition rates of the test data for Type 1 decision trees and those of the conventional one-against-all and fuzzy support vector machines using the data sets listed in Table 1.3. We used the polynomial kernel with degrees 2 to 4. The ranges of the input variables were normalized into $[0, 1]$. We trained the support vector machine by the primal–dual interior-point method combined with the decomposition technique.

We set $C = 20{,}000$ for the thyroid data set; $C = 10{,}000$ for the MNIST data set; and $C = 2{,}000$ for the remaining data sets. We used an Athlon MP 2000 personal computer.

Table 3.1 shows the recognition rates of the test data for the conventional one-against-all support vector machine, fuzzy support vector machine (FSVM), and Type 1 decision tree. The best recognition rate in a row is shown in bold. Column "Train." on the left lists the training time for SVM and FSVM, and the column on the right lists the training time for the Type 1 decision tree.

**Table 3.1** Performance of decision-tree support vector machines

| Data | Kernel | SVM (%) | FSVM (%) | Train. (s) | Type 1 (%) | Train. (s) |
|------|--------|---------|----------|------------|------------|------------|
| Iris | d2 | 92.00 | **94.67** | – | 93.33 | – |
|      | d3 | 93.33 | **94.67** | – | 93.33 | – |
| Numeral | d2 | 99.02 | 99.39 | 0.5 | **99.76** | 0.2 |
|         | d3 | 98.90 | 99.51 | 0.5 | **99.76** | 0.3 |
| Thyroid | d2 | 95.13 | 97.20 | 245 | **97.78** | 5 |
|         | d3 | 95.51 | 97.40 | 290 | **97.84** | 5 |
|         | d4 | 95.57 | 97.58 | 21 | **97.72** | 5 |
| Blood cell | d2 | 88.77 | **93.03** | 24 | 92.45 | 9 |
|            | d3 | 88.84 | **93.10** | 23 | 92.19 | 8 |
|            | d4 | 86.61 | **92.68** | 22 | 91.48 | 7 |
| Hiragana-50 | d2 | 95.73 | **99.07** | 126 | 97.74 | 50 |
|             | d3 | 96.20 | **99.35** | 123 | 98.00 | 52 |
|             | d4 | 96.33 | **99.37** | 136 | 98.07 | 53 |
| Hiragana-105 | d2 | 99.99 | **100** | 530 | 99.99 | 247 |
|              | d3 | **100** | **100** | 560 | **100** | 237 |
| Hiragana-13 | d2 | 96.25 | **99.50** | 238 | 98.17 | 64 |
|             | d3 | 96.09 | **99.35** | 229 | 98.28 | 63 |
|             | d4 | 96.12 | **99.34** | 231 | 98.44 | 64 |
| MNIST | d2 | 96.06 | **98.17** | 2,760 | 97.71 | 870 |
|       | d3 | 96.56 | **98.38** | 4,166 | 97.74 | 967 |

From the table, the recognition rate of the Type 1 decision tree is, in most cases, better than that of the SVM, but except for the numeral and thyroid data sets, it is lower than that of the FSVM, although the difference is small.

The training times of the SVM and the FSVM are the same. From the table, the training time of Type 1 decision tree is usually two to four times shorter than that of the SVM. This is because in training a Type 1 decision tree, the number of training data decreases as training proceeds from the top node to the leaf nodes, but for the SVM all the training data are used to determine $n$ decision functions.

## 3.2 Pairwise Support Vector Machines

In this section, we discuss pairwise support vector machines and their variants. Pairwise support vector machines reduce the unclassifiable regions that occur for one-against-all support vector machines. But unclassifiable regions still exist. To resolve unclassifiable regions, we discuss fuzzy support vector machines and decision-tree-based support vector machines.

### 3.2.1 Conventional Support Vector Machines

In pairwise support vector machines, we determine the decision functions for all the combinations of class pairs. In determining a decision function for a class pair, we use the training data for the corresponding two classes. Thus, in each training session, the number of training data is reduced considerably compared to one-against-all support vector machines, which use all the training data. But the number of decision functions is $n(n-1)/2$, compared to $n$ for one-against-all support vector machines, where $n$ is the number of classes.

Let the decision function for class $i$ against class $j$, with the maximum margin, be

$$D_{ij}(\mathbf{x}) = \mathbf{w}_{ij}^\top \boldsymbol{\phi}(\mathbf{x}) + b_{ij}, \tag{3.28}$$

where $\mathbf{w}_{ij}$ is the $l$-dimensional vector, $\boldsymbol{\phi}(\mathbf{x})$ is a mapping function that maps $\mathbf{x}$ into the $l$-dimensional feature space, $b_{ij}$ is the bias term, and $D_{ij}(\mathbf{x}) = -D_{ji}(\mathbf{x})$.

The regions

$$R_i = \{\mathbf{x} \mid D_{ij}(\mathbf{x}) > 0, j = 1, \ldots, n, j \neq i\} \quad \text{for } i = 1, \ldots, n \tag{3.29}$$

do not overlap, and if $\mathbf{x}$ is in $R_i$, $\mathbf{x}$ is considered to belong to class $i$. The problem is that $\mathbf{x}$ may not be in any of $R_i$. We classify $\mathbf{x}$ by voting, namely, for the input vector $\mathbf{x}$ we calculate

$$D_i(\mathbf{x}) = \sum_{j \neq i, j=1}^{n} \operatorname{sign}(D_{ij}(\mathbf{x})), \tag{3.30}$$

where

$$\operatorname{sign}(x) = \begin{cases} 1 & \text{for } x \geq 0, \\ -1 & \text{for } x < 0, \end{cases} \tag{3.31}$$

and we classify $\mathbf{x}$ into the class

$$\arg\max_{i=1,\ldots,n} D_i(\mathbf{x}). \tag{3.32}$$

If $\mathbf{x} \in R_i$, $D_i(\mathbf{x}) = n - 1$ and $D_k(\mathbf{x}) < n - 1$ for $k \neq i$. Thus $\mathbf{x}$ is classified into $i$. But if any of $D_i(\mathbf{x})$ is not $n - 1$, (3.32) may be satisfied for plural $i$'s. In this case, $\mathbf{x}$ is unclassifiable. In the shaded region in Fig. 3.6, $D_i(\mathbf{x}) = 0$ ($i = 1, 2$, and 3). Thus the shaded region is unclassifiable, although the unclassifiable region is much smaller than that for the one-against-all support vector machine shown in Fig. 3.1.

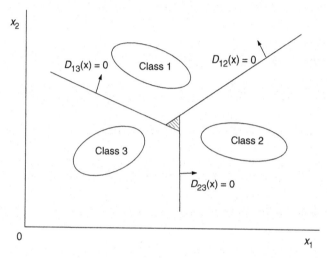

**Fig. 3.6** Unclassifiable regions by the pairwise formulation. Reprinted from [4, p. 115]

## 3.2.2 Fuzzy Support Vector Machines

### 3.2.2.1 Architecture

Similar to the one-against-all formulation, we introduce the membership function to resolve unclassifiable regions while realizing the same classification results with those of the conventional pairwise classification for the classifiable regions. To do this, for the optimal separating hyperplane $D_{ij}(\mathbf{x}) = 0$ ($i \neq j$) we define one-dimensional membership functions $m_{ij}(\mathbf{x})$ in the directions orthogonal to $D_{ij}(\mathbf{x}) = 0$ as follows:

$$m_{ij}(\mathbf{x}) = \begin{cases} 1 & \text{for} \quad D_{ij}(\mathbf{x}) \geq 1, \\ D_{ij}(\mathbf{x}) & \text{otherwise.} \end{cases} \tag{3.33}$$

We define the class $i$ membership function of $\mathbf{x}$ by the minimum operation for $m_{ij}(\mathbf{x})$ $(j \neq i, j = 1, \ldots, n)$:

$$m_i(\mathbf{x}) = \min_{\substack{j \neq i, \\ j=1,\ldots,n}} m_{ij}(\mathbf{x}), \tag{3.34}$$

or the average operation:

$$m_i(\mathbf{x}) = \frac{1}{n-1} \sum_{\substack{j \neq i, \\ j=1}}^{n} m_{ij}(\mathbf{x}). \tag{3.35}$$

Now an unknown data sample $\mathbf{x}$ is classified into the class

$$\arg \max_{i=1,\ldots,n} m_i(\mathbf{x}). \tag{3.36}$$

Equation (3.34) is equivalent to

$$m_i(\mathbf{x}) = \min \left( 1, \min_{\substack{j \neq i, \\ j=1,\ldots,n}} D_{ij}(\mathbf{x}) \right). \tag{3.37}$$

Because $m_i(\mathbf{x}) = 1$ holds for only one class, classification with the minimum operator is equivalent to classifying $\mathbf{x}$ into the class

$$\arg \max_{i=1,\ldots,n} \min_{\substack{j \neq i, \\ j=1,\ldots,n}} D_{ij}(\mathbf{x}). \tag{3.38}$$

Thus, the unclassifiable region shown in Fig. 3.6 is resolved as shown in Fig. 3.7 for the fuzzy support vector machine with the minimum operator.

### 3.2.3 Performance Comparison of Fuzzy Support Vector Machines

In this section we compare L1 and L2 support vector machines for one-against-all and pairwise classification for the data sets listed in Table 1.3. We scaled the input ranges into $[0, 1]$. Except for the MNIST data set, we determined the kernels and parameters by fivefold cross-validation. We used linear kernels; polynomial kernels with degrees 2, 3, and 4; and RBF kernels with $\gamma = 0.1, 1$, and 10. The values of $C$ were selected from 1, 50, 100, 1,000, 3,000, 5,000, 7,000, 10,000, and 100,000. We selected the kernel and parameter values with the highest recognition rate for the validation data set. If the

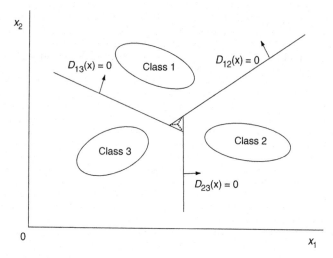

**Fig. 3.7** Extended generalization regions. Reprinted from [4, p. 116]

same recognition rate was obtained, we broke the tie by selecting the simplest structure as follows:

1. Select the kernel and parameter values with the highest recognition rate for the training data.
2. Select polynomial kernels from polynomial and RBF kernels.
3. Select the polynomial kernel with the smallest degree from polynomial kernels.
4. Select the RBF kernel with the smallest value of $\gamma$ from RBF kernels.
5. Select the value with the largest value of $C$ from different values of $C$.

For the MNIST data set, we set $C = 10,000$ and selected the kernel that realized the maximum recognition rate for the test data.

Table 3.2 lists the results for one-against-all L1 and L2 SVMs and pairwise L1 and L2 SVMs. The "Parm" row lists the kernel and the value of $C$ selected. If the values of $C$ selected by cross-validation are different for the minimum and average operators, we only show the value for the minimum operator. The "Dis.," "Min," and "Avg." rows list the recognition rates of the test data, those of the training data in parentheses if lower than 100%, with discrete functions, minimum, and average operators, respectively. The "SVs" row lists the number of average support vectors per decision function and the numeral in parentheses shows the number of bounded support vectors for L1 SVMs. The maximum recognition rate of the test data for a data set is shown in bold.

For some cases, such as pairwise SVMs for the iris data, the recognition rates of the discrete SVMs and fuzzy SVMs are the same. This means that no test data were in the unclassifiable regions. Except for these cases, the

**Table 3.2** Performance comparison of support vector machines

| Data | Item | One-against-all | | Pairwise | |
|------|------|------|------|------|------|
| | | L1 SVM | L2 SVM | L1 SVM | L2 SVM |
| Iris | Parm | $\gamma 0.1, C5,000$ | $d1, C2,000$ | $\gamma 1, C100$ | $\gamma 0.1, C100$ |
| | Dis. | 92.00 (97.33) | 69.33 (74.67) | **97.33** (98.67) | **97.33** |
| | Min | 94.67 | 94.67 | **97.33** (98.67) | **97.33** |
| | Avg. | 94.67 | 94.67 | **97.33** (98.67) | **97.33** |
| | SVs | 10 (5) | 25 | 10 (7) | 21 |
| Numeral | Parm | $\gamma 1, C50$ | $d3, C1$ | $d2, C1$ | $\gamma 0.1, C1,000$ |
| | Dis. | 99.02 (99.51) | 99.15 | 99.63 | **99.76** |
| | Min | 99.27 (99.88) | 99.63 | 99.63 | **99.76** |
| | Avg. | 99.27 (99.88) | 99.63 | 99.63 | **99.76** |
| | SVs | 15 (3) | 47 | 6 | 13 |
| Thyroid | Parm | $d4, C10^5$ | $d4, C10^5$ | $d1, C10^5$ | $d3, C10^4$ |
| | Dis. | 95.97 (99.84) | 96.09 (99.89) | 97.29 (98.59) | 97.67 (99.95) |
| | Min | **97.93** (99.97) | 97.81 | 97.61 (98.75) | **97.93** (99.95) |
| | Avg. | **97.93** (99.97) | 97.81 | 97.64 (98.73) | **97.93** (99.95) |
| | SVs | 87 (7) | 96 | 68 (52) | 55 |
| Blood cell | Parm | $d2, C3,000$ | $\gamma 10, C100$ | $d1, C50$ | $d2, C10$ |
| | Dis. | 86.87 (94.93) | 90.39 (94.80) | 91.58 (95.41) | 92.87 (96.51) |
| | Min | 93.16 (97.68) | **93.58** (97.19) | 92.03 (95.58) | 92.97 (96.61) |
| | Avg. | 93.16 (97.68) | **93.58** (97.19) | 92.06 (96.03) | 92.94 (96.58) |
| | SVs | 92 (29) | 188 | 19 (11) | 34 |
| H-50 | Parm | $\gamma 10, C5,000$ | $\gamma 10, C1,000$ | $\gamma 10, C10^4$ | $\gamma 10, C10^4$ |
| | Dis. | 97.72 | 97.74 | 99.00 | 99.09 |
| | Min | 99.26 | **99.28** | 99.11 | 99.11 |
| | Avg. | 99.26 | **99.28** | 99.11 | 99.11 |
| | SVs | 71 | 97 | 21 | 21 |
| H-13 | Parm | $\gamma 10, C3,000$ | $\gamma 10, C500$ | $\gamma 10, C7,000$ | $\gamma 10, C2,000$ |
| | Dis. | 98.10 | 98.68 (99.83) | 99.63 | 99.70 |
| | Min | 99.63 | 99.69 (99.96) | 99.74 | **99.76** |
| | Avg. | 99.63 | 99.69 (99.96) | 99.70 | 99.72 |
| | SVs | 39 (1) | 71 | 10 | 11 |
| H-105 | Parm | $\gamma 10, C10^4$ | $\gamma 10, C10^4$ | $\gamma 10, C10^4$ | $\gamma 10, C10^4$ |
| | Dis. | **100** | **100** | 99.93 | **100** |
| | Min | **100** | **100** | 99.95 | **100** |
| | Avg. | **100** | **100** | 99.94 | **100** |
| | SVs | 91 | 91 | 13 | 26 |
| MNIST | Parm | $\gamma 10, C10^4$ | $\gamma 10, C10^4$ | $\gamma 10, C10^4$ | $\gamma 10, C10^4$ |
| | Min | **98.55** | **98.55** | 98.32 | 98.32 |
| | SVs | 2,287 | 2,295 | 602 | 603 |

recognition rates of the fuzzy SVMs are better. As the theory tells us, the recognition rates by minimum operators and average operators are the same for the one-against-all SVMs. In addition, for pairwise SVMs there is not much difference.

Comparing L1 and L2 SVMs, there is not much difference in the recognition rates of the test data, but usually L2 SVMs require more support vectors.

Comparing one-against-all and pairwise SVMs, we can see the following observations:

1. The recognition improvement of pairwise classification by the introduction of membership functions is small. Thus, the unclassifiable regions by pairwise SVMs are smaller than those by one-against-all SVMs.
2. In most cases, the recognition rates of the test data by the pairwise fuzzy SVMs are almost the same as those by the one-against-all SVMs.
3. The number of support vectors of pairwise SVMs is smaller than that of one-against-all SVMs. This is because in pairwise SVMs one class needs to be separated from another class but in one-against-all SVMs, one class needs to be separated from the remaining classes; thus more support vectors are necessary. But the total number of support vectors for the pairwise SVMs may be larger for $n > 3$ because $n(n-1)/2$ decision functions are necessary.

### 3.2.4 Cluster-Based Support Vector Machines

In a one-against-all support vector machine, all the training data are used for training, but for a pairwise support vector machine, training data for two classes are used at a time. Thus for a large problem, a pairwise support vector machine handles a much smaller number of training data than a one-against-all support vector machine. But if the number of data for one class is very large, training becomes prohibitive, even for a pairwise support vector machine. To solve the problem in such a situation, Lu et al. [26] proposed dividing the training data for each class into clusters and determining, like pairwise support vector machines, decision functions for cluster pairs. We call this a *cluster-based support vector machine*. In the following, we discuss the architecture of cluster-based support vector machines with minimum operators.

Assume that class $i$ $(i = 1, \ldots, n)$ is divided into $N_i$ clusters and we denote the $j$th cluster for class $i$ cluster $ij$. Let the decision function for cluster $ij$ and cluster $op$ be

$$D_{ij-op}(\mathbf{x}) = \mathbf{w}_{ij-op}^{\top}\phi(\mathbf{x}) + b_{ij-op}, \qquad (3.39)$$

where $\phi(\mathbf{x})$ is a mapping function from $\mathbf{x}$ to the $l$-dimensional feature space, $\mathbf{w}_{ij-op}$ is an $l$-dimensional vector, $b_{ij-op}$ is a bias term, and $D_{ij-op}(\mathbf{x}) = -D_{op-ij}(\mathbf{x})$. For cluster $ij$, we determine the decision function $D_{ij}(\mathbf{x})$ using $D_{ij-op}$ $(o \neq i, o = 1, \ldots, n, j = 1, \ldots, N_o)$:

$$D_{ij}(\mathbf{x}) = \min_{\substack{o \neq i, o = 1, \ldots, n, \\ p = 1, \ldots, N_o}} D_{ij-op}(\mathbf{x}). \tag{3.40}$$

Now if $\mathbf{x}$ belongs to cluster $ij$, $\mathbf{x}$ needs to be classified into class $i$. Thus, we define the decision function for class $i$ by

$$D_i(\mathbf{x}) = \max_{j=1,\ldots,N_i} D_{ij}(\mathbf{x}). \tag{3.41}$$

Then unknown $\mathbf{x}$ is classified into

$$\arg \max_{i=1,\ldots,n} D_i(\mathbf{x}). \tag{3.42}$$

For the special case where each class consists of one cluster, (3.42) reduces to (3.38), which is equivalent to a pairwise fuzzy support vector machine.

*Example 3.1.* Consider the case where each of two classes consists of two clusters as shown in Fig. 3.8. Cluster 11 is separated from Clusters 21 and 22 by $D_{11-21}(\mathbf{x}) = 0$ and $D_{11-22}(\mathbf{x}) = 0$, respectively. Thus the region for Cluster 11, where $D_{11}(\mathbf{x}) > 0$, is $\{\mathbf{x} \mid D_{11-21}(\mathbf{x}) > 0, D_{11-22}(\mathbf{x}) > 0\}$. Likewise, the region for Cluster 12, where $D_{12}(\mathbf{x}) > 0$, is $\{\mathbf{x} \mid D_{12-21}(\mathbf{x}) > 0, D_{12-22}(\mathbf{x}) > 0\}$. These regions are disjoint, and if $\mathbf{x}$ is in any of the regions, it is classified into Class 1. On the other hand, as seen from the figure, the regions for Clusters 21 and 22 are overlapped.

Similar to pairwise support vector machines, unclassifiable regions that may appear using discrete decision functions are resolved using the continuous decision functions given by (3.40). But in this example, unclassifiable regions do not exist, even if we use discrete decision functions.

## 3.2.5 Decision-Tree-Based Support Vector Machines

Similar to decision-tree-based support vector machines discussed for one-against-all formulation, we can formulate decision-tree-based support vector machines for pairwise classification [5–7, 27]. In this section, we discuss two types of decision-tree-based support vector machines and then we discuss how to optimize structures of decision trees [9, 10].

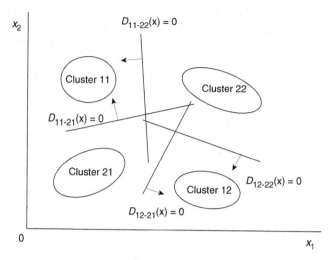

**Fig. 3.8** Decision boundaries by cluster-based classification

### 3.2.5.1 Decision Directed Acyclic Graph Support Vector Machines

To resolve unclassifiable regions for pairwise support vector machines, Platt et al. [6] proposed decision-tree-based pairwise support vector machines called *decision directed acyclic graph (DDAG) support vector machines*. In the following we call them DDAGs for short. Figure 3.9 shows the decision tree for the three classes shown in Fig. 3.6. In the figure, $\bar{i}$ shows that $\mathbf{x}$ does not belong to class $i$. As the top-level classification, we can choose any pair of classes. And except for the leaf node if $D_{ij}(\mathbf{x}) > 0$, we consider that $\mathbf{x}$ does not belong to class $j$, and if $D_{ij}(\mathbf{x}) < 0$ not class $i$.[1] Thus if $D_{12}(\mathbf{x}) > 0$, $\mathbf{x}$ does not belong to Class 2. Thus it belongs to either Class 1 or Class 3, and the next classification pair is Classes 1 and 3. The generalization regions become as shown in Fig. 3.10. Unclassifiable regions are resolved, but clearly the generalization regions depend on the tree formation.

Figure 3.11 shows a DDAG for four classes. At the top level, Classes 1 and 2 are selected. At the second level, Classes 1 and 4, and 2 and 3, which cover all four classes, are selected. But we can select any pair from Classes 1, 3, and 4 at the left node and from 2, 3, and 4 at the right node. Thus we may select Classes 3 and 4 for both nodes as shown in Fig. 3.12. This is an extension of the DDAG originally defined.

Classification by an original DDAG is executed by list processing, namely, first we generate a list with class numbers as elements. Then we calculate the decision function, for the input $\mathbf{x}$, corresponding to the first and last elements. Let these classes be $i$ and $j$ and $D_{ij}(\mathbf{x}) > 0$. We delete the element

---

[1] We may resolve the tie $D_{ij}(\mathbf{x}) = 0$ by $D_{ij}(\mathbf{x}) \geq 0$.

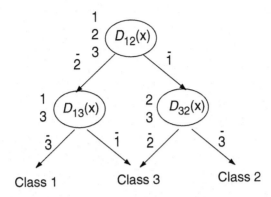

Fig. 3.9 Decision-tree-based pairwise classification

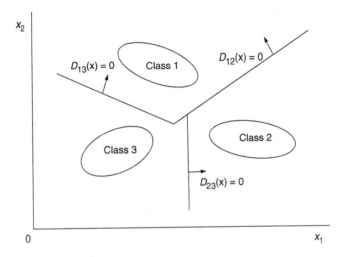

Fig. 3.10 Generalization region by decision-tree-based pairwise classification

$j$ from the list. We repeat the procedure until one element is left. Then we classify $\mathbf{x}$ into the class that corresponds to the element number. For Fig. 3.11, we generate the list $\{1, 3, 4, 2\}$. If $D_{12}(\mathbf{x}) > 0$, we delete element 2 from the list; we obtain $\{1, 3, 4\}$. Then if $D_{14}(\mathbf{x}) > 0$, we delete element 4 from the list; $\{1, 3\}$. If $D_{13}(\mathbf{x}) > 0$, we delete element 3 from the list. Because only one remains in the list, we classify $\mathbf{x}$ into Class 1.

Training of a DDAG is the same as conventional pairwise support vector machines, namely, we need to determine $n(n-1)/2$ decision functions for an $n$-class problem. The advantage of DDAGs is that classification is faster than by conventional pairwise support vector machines or pairwise fuzzy support

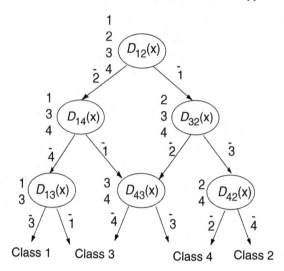

**Fig. 3.11** A DDAG for four classes

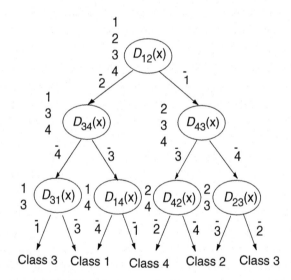

**Fig. 3.12** An extended DDAG equivalent to the ADAG shown in Fig. 3.13

vector machines. In a DDAG, classification can be done by calculating $(n-1)$ decision functions.

### 3.2.5.2 Adaptive Directed Acyclic Graphs

Pontil and Verri [5] proposed using rules of a tennis tournament to resolve unclassifiable regions. Not knowing their work, Kijsirikul and Ussivakul [7] proposed the same method and called it *adaptive directed acyclic graph (ADAG)*. For three-class problems, the ADAG is equivalent to the DDAG. Reconsider the example shown in Fig. 3.6. Let the first-round matches be {Class 1, Class 2} and {Class 3}. Then for an input **x**, in the first match, **x** is classified into Class 1 or Class 2, and in the second match **x** is classified into Class 3. Then the second-round match is either {Class 1, Class 3} or {Class 2, Class 3} according to the outcome of the first-round match. The resulting generalization regions for classes are the same as those shown in Fig. 3.10. Thus for a three-class problem there are three different ADAGs, each having an equivalent DDAG.

When there are more than three classes, the set of ADAGs is included in the set of extended DDAGs. Consider a four-class problem and let the ADAG be as shown in Fig. 3.13, namely, the first-round matches are {Class 1, Class 2} and {Class 3, Class 4}. An equivalent DDAG is shown in Fig. 3.12. The order of matches {Class 1, Class 2} and {Class 3, Class 4} is irrelevant, namely, they are independent. This can be realized in a DDAG by setting the match {Class 1, Class 2} at all the nodes of one level and the match {Class 3, Class 4} at all the nodes of another level of the tree. Here, we set the match {Class 1, Class 2} at the top of the tree. Thus we set the match {Class 3, Class 4} at the two nodes of the second level. The DDAG obtained by this method is an extension of the original DDAG. In this way, for an ADAG including $n$ classes, we can generate an equivalent DDAG.

However, classification cannot be done by the list processing discussed previously, namely, for the list $\{1, 3, 4, 2\}$ if $D_{12}(\mathbf{x}) > 0$, we delete element 2 and obtain $\{1, 3, 4\}$. Then we calculate $D_{14}(\mathbf{x})$, but this does not correspond to Fig. 3.12.

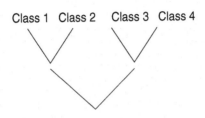

**Fig. 3.13** An ADAG for a four-class problem

Any ADAG can be converted to an equivalent DDAG, but the reverse is not true for $n \geq 4$. The DDAG shown in Fig. 3.11 cannot be converted to an ADAG because the second-level decision functions are different and do not constitute matches. But according to the computer simulations [28, 7], classification performance of the two methods is almost identical.

### 3.2.5.3 Optimizing Decision Trees

Classification by DDAGs or ADAGs is faster than by pairwise fuzzy support vector machines. But the problem is that the generalization ability depends on the structure of decision trees. To solve this problem for ADAGs, in [29], ADAGs are reordered so that the sum of $\|\mathbf{w}_{ij}\|$ associated with the leaf nodes is minimized. Here we discuss optimizing structures of DDAGs and ADAGs according to [9, 10].

In DDAGs, the unclassifiable regions are assigned to the classes associated with the leaf nodes. By the DDAG for the three-class problem shown in Fig. 3.9, the unclassifiable region is assigned to Class 3, as shown in Fig. 3.10, which corresponds to the leaf node $D_{32}(\mathbf{x})$.

Suppose that the decision boundaries for a four-class problem are given by Fig. 3.14. Then the unclassifiable regions by the conventional pairwise support vector machine are as shown in Fig. 3.15. The shaded regions show the unclassifiable regions, and the thick lines show class boundaries.

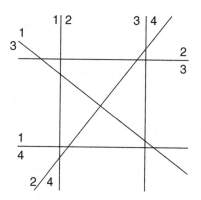

**Fig. 3.14** Decision boundaries for a four-class problem. Reprinted from [10, p. 124]

If the DDAG for the four-class problem is given by Fig. 3.11, the class boundaries are as shown in Fig. 3.16. Region A in the figure is classified into Class 3 by $D_{34}(\mathbf{x})$, which is at a leaf node of the DDAG.

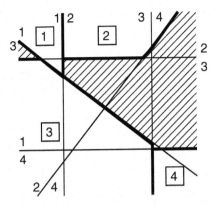

**Fig. 3.15** Unclassifiable regions for Fig. 3.14. Reprinted from [10, p. 125]

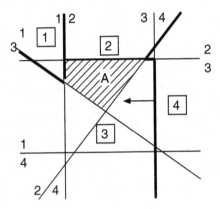

**Fig. 3.16** Class regions for Fig. 3.14 using the DDAG given by Fig. 3.11. Reprinted from [10, p. 125]

Because any ADAG is converted into a DDAG, the preceding discussions hold for ADAGs, namely, the unclassifiable regions are assigned to the classes associated with the leaf nodes of the equivalent DDAG.

The unclassifiable regions caused by the conventional pairwise classification are assigned to the classes associated with the leaf nodes of a DDAG. Thus, if we put the class pairs that are easily separated in the upper nodes, unclassifiable regions are assigned to the classes that are difficult to separate. This means that the class pairs that are difficult to separate are classified by the decision boundaries that are determined by these pairs.

In forming a DDAG or an ADAG, we need to train support vector machines for all pairs of classes. Thus, to determine the optimal structure, we can use any of the measures that are developed for estimating the generalization ability (see Section 2.8.2).

Therefore, the algorithm to determine the DDAG structure for an $n$-class problem is as follows:

1. Generate the initial list: $\{1, \ldots, n\}$.
2. If there are no generated lists, terminate the algorithm. Otherwise, select a list and select the class pair $(i, j)$ with the highest generalization ability from the list.
3. If the list selected at Step 2 has more than two elements, generate two lists deleting $i$ or $j$ from the list. Go to Step 2.

Figure 3.17 shows an example of a four-class problem. First we generate the list $\{1, 2, 3, 4\}$. At the top level we select a pair of classes that has the highest generalization ability from Classes 1 to 4. Let them be Classes 1 and 2. Then we generate the two lists $\{2, 3, 4\}$ and $\{1, 3, 4\}$. We iterate this procedure for the two lists.

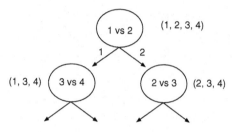

**Fig. 3.17** Determination of a DDAG for a four-class problem. Reprinted from [10, p. 126]

This procedure determines the structure off-line. We can determine the structure while classifying $\mathbf{x}$ as follows:

1. Generate the initial list: $\{1, \ldots, n\}$.
2. Select the class pair $(i, j)$ with the highest generalization ability from the list. Let $\mathbf{x}$ be on the Class $i$ side of the decision function. Delete $j$ from the list. Otherwise, delete $i$.
3. If the list selected at Step 2 has more than one element, go to Step 2. Otherwise, classify $\mathbf{x}$ into the class associated with the element and terminate the algorithm.

Because any ADAG is converted into an equivalent DDAG, we can determine the tree structure selecting class pairs with the highest generalization ability. Figure 3.18 shows an example of an eight-class problem. At the first level, we select the class pair with the highest generalization ability; let them be Classes 1 and 4. We iterate the procedure for the remaining classes and let the class pairs be Classes 6 and 7, 2 and 5, and 3 and 8. In the second level, the pairs of classes are determined according to the input $\mathbf{x}$. Let the candidate classes be 1, 6, 5, and 3 for $\mathbf{x}$. Then, we choose the pair of classes with the highest generalization ability; let it be (1, 5). Then the remaining pair is (3, 6). We iterate the procedure until **x** is classified into a definite class.

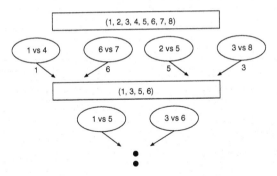

**Fig. 3.18** Determination of an ADAG for a four-class problem. Reprinted from [10, p. 126]

#### 3.2.5.4 Performance Evaluation

Because the generalization abilities of the ADAGs and DDAGs do not differ very much, in the following we show only the results for DDAGs. As the measure for estimating the generalization ability, we evaluated the VC dimension given by (2.317), the LOO error estimator given by (2.318), and Joachims' $\xi\alpha$ LOO estimator given by (2.319), but there was not much difference [10]. Therefore, in the following we show the results using the LOO error estimator given by (2.318), namely the number of support vectors divided by the number of training data.

We compared the maximum, minimum, and average recognition rates of the test data for DDAGs and the recognition rate of the pairwise fuzzy support vector machine with minimum operators, using the data sets listed in Table 1.3. We used the polynomial kernel with degree 3 and RBF kernel with $\gamma = 1$. The input range was normalized into $[0, 1]$. We trained support vector machines by the primal–dual interior-point method combined with the decomposition technique. For the thyroid and MNIST data sets we set $C = 10,000$ and for other data sets $C = 1,000$. We used an Athlon MP 2000 personal computer.

Table 3.3 shows the recognition rates of the test data for the conventional pairwise support vector machine (SVM), pairwise fuzzy support vector machine (FSVM), and DDAGs. Column "OPT" lists the recognition rate for the optimum DDAG. To compare the recognition rate of the DDAGs, the maximum, minimum, and average recognition rates for the DDAGs are also listed. In column "Kernel," for instance, $d3$ denotes the polynomial kernel with degree 3 and $\gamma1$ denotes the RBF kernel with $\gamma = 1$.

The number of DDAGs for a three-class problem is 3, but it explodes as $n$ increases. Thus, except for the iris and thyroid data, we randomly generated 10,000 DDAGs and calculated the maximum, minimum, and average recognition rates of the test data.

From the table, for the iris data, all the recognition rates are the same; this means that there are no test data in unclassifiable regions. The recognition rate of the optimum DDAG is comparable to the average recognition rate of DDAGs; in 9 cases out of 16, the recognition rates of the optimum DDAGs are better than or equal to the average recognition rates.

The recognition rates of FSVMs are equal to or better than the average recognition rates of DDAGs and comparable to the maximum recognition rates of DDAGs.

**Table 3.3** Performance of pairwise support vector machines

| Data | Kernel | SVM (%) | FSVM (%) | DDAG | | | |
|---|---|---|---|---|---|---|---|
| | | | | Max. (%) | Min. (%) | Ave (%) | OPT (%) |
| Iris | $d3$ | 93.33 | 93.33 | 93.33 | 93.33 | 93.33 | 93.33 |
| | $\gamma1$ | 97.33 | 97.33 | 97.33 | 97.33 | 97.33 | 97.33 |
| Numeral | $d3$ | 99.76 | 100 | 100 | 99.76 | 99.84 | 99.76 |
| | $\gamma1$ | 99.63 | 99.63 | 99.88 | 99.63 | 99.72 | 99.88 |
| Thyroid | $d3$ | 97.81 | 97.87 | 97.89 | 97.81 | 97.86 | 97.81 |
| | $\gamma1$ | 97.26 | 97.37 | 97.40 | 97.26 | 97.34 | 97.26 |
| Blood cell | $d3$ | 92.07 | 92.77 | 93.00 | 92.32 | 92.47 | 92.65 |
| | $\gamma1$ | 91.93 | 92.23 | 92.41 | 91.84 | 92.08 | 92.00 |
| Hiragana-50 | $d3$ | 98.57 | 98.89 | 99.37 | 98.52 | 98.75 | 98.61 |
| | $\gamma1$ | 98.37 | 98.85 | 99.22 | 98.29 | 98.56 | 98.50 |
| Hiragana-105 | $d3$ | 100 | 100 | 100 | 100 | 100 | 100 |
| | $\gamma1$ | 99.98 | 100 | 100 | 99.98 | 99.99 | 99.99 |
| Hiragana-13 | $d3$ | 99.63 | 99.66 | 99.72 | 99.58 | 99.64 | 99.66 |
| | $\gamma1$ | 99.61 | 99.65 | 99.70 | 99.55 | 99.60 | 99.63 |
| MNIST | $d3$ | 97.85 | 98.01 | 97.99 | 97.91 | 97.96 | 97.91 |
| | $\gamma1$ | 97.18 | 97.78 | 97.49 | 97.31 | 97.41 | 97.54 |

For pairwise classification, training time is the same for fuzzy SVMs and DDAGs, but classification time of DDAGs is faster. Table 3.4 lists classification time of the blood cell and hiragana-50 test data sets. DDAGs are much faster than SVMs and FSVMs.

**Table 3.4** Classification time comparison

| Data | Kernel | DDAG (s) | SVM (s) | FSVM (s) |
|------|--------|----------|---------|----------|
| Blood cell | $d3$ | 0.9 | 1.2 | 2.4 |
| | $\gamma1$ | 0.7 | 2.6 | 5.2 |
| Hiragana-50 | $d3$ | 15 | 66 | 114 |
| | $\gamma1$ | 17 | 76 | 159 |

## 3.2.6 Pairwise Classification with Correcting Classifiers

In pairwise classification, let the classifier for classes $i$ and $j$ generate the estimate of the probability that the input $\mathbf{x}$ belongs to class $i$, $p_{ij}(\mathbf{x})$, where $p_{ji}(\mathbf{x}) = 1 - p_{ij}(\mathbf{x})$. Then we can estimate the probability that $\mathbf{x}$ belongs to class $i$, $p_i(\mathbf{x})$, by

$$p_i(\mathbf{x}) = \frac{2}{n\,(n-1)} \sum_{\substack{j \neq i, \\ j=1}}^{n} p_{ij}(\mathbf{x}), \qquad (3.43)$$

where $2/(n\,(n-1))$ is to ensure $\sum_{i=1,\ldots,n} p_i(\mathbf{x}) = 1$, and we can classify $\mathbf{x}$ into

$$\arg\max_i p_i(\mathbf{x}). \qquad (3.44)$$

This classification method is called *pairwise coupling classification* [30]. Because $p_i(\mathbf{x})$ is continuous, unclassifiable regions do not occur.

In pairwise coupling classification, $p_{ij}(\mathbf{x})$ is estimated using the training data belonging to classes $i$ and $j$. Thus if $\mathbf{x}$ belongs to class $k$, a large value of $p_{ij}(\mathbf{x})$ or $p_{ji}(\mathbf{x})$ $(k \neq i, j)$ may mislead classification. To avoid this situation, Moreira and Mayoraz [31] introduced correcting classifiers. In addition to a pairwise classifier that separates class $i$ from class $j$, we generate a classifier that separates classes $i$ and $j$ from the remaining classes. Let the output of the classifier be $q_{ij}(\mathbf{x})$, where $q_{ij}(\mathbf{x})$ is the probability that $\mathbf{x}$ belongs to class $i$ or $j$. Then we calculate $p_i(\mathbf{x})$ by

$$p_i(\mathbf{x}) = \frac{2}{n\,(n-1)} \sum_{\substack{j \neq i, \\ j=1}}^{n} p_{ij}(\mathbf{x})\, q_{ij}(\mathbf{x}). \qquad (3.45)$$

By multiplying $p_{ij}(\mathbf{x})$ by $q_{ij}(\mathbf{x})$, $p_i(\mathbf{x})$ is decreased when $\mathbf{x}$ does not belong to class $i$. Thus the problem with pairwise coupling classification can be avoided.

According to the computer experiments [31], the generalization ability of pairwise coupling with correcting classifiers was better than that of pairwise coupling. In addition, if correcting classifiers alone were used as classifiers, their performance was comparable with that of pairwise coupling with correcting classifiers.

In correcting classifiers, there are $_nC_2$ decision functions. For three-class classification, separating two classes from the remaining class is equivalent to separating one class from the remaining classes, namely one-against-all classification. The target values for four classes are shown in a matrix form as follows:

$$
\begin{vmatrix}
1 & 1 & -1 & -1 \\
1 & -1 & 1 & -1 \\
1 & -1 & -1 & 1 \\
-1 & 1 & 1 & -1 \\
-1 & 1 & -1 & 1 \\
-1 & -1 & 1 & 1
\end{vmatrix},
\tag{3.46}
$$

where the $i$th row corresponds to the $i$th decision function and $j$th column, class $j$. Each decision function has a complementary counterpart. For example, the first and sixth rows are complementary. But for $n > 4$, each decision function is distinct. Let the target value of class $i$ for the $j$th decision function be $g_{ij}$ and the output of the $j$th decision function be $D_j(\mathbf{x})$. Then input $\mathbf{x}$ is classified into class

$$
\arg\max_i \sum_{j=1}^{nC_2} a(i,j),
\tag{3.47}
$$

where

$$
a(i,j) = \begin{cases}
D_j(\mathbf{x}) & \text{for } g_{ij} = 1, \\
1 - D_j(\mathbf{x}) & \text{for } g_{ij} = -1.
\end{cases}
\tag{3.48}
$$

To apply pairwise coupling to support vector machines we need to calculate $p_{ij}(\mathbf{x})$. In [30], $p_{ij}(\mathbf{x})$ and $p_{ji}(\mathbf{x})$ are determined by approximating the normal distributions in the direction orthogonal to the decision function. In [32], in addition to training support vector machines, sigmoid functions are trained [33] (see Section 4.12).

Pairwise coupling support vector machines and pairwise fuzzy support vector machines do not differ very much, either posteriori probabilities or degrees of membership are used. Thus, the idea of correcting classifiers can be readily introduced to pairwise fuzzy support vector machines. But we need to train correcting classifiers using all the training data. This leads to long training times compared to pairwise classification.

## 3.3 Error-Correcting Output Codes

Error-correcting codes, which detect and correct errors in data transmission channels, are used to encode classifier outputs to improve generalization ability. The codes are called *error-correcting output codes (ECOCs)*. For support vector machines, in addition to generalization improvement they can be used to resolve unclassifiable regions. In this section, we first discuss how error-correcting codes can be used for pattern classification. Next, by introducing "don't care" outputs, we discuss a unified scheme for output coding that includes one-against-all and pairwise formulations. Then we show the equivalence of the error-correcting codes with the membership functions. Finally, we compare performance of ECOC support vector machines with one-against-all support vector machines.

### 3.3.1 Output Coding by Error-Correcting Codes

Dietterich and Bakiri [11] proposed using error-correcting output codes for multiclass problems. Let $g_{ij}$ be the target value of the $j$th decision function $D_j(\mathbf{x})$ for class $i$ $(i = 1, \ldots, n)$:

$$g_{ij} = \begin{cases} 1 & \text{if } D_j(\mathbf{x}) > 0 \text{ for class } i, \\ -1 & \text{otherwise.} \end{cases} \qquad (3.49)$$

The $j$th column vector $\mathbf{g}_j = (g_{1j}, \ldots, g_{nj})^\top$ is the target vector for the $j$th decision function. If all the elements of a column are 1 or $-1$, classification is not performed by this decision function and two column vectors with $\mathbf{g}_i = -\mathbf{g}_j$ result in the same decision function. Thus the maximum number of distinct decision functions is $2^{n-1} - 1$.

The $i$th row vector $(g_{i1}, \ldots, g_{ik})$ corresponds to a code word for class $i$, where $k$ is the number of decision functions. In error-correcting codes, if the minimum Hamming distance between pairs of code words is $h$, the code can correct at least $\lfloor (h - 1)/2 \rfloor$-bit errors, where $\lfloor a \rfloor$ gives the maximum integer value that does not exceed $a$. For three-class problems, there are three decision functions in maximum as shown in Table 3.5, which is equivalent to one-against-all formulation, and there is no error-correcting capability. Thus ECOC is considered to be a variant of one-against-all classification.

**Table 3.5** Error-correcting codes for three classes (one-against-all)

| Class | $\mathbf{g}_1$ | $\mathbf{g}_2$ | $\mathbf{g}_3$ |
|-------|------|------|------|
| 1 | 1 | −1 | −1 |
| 2 | −1 | 1 | −1 |
| 3 | −1 | −1 | 1 |

## 3.3.2 Unified Scheme for Output Coding

Introducing "don't care" outputs, Allwein et al. [34] unified output codes that include one-against-all, pairwise, and ECOC schemes. Denoting a "don't care" output by 0, pairwise classification for three classes can be shown as in Table 3.6.

**Table 3.6** Extended error-correcting codes for pairwise classification with three classes

| Class | $g_1$ | $g_2$ | $g_3$ |
|-------|-------|-------|-------|
| 1     | 1     | 0     | -1    |
| 2     | -1    | 1     | 0     |
| 3     | 0     | -1    | 1     |

To calculate the distance of $\mathbf{x}$ from the $j$th decision function for class $i$, we define the error $\varepsilon_{ij}(\mathbf{x})$ by

$$\varepsilon_{ij}(\mathbf{x}) = \begin{cases} 0 & \text{for } g_{ij} = 0, \\ \max(1 - g_{ij}D_j(\mathbf{x}), 0) & \text{otherwise.} \end{cases} \tag{3.50}$$

If $g_{ij} = 0$, we need to skip this case. Thus, $\varepsilon_{ij}(\mathbf{x}) = 0$. If $g_{ij}D_j(\mathbf{x}) \geq 1$, $\mathbf{x}$ is on the correct side of the $j$th decision function with more than or equal to the margin of 1. Thus, $\varepsilon_{ij}(\mathbf{x}) = 0$. If $g_{ij}D_j(\mathbf{x}) < 1$, $\mathbf{x}$ is on the wrong side or even if it is on the correct side, the margin is smaller than 1. We evaluate this disparity by $1 - g_{ij}D_i(\mathbf{x})$.

Then the distance of $\mathbf{x}$ from class $i$ is given by

$$d_i(\mathbf{x}) = \sum_{j=1}^{k} \varepsilon_{ij}(\mathbf{x}). \tag{3.51}$$

Using (3.51), $\mathbf{x}$ is classified into

$$\arg \min_{i=1,\dots,n} d_i(\mathbf{x}). \tag{3.52}$$

Instead of (3.50), if we use the discrete function

$$\varepsilon_{ij}(\mathbf{x}) = \begin{cases} 0 & \text{for } g_{ij} = 0, \\ 0 & \text{for } g_{ij} = \pm 1, \ g_{ij}D_i(\mathbf{x}) > 0, \\ 1 & \text{otherwise;} \end{cases} \tag{3.53}$$

(3.51) gives the Hamming distance. But by this formulation, as seen in Sections 3.1 and 3.2, unclassifiable regions occur.

### 3.3.3 Equivalence of ECOC with Membership Functions

Here we discuss the relationship between ECOC and membership functions. For $g_{ij} = \pm 1$, the error $\varepsilon_{ij}(\mathbf{x})$ is expressed by the one-dimensional membership function $m_{ij}(\mathbf{x})$:

$$
\begin{aligned}
m_{ij}(\mathbf{x}) &= \min(g_{ij} D_j(\mathbf{x}), 1) \\
&= 1 - \varepsilon_{ij}(\mathbf{x}).
\end{aligned}
\tag{3.54}
$$

Thus, if we define the membership function for class $i$ by

$$
m_i(\mathbf{x}) = \frac{1}{\sum_{j=1}^{k} |g_{ij}|} \sum_{\substack{g_{ij} \neq 0, \\ j=1}}^{k} m_{ij}(\mathbf{x})
\tag{3.55}
$$

and classify $\mathbf{x}$ into

$$
\arg \max_{i=1,\dots,n} m_i(\mathbf{x}),
\tag{3.56}
$$

we obtain the same recognition result as that by (3.52). This is equivalent to a fuzzy support vector machine with average operators.

Similarly, instead of (3.51), if we use

$$
d_i(\mathbf{x}) = \max_{j=1,\dots,n} \varepsilon_{ij}(\mathbf{x}),
\tag{3.57}
$$

the resulting classifier is equivalent to a fuzzy support vector machine with minimum operators.

According to the discussions in Section 3.1.3, one-against-all fuzzy support vector machines with average and minimum operators give the same decision boundaries. For pairwise classification, they are different, but according to Section 3.2 the difference is small.

### 3.3.4 Performance Evaluation

We compared recognition performance of ECOC support vector machines with one-against-all support vector machines using the blood cell data and hiragana-50 data listed in Table 1.3 [35, 22]. As error-correcting codes we used the BCH (Bose–Chaudhuri–Hochquenghem) codes, which belong to one type of cyclic codes. We used four BCH codes with 15, 31, 63, and 127 word lengths, properly setting the minimum Hamming distances. For each word length we randomly assigned the class labels to generated codes 10 times, and using the assigned codes we trained 10 ECOC support vector machines with $C = 5,000$.

**Table 3.7** Performance of blood cell data with polynomial kernels ($d = 3$)

| Code | Hamming (%) | Average (%) | Minimum (%) |
|---|---|---|---|
| One-against-all | 87.13 (92.41) | **92.84** (96.09) | **92.84** (96.09) |
| (15, 5, 7) | 90.17 (93.34) | 91.56 (94.45) | 91.19 (93.95) |
| (31, 11, 11) | 90.86 (93.60) | 91.90 (94.59) | 91.80 (94.16) |
| (63, 7, 31) | **91.82** (94.64) | 92.20 (94.98) | 92.23 (94.32) |
| (127, 8, 63) | 91.80 (94.58) | 92.01 (94.82) | 91.93 (96.09) |

Table 3.7 shows the results for the blood cell data with polynomial kernels with degree 3. In the "Code" column, e.g., (15, 5, 7) means that the word length is 15 bits, the number of information bits is 5, and the minimum Hamming distance is 7. The "Hamming," "Average," and "Minimum" columns list the average recognition rates of the test and the training data (in parentheses) using the Hamming distance, the average operator, and the minimum operator, respectively. The numeral in boldface shows the maximum recognition rate among the different codes.

From the table, using the Hamming distance, the recognition rates of both training and test data improved as the word length was increased, and they reached the maximum at the word length of 63. But because by the Hamming distance unclassifiable regions existed, the recognition rates were lower than by average and minimum operators. By the average and minimum operators, however, the one-against-all support vector machines showed the best recognition rates. This may be caused by the lower recognition rates of the training data by the ECOC support vector machines than by the one-against-all support vector machines.

Thus, to improve the recognition rate of the test data, we used the RBF kernels. Table 3.8 shows the results for the RBF kernels with $\gamma = 1$. The ECOC support vector machines showed better recognition performance than the one-against-all support vector machines. In addition, the average operator showed better recognition performance than the minimum operator.

Table 3.9 shows the results of the hiragana-50 data for the polynomial kernels with degree 3. The ECOC support vector machine with the average

**Table 3.8** Performance of blood cell data with RBF kernels ($\gamma = 1$)

| Code | Hamming (%) | Average (%) | Minimum (%) |
|---|---|---|---|
| One-against-all | 86.68 (98.58) | 92.94 (99.29) | 92.94 (99.29) |
| (15, 5, 7) | 92.43 (98.27) | 93.47 (98.49) | 93.07 (98.18) |
| (31, 11, 11) | 92.88 (98.36) | 93.85 (98.59) | 93.53 (98.13) |
| (63, 10, 27) | **93.68** (98.64) | **94.05** (98.68) | **93.75** (98.37) |
| (127, 15, 55) | **93.68** (98.60) | 93.96 (98.61) | 93.63 (97.94) |

**Table 3.9** Performance of hiragana-50 data with polynomial kernels $(d = 3)$

| Code | Hamming (%) | Average (%) | Minimum (%) |
|---|---|---|---|
| One-against-all | 97.55 (100) | 99.28 (100) | **99.28** (100) |
| (15, 5, 7) | 95.50 (99.96) | 97.63 (99.85) | 96.93 (99.97) |
| (31, 11, 11) | 98.38 (99.99) | 99.01 (100) | 98.56 (99.97) |
| (63, 7, 31) | 99.01 (100) | 99.30 (100) | 99.17 (99.97) |
| (127, 8, 63) | **99.31** (100) | **99.46** (100) | 99.26 (99.97) |

operator and the word length of 127 showed the best recognition performance. But some ECOC support vector machines showed lower recognition performance than the one-against-all support vector machine. Thus the performance of ECOC support vector machines was not stable.

From the preceding computer experiments, it is shown that ECOC support vector machines do not always perform better than one-against-all support vector machines. Thus, to obtain good recognition performance, we need to optimize the structure of ECOC support vector machines [36, 37]. To do this, Pérez-Cruz and Artés-Rodríguez [36] proposed an iterative pruning of worst classifier from redundant classifiers.

## 3.4 All-at-Once Support Vector Machines

In this section, we resolve unclassifiable regions of multiclass problems by determining all the decision functions simultaneously [12–14, 20, 38–40, 15, 41]. In [42, pp. 174–176], a multiclass problem is converted into a two-class problem by expanding the $m$-dimensional input data into $(m \times n)$-dimensional data.[2] Here, we do not use this method. We discuss an all-at-once support vector machine that uses $(M \times n)$ slack variables [12].

For an $n$-class problem, we define the decision function for class $i$ by

$$D_i(\mathbf{x}) = \mathbf{w}_i^\top \phi(\mathbf{x}) + b_i, \qquad (3.58)$$

where $\mathbf{w}_i$ is the weight vector for class $i$ in the feature space, $\phi(\mathbf{x})$ is the mapping function, and $b_i$ is the bias term. For class $i$ data $\mathbf{x}$ to be correctly classified, $D_i(\mathbf{x})$ needs to be the largest among $D_j(\mathbf{x})$ $(j = 1, \ldots, n)$, namely, the following inequalities must hold:

$$\mathbf{w}_i^\top \phi(\mathbf{x}) + b_i > \mathbf{w}_j^\top \phi(\mathbf{x}) + b_j \quad \text{for } j \neq i, j = 1, \ldots, n. \qquad (3.59)$$

---

[2] A similar method is discussed in [43].

We consider determining the $n$ decision functions at the same time so that the margins between classes are maximized. The resulting L1 soft-margin support vector machine is as follows:

$$\text{minimize} \quad Q(\mathbf{w}, \mathbf{b}, \boldsymbol{\xi}) = \frac{1}{2} \sum_{i=1}^{n} \|\mathbf{w}_i\|^2 + C \sum_{i=1}^{M} \sum_{\substack{j \neq y_i, \\ j=1}}^{n} \xi_{ij} \quad (3.60)$$

$$\text{subject to} \quad (\mathbf{w}_{y_i} - \mathbf{w}_j)^\top \boldsymbol{\phi}(\mathbf{x}_i) + b_{y_i} - b_j \geq 1 - \xi_{ij}$$
$$\text{for } j \neq y_i, j = 1, \ldots, n, i = 1, \ldots, M, \quad (3.61)$$

where $y_i (\in \{1, \ldots, n\})$ is the class label for $\mathbf{x}_i$, $C$ is the margin parameter that determines the trade-off between the maximization of the margin and minimization of the classification error, $\xi_{ij} (\geq 0)$ is the slack variable associated with $\mathbf{x}_i$ and class $j$, $\boldsymbol{\xi} = (\ldots, \xi_{ij}, \ldots)^\top$, $\mathbf{w} = (\mathbf{w}_1, \ldots, \mathbf{w}_n)$, and $\mathbf{b} = (b_1, \ldots, b_n)^\top$. The first term in the objective function gives the reciprocal of the sum of the squared margins and the second term is concerned with the classification error.

Introducing the nonnegative Lagrange multipliers $\alpha_{ij}$ and $\beta_{ij}$, we obtain

$$Q(\mathbf{w}, \mathbf{b}, \boldsymbol{\xi}, \boldsymbol{\alpha}, \boldsymbol{\beta}) = \frac{1}{2} \sum_{i=1}^{n} \|\mathbf{w}_i\|^2 + C \sum_{i=1}^{M} \sum_{\substack{j \neq y_i, \\ j=1}}^{n} \xi_{ij}$$
$$- \sum_{i=1}^{M} \sum_{\substack{j \neq y_i, \\ j=1}}^{n} \alpha_{ij} \left( (\mathbf{w}_{y_i} - \mathbf{w}_j)^\top \boldsymbol{\phi}(\mathbf{x}_i) + b_{y_i} - b_j - 1 + \xi_{ij} \right)$$
$$- \sum_{i=1}^{M} \sum_{\substack{j \neq y_i, \\ j=1}}^{n} \beta_{ij} \, \xi_{ij}$$
$$= \frac{1}{2} \sum_{i=1}^{n} \|\mathbf{w}_i\|^2 - \sum_{i=1}^{M} \sum_{j=1}^{n} z_{ij} \left( \mathbf{w}_j^\top \boldsymbol{\phi}(\mathbf{x}_i) + b_j - 1 \right)$$
$$- \sum_{i=1}^{M} \sum_{\substack{j \neq y_i, \\ j=1}}^{n} (\alpha_{ij} + \beta_{ij} - C) \xi_{ij}, \quad (3.62)$$

where

$$z_{ij} = \begin{cases} \displaystyle\sum_{\substack{k \neq y_i, \\ k=1}}^{n} \alpha_{ik} & \text{for } j = y_i, \\ -\alpha_{ij} & \text{otherwise.} \end{cases} \quad (3.63)$$

The conditions of optimality are given by

$$\frac{\partial Q(\mathbf{w}, \mathbf{b}, \boldsymbol{\xi}, \boldsymbol{\alpha}, \boldsymbol{\beta})}{\partial \mathbf{b}} = \mathbf{0}, \quad (3.64)$$

$$\frac{\partial Q(\mathbf{w}, \mathbf{b}, \boldsymbol{\xi}, \boldsymbol{\alpha}, \boldsymbol{\beta})}{\partial \mathbf{w}} = \mathbf{0}, \tag{3.65}$$

$$\frac{\partial Q(\mathbf{w}, \mathbf{b}, \boldsymbol{\xi}, \boldsymbol{\alpha}, \boldsymbol{\beta})}{\partial \boldsymbol{\xi}} = \mathbf{0}, \tag{3.66}$$

$$\alpha_{ij} \left( (\mathbf{w}_{y_i} - \mathbf{w}_j)^\top \phi(\mathbf{x}_i) + b_{y_i} - b_j - 1 + \xi_{ij} \right) = 0$$
$$\text{for } j \neq y_i, j = 1, \ldots, n, i = 1, \ldots, M, \tag{3.67}$$

$$\beta_{ij} \, \xi_{ij} = 0 \quad \text{for } j \neq y_i, j = 1, \ldots, n, i = 1, \ldots, M, \tag{3.68}$$

where (3.67) and (3.68) are the KKT (complementarity) conditions.
Using (3.62), (3.64), (3.65), and (3.66) reduce, respectively, to

$$\sum_{i=1}^{M} z_{ij} = 0 \quad \text{for } j = 1, \ldots, n, \tag{3.69}$$

$$\mathbf{w}_j = \sum_{i=1}^{M} z_{ij} \, \phi(\mathbf{x}_i) \quad \text{for } j = 1, \ldots, n, \tag{3.70}$$

$$\alpha_{ij} + \beta_{ij} = C, \quad \alpha_{ij} \geq 0, \quad \beta_{ij} \geq 0$$
$$\text{for } i = 1, \ldots, M, \quad j \neq y_i, j = 1, \ldots, n. \tag{3.71}$$

Thus we obtain the following dual problem:

$$\text{maximize} \quad Q(\boldsymbol{\alpha}) = \sum_{\substack{i=1 \\ }}^{M} \sum_{\substack{j \neq y_i, \\ j=1}}^{n} \alpha_{ij} - \frac{1}{2} \sum_{i,k=1}^{M} \sum_{j=1}^{n} z_{ij} \, z_{kj} \, K(\mathbf{x}_i, \mathbf{x}_j) \tag{3.72}$$

$$\text{subject to} \quad \sum_{i=1}^{M} z_{ij} = 0 \quad \text{for } j = 1, \ldots, n, \tag{3.73}$$

$$0 \leq \alpha_{ij} \leq C \quad \text{for } i = 1, \ldots, M, j \neq y_i, j = 1, \ldots, n. \tag{3.74}$$

The decision function is given by

$$D_i(\mathbf{x}) = \sum_{j=1}^{M} z_{ji} \, K(\mathbf{x}_j, \mathbf{x}) + b_i. \tag{3.75}$$

Because $\alpha_{ji}$ are nonzero for the support vectors, the summation in (3.75) is added only for nonzero $z_{ji}$.

Then the data sample $\mathbf{x}$ is classified into the class

$$\arg \max_{i=1,\ldots,n} D_i(\mathbf{x}). \tag{3.76}$$

If the maximum is reached for plural classes, the data sample is on the class boundary and is unclassifiable.

## 3.5 Comparisons of Architectures

We have discussed four types of support vector machines: one-against-all, pairwise, ECOC, and all-at-once support vector machines. In this section we summarize their characteristics and compare their trainability.

### 3.5.1 One-Against-All Support Vector Machines

To resolve unclassifiable regions of the original one-against-all support vector machines with discrete decision functions, the following extensions have been discussed in this chapter:

1. Support vector machines with continuous decision functions. By taking the maximum value among decision functions, unclassifiable regions are resolved.
2. Fuzzy support vector machines with minimum and average operators are proved to be equivalent to support vector machines with continuous decision functions.
3. Decision-tree-based support vector machines. Training of the preceding two types of support vector machines is the same as of support vector machines with discrete decision functions. But for an $n$-class problem, because a decision-tree-based support vector machine determines $n - 1$ decision functions using smaller numbers of training data, training is faster.

For the data sets tested, an improvement of generalization ability of the fuzzy support vector machines over support vector machines with discrete decision functions was large.

### 3.5.2 Pairwise Support Vector Machines

By pairwise support vector machines with voting schemes unclassifiable regions are reduced but they still exist. Thus to resolve unclassifiable regions the following extensions have been discussed:

1. Fuzzy support vector machines with minimum and average operators. The improvement by introducing the membership functions was smaller than for one-against-all support vector machines because the size of unclassifiable regions is smaller. In addition, there is not much difference of recognition rates using minimum and average operators.
2. Decision-tree-based support vector machines: decision directed acyclic graphs (DDAGs) and adaptive directed acyclic graphs (ADAGs). DDAGs are more general than ADAGs but the difference of generalization ability

is very small. The generalization ability of fuzzy support vector machines is better than the average generalization ability of DDAGs and ADAGs, but the difference is small for the tested data sets. DDAGs and ADAGs are suited for the problem where high-speed classification is necessary. Optimization of DDAGs and ADAGs was discussed, and the recognition rates of the optimized DDAGs were comparable or better than the average recognition rates of DDAGs.

### 3.5.3 ECOC Support Vector Machines

Error-correcting output codes are introduced to improve generalization ability of support vector machines. By ECOC, resolution of unclassifiable regions is also achieved. ECOC support vector machines are equivalent to fuzzy support vector machines with average operators.

ECOC support vector machines are an extension of one-against-all support vector machines and by introducing "don't care" outputs, any classification scheme including pairwise classification can be realized.

According to the performance evaluation, ECOC support vector machines did not always outperform one-against-all fuzzy support vector machines, and there was an optimal code word length. Thus, optimization of ECOC structures is necessary.

### 3.5.4 All-at-Once Support Vector Machines

By determining all the decision functions at once, unclassifiable regions are resolved. But the number of variables is $M \times (n-1)$, where $M$ is the number of training data and $n$ is the number of classes. Thus training becomes difficult as the number of training data becomes large.

### 3.5.5 Training Difficulty

Table 3.10 summarizes the characteristics of four types of support vector machines from the number of decision functions to be determined for $n$ class problems, the number of variables solved simultaneously for the $M$ training data, and the number of equalities in the dual optimization problem. In the table, $M_i$ is the number of training data belonging to class $i$ and $k$ is the length of code words.

In the following we discuss the difference of support vector machines from the standpoint of separability. We say that training data are separable by

**Table 3.10** Comparison of support vector machines

|                    | One-against-all | Pairwise    | ECOC | All-at-once |
|--------------------|-----------------|-------------|------|-------------|
| Decision functions | $n$             | $n(n-1)/2$  | $k$  | $n$         |
| Variables          | $M$             | $M_i + M_j$ | $M$  | $M(n-1)$    |
| Equalities         | 1               | 1           | 1    | $n$         |

a support vector machine if each decision function for the support vector machine separates training data in the feature space. This means that training data are separated correctly 100% by the support vector machine.

*Example 3.2.* Consider a one-dimensional case shown in Fig. 3.19, where Class 1 is in $(a, b)$, Class 2 is in $(-\infty, a)$, and Class 3 is in $(b, \infty)$. We consider linear decision functions for three classes [44].

Because Class 1 is not separated from Classes 2 and 3 by a linear decision function, the problem is not separable by one-against-all formulation. But by pairwise formulation, by setting

$$g_{12}(x) = x - a, \tag{3.77}$$

$$g_{13}(x) = -x + b, \tag{3.78}$$

$$g_{23}(x) = -x + c, \tag{3.79}$$

where $b > c > a$, the problem is separable.

By all-at-once formulation, the problem is also separable. Figure 3.19 shows an example of $g_i(x)$. Because $g_1(x) > g_2(x)$ and $g_1(x) > g_3(x)$ in the interval $(a, b)$, the data in this interval are classified into Class 1.

Therefore, the separation power of one-against-all formulation is lower than pairwise or all-at-once formulation.

**Theorem 3.1.** *If training data are separable by a one-against-all support vector machine, the training data are separable by a pairwise support vector machine. But the reverse is not always true.*

*Proof.* Suppose for an $n$-class problem, training data are separable by a one-against-all support vector machine. Then class $i$ is separated from the set of classes $\{j \mid j \neq i, j = 1, \ldots, n\}$. This means that class $i$ is separated from class $j$ $(j \neq i, j = 1, \ldots, n)$. Therefore, any pair of classes $i$ and $j$ is separable in the feature space.

We prove that the reverse does not hold by a counterexample. Example 3.2 shows one counterexample, but here we consider a three-class case shown in Fig. 3.20, where each pair of classes is separated linearly. Thus the training data are separable by a pairwise support vector machine. But as shown in Fig. 3.21, the training data are not separable by a one-against-all support vector machine. Thus the reverse is not always true. ∎

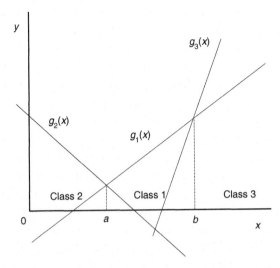

**Fig. 3.19** Class boundaries for a one-dimensional case by all-at-once formulation

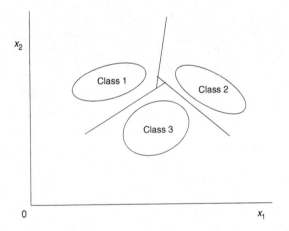

**Fig. 3.20** Class boundaries by a pairwise support vector machine

Theorem 3.1 tells us that for a given kernel, training data are more difficult to separate by one-against-all support vector machines than by pairwise support vector machines. In addition, for one-against-all support vector machines, the number of training data for determining a decision function is the total number of training data compared to the sum of two-class training data for pairwise support vector machines. Thus, training of a one-against-all support vector machine is more difficult than that of a pairwise support vector machine.

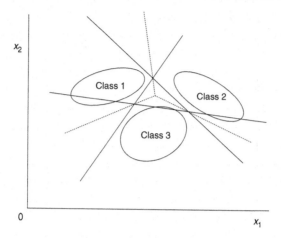

**Fig. 3.21** Class boundaries by a one-against-all support vector machine

The above discussions do not mean that the generalization ability of pairwise support vector machines is better than that of one-against-all support vector machines. According to the computer experiments shown in Table 3.2, they are compatible. This is also claimed in [45] by an extensive computer experiment using the data sets from the UCI repository.

**Theorem 3.2.** *If training data are separable by an ECOC support vector machine, the training data are separable by a pairwise support vector machine. But the reverse is not always true.*

*Proof.* Suppose for an $n$-class problem, training data are separable by an ECOC support vector machine without don't care outputs. Then for the $k$th decision function a set of classes $L_1$ is separated from a set of classes $L_2$, where $L_1 \cup L_2 = \{1, \ldots, n\}$. Thus any class in $L_1$ is separated from any class in $L_2$. Assume that there is a pair of classes $i$ and $j$ that are not separable. This means that classes $i$ and $j$ are on the same sides of all the decision functions. This does not happen because the ECOC does not classify classes $i$ and $j$. The above discussions also hold for the ECOC with don't care outputs.

Because ECOC support vector machines are an extension of one-against-all support vector machines, the reverse is not always true. ∎

Now compare one-against-all and all-at-once support vector machines when a classification problem is separable by the one-against-all support vector machine. In the one-against-all support vector machine, the decision function for class $i$ is determined so that $D_i(\mathbf{x}) \geq 1$ for $\mathbf{x}$ belonging to class $i$ and $D_i(\mathbf{x}) \leq -1$, otherwise. But by classification using continuous decision functions or fuzzy membership functions, if

$$D_i(\mathbf{x}) > D_j(\mathbf{x}) \quad \text{for } j \neq i, j = 1, \ldots, n, \tag{3.80}$$

**x** is classified into class $i$. Equation (3.80) is the same constraint as that of the all-at-once support vector machine. Thus it is estimated that the decision boundaries obtained by one-against-all and all-at-once support vector machines are quite similar (compare the decision boundaries in Figs. 3.21 and 3.22).

As shown in Figs. 3.20 and 3.22, for three classes, training data that are separable by an all-at-once support vector machine are separable by a pairwise support vector machine, and vice versa. But in general, separation of a smaller number of data with a smaller number of constraints is easier than that by a larger number of data. Thus, training data are more separable by pairwise support vector machines than by all-at-once support vector machines.

Considering the fact that an all-at-once support vector machine has similar decision functions with a one-against-all support vector machine and that its training is more difficult, an all-at-once support vector machine should be a last choice as a classifier.

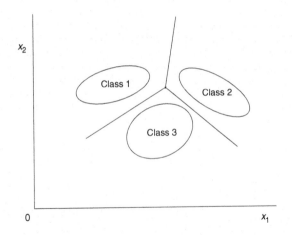

**Fig. 3.22** Class boundaries by an all-at-once support vector machine

## 3.5.6 Training Time Comparison

We evaluated the training time of support vector machines using the data sets listed in Table 1.3. We solved the optimization problem repeatedly, adding 200 data for the MNIST data set and 50 data except for the set at a time. We used the primal–dual interior-point method [46] combined with the decomposition technique to solve the quadratic programming problem. We set the value of the margin parameter $C$ to 5,000. We used an Athlon (1 GHz) computer for the MNIST data set and a Pentium III (933 MHz) computer for the other data sets.

Table 3.11 shows the training time for one-against-all and pairwise support vector machines. For the hiragana-105 data with 38 classes, 703 decision functions need to be determined for pairwise classification, but training a pairwise support vector machine is six times faster than training a one-against-all support vector machine. For the data sets tried, training speedup by pairwise classification is 3–9.

**Table 3.11** Training time comparison

| Data | Kernel | One-against-all (s) | Pairwise (s) | Ratio |
|------|--------|---------------------|--------------|-------|
| Thyroid | $d4$ | 44 | 5 | 8.8 |
| Blood cell | $d4$ | 32 | 5 | 6.4 |
| Hiragana-50 | $d2$ | 128 | 40 | 3.2 |
| Hiragana-105 | $d2$ | 600 | 98 | 6.1 |
| Hiragana-13 | $d2$ | 271 | 43 | 6.3 |
| MNIST | $d3$ | 3,153 | 554 | 5.7 |

# References

1. V. N. Vapnik. *The Nature of Statistical Learning Theory.* Springer-Verlag, New York, NY, 1995.
2. U. H.-G. Kreßel. Pairwise classification and support vector machines. In B. Schölkopf, C. J. C. Burges, and A. J. Smola, editors, *Advances in Kernel Methods: Support Vector Learning*, pages 255–268. MIT Press, Cambridge, MA, 1999.
3. T. Inoue and S. Abe. Fuzzy support vector machines for pattern classification. In *Proceedings of International Joint Conference on Neural Networks (IJCNN '01)*, volume 2, pages 1449–1454, Washington, DC, 2001.
4. S. Abe and T. Inoue. Fuzzy support vector machines for multiclass problems. In *Proceedings of the Tenth European Symposium on Artificial Neural Networks (ESANN 2002)*, pages 113–118, Bruges, Belgium, 2002.
5. M. Pontil and A. Verri. Support vector machines for 3-D object recognition. *IEEE Transactions on Pattern Analysis and Machine Intelligence*, 20(6):637–646, 1998.
6. J. C. Platt, N. Cristianini, and J. Shawe-Taylor. Large margin DAGs for multiclass classification. In S. A. Solla, T. K. Leen, and K.-R. Müller, editors, *Advances in Neural Information Processing Systems 12*, pages 547–553. MIT Press, Cambridge, MA, 2000.
7. B. Kijsirikul and N. Ussivakul. Multiclass support vector machines using adaptive directed acyclic graph. In *Proceedings of the 2002 International Joint Conference on Neural Networks (IJCNN '02)*, volume 1, pages 980–985, Honolulu, Hawaii, 2002.
8. F. Takahashi and S. Abe. Decision-tree-based multiclass support vector machines. In *Proceedings of the Ninth International Conference on Neural Information Processing (ICONIP '02)*, volume 3, pages 1418–1422, Singapore, 2002.

9. F. Takahashi and S. Abe. Optimizing directed acyclic graph support vector machines. In *Proceedings of Artificial Neural Networks in Pattern Recognition (ANNPR 2003)*, pages 166–170, Florence, Italy, 2003.

10. F. Takahashi and S. Abe. Optimal structure of decision-tree-based pairwise support vector machines. *Transactions of the Institute of Systems, Control and Information Engineers*, 17(3):122–130, 2004 (in Japanese).

11. T. G. Dietterich and G. Bakiri. Solving multiclass learning problems via error-correcting output codes. *Journal of Artificial Intelligence Research*, 2:263–286, 1995.

12. V. N. Vapnik. *Statistical Learning Theory*. John Wiley & Sons, New York, NY, 1998.

13. J. Weston and C. Watkins. Multi-class support vector machines. Technical Report CSD-TR-98-04, Royal Holloway, University of London, London, UK, 1998.

14. J. Weston and C. Watkins. Support vector machines for multi-class pattern recognition. In *Proceedings of the Seventh European Symposium on Artificial Neural Networks (ESANN 1999)*, pages 219–224, Bruges, Belgium, 1999.

15. K. Crammer and Y. Singer. On the learnability and design of output codes for multi-class problems. *Machine Learning*, 47(2–3):201–233, 2002.

16. K. Crammer and Y. Singer. Improved output coding for classification using continuous relaxation. In T. K. Leen, T. G. Dietterich, and V. Tresp, editors, *Advances in Neural Information Processing Systems 13*, pages 437–443. MIT Press, Cambridge, MA, 2001.

17. G. Rätsch, A. J. Smola, and S. Mika. Adapting codes and embeddings for polychotomies. In S. Becker, S. Thrun, and K. Obermayer, editors, *Advances in Neural Information Processing Systems 15*, pages 529–536. MIT Press, Cambridge, MA, 2003.

18. S. Abe. Analysis of multiclass support vector machines. In *Proceedings of International Conference on Computational Intelligence for Modelling, Control and Automation (CIMCA 2003)*, pages 385–396, Vienna, Austria, 2003.

19. T. C. Mota and A. C. G. Thomé. One-against-all-based multiclass SVM strategies applied to vehicle plate character recognition. In *Proceedings of the 2009 International Joint Conference on Neural Networks (IJCNN 2009)*, pages 2153–2159, Atlanta, GA, 2009.

20. K. P. Bennett. Combining support vector and mathematical programming methods for classification. In B. Schölkopf, C. J. C. Burges, and A. J. Smola, editors, *Advances in Kernel Methods: Support Vector Learning*, pages 307–326. MIT Press, Cambridge, MA, 1999.

21. E. Mayoraz and E. Alpaydin. Support vector machines for multi-class classification. In J. Mira and J. V. Sánchez-Andrés, editors, *Engineering Applications of Bio-Inspired Artificial Neural Networks (IWANN'99)—Proceedings of International Work—Conference on Artificial and Natural Neural Networks, Alicante, Spain*, volume 2, pages 833–842, 1999.

22. T. Kikuchi and S. Abe. Comparison between error correcting output codes and fuzzy support vector machines. *Pattern Recognition Letters*, 26(12):1937–1945, 2005.

23. A. Juneja and C. Espy-Wilson. Speech segmentation using probabilistic phonetic feature hierarchy and support vector machines. In *Proceedings of International Joint Conference on Neural Networks (IJCNN 2003)*, volume 1, pages 675–679, Portland, OR, 2003.

24. F. Schwenker. Hierarchical support vector machines for multi-class pattern recognition. In *Proceedings of the Fourth International Conference on Knowledge-Based Intelligent Engineering Systems and Allied Technologies (KES 2000)*, volume 2, pages 561–565, Brighton, UK 2000.

25. F. Schwenker. Solving multi-class pattern recognition problems with tree structured support vector machines. In B. Radig and S. Florczyk, editors, *Pattern Recognition: Proceedings of Twenty-Third DAGM Symposium, Munich, Germany*, pages 283–290. Springer-Verlag, Berlin, Germany, 2001.

26. B.-L. Lu, K.-A. Wang, M. Utiyama, and H. Isahara. A part-versus-part method for massively parallel training of support vector machines. In *Proceedings of International*

*Joint Conference on Neural Networks (IJCNN 2004)*, volume 1, pages 735–740, Budapest, Hungary, 2004.

27. G. Guo, S. Z. Li, and K. L. Chan. Support vector machines for face recognition. *Image and Vision Computing*, 19(9–10):631–638, 2001.

28. C. Nakajima, M. Pontil, and T. Poggio. People recognition and pose estimation in image sequences. In *Proceedings of International Joint Conference on Neural Networks (IJCNN 2000)*, volume IV, pages 189–194, Como, Italy, 2000.

29. T. Phetkaew, B. Kijsirikul, and W. Rivepiboon. Reordering adaptive directed acyclic graphs: An improved algorithm for multiclass support vector machines. In *Proceedings of International Joint Conference on Neural Networks (IJCNN 2003)*, volume 2, pages 1605–1610, Portland, OR, 2003.

30. T. Hastie and R. Tibshirani. Classification by pairwise coupling. In M. I. Jordan, M. J. Kearns, and S. A. Solla, editors, *Advances in Neural Information Processing Systems 10*, pages 507–513. MIT Press, Cambridge, MA, 1998.

31. M. Moreira and E. Mayoraz. Improved pairwise coupling classification with correcting classifiers. In C. Nédellec and C. Rouveirol, editors, *Machine Learning: ECML-98: Proceedings of Tenth European Conference on Machine Learning, Chemnitz, Germany*. Springer-Veralg, Berlin, Germany, 1998.

32. Z. Li and S. Tang. Face recognition using improved pairwise coupling support vector machines. In *Proceedings of the Ninth International Conference on Neural Information Processing (ICONIP '02)*, volume 2, pages 876–880, Singapore, 2002.

33. J. C. Platt. Probabilities for SV machines. In A. J. Smola, P. L. Bartlett, B. Schölkopf, and D. Schuurmans, editors, *Advances in Large Margin Classifiers*, pages 61–73. MIT Press, Cambridge, MA, 2000.

34. E. L. Allwein, R. E. Schapire, and Y. Singer. Reducing multiclass to binary: A unifying approach for margin classifiers. *Journal of Machine Learning Research*, 1:113–141, 2000.

35. T. Kikuchi and S. Abe. Error correcting output codes vs. fuzzy support vector machines. In *Proceedings of Artificial Neural Networks in Pattern Recognition (ANNPR 2003)*, pages 192–196, Florence, Italy, 2003.

36. F. Pérez-Cruz and A. Artés-Rodríguez. Puncturing multi-class support vector machines. In J. R. Dorronsoro, editor, *Artificial Neural Networks (ICANN 2002)— Proceedings of International Conference, Madrid, Spain*, pages 751–756. Springer-Verlag, Berlin, Germany, 2002.

37. A. Passerini, M. Pontil, and P. Frasconi. New results on error correcting output codes of kernel machines. *IEEE Transactions on Neural Networks*, 15(1):45–54, 2004.

38. E. J. Bredensteiner and K. P. Bennett. Multicategory classification by support vector machines. *Computational Optimization and Applications*, 12(1–3):53–79, 1999.

39. Y. Guermeur, A. Elisseeff, and H. Paugam-Moisy. A new multi-class SVM based on a uniform convergence result. In *Proceedings of the IEEE-INNS-ENNS International Joint Conference on Neural Networks (IJCNN 2000)*, volume 4, pages 183–188, Como, Italy, 2000.

40. S. Borer and W. Gerstner. Support vector representation of multi-categorical data. In J. R. Dorronsoro, editor, *Artificial Neural Networks (ICANN 2002)—Proceedings of International Conference, Madrid, Spain*, pages 733–738. Springer-Verlag, Berlin, Germany, 2002.

41. C. Angulo, X. Parra, and A. Català. An [sic] unified framework for 'all data at once' multi-class support vector machines. In *Proceedings of the Tenth European Symposium on Artificial Neural Networks (ESANN 2002)*, pages 161–166, Bruges, Belgium, 2002.

42. R. O. Duda and P. E. Hart. *Pattern Classification and Scene Analysis*. John Wiley & Sons, New York, NY, 1973.

43. D. Anguita, S. Ridella, and D. Sterpi. A new method for multiclass support vector machines. In *Proceedings of International Joint Conference on Neural Networks (IJCNN 2004)*, volume 1, pages 407–412, Budapest, Hungary, 2004.

44. D. Tsujinishi, Y. Koshiba, and S. Abe. Why pairwise is better than one-against-all or all-at-once. In *Proceedings of International Joint Conference on Neural Networks (IJCNN 2004)*, volume 1, pages 693–698, Budapest, Hungary, 2004.

45. R. Rifkin and A. Klautau. In defense of one-vs-all classification. *Journal of Machine Learning Research*, 5:101–141, 2004.

46. R. J. Vanderbei. LOQO: An interior point code for quadratic programming. Technical Report SOR-94-15, Princeton University, 1998.

# Chapter 4
# Variants of Support Vector Machines

Since the introduction of support vector machines, numerous variants have been developed. In this chapter, we discuss some of them: least-squares support vector machines, linear programming support vector machines, sparse support vector machines, etc. We also discuss learning paradigms: incremental training, learning using privileged information, semi-supervised learning, multiple classifier systems, multiple kernel learning, and other topics: confidence level and visualization of support vector machines.

## 4.1 Least-Squares Support Vector Machines

Suykens et al. [1–3] proposed least-squares (LS) support vector machines for two-class problems, in which the inequality constraints in L2 soft-margin support vector machines are converted into equality constraints. The training of the LS support vector machine is done by solving a set of linear equations, instead of a quadratic programming problem.

We can extend the two-class LS support vector machines to multiclass LS support vector machines [4, 2] in a way similar to that of support vector machines. In [5], the effect of coding methods for multiclass classification on generalization ability was evaluated by computer experiment. Three coding methods—one-against-all classification, a variant of pairwise classification, i.e., correcting classification, and classification based on error-correcting output codes—were tested for two benchmark data sets and the last two outperformed the first one.

In the following, we first discuss two-class LS support vector machines. Then we discuss one-against-all, pairwise, and all-at-once LS support vector machines and finally we compare their classification and training performance.

S. Abe, *Support Vector Machines for Pattern Classification*,
Advances in Pattern Recognition, DOI 10.1007/978-1-84996-098-4_4,
© Springer-Verlag London Limited 2010

## 4.1.1 Two-Class Least-Squares Support Vector Machines

For a two-class problem, we consider the following decision function:

$$D(\mathbf{x}) = \mathbf{w}^{\top}\phi(\mathbf{x}) + b, \tag{4.1}$$

where $\mathbf{w}$ is the $l$-dimensional vector, $b$ is the bias term, and $\phi(\mathbf{x})$ is the $l$-dimensional vector that maps $m$-dimensional vector $\mathbf{x}$ into the feature space. If $D(\mathbf{x}) > 0$, $\mathbf{x}$ is classified into Class 1 and if $D(\mathbf{x}) < 0$, Class 2.

The LS support vector machine is formulated as follows:

$$\text{minimize} \quad \frac{1}{2}\mathbf{w}^{\top}\mathbf{w} + \frac{C}{2}\sum_{i=1}^{M}\xi_i^2 \tag{4.2}$$

$$\text{subject to} \quad y_i\left(\mathbf{w}^{\top}\phi(\mathbf{x}_i) + b\right) = 1 - \xi_i \quad \text{for } i = 1, \ldots, M, \tag{4.3}$$

where $(\mathbf{x}_i, y_i)$ $(i = 1, \ldots, M)$ are $M$ training input–output pairs, $y_i = 1$ or $-1$ if $\mathbf{x}_i$ belongs to Class 1 or 2, respectively, $\xi_i$ are the slack variables for $\mathbf{x}_i$, and $C$ is the margin parameter. Here, if $\xi_i \geq 1$, $\mathbf{x}_i$ is misclassified and otherwise, $\mathbf{x}_i$ is correctly classified. Unlike L1 or L2 support vector machines, $\xi_i$ can be negative. The first term in the objective function is the reciprocal of the squared margin divided by 2, the second term is to control the number of misclassifications, and $C$ controls the trade-off between the two terms.

Multiplying $y_i$ to both sides of the equation in (4.3), we obtain

$$y_i - \mathbf{w}^{\top}\phi(\mathbf{x}_i) - b = y_i\,\xi_i. \tag{4.4}$$

Because $\xi_i$ takes either a positive or a negative value and $|y_i| = 1$, instead of (4.3), we can use

$$y_i - \mathbf{w}^{\top}\phi(\mathbf{x}_i) - b = \xi_i. \tag{4.5}$$

Introducing the Lagrange multipliers $\alpha_i$ into (4.2) and (4.5), we obtain the unconstrained objective function:

$$Q(\mathbf{w}, b, \boldsymbol{\alpha}, \boldsymbol{\xi})$$
$$= \frac{1}{2}\mathbf{w}^{\top}\mathbf{w} + \frac{C}{2}\sum_{i=1}^{M}\xi_i^2 - \sum_{i=1}^{M}\alpha_i\left(\mathbf{w}^{\top}\phi(\mathbf{x}_i) + b - y_i + \xi_i\right), \tag{4.6}$$

where $\boldsymbol{\alpha} = (\alpha_1, \ldots, \alpha_M)^{\top}$ and $\boldsymbol{\xi} = (\xi_1, \ldots, \xi_M)^{\top}$.

Taking the partial derivatives of (4.6) with respect to $\mathbf{w}$, $b$, and $\boldsymbol{\xi}$ and equating them to zero, together with the equality constraint (4.5), we obtain the optimal conditions as follows:

$$\mathbf{w} = \sum_{i=1}^{M}\alpha_i\,\phi(\mathbf{x}_i), \tag{4.7}$$

$$\sum_{i=1}^{M} \alpha_i = 0, \tag{4.8}$$

$$\alpha_i = C\,\xi_i \quad \text{for } i = 1, \ldots, M, \tag{4.9}$$

$$\mathbf{w}^\top \boldsymbol{\phi}(\mathbf{x}_i) + b - y_i + \xi_i = 0 \quad \text{for } i = 1, \ldots, M. \tag{4.10}$$

From (4.9), unlike L1 or L2 support vector machines, $\alpha_i$ can be negative.

Substituting (4.7) and (4.9) into (4.10) and expressing it and (4.8) in matrix form, we obtain

$$\begin{pmatrix} \Omega & \mathbf{1} \\ \mathbf{1}^\top & 0 \end{pmatrix} \begin{pmatrix} \boldsymbol{\alpha} \\ b \end{pmatrix} = \begin{pmatrix} \mathbf{y} \\ 0 \end{pmatrix} \tag{4.11}$$

or

$$\Omega\boldsymbol{\alpha} + \mathbf{1}b = \mathbf{y}, \tag{4.12}$$

$$\mathbf{1}^\top \boldsymbol{\alpha} = 0, \tag{4.13}$$

where $\mathbf{1}$ is the $M$-dimensional vector and

$$\Omega_{ij} = \boldsymbol{\phi}^\top(\mathbf{x}_i)\,\boldsymbol{\phi}(\mathbf{x}_j) + \frac{\delta_{ij}}{C}, \tag{4.14}$$

$$\delta_{ij} = \begin{cases} 1 & i = j, \\ 0 & i \neq j, \end{cases} \tag{4.15}$$

$$\mathbf{y} = (y_1, \ldots, y_M)^\top, \tag{4.16}$$

$$\mathbf{1} = (1, \ldots, 1)^\top. \tag{4.17}$$

Like the support vector machine, setting $K(\mathbf{x}, \mathbf{x}') = \boldsymbol{\phi}^\top(\mathbf{x})\,\boldsymbol{\phi}(\mathbf{x}')$, we can avoid the explicit treatment of variables in the feature space.

The original minimization problem is solved by solving (4.11) for $\boldsymbol{\alpha}$ and $b$ as follows. Because of $1/C\,(> 0)$ in the diagonal elements, $\Omega$ is positive definite. Therefore,

$$\boldsymbol{\alpha} = \Omega^{-1}(\mathbf{y} - \mathbf{1}\,b). \tag{4.18}$$

Substituting (4.18) into (4.13), we obtain

$$b = (\mathbf{1}^\top \Omega^{-1} \mathbf{1})^{-1} \mathbf{1}^\top \Omega^{-1} \mathbf{y}. \tag{4.19}$$

Thus, substituting (4.19) into (4.18), we obtain $\boldsymbol{\alpha}$.

By changing the inequality constraints into the equality constraints, training of support vector machines reduces to solving a set of linear equations instead of a quadratic programming problem. But by this formulation, sparsity of $\boldsymbol{\alpha}$ is not guaranteed. To avoid this, Suykens et al. [2, 6] proposed pruning the data whose associated $\alpha_i$ have small absolute values, namely, first we solve (4.11) using all the training data. Next, we sort $\alpha_i$ according to their absolute values and delete a portion of the training data set (say 5% of the set) starting from the data with the minimum absolute value in order. Then we solve (4.11) using the reduced training data set and iterate this procedure while the user-defined performance index is not degraded.

Cawley and Talbot [7] proposed a greedy algorithm. By assuming that the weight vector is expressed by

$$\mathbf{w} = \sum_{i \in S} \alpha_i \, y_i \, \phi(\mathbf{x}_i), \tag{4.20}$$

where $S \subset \{1, \ldots, M\}$, (4.2) becomes

$$\frac{1}{2} \sum_{i,j \in S} \alpha_i \, \alpha_j \, y_i \, y_j \, K(\mathbf{x}_i, \mathbf{x}_j)$$

$$+ \frac{C}{2} \sum_{i=1}^{M} \left( \sum_{j \in S} y_i \, (y_j \, \alpha_j \, K(\mathbf{x}_i, \mathbf{x}_j) + b) - 1 \right)^2 . \tag{4.21}$$

Then starting with only a bias term, i.e., $S = \emptyset$, the training pattern that minimizes the objective function (4.21) is added to $S$ until some convergence test is satisfied.

For linear kernels, saving calculated $\mathbf{w}$, we can speed up classification [8]. Unfortunately, as discussed in Section 2.6.5, a similar method is not applicable to nonlinear kernels.

## 4.1.2 One-Against-All Least-Squares Support Vector Machines

For a one-against-all LS support vector machine, we determine $n$ decision functions that separate one class from the remaining classes. The $i$th decision function

$$D_i(\mathbf{x}) = \mathbf{w}_i^\top \, \phi(\mathbf{x}) + b_i \tag{4.22}$$

separates class $i$ from the remaining classes with the maximum margin, where $\mathbf{w}_i$ is the $l$-dimensional weight vector and $b_i$ is the bias term.

The hyperplane $D_i(\mathbf{x}) = 0$ forms the optimal separating hyperplane. In classification, if for the input vector $\mathbf{x}$

$$D_i(\mathbf{x}) > 0 \tag{4.23}$$

is satisfied for one $i$, $\mathbf{x}$ is classified into class $i$. Because only the sign of the decision function is used, the decision is discrete.

If (4.23) is satisfied for plural $i$'s, or if there is no $i$ that satisfies (4.23), $\mathbf{x}$ is unclassifiable.

To avoid this, instead of the discrete decision functions, continuous decision functions can be used, namely, data sample $\mathbf{x}$ is classified into the class

$$\arg \max_{i=1,\ldots,n} D_i(\mathbf{x}). \tag{4.24}$$

Another way of resolving unclassifiable regions is to introduce membership functions. For class $i$ we define one-dimensional membership functions $m_{ij}(\mathbf{x})$ in the directions orthogonal to the optimal separating hyperplanes $D_j(\mathbf{x}) = 0$ as follows:

1. For $i = j$

$$m_{ii}(\mathbf{x}) = \begin{cases} 1 & \text{for} \quad D_i(\mathbf{x}) \geq 1, \\ D_i(\mathbf{x}) & \text{otherwise.} \end{cases} \tag{4.25}$$

2. For $i \neq j$

$$m_{ij}(\mathbf{x}) = \begin{cases} 1 & \text{for} \quad D_j(\mathbf{x}) \leq -1, \\ -D_j(\mathbf{x}) & \text{otherwise.} \end{cases} \tag{4.26}$$

For $i \neq j$, class $i$ is on the negative side of $D_j(\mathbf{x}) = 0$.

We define the class $i$ membership function of $\mathbf{x}$ by the minimum operation for $m_{ij}(\mathbf{x})$ $(j = 1, \ldots, n)$:

$$m_i(\mathbf{x}) = \min_{j=1,\ldots,n} m_{ij}(\mathbf{x}), \tag{4.27}$$

or the average operation:

$$m_i(\mathbf{x}) = \frac{1}{n} \sum_{j=1,\ldots,n} m_{ij}(\mathbf{x}). \tag{4.28}$$

The data sample $\mathbf{x}$ is classified into the class

$$\arg \max_{i=1,\ldots,n} m_i(\mathbf{x}). \tag{4.29}$$

In Section 3.1.3, one-against-all support vector machines with continuous decision functions and one-against-all fuzzy support vector machines with minimum or average operators are proved to be equivalent. This also holds for LS support vector machines.

## 4.1.3 Pairwise Least-Squares Support Vector Machines

In pairwise classifications, unclassifiable regions exist. Thus, similar to Section 3.2.2, we introduce fuzzy membership functions to resolve unclassifiable regions in pairwise classification [9, 10].

In pairwise classification we require a binary classifier for each possible pair of classes and the number of the total pairs is $n(n-1)/2$ for an $n$-class problem. The decision function for the pair of classes $i$ and $j$ is given by

$$D_{ij}(\mathbf{x}) = \mathbf{w}_{ij}^{\top}\boldsymbol{\phi}(\mathbf{x}) + b_{ij}, \tag{4.30}$$

where $\mathbf{w}_{ij}$ is the $l$-dimensional weight vector, $\boldsymbol{\phi}(\mathbf{x})$ maps $\mathbf{x}$ into the $l$-dimensional feature space, $b_{ij}$ is the bias term, and $D_{ij}(\mathbf{x}) = -D_{ji}(\mathbf{x})$. Then for data sample $\mathbf{x}$ we calculate

$$D_i(\mathbf{x}) = \sum_{j\neq i, j=1}^{n} \text{sign}(D_{ij}(\mathbf{x})), \tag{4.31}$$

where

$$\text{sign}(a) = \begin{cases} 1 & a \geq 0, \\ -1 & \text{otherwise,} \end{cases}$$

and this data sample is classified into the class

$$\arg \max_{i=1,\dots,n} D_i(\mathbf{x}). \tag{4.32}$$

If (4.32) is satisfied for one $i$, $\mathbf{x}$ is classified into class $i$. But if (4.32) is satisfied for plural $i$'s, $\mathbf{x}$ is unclassifiable.

To avoid this, we introduce membership functions. First, we define the one-dimensional membership function, $m_{ij}(\mathbf{x})$, in the direction orthogonal to the optimal separating hyperplane $D_{ij}(\mathbf{x})$ as follows:

$$m_{ij} = \begin{cases} 1 & \text{for} \quad D_{ij}(\mathbf{x}) \geq 1, \\ D_{ij}(\mathbf{x}) & \text{otherwise.} \end{cases} \tag{4.33}$$

Here, we allow a negative degree of membership to make any data except those on the decision boundary be classified.

Using the minimum operation the membership function, $m_i(\mathbf{x})$, of $\mathbf{x}$ for class $i$ is given by

$$m_i(\mathbf{x}) = \min_{j=1,\dots,n} m_{ij}(\mathbf{x}). \tag{4.34}$$

The shape of the resulting membership function is a truncated polyhedral pyramid in the feature space, and the contour surface, in which the degree of membership is the same, is parallel to the decision function.

Using the average operation the membership function, $m_i(\mathbf{x})$, of $\mathbf{x}$ for class $i$ is given by

$$m_i(\mathbf{x}) = \frac{1}{n-1} \sum_{j \neq i, j=1}^{n} m_{ij}(\mathbf{x}). \tag{4.35}$$

The shape of the resulting membership function is a truncated polyhedral pyramid but some part of the contour surface is not parallel to the decision function.

Using either (4.34) or (4.35), data sample $\mathbf{x}$ is classified into the class

$$\arg \max_{i=1,\ldots,n} m_i(\mathbf{x}). \tag{4.36}$$

Comparing the minimum and average operations, the regions where $m_i(\mathbf{x}) = 1$ are the same, but the regions where $m_i(\mathbf{x}) < 1$ are different. We can show that the decision boundaries with the minimum operation are the same as those given by (4.31) for classifiable regions but the decision boundaries for the average operation are not. Thus the recognition rate using the minimum operation is always better than or equal to that by the conventional pairwise LS support vector machine. But this does not hold for the average operation.

### 4.1.4 All-at-Once Least-Squares Support Vector Machines

Similar to all-at-once support vector machines discussed in Section 3.4, we can define all-at-once LS support vector machines.

The maximum-margin classifier is obtained by solving

$$\text{minimize} \quad \frac{1}{2} \sum_{j=1}^{n} \|\mathbf{w}_j\|^2 + \frac{C}{2} \sum_{i=1}^{M} \sum_{\substack{j \neq y_i, \\ j=1}}^{n} \xi_{ij}^2 \tag{4.37}$$

$$\text{subject to} \quad \mathbf{w}_j^\top \boldsymbol{\phi}(\mathbf{x}_i) + b_{y_i} - b_j = 1 - \xi_{ij}$$
$$\text{for} \quad j \neq y_i, j = 1, \ldots, n, i = 1, \ldots, M, \tag{4.38}$$

where $\mathbf{w}_j$ is the $l$-dimensional weight vector for class $j$, $b_j$ is the bias term for class $j$, $\xi_{ij}$ are slack variables associated with $\mathbf{x}_i$ and class $j$, $y_i \, (\in \{1, \ldots, n\})$ is the class label for $\mathbf{x}_i$, and $C$ is the margin parameter that determines the trade-off between the maximization of the margin and minimization of the classification error.

Introducing the Lagrange multipliers $\alpha_{ij}$, we obtain

$$Q(\mathbf{w}, \mathbf{b}, \boldsymbol{\xi}, \boldsymbol{\alpha}, \boldsymbol{\beta}) = \frac{1}{2} \sum_{j=1}^{n} \|\mathbf{w}_j\|^2 + \frac{C}{2} \sum_{i=1}^{M} \sum_{\substack{j \neq y_i, \\ j=1}}^{n} \xi_{ij}^2$$

$$-\sum_{i=1}^{M} \sum_{\substack{j \neq y_i, \\ j=1}}^{n} \alpha_{ij} \left( (\mathbf{w}_{y_i} - \mathbf{w}_j)^{\top} \phi(\mathbf{x}_i) + b_{y_i} - b_j - 1 + \xi_{ij} \right). \quad (4.39)$$

Taking the partial derivatives of (4.39) with respect to $\mathbf{w}_j, b_j, \alpha_{ij}$, and $\xi_{ij}$ and equating them to zero, we obtain the optimal conditions as follows:

$$\mathbf{w}_j = \sum_{i=1}^{M} z_{ij}\, \phi(\mathbf{x}_i) \quad \text{for } j = 1, \ldots, n, \qquad (4.40)$$

$$\sum_{i=1}^{M} z_{ij} = 0 \quad \text{for } j = 1, \ldots, n, \qquad (4.41)$$

$$\alpha_{ij} = C\xi_{ij}, \quad \alpha_{ij} \geq 0$$
$$\text{for } i = 1, \ldots, M, \, j \neq y_i, \, j = 1, \ldots, n, \qquad (4.42)$$

$$(\mathbf{w}_{y_i} - \mathbf{w}_j)^{\top} \phi(\mathbf{x}_i) + b_{y_i} - b_j - 1 + \xi_{ij} = 0$$
$$\text{for } i = 1, \ldots, M, \, j \neq y_i, \, j = 1, \ldots, n, \qquad (4.43)$$

where

$$z_{ij} = \begin{cases} \displaystyle\sum_{\substack{k \neq y_i, \\ k=1}}^{n} \alpha_{ik} & \text{for } j = y_i, \\ -\alpha_{ij} & \text{otherwise.} \end{cases} \qquad (4.44)$$

Similar to a two-class problem, substituting (4.40) and (4.42) into (4.43), we can solve the resulting equation and (4.41) for $\alpha_{ij}$ and $b_i$.

The decision function for class $i$ is given by

$$D_i(\mathbf{x}) = \sum_{j=1}^{M} z_{ji}\, \phi^{\top}(\mathbf{x}_j)\, \phi(\mathbf{x}) + b_i. \qquad (4.45)$$

As usual, to avoid explicit treatment of variables in the feature space, we use kernel tricks: $K(\mathbf{x}, \mathbf{x}') = \phi^{\top}(\mathbf{x})\, \phi(\mathbf{x}')$.

## 4.1.5 Performance Comparison

### 4.1.5.1 Condition of Experiments

Using the data sets listed in Table 1.3, we compared the performance of the fuzzy one-against-all LS support vector machine (LS SVM), the fuzzy pairwise LS SVM with minimum and average operators, and the all-at-once LS SVM. Each input variable was scaled into $[0, 1]$.

Among linear kernels, polynomial kernels with degrees 2, 3, and 4, and RBF kernels with $\gamma = 0.1, 1$, and 10 we selected the optimum kernel and the value of $C$ from 1 to 100,000 by fivefold cross-validation. The experiments were done on an AthlonMP 2 GHz personal computer.

### 4.1.5.2 Classification Performance

Table 4.1 shows the recognition performance of the fuzzy one-against-all LS SVM, the fuzzy pairwise LS SVMs with minimum and average operators, and the all-at-once LS SVM. The "Parm" row lists the kernel type and its parameter value and the $C$ value optimized by the fivefold cross-validation. The "Min" and "Avg." rows show the recognition rates with the decision functions with minimum operators and with average operators, respectively. If the recognition rates of the training data were not 100%, we list the recognition rates in parentheses. For the all-at-once LS SVM, we list the recognition rates in "Min" rows. The highest recognition rates of the test data are shown in boldface.

We could not get the results of the all-at-once LS SVM other than for the iris, numeral, and thyroid data sets due to memory overflow. For the same reason, for the MNIST data set, we set $C = 10,000$ and used RBF kernels for the pairwise LS SVM.

In the following, the recognition rate means that of the test data.

For all the data sets, the fuzzy pairwise LS SVMs performed best, and except for the hiragana-13 data set the average operator performed better than or equal to the minimum operator.

Now compare the recognition rates of LS SVMs with those of SVMs listed in Table 3.2. For the SVMs, the performance depends on the data sets; no single architecture performed best. But for the LS SVMs, the pairwise fuzzy LS SVMs performed best. Except for the thyroid data set, the pairwise LS SVM with the average operators showed the same or higher recognition rates than the SVMs. Especially for the blood cell data set, the difference was large. But for the thyroid data set, the recognition rate of the LS SVMs was very poor. Therefore, the performance of LS SVMs is more dependable on the data sets than that of the SVMs.

### 4.1.5.3 Training Speed

Table 4.2 shows the training time of the one-against-all, pairwise, and all-at-once LS SVMs for the polynomial kernels with degree 2. In training the LS SVM, we used the Cholesky factorization to solve the set of linear equations. For all the cases, training of the pairwise LS SVM was the fastest and training of the all-at-once LS SVM was the slowest.[1] For the blood cell and hiragana data sets, we could not train the all-at-once LS SVM because of the memory

---

[1] For regular SVMs, this fact was shown in [11].

**Table 4.1** Performance comparison of LS support vector machines

| Data | Item | One-against-all | Pairwise | All-at-once |
|---|---|---|---|---|
| Iris | Parm | $\gamma 1,\ C10^4$ | $d1,\ C50$ | $d3,\ C10^4$ |
| | Min | 96.00 | 97.33 | 92.00 |
| | Avg. | 96.00 | **98.67** | – |
| Numeral | Parm | $\gamma 10,\ C50$ | $d1,\ C10$ | $\gamma 0.1,\ C500$ |
| | Min | 99.39 | 99.27 (99.75) | 99.02 (99.75) |
| | Avg. | 99.39 | **99.76** | – |
| Thyroid | Parm | $\gamma 10,\ C10^5$ | $\gamma 10,\ C10^5$ | $\gamma 10,\ C10^5$ |
| | Min | 94.22 (97.38) | 95.48 (98.49) | 94.25 (97.91) |
| | Avg. | 94.22 (97.38) | **95.57** (98.28) | – |
| Blood cell | Parm | $\gamma 10,\ C500$ | $d2,\ C10^5$ | – |
| | Min | 93.58 (96.48) | 93.55 (97.87) | – |
| | Avg. | 93.58 (96.48) | **94.32** (98.10) | – |
| Hiragana-50 | Parm | $\gamma 10,\ C10^4$ | $\gamma 1,\ C3,000$ | – |
| | Min | 99.22 | 99.02 | – |
| | Avg. | 99.22 | **99.33** | – |
| Hiragana-13 | Parm | $\gamma 10,\ C10^4$ | $\gamma 10,\ C10^5$ | – |
| | Min | 99.64 (99.77) | **99.90** | – |
| | Avg. | 99.64 (99.77) | 99.88 | – |
| Hiragana-105 | Parm | $\gamma 10,\ C10^4$ | $\gamma 10,\ C7,000$ | – |
| | Min | **100** | **100** | – |
| | Avg. | **100** | **100** | – |
| MNIST | Parm | – | $\gamma 10,\ C10^4$ | – |
| | Min | – | **98.79** | – |
| | Avg. | – | **98.79** | – |

overflow. Therefore, as indicated in [12], we need to use iterative methods for speedup and efficient memory use.

**Table 4.2** Training time in seconds

| Data | One-against-all | Pairwise | All-at-once |
|---|---|---|---|
| Numeral | 25 | 1 | 2,026 |
| Thyroid | 716 | 409 | 1,565 |
| Blood cell | 1,593 | 59 | – |
| Hiragana-50 | 15,130 | 129 | – |

## 4.1.5.4 Classification Speed

Table 4.3 lists the classification time for linear kernels with the values of $C$ determined by fivefold cross-validation. The conventional and calculated methods mean that the weights are calculated for each data sample and the weights are calculated once before classification, respectively. Except for the iris data set, which is very small, speedup by the calculated method is evident.

**Table 4.3** Classification time in seconds

| Data | Method | One-against-all | Pairwise | All-at-once |
|------|--------|-----------------|----------|-------------|
| Iris | Conventional | 0.01 | 0.00 | 0.01 |
|      | Calculated | 0.00 | 0.00 | 0.00 |
| Numeral | Conventional | 3.1 | 2.7 | 3.2 |
|         | Calculated | 0.01 | 0.02 | 0.00 |
| Thyroid | Conventional | 43 | 28 | 43 |
|         | Calculated | 0.02 | 0.02 | 0.02 |
| Blood cell | Conventional | 80 | 52 | – |
|            | Calculated | 0.02 | 0.26 | – |

#### 4.1.5.5 Influence of Outliers

Because LS SVMs use equality constraints instead of inequality constraints, they are vulnerable to outliers [12]. The only difference between LS SVMs and L2 SVMs is that the former uses equality constraints whereas the latter uses the inequality constraints. Thus, we compared their recognition performance when outliers were included.

For evaluation, we used the blood cell data belonging to Classes 2 and 3, which overlap heavily and are difficult to classify. As outliers, we added 10 data belonging to classes other than 2 and 3 to Class 2 training data. We used the polynomial kernel with degree 2.

Figures 4.1 and 4.2 show the recognition rates against the margin parameter $C$ for the LS SVM and L2 SVM, respectively. In the figures, the dotted lines show the recognition rates of the training data and the solid lines show those of the test data. We calculated the recognition rate of the training data excluding outliers.

In Fig. 4.1, the recognition rates of the training data for the LS SVM did not change much for $100 < C < 10,000$ even if the outliers were included. But the recognition rate of test data dropped rapidly, especially when outliers were included.

In Fig. 4.2, the recognition rates of the training data for the L2 SVM increased as the value of $C$ was increased, and there is not much difference between the recognition rates with and those without outliers. In addition, the recognition rate of the test data with outliers was almost constant for the change of $C$, and for a large value of $C$ it is better than without outliers.

Comparing Figs. 4.1 and 4.2, we can see that the L2 SVM is more robust than the LS SVM for outliers.

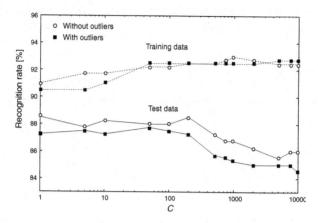

**Fig. 4.1** Influence of outliers to the LS support vector machine. Reprinted from [10, p. 791] with permission from Elsevier

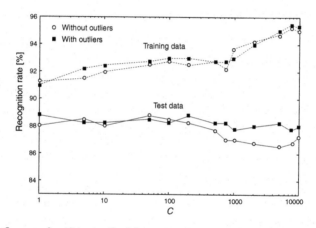

**Fig. 4.2** Influence of outliers to the L2 support vector machine. Reprinted from [10, p. 791] with permission from Elsevier

## 4.2 Linear Programming Support Vector Machines

In the original formulation of support vector machines, quadratic programming problems need to be solved. But we can formulate classification problems by linear programming, replacing the quadratic objective function with a linear function [13–16]. In this section, we discuss linear programming (LP) support vector machines, their characteristics, and their classification performance for benchmark data sets.

## 4.2.1 Architecture

In the L1 soft-margin support vector machine, replacing the L2 norm $\|\mathbf{w}\|_2^2 = w_1^2 + w_2^2 + \cdots + w_l^2$ in the objective function with an L1 norm $\|\mathbf{w}\|_1 = |w_1| + |w_2| + \cdots + |w_l|$, where $\mathbf{w}$ is the coefficient vector of the separating hyperplane, we obtain the following LP support vector machine:

$$\text{minimize} \quad Q(\mathbf{w}, \boldsymbol{\xi}) = \sum_i^l |w_i| + C \sum_{i=1}^M \xi_i \tag{4.46}$$

$$\text{subject to} \quad y_i \left(\mathbf{w}^\top \boldsymbol{\phi}(\mathbf{x}_i) + b\right) \geq 1 - \xi_i \quad \text{for } i = 1, \ldots, M, \tag{4.47}$$

where $C$ is the margin parameter, $\xi_i$ are slack variables associated with the training data $\mathbf{x}_i$, $y_i$ are class labels and are 1 if $\mathbf{x}_i$ belong to Class 1 and $-1$ otherwise, $\boldsymbol{\phi}(\mathbf{x})$ is the $l$-dimensional mapping function that maps $\mathbf{x}$ into the feature space, and $b$ is the bias term.

By this formulation, for the linear kernel, i.e., $\boldsymbol{\phi}(\mathbf{x}) = \mathbf{x}$, we can solve the problem by linear programming. However, for the kernels other than linear kernels, we need to treat the feature space explicitly.

To formulate an LP support vector machine in the feature space, instead of the mapping function to the feature space, $\boldsymbol{\phi}(\mathbf{x})$, we define the mapping function to a subspace of the feature space by [17]

$$\boldsymbol{\psi}(\mathbf{x}) = (K(\mathbf{x}_1, \mathbf{x}), \ldots, K(\mathbf{x}_M, \mathbf{x}))^\top. \tag{4.48}$$

This is similar to, but different from, the mapping function to the empirical feature space given by (2.151) on p. 51. Therefore, in general

$$K(\mathbf{x}, \mathbf{x}') = \boldsymbol{\phi}^\top(\mathbf{x}) \boldsymbol{\phi}(\mathbf{x}') \neq \boldsymbol{\psi}^\top(\mathbf{x}) \boldsymbol{\psi}(\mathbf{x}'). \tag{4.49}$$

Using (4.48), the decision function becomes [18]

$$D(\mathbf{x}) = \boldsymbol{\alpha}^\top \boldsymbol{\psi}(\mathbf{x}) + b = \sum_{i=1}^M \alpha_i K(\mathbf{x}, \mathbf{x}_i) + b, \tag{4.50}$$

where $\boldsymbol{\alpha} = (\alpha_1, \ldots, \alpha_M)^\top$ and $\alpha_i$ take on real values. Thus, unlike L1 support vector machines, in (4.50) we need not multiply $\alpha_i K(\mathbf{x}, \mathbf{x}_i)$ by $y_i$. Then the resulting LP support vector machine becomes as follows:

$$\text{minimize} \quad Q(\boldsymbol{\alpha}, \boldsymbol{\xi}) = \sum_{i=1}^M (|\alpha_i| + C \xi_i) \tag{4.51}$$

$$\text{subject to} \quad y_j \left(\sum_{i=1}^M \alpha_i K(\mathbf{x}_j, \mathbf{x}_i) + b\right) \geq 1 - \xi_j \quad \text{for } j = 1, \ldots, M. \tag{4.52}$$

Letting $\alpha_i = \alpha_i^+ - \alpha_i^-$ and $b = b^+ - b^-$, where $\alpha_i^+ \geq 0$, $\alpha_i^- \geq 0$, $b^+ \geq 0$, and $b^- \geq 0$, we can solve (4.51) and (4.52) for $\boldsymbol{\alpha}$, $b$, and $\boldsymbol{\xi}$ by linear programming. But because

$$\mathbf{w} = \sum_{i=1}^{M} \alpha_i \, \phi(\mathbf{x}_i), \tag{4.53}$$

minimization of the sum of $|\alpha_i|$ does not lead to maximization of the margin measured in the L1 norm.

By this formulation, the number of variables is $3M + 2$ and the number of inequality constraints is $M$. Thus for a large number of training data, training becomes very slow even by linear programming. Therefore, we need to use decomposition techniques [19, 20] as discussed in Section 5.9.

*Example 4.1.* Consider solving the problem in Example 2.8 on p. 74 by the LP support vector machine. Because the problem is linearly separable, we consider a solution with $\xi_i = 0$ for $i = 1, \ldots, 4$. Then, the objective function in (4.51) becomes

$$Q(\boldsymbol{\alpha}, \boldsymbol{\xi}) = |\alpha_1| + |\alpha_2| + |\alpha_3| + |\alpha_4| \tag{4.54}$$

and the constraints in (4.52) become

$$\begin{align}
\alpha_1 - \alpha_4 + b \geq 1 \quad &\text{for Sample 1,} \tag{4.55} \\
\alpha_2 - \alpha_3 + b \geq 1 \quad &\text{for Sample 2,} \tag{4.56} \\
\alpha_2 - \alpha_3 - b \geq 1 \quad &\text{for Sample 3,} \tag{4.57} \\
\alpha_1 - \alpha_4 - b \geq 1 \quad &\text{for Sample 4.} \tag{4.58}
\end{align}$$

Because (4.54) is minimized, from (4.55) and (4.58) or (4.56) and (4.57), $b$ needs to be 0. Thus, (4.55), (4.56), (4.57), and (4.58) reduce to

$$\alpha_1 - \alpha_4 \geq 1, \tag{4.59}$$
$$\alpha_2 - \alpha_3 \geq 1. \tag{4.60}$$

Therefore, (4.54) is minimized when

$$\begin{align}
\alpha_1 = 1 - \beta_1, \qquad &\alpha_4 = -\beta_1, \tag{4.61} \\
\alpha_2 = 1 - \beta_2, \qquad &\alpha_3 = -\beta_2, \tag{4.62}
\end{align}$$

where $1 \geq \beta_1 \geq 0$ and $1 \geq \beta_2 \geq 0$.

Similar to the solution of the L1 support vector machine, the solution is nonunique and the decision boundary is given by

$$D(\mathbf{x}) = x_1 + x_2 = 0, \tag{4.63}$$

which is the same as that for the L1 support vector machine. Notice that for $\beta_1 = \beta_2 = 0$, $\alpha_1$ and $\alpha_2$ are 1, and $\alpha_3$ and $\alpha_4$ are zero, namely, the dual variables associated with Class 2 are all zero. Thus, the definition of support vectors does not hold for LP support vector machines.

Similar to L1 and L2 support vector machines, LP support vector machines have degenerate solutions, namely, $\alpha_i$ are all zero. The difference is that LP support vector machines have degenerate solutions when the value of $C$ is small as the following theorem shows.

**Theorem 4.1.** *For the LP support vector machine, there exists a positive $C_0$ such that for $0 \leq C \leq C_0$ the solution is degenerate.*

*Proof.* Because of the slack variables $\xi_i$, (4.52) has a feasible solution. Thus, for some $C$, (4.51) and (4.52) have the optimal solution with some $\alpha_i$ being nonzero.

For $\boldsymbol{\alpha} = \mathbf{0}$, (4.52) reduces to

$$y_i \, b \geq 1 - \xi_i. \tag{4.64}$$

For $b = 0$, (4.64) is satisfied for $\xi_i = 1$. Then (4.51) is

$$Q(\boldsymbol{\alpha}, \boldsymbol{\xi}) = MC. \tag{4.65}$$

Thus, by decreasing the value of $C$ from a large value, we can find a maximum value of $C$, $C_0$, in which (4.51) is minimized for $\boldsymbol{\alpha} = \mathbf{0}$. For $0 < C \leq C_0$, it is evident that $\boldsymbol{\alpha} = \mathbf{0}$ is the optimal solution for (4.51) and (4.52). ∎

Zhou et al. [16] proposed a slightly different linear programming support vector machine as follows:

$$\text{minimize} \quad Q(r, \boldsymbol{\alpha}, \boldsymbol{\xi}) = -r + C \sum_{i=1}^{M} \xi_i \tag{4.66}$$

$$\text{subject to} \quad y_i \left( \sum_{j=1}^{M} \alpha_j \, K(\mathbf{x}_j, \mathbf{x}_i) + b \right) \geq r - \xi_i \quad \text{for } i = 1, \ldots, M, \tag{4.67}$$

$$r \geq 0, \tag{4.68}$$

$$-1 \leq \alpha_i \leq 1 \quad \text{for } i = 1, \ldots, M, \tag{4.69}$$

$$\xi_i \geq 0 \quad \text{for } i = 1, \ldots, M. \tag{4.70}$$

In [16], $\alpha_j \, y_j$ is used instead of $\alpha_j$ in (4.67). But this is not necessary because $\alpha_j$ may take on negative values. In regular support vector machines $r$ in (4.67) is 1, but here $r$ is maximized.

Similar to two-class support vector machines, we can easily extend two-class LP support vector machines to multiclass LP support vector machines, such as one-against-all, pairwise, and ECOC LP support vector machines. But because the extension is straightforward [21], we will not discuss the details here.

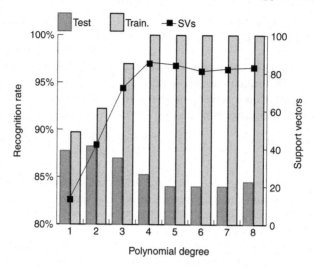

**Fig. 4.3** Recognition rates and support vectors against a polynomial degree

## 4.2.2 Performance Evaluation

In this section, we first show the performance of LP support vector machines (LP SVMs) for the two-class blood cell data used in Section 2.6.9. Then we show the performance for the multiclass data sets listed in Table 1.3.

### 4.2.2.1 Performance of Two-Class LP Support Vector Machines

Figure 4.3 shows the recognition rates and the number of support vectors against the polynomial degree with $C = 5,000$. As the polynomial degree becomes higher, the recognition rate of the training data increases and reaches 100% for the polynomial degree of 4. But the recognition rate of the test data reaches the maximum at the polynomial degree of 2 and decreases afterward.

Figure 4.4 shows the recognition rates and the number of support vectors against the RBF kernel parameter $\gamma$ with $C = 5,000$. As the value of $\gamma$ becomes larger, the recognition rate of the training data increases. But the recognition rate of the test data reaches the maximum at $\gamma = 3$. The number of support vectors increases as the value of $\gamma$ increases.

Figure 4.5 shows the recognition rates and the number of support vectors against $C$ for RBF kernels with $\gamma = 1$. For $C = 0.1$, the recognition rates are very low due to the effect of the degenerate solution. For $C = 0.01$, the degenerate solution was obtained. The recognition rates of the test data do not vary very much for $C$ larger than or equal to 1.

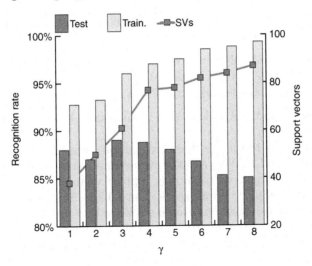

**Fig. 4.4** Recognition rates and support vectors against $\gamma$

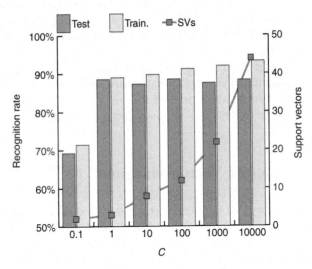

**Fig. 4.5** Recognition rates and support vectors against $C$

## 4.2.2.2 Performance of Multiclass LP Support Vector Machines

Here we show the performance of one-against-all and pairwise fuzzy LP SVMs. We determined the kernels and parameters by fivefold cross-validation. For each following kernel, we set the value of $C$ from 1, 10, 50, 100, 500, 1,000,

2,000, 3,000, 5,000, 7,000, 10,000, and 100,000 and determined the optimal
kernel and the parameters:

1. Linear kernels and polynomial kernels with degrees 2, 3, and 4
2. RBF kernels with $\gamma = 0.1$, 1, and 10
3. Neural network kernels with $\nu = 0.2$ and $a = 0.1$

Here, we used neural network kernels because kernels need not be positive
semidefinite.

Table 4.4 shows the parameters, recognition rates of the one-against-all
and pairwise fuzzy LP SVM, and the average number of support vectors per
decision function, where we count only support vectors with nonzero $\alpha_i$ (for
the definition of support vectors, see Section 5.9.2.1). The recognition rates of
the training data are shown in parentheses when they are not 100%. For each
data set, the maximum recognition rate of the test data is shown in boldface.
From the tables, the recognition rates of the test data of both methods are
comparable.

Similar to the one-against-all fuzzy SVMs, the recognition rates of the
fuzzy LP SVMs with minimum and average operators were the same. Comparing with Table 3.2, the recognition rates of the test data are comparable
but a little lower. But the numbers of support vectors of LP SVMs are smaller.
This is because in LP SVMs, the sum of $|\alpha_i|$ is minimized. Thus it leads to
fewer support vectors.

Similar to the pairwise fuzzy SVMs, the recognition rates of the fuzzy LP
SVMs with minimum and average operators are almost the same. Comparing
with Table 3.2, the recognition rates of the test data by fuzzy LP SVMs are
slightly lower but the numbers of support vectors are smaller.

## 4.3 Sparse Support Vector Machines

One of the advantages of support vector machines is that the solution is
sparse, namely, the solution is expressed by a small portion of training data.
In extreme cases, the solution of a two-class problem can be expressed by two
data belonging to opposite classes. But, for a very difficult classification problem, the number of support vectors increases and thus sparsity is decreased.
For LS support vector machines, the situation is still worse: all the training
data become support vectors. Because the decision function is expressed by
the weighted sum of kernels plus the bias term usually called *kernel expansion*, an increase of support vectors results in slowing down classification. In
this section, we discuss sparse L1, L2, and LS support vector machines using
the idea of the empirical feature space discussed in Section 2.3.6.

**Table 4.4** Performance comparison of LP support vector machines

| Data | Item | One-against-all LP SVM | Pairwise LP SVM |
|---|---|---|---|
| Iris | Parm | $\gamma 0.1$, $C100$ | $\gamma 1$, $C5,000$ |
| | Dis. | 92.00 (94.67) | 92.00 |
| | Min | **94.67** (97.33) | 92.00 |
| | Avg. | **94.67** (97.33) | 92.00 |
| | SVs | 3 | 3 |
| Numeral | Parm | $d2$, $C1$ | $\gamma 10$, $C5,000$ |
| | Dis. | 99.27 (99.63) | 99.39 |
| | Min | **99.51** (99.75) | 99.39 |
| | Avg. | **99.51** (99.75) | 99.39 |
| | SVs | 7 | 4 |
| Thyroid | Parm | $d2$, $C5,000$ | Linear, $C5,000$ |
| | Dis. | 96.00 (99.28) | 97.35 (98.59) |
| | Min | 97.61 (99.58) | 97.64 (98.78) |
| | Avg. | 97.61 (99.58) | **97.67** (98.78) |
| | SVs | 69 | 14 |
| Blood cell | Parm | $d4$, $C1$ | $\gamma 10$, $C50$ |
| | Dis. | 87.84 (93.38) | 91.90 (97.16) |
| | Min | **92.94** (96.42) | 92.52 (97.26) |
| | Avg. | **92.94** (96.42) | 92.65 (97.32) |
| | SVs | 27 | 6 |
| Hiragana-50 | Parm | $\gamma 10$, $C50$ | $\gamma 10$, $C50$ |
| | Dis. | 94.62 | 97.44 (99.93) |
| | Min | **98.11** | 97.87 (99.98) |
| | Avg. | **98.11** | 97.79 |
| | SVs | 32 | 8 |
| Hiragana-13 | Parm | $\gamma 10$, $C100$ | nn, $C5,000$ |
| | Dis. | 97.01 (99.63) | 98.79 (99.81) |
| | Min | **99.34** (99.90) | 99.13 (99.87) |
| | Avg. | **99.34** (99.90) | 99.16 (99.93) |
| | SVs | 27 | 4 |
| Hiragana-105 | Parm | $\gamma 10$, $C10$ | $\gamma 10$, $C50$ |
| | Dis. | 99.84 (99.99) | **100** |
| | Min | **100** | **100** |
| | Avg. | **100** | **100** |
| | SVs | 38 | 9 |

## 4.3.1 Several Approaches for Sparse Support Vector Machines

There are many papers discussing sparse support vector machines, e.g., [22–25]. The solution of support vector machines is expressed by the weighted sum of kernels plus the bias term

$$D(\mathbf{x}) = \sum_{i \in S} \alpha_i K(\mathbf{x}_i, \mathbf{x}) + b, \tag{4.71}$$

where $S$ is a set of support vector indices, $\alpha_i$ are nonzero constants, $K(\mathbf{x}, \mathbf{x}')$ is the kernel, and $b$ is the bias term. The purpose of sparse support vector machines is to minimize the number of support vectors while retaining the generalization ability comparable with that of the regular support vector machines. The right-hand side of (4.71) is called *kernel expansion*. Thus the problem is how to select the smallest number of support vectors, i.e., the smallest kernel expansion. There are roughly three approaches:

1. Simplify the decision function after training a support vector machine [22].
2. Reformulate the formulation of support vector machines so that sparsity is increased.

   Wu et al. [26] impose, as a constraint, the weight vector that is expressed by a fixed number of kernel functions and solved the optimization problem by the steepest descent method.
3. Sequentially build up the kernel expansion during training [23] or determine the kernel expansion before training [27–30].

   Keerthi et al. [23] train L2 support vector machines in the primal form. The idea is to select basis vectors by forward selection and for the selected basis vectors train support vector machines by Newton's method. This process is iterated until some stopping condition is satisfied.

   In [27–30] linearly independent data in the empirical feature space are selected either by the Cholesky factorization or kernel discriminant analysis and in the empirical feature space generated by these data a support vector machine is trained. This method is first derived for LS support vector machines and then extended to regular support vector machines.

In addition to the above methods, some use different architectures to realize sparse solutions: relevance vector machines based on a Bayesian learning framework [31] and their variants [32]; and import vector machines [33, 34] based on kernel logistic regression [35].

Because for LS support vector machines, all the training data become support vectors, several methods classified into Approach 3 have been developed. Suykens et al.'s method [12, 3] prunes support vectors with small absolute values of the associated dual variables and retrain an LS support vector machine using the reduced training data set. This process is iterated until sufficient sparsity is realized. Cawley and Talbot [36] reformulate LS support vector machines using the kernel expansion. And the above pruning method is used to reduce support vectors. Because the training data are reduced during pruning, information for the deleted training data is lost for the trained LS support vector machine. To overcome this problem, Valyon and Horváth [37] select linearly independent data in the feature space from the training data, and using the selected training data as support vectors obtain the solution by least-squares method using all the training data. Jiao et al.

[38] select linearly independent data by forward selection that minimize the objective function.

In the following we discuss sparse support vector machines based on [27–30], namely, first we select linearly independent data in the empirical feature space and then train support vector machines in the empirical feature space. To select linearly independent data in the empirical feature space we use the Cholesky factorization [27] or forward selection based on the class separability calculated by the linear discriminant analysis (LDA) in the empirical feature space.

## 4.3.2 Idea

From the discussions in Section 2.3.6, the empirical feature space, which is mapped by the mapping function given by (2.151), is equivalent to the feature space in that the associated kernel functions give the same values. Thus, instead of training support vector machines in the feature space, we can equivalently train ones in the finite empirical feature space. But to do so we need to transform input variables into variables in the empirical feature space by (2.151). To speed up generating the empirical feature space we select linearly independent training data that span the empirical feature space. Let the $N (\leq M)$ training data $\mathbf{x}_{i_1}, \ldots, \mathbf{x}_{i_N}$ be linearly independent in the empirical feature space, where $\mathbf{x}_{i_j} \in \{\mathbf{x}_1, \ldots, \mathbf{x}_M\}$ and $j = 1, \ldots, N$. Then, we use the following mapping function:

$$\mathbf{h}(\mathbf{x}) = (K(\mathbf{x}_{i_1}, \mathbf{x}), \ldots, K(\mathbf{x}_{i_N}, \mathbf{x}))^\top, \tag{4.72}$$

where $K(\mathbf{x}, \mathbf{x}')$ is a kernel. Because a support vector machine is trained in the empirical feature space, obtained support vectors are expressed by a linear combination of $\mathbf{x}_{i_1}, \ldots, \mathbf{x}_{i_N}$. According to the original definition, support vectors are the smallest subset of training data that generates the decision function. Therefore, support vectors in this case are the support vectors plus $\mathbf{x}_{i_1}, \ldots, \mathbf{x}_{i_N}$. But because in classification, we need only $\mathbf{x}_{i_1}, \ldots, \mathbf{x}_{i_N}$ and support vectors are not necessary, in the following we consider $\mathbf{x}_{i_1}, \ldots, \mathbf{x}_{i_N}$ as support vectors.

According to this definition, support vectors do not change even if the value of the margin parameter changes. And the number of support vectors is the number of selected independent training data that span the empirical feature space. Then reducing $N$ without deteriorating the generalization ability we can realize sparse support vector machines.

Then which is sparser, the original support vector machine trained in the feature space or the support vector machine trained in the empirical feature space? That depends on classification problems for the L1 and L2 support vector machines but for LS support vector machines, the LS support vector machine trained in the empirical feature space is sparser.

If we use linear kernels we do not need to select linearly independent variables if the number of input variables is very small compared to the number of training data. In that case, instead of (4.72), we use

$$h(\mathbf{x}) = \mathbf{x}. \tag{4.73}$$

This is equivalent to using $\mathbf{e}_i$ $(i = 1, \ldots, m)$, where $\mathbf{e}_i$ are the basis vectors in the input space, in which the $i$th element is 1 and other elements 0. If the training data span the input space, the support vector machine trained in the feature space and the support vector machine using (4.73) are equivalent but if the training data do not span the input space, they are different.

In the following first we discuss training L1, L2, and LS support vector machines in the empirical feature space. Then, we discuss two methods in selecting independent data and finally, we show performance evaluation results using some benchmark data sets.

### 4.3.3 Support Vector Machines Trained in the Empirical Feature Space

First we discuss L1 support vector machines trained in the empirical feature space.

Let the $M$ pairs of training inputs and outputs be $(\mathbf{x}_i, y_i)$ $(i = 1, \ldots, M)$ and $y_i = 1$ for Class 1 and $y_i = -1$ for Class 2. And let the decision function in the feature space be given by

$$D(\mathbf{x}) = \mathbf{w}^\top \boldsymbol{\phi}(\mathbf{x}) + b, \tag{4.74}$$

where $\mathbf{w}$ is the $l$-dimensional vector in the feature space, $\boldsymbol{\phi}(\mathbf{x})$ is the mapping function that maps the input space into the feature space, and $b$ is the bias term. Then the optimal separating hyperplane of the L1 support vector machine in the feature space is determined by solving

$$\text{minimize} \quad Q(\mathbf{w}, b, \boldsymbol{\xi}_i) = \frac{1}{2}\mathbf{w}^\top \mathbf{w} + C \sum_{i=1}^{M} \xi_i \tag{4.75}$$

$$\text{subject to} \quad y_i (\mathbf{w}^\top \boldsymbol{\phi}(\mathbf{x}_i) + b) \geq 1 - \xi_i \quad \text{for } i = 1, \ldots, M, \tag{4.76}$$

where $\xi_i$ is the slack variable for $\mathbf{x}_i$ and $C$ is the margin parameter.

Let the optimal separating hyperplane in the empirical feature space be

$$D(\mathbf{x}) = \mathbf{v}^\top \mathbf{h}(\mathbf{x}) + b_{\mathrm{e}}, \tag{4.77}$$

where $\mathbf{v}$ is the $N$-dimensional vector in the empirical feature space and $b_e$ is the bias term. Then the optimal separating hyperplane in the empirical feature space is determined by solving

$$\text{minimize} \quad Q(\mathbf{v}, b_e, \boldsymbol{\xi}) = \frac{1}{2}\mathbf{v}^{\top}\mathbf{v} + C\sum_{i=1}^{M}\xi_i \tag{4.78}$$

$$\text{subject to} \quad y_i\left(\mathbf{v}^{\top}\mathbf{h}(\mathbf{x}_i) + b_e\right) \geq 1 - \xi_i \quad \text{for } i = 1,\ldots,M. \tag{4.79}$$

Because the dimension of $\mathbf{v}$ is at most $M$, the optimization problem given by (4.78) and (4.79) is solvable in its primal form although the optimization problem given by (4.75) and (4.76) is not if the dimension of the feature space is infinite. But here we consider the dual problem derived from (4.78) and (4.79) as follows:

$$\text{maximize} \quad Q(\boldsymbol{\alpha}) = \sum_{i=1}^{M}\alpha_i - \frac{1}{2}\sum_{i,j=1}^{M}\alpha_i\,\alpha_j\,y_i\,y_j\,K_e(\mathbf{x}_i,\mathbf{x}_j) \tag{4.80}$$

$$\text{subject to} \quad \sum_{i=1}^{M}y_i\,\alpha_i = 0, \quad C \geq \alpha_i \geq 0 \quad \text{for } i = 1,\ldots,M, \tag{4.81}$$

where $K_e(\mathbf{x},\mathbf{x}')$ is the kernel for the empirical feature space, $K_e(\mathbf{x},\mathbf{x}') = \mathbf{h}^{\top}(\mathbf{x})\,\mathbf{h}(\mathbf{x}')$, and $\alpha_i\,(i = 1,\ldots,M)$ are Lagrange multipliers associated with $\mathbf{x}_i$.

Similar to the above discussions, the L2 support vector machine in the empirical feature space is given by

$$\text{maximize} \quad Q(\boldsymbol{\alpha}) = \sum_{i=1}^{M}\alpha_i - \frac{1}{2}\sum_{i,j=1}^{M}\left(\alpha_i\,\alpha_j\,y_i\,y_j K_e(\mathbf{x}_i,\mathbf{x}_j) + \frac{\delta_{ij}}{C}\right) \tag{4.82}$$

$$\text{subject to} \quad \sum_{i=1}^{M}y_i\,\alpha_i = 0, \quad \alpha_i \geq 0 \quad \text{for } i = 1,\ldots,M. \tag{4.83}$$

The difference of training support vector machines in the feature space or in the empirical feature space is whether we use $K(\mathbf{x},\mathbf{x}')$ or $K_e(\mathbf{x},\mathbf{x}')$. Therefore, by a slight modification of the software for support vector machines, we obtain software for training support vector machines in the empirical feature space.

The LS support vector machine in the feature space is trained by

$$\text{minimize} \quad \frac{1}{2}\mathbf{w}^{\top}\mathbf{w} + \frac{C}{2}\sum_{i=1}^{M}\xi_i^2 \tag{4.84}$$

$$\text{subject to} \quad \mathbf{w}^{\top}\phi(\mathbf{x}_i) + b = y_i - \xi_i \quad \text{for } i = 1,\ldots,M. \tag{4.85}$$

Similarly the LS support vector machine in the empirical feature space is trained by

$$\text{minimize} \quad Q(\mathbf{v}, \boldsymbol{\xi}, b_e) = \frac{1}{2} \mathbf{v}^\top \mathbf{v} + \frac{C}{2} \sum_{i=1}^{M} \xi_i^2 \tag{4.86}$$

$$\text{subject to} \quad \mathbf{v}^\top \mathbf{h}(\mathbf{x}_i) + b_e = y_i - \xi_i \quad \text{for } i = 1, \ldots, M. \tag{4.87}$$

We can solve (4.86) and (4.87) either in the primal or dual form. But because the primal form is faster for $N < M$ [39], in the following we discuss the primal form.

Substituting (4.87) into (4.86), we obtain

$$Q(\mathbf{v}, \boldsymbol{\xi}, b_e) = \frac{1}{2} \mathbf{v}^\top \mathbf{v} + \frac{C}{2} \sum_{i=1}^{M} (y_i - \mathbf{v}^\top \mathbf{h}(\mathbf{x}_i) - b_e)^2. \tag{4.88}$$

Equation (4.88) is minimized when the following equations are satisfied:

$$\frac{\partial Q(\mathbf{v}, \boldsymbol{\xi}, b_e)}{\partial \mathbf{v}} = \mathbf{v} - C \sum_{i=1}^{M} (y_i - \mathbf{v}^\top \mathbf{h}(\mathbf{x}_i) - b_e) \mathbf{h}(\mathbf{x}_i) = \mathbf{0} \tag{4.89}$$

$$\frac{\partial Q(\mathbf{v}, \boldsymbol{\xi}, b_e)}{\partial b} = -C \sum_{i=1}^{M} (y_i - \mathbf{v}^\top \mathbf{h}(\mathbf{x}_i) - b_e) = 0. \tag{4.90}$$

From (4.90)

$$b_e = \frac{1}{M} \sum_{i=1}^{M} (y_i - \mathbf{v}^\top \mathbf{h}(\mathbf{x}_i)). \tag{4.91}$$

Substituting (4.91) into (4.89), we obtain

$$\left( \frac{1}{C} + \sum_{i=1}^{M} \mathbf{h}(\mathbf{x}_i) \mathbf{h}^\top(\mathbf{x}_i) - \frac{1}{M} \sum_{i,j=1}^{M} \mathbf{h}(\mathbf{x}_i) \mathbf{h}^\top(\mathbf{x}_j) \right) \mathbf{v}$$

$$= \sum_{i=1}^{M} y_i \, \mathbf{h}(\mathbf{x}_i) - \frac{1}{M} \sum_{i,j=1}^{M} y_i \, \mathbf{h}(\mathbf{x}_j). \tag{4.92}$$

The coefficient matrix of $\mathbf{v}$ is proved to be positive definite [39]. Therefore, from (4.92) and (4.91) we obtain $\mathbf{v}$ and $b_e$.

## 4.3.4 Selection of Linearly Independent Data

### 4.3.4.1 Idea

In this section we discuss two methods for selecting linearly independent data. The first method uses the Cholesky factorization of the kernel matrix. During factorization, if the argument of the square root associated with the diagonal element is smaller than the prescribed threshold value, we delete the associated row and column and continue decomposing the matrix. By increasing the threshold value, we can increase the sparsity of the support vector machine.

So long as we select the linearly independent data that span the empirical feature space, different sets of linearly independent data do not affect the generalization ability of the support vector machine, because the different sets span the same empirical feature space.[2]

But the different sets of the linearly independent data for the reduced empirical feature space span different reduced empirical feature spaces. Thus, the processing order of the training data affects the generalization ability of the support vector machine.

To overcome this problem, we consider selecting independent data that maximally separate two classes using linear discriminant analysis (LDA) calculated in the reduced empirical feature space. This idea of selecting data by LDA is essentially the same with that used in [40].

### 4.3.4.2 Selection of Independent Data by the Cholesky Factorization

Let the kernel matrix $K = \{K(\mathbf{x}_i, \mathbf{x}_j)\}$ $(i, j = 1, \ldots, M)$ be positive definite. Then $K$ is decomposed by the Cholesky factorization into

$$K = L L^\top, \tag{4.93}$$

where $L$ is the regular lower triangular matrix and each element $L_{ij}$ is given by

$$L_{op} = \frac{K_{op} - \sum_{n=1}^{p-1} L_{pn} L_{on}}{L_{pp}} \quad \text{for } o = 1, \ldots, M, \quad p = 1, \ldots, o-1, \tag{4.94}$$

---

[2] Strictly speaking this statement is wrong, because support vector machines are not linear-transformation invariant. But by cross-validation, the difference of the generalization abilities may be small.

$$L_{aa} = \sqrt{K_{aa} - \sum_{n=1}^{a-1} L_{an}^2} \quad \text{for } a = 1, 2, \ldots, M. \tag{4.95}$$

Here, $K_{ij} = K(\mathbf{x}_i, \mathbf{x}_j)$.

Then during the Cholesky factorization, if the argument of the square root associated with the diagonal element is smaller than the prescribed value $\eta_C \ (> 0)$

$$H_{aa} - \sum_{n=1}^{a-1} L_{an}^2 \leq \eta_C, \tag{4.96}$$

we delete the associated row and column and continue decomposing the matrix. The training data that are not deleted in the Cholesky factorization are linearly independent.

The above Cholesky factorization can be done incrementally [41], namely, instead of calculating the full kernel matrix in advance, if (4.96) is not satisfied, we overwrite the $a$th column and row elements with those newly calculated using the previously selected data and $\mathbf{x}_{a+1}$. Thus the dimension of $L$ is the number of selected training data, not the number of training data.

To increase sparsity of support vector machines, we increase the value of $\eta_C$. The optimal value is determined by cross-validation.

### 4.3.4.3 Forward Selection by Linear Discriminant Analysis in the Empirical Feature Space

Using LDA in the empirical feature space, which will be discussed in Section 6.5.2, we select linearly independent data by forward selection. Starting from an empty set we add one data sample at a time that maximizes (6.81) if the data sample is added. Let the set of selected data indices be $S^k$ and the set of remaining data indices be $T^k$, where $k$ denotes that $k$ data are selected. Initially $S^0 = \phi$ and $T^0 = \{1, \ldots, M\}$. Let $S_j^k$ denote that $\mathbf{x}_j$ for $j \in T^k$ is temporarily added to $S^k$. Let $\mathbf{h}^{k,j}(\mathbf{x})$ be the mapping function with $\mathbf{x}_j$ temporarily added to the selected data:

$$\mathbf{h}^{k,j}(\mathbf{x}) = (K(\mathbf{x}_{i_1}, \mathbf{x}), \ldots, K(\mathbf{x}_{i_k}, \mathbf{x}), K(\mathbf{x}_j, \mathbf{x}))^\top, \tag{4.97}$$

where $\{i_1, \ldots, i_k\} = S^k$. And let $J_{\text{opt}}^{k,j}$ be the optimum value of the objective function with the mapping function $\mathbf{h}^{k,j}(\mathbf{x})$. Then we calculate

$$j_{\text{opt}} = \arg_j J_{\text{opt}}^{k,j} \quad \text{for } j \in T^k \tag{4.98}$$

and if the addition of $\mathbf{x}_{j_{\text{opt}}}$ results in a sufficient increase in the objective function

$$\left( J_{\text{opt}}^{k,j_{\text{opt}}} - J_{\text{opt}}^k \right) / J_{\text{opt}}^{k,j_{\text{opt}}} \geq \eta_L, \tag{4.99}$$

where $\eta_L$ is a positive parameter, we increment $k$ by 1 and add $j_{\text{opt}}$ to $S^k$ and delete it from $T^k$. If the above equation does not hold we stop forward selection. We must notice that $J_{\text{opt}}^{k,j}$ is non-decreasing for the addition of data [42]. Thus the left-hand side of (4.99) is nonnegative.

If the addition of a data sample results in the singularity of $Q_{\mathbf{T}}^{k,j}$, where $Q_{\mathbf{T}}^{k,j}$ is the total scatter matrix evaluated using the data with $S^{k,j}$ indices, we consider that the data sample does not give useful information in addition to the already selected data. Thus, instead of adding a small value to the diagonal elements of $Q_{\mathbf{T}}^{k,j}$ to avoid singularity, we do not consider this data sample for a candidate of addition. This is equivalent to calculating the pseudo-inverse of $Q_{\mathbf{T}}^{k,j}$.

The necessary and sufficient condition of a matrix being positive definite is that all the principal minors are positive. And notice that the exchange of two rows and then the exchange of the associated two columns do not change the singularity of the matrix. Thus, if $\mathbf{x}_j$ causes the singularity of $Q_{\mathbf{T}}^{k,j}$, later addition will always cause singularity of the matrix, namely, we can delete $j$ from $T^k$ permanently. If there are many training data that cause singularity of the matrix, forward selection becomes efficient.

Thus the procedure of independent data selection is as follows:

1. Set $S^0 = \phi$, $T^0 = \{1, \ldots, M\}$, and $k = 0$. Calculate $j_{\text{opt}}$ given by (4.98) and set $S^1 = \{j_{\text{opt}}\}$, $T^1 = T^0 - \{j_{\text{opt}}\}$, and $k = 1$.
2. If for some $j \in T^k$, $Q_{\mathbf{T}}^{k,j}$ is singular, permanently delete $j$ from $T^k$ and calculate $j_{\text{opt}}$ given by (4.98). If (4.99) is satisfied, go to Step 3. Otherwise terminate the algorithm.
3. Set $S^{k+1} = S^k \cup \{j_{\text{opt}}\}$ and $T^{k+1} = T^k - \{j_{\text{opt}}\}$. Increment $k$ by 1 and go to Step 2.

Keeping the Cholesky factorization of $Q_{\mathbf{T}}^k$, the Cholesky factorization of $Q_{\mathbf{T}}^{k,j}$ is done incrementally; namely, using the factorization of $Q_{\mathbf{T}}^k$, the factorization of $Q_{\mathbf{T}}^{k,j}$ is obtained by calculating the $(k+1)$th diagonal element and column elements. This accelerates the calculation of the inverse of the within-class scatter matrix.

For a large training data set, forward selection will be inefficient. To speed up selecting linearly independent data in such a situation, we may divide the training data set into several subsets and repeat forward selection for each subset.

### 4.3.5 Performance Evaluation

We evaluated the generalization ability and sparsity of sparse L1, L2, and LS SVMs using the two-class data sets listed in Table 1.1. Because in most cases forward selection by LDA worked better than the Cholesky factorization, we selected linearly independent data by forward selection based on LDA with

$\eta_L = 10^{-3}$ for L1 and L2 SVMs and $\eta_L = 10^{-4}$ for the LS SVM. We judged singularity of a matrix when a diagonal element in the Cholesky factorization was smaller than $10^{-6}$. In the following we use SL1 SVM, SL2 SVM, and SLS SVM to denote sparse support vector machines according to the preceding procedure.

We normalized the input ranges into $[0, 1]$ and used RBF kernels and determined the parameter values of $C$ and $\gamma$ using the first five training data sets by fivefold cross-validation; the value of $C$ was selected from among $\{1, 10, 50, 100, 500, 1,000, 2,000, 3,000, 5,000, 8,000, 10,000, 50,000, 100,000\}$ and the value of $\gamma$ from among $\{0.1, 0.5, 1, 5, 10, 15\}$.

Table 4.5 lists the determined parameter values. The values of $\gamma$ for L1 and SL1 SVMs are the same for each problem. This is also true for L2 and SL2 SVMs but for LS and SLS SVMs the values are the same for only 3 problems out of 13. In most cases, the values of $C$ for sparse SVMs are larger than those of the associated non-sparse SVMs.

**Table 4.5** Parameter setting for two-class problems

| Data | L1 SVM $\gamma$ | $C$ | SL1 SVM $\gamma$ | $C$ | L2 SVM $\gamma$ | $C$ | SL2 SVM $\gamma$ | $C$ | LS SVM $\gamma$ | $C$ | SLS SVM $\gamma$ | $C$ |
|---|---|---|---|---|---|---|---|---|---|---|---|---|
| Banana | 15 | 100 | 15 | 8,000 | 10 | 3,000 | 10 | $10^5$ | 10 | 500 | 15 | $10^5$ |
| B. cancer | 0.1 | 500 | 0.1 | 1,000 | 0.1 | 500 | 0.1 | 2,000 | 0.5 | 10 | 1 | 100 |
| Diabetes | 0.1 | 3,000 | 0.1 | 50,000 | 0.1 | 3,000 | 0.1 | 50,000 | 10 | 1 | 5 | 10 |
| Flare-solar | 0.5 | 10 | 0.5 | 10 | 0.5 | 1 | 0.5 | 50 | 0.1 | 100 | 0.5 | 1,000 |
| German | 0.1 | 50 | 0.1 | 3,000 | 0.1 | 50 | 0.1 | 3,000 | 0.5 | 50 | 1 | 3,000 |
| Heart | 0.1 | 50 | 0.1 | $10^4$ | 0.1 | 50 | 0.1 | $10^4$ | 0.1 | 10 | 0.5 | 50 |
| Image | 15 | 500 | 15 | $10^5$ | 15 | 500 | 15 | $10^5$ | 10 | 3,000 | 10 | $10^5$ |
| Ringnorm | 15 | 1 | 15 | 1 | 15 | 1,000 | 15 | 50 | 10 | 1 | 0.5 | 1 |
| Splice | 10 | $10^5$ | 10 | $10^5$ | 10 | 1,000 | 10 | $10^4$ | 5 | 50 | 0.5 | 1 |
| Thyroid | 15 | 100 | 15 | 500 | 15 | 100 | 15 | 500 | 10 | 50 | 15 | $10^4$ |
| Titanic | 0.5 | 50 | 0.5 | 1,000 | 0.5 | 10 | 0.5 | 100 | 0.5 | 500 | 5 | 3,000 |
| Twonorm | 0.5 | 1 | 0.5 | 10 | 0.5 | 1 | 0.5 | 50,000 | 0.1 | 10 | 0.1 | 50,000 |
| Waveform | 10 | 1 | 10 | 1 | 15 | 1 | 15 | 1 | 10 | 1 | 10 | 10 |

Table 4.6 shows the average classification errors and standard deviations. Comparing L1 and SL1 SVMs, for the image and splice problems, the average errors of the SL1 SVM are inferior. This may be caused by excessive deletion of training data. For L2 and SL2 SVMs, the L2 SVM for the flare-solar problem shows inferior performance. For the LS and SLS SVMs, the LS SVM shows inferior performance for the waveform problem and the SLS SVM does for the image and ringnorm problems. For the ringnorm problem, changing the value of $C$ to 10, the classification error was $5.57 \pm 1.1$ with support vectors $6.5 \pm 0.9$. Therefore, inferior solution was caused by improper selection of parameter values. For the other problems the performance is comparable. Therefore, conventional and sparse SVMs show comparable classification performance.

**Table 4.6** Comparison of the average classification errors and the standard deviations of the errors in percentage

| Data | L1 SVM | SL1 SVM | L2 SVM | SL2 SVM | LS SVM | SLS SVM |
|------|--------|---------|--------|---------|--------|---------|
| Banana | 10.7±0.52 | 10.9±0.60 | 10.6±0.51 | 10.7±0.61 | 10.7±0.52 | 10.8±0.52 |
| B. cancer | 27.6±4.7 | 28.6±4.5 | 25.3±4.1 | 27.5±4.8 | 25.7±4.5 | 25.4±4.5 |
| Diabetes | 23.7±1.8 | 24.4±1.8 | 23.5±1.9 | 24.0±1.8 | 23.2±1.7 | 22.9±1.8 |
| Flare-solar | 32.4±1.7 | 32.6±1.7 | 35.7±3.2 | 32.6±1.7 | 33.3±1.6 | 33.4±1.5 |
| German | 23.8±2.3 | 23.8±2.3 | 25.4±2.3 | 25.3±2.1 | 23.3±2.1 | 24.0±2.2 |
| Heart | 16.3±3.4 | 16.7±3.5 | 16.4±2.9 | 16.5±3.2 | 16.0±3.4 | 16.2±3.3 |
| Image | 2.7±0.41 | 3.8±0.59 | 3.6±0.72 | 3.8±0.59 | 2.66±0.37 | 3.80±0.65 |
| Ringnorm | 2.2±0.30 | 1.8±0.22 | 1.8±0.20 | 1.8±0.22 | 4.08±0.58 | 7.56±5.2 |
| Splice | 10.8±0.71 | 15.5±0.70 | 14.1±1.3 | 15.5±0.70 | 11.3±0.51 | 11.6±0.61 |
| Thyroid | 3.9±2.1 | 4.3±1.9 | 3.8±1.9 | 4.3±2.0 | 4.84±2.5 | 4.68±2.5 |
| Titanic | 22.5±0.55 | 23.0±1.2 | 23.2±1.4 | 23.2±1.8 | 22.5±0.97 | 22.4±1.1 |
| Twonorm | 2.4±0.14 | 3.1±0.36 | 2.8±0.66 | 2.9±0.31 | 1.90±0.61 | 2.63±0.20 |
| Waveform | 10.0±0.44 | 10.5±0.43 | 9.8±0.40 | 10.5±0.43 | 14.9±0.98 | 9.72±0.37 |

Table 4.7 lists the number of support vectors. Except for the thyroid problem, the number of support vectors for SL1 and SL2 SVMs are smaller than L1 and L2 SVMs, respectively. And except for the banana problem, the number of support vectors for SL1 and SL2 SVMs is the same. This is because the $\gamma$ values for SL1 and SL2 SVMs are the same except for the banana problem as seen from Table 4.5. For LS and SLS SVMs, the number of support vectors for SLS SVMs is smaller, because for the LS SVM all the training data become support vectors.

Although SLS SVMs requires more support vectors than L1 and L2 SVMs, their classification performance is comparable.

**Table 4.7** Comparison of support vectors

| Data | L1 SVM | SL1 SVM | L2 SVM | SL2 SVM | LS SVM | SLS SVM |
|------|--------|---------|--------|---------|--------|---------|
| Banana | 101±10 | 22.9±8.9 | 168±21 | 20.1±2.3 | 400 | 27.8±1.6 |
| B. cancer | 124±11 | 9.8±1.9 | 176±6.4 | 9.8±1.9 | 200 | 42.1±1.9 |
| Diabetes | 255±12 | 6.0±0.61 | 393±9.1 | 6.0±0.62 | 468 | 65.7±4.1 |
| Flare-solar | 530±14 | 12±3.1 | 322±96 | 12±3.1 | 666 | 16.0±1.5 |
| German | 398±6.1 | 13.1±2.2 | 575±12 | 13.1±2.2 | 700 | 95.2±3.9 |
| Heart | 73.9±5.6 | 9.7±1.7 | 123±7.8 | 9.7±1.7 | 170 | 32.5±2.8 |
| Image | 151±8.0 | 66.2±7.7 | 221±15 | 66.2±7.7 | 1,300 | 132±7.7 |
| Ringnorm | 130±5.5 | 32.2±24 | 249±7.5 | 32.2±24 | 400 | 10.3±1.2 |
| Splice | 741±14 | 241±9.0 | 376±125 | 241±9.0 | 1,000 | 619±10 |
| Thyroid | 14.1±2.0 | 27.1±5.0 | 18.0±4.5 | 27.1±5.0 | 140 | 38.3±2.5 |
| Titanic | 139±10 | 5.7±1.4 | 149±3.5 | 5.7±1.4 | 150 | 9.97±1.3 |
| Twonorm | 255±8.0 | 7.6±1.0 | 259±45 | 7.6±1.0 | 400 | 11.6±1.1 |
| Waveform | 153±8.9 | 112±24 | 208±9.9 | 112±24 | 400 | 278±9.4 |

## 4.4 Performance Comparison of Different Classifiers

In this section, we compare performance of L1, L2, LS, SLS, and LP SVMs using the two-class data sets, microarray data sets, and multiclass data sets shown in Section 1.3. We have already shown or will show their performance and the results shown there may be different from those shown here. The reason is due to the parameter ranges used for evaluation. Therefore, to compare different classifiers, we need to be careful in evaluation conditions and we need to notice that the performance difference is only valid for the conditions that we used in evaluation.

For the two-class problems we used RBF kernels. Table 4.8 shows the parameter values and the average number of support vectors and their standard deviations for the two-class problems. In the table, $\gamma$ and $C$ values were determined by fivefold cross-validation and the $\gamma$ value was selected from $\{0.1, 0.5, 1, 5, 10, 15\}$ and the $C$ value from $\{1, 10, 50, 100, 500, 1,000, 2,000\}$. The numerals in the "SVs" column of the LP SVM show the average number of nonzero support vectors and its standard deviation and average number of all the support vectors and its standard deviation in parentheses. For the LS SVM, all the training data become support vectors and thus the number of support vectors is the largest. For most cases, the numbers of support vectors for the L1 SVM and LP SVM are almost the same. But because the numbers of nonzero support vectors for the LP SVM decrease drastically, the classification time of the LP SVM is much shorter.

**Table 4.8** Parameter values and support vectors for two-class problems

| Data | L1 SVM | | | LS SVM | | | LP SVM | | |
|------|--------|------|------|--------|------|------|--------|------|------|
| | $\gamma$ | $C$ | SVs | $\gamma$ | $C$ | SVs | $\gamma$ | $C$ | SVs |
| Banana | 15 | 100 | 101±10 | 10 | 500 | 400 | 15 | 100 | 18±1.5 (106±11) |
| B. cancer | 1 | 10 | 114±5.5 | 0.5 | 10 | 200 | 1 | 50 | 24±2.1 (115±6.2) |
| Diabetes | 0.5 | 50 | 250±7.9 | 10 | 1 | 468 | 0.1 | 100 | 8.6±0.73 (255±7.8) |
| Flare-solar | 1 | 10 | 448±13 | 0.1 | 100 | 666 | 5 | 1 | 13±3.1 (454±13) |
| German | 0.5 | 50 | 388±11 | 0.5 | 50 | 700 | 5 | 1 | 28±2.7 (410±11) |
| Heart | 0.1 | 50 | 74±5.8 | 0.1 | 10 | 170 | 0.1 | 10 | 6.7±1.0 (86±6.1) |
| Image | 15 | 500 | 146±2.1 | 10 | 3,000 | 1,300 | 15 | 100 | 68±3.6 (198±10) |
| Ringnorm | 15 | 1 | 131±5.6 | 10 | 1 | 400 | 15 | 1 | 16±2.1 (57±4.9) |
| Splice | 10 | 2,000 | 740±13 | 5 | 50 | 1,000 | 5 | 10 | 346±17 (399±17) |
| Thyroid | 5 | 1,000 | 13±1.9 | 10 | 50 | 140 | 5 | 10 | 7.7±1.1 (20±2.8) |
| Titanic | 5 | 2,000 | 67±9.8 | 0.5 | 500 | 150 | 10 | 10 | 8.0±1.7 (68±9.9) |
| Twonorm | 0.5 | 1 | 255±8.0 | 0.1 | 10 | 400 | 1 | 10 | 16±1.9 (59±5.0) |
| Waveform | 15 | 1 | 149±8.7 | 10 | 1 | 400 | 1 | 1 | 7.1±1.5 (186±9.9) |

Table 4.9 shows the average classification errors and their standard deviations of the test data sets for the three classifiers. For the heart problem,

although $C = 1$ was selected for the LP SVP, the degenerate solutions were obtained for some data sets. Thus, we obtained solutions setting $C = 10$.

We performed Welch $t$-test for a pair of the average classification errors and their standard deviations with the significance level of 0.05. We show the average classification error and/or its standard deviation in boldface if they are statistically better than the remaining two and there is no statistical difference between the remaining two. But if the remaining two are statistically different, we show the worst average classification error and/or its standard deviation in italic. For instance, for the waveform problem, the average classification error and its standard deviation of the L1 SVM is the best and the average classification error of the LS SVM is the worst although the standard deviations of the LS SVM and LP SVM are statistically not different.

From the table, classification performance of the L1 SVM is better than that of the LS SVM or LP SVM and the LS SVM and LP SVM show comparable classification performance.

**Table 4.9** Comparison of average classification errors and their standard deviations in percentage for two-class problems

| Data | L1 SVM | LS SVM | LP SVM |
|------|--------|--------|--------|
| Banana | **10.5±0.46** | 10.7±0.52 | **10.6±0.49** |
| B. cancer | 25.6±4.4 | 25.7±4.5 | 25.9±4.6 |
| Diabetes | **23.3±1.8** | **23.2±1.7** | 23.7±1.8 |
| Flare-solar | **32.5±1.7** | 33.3±1.6 | **32.4±1.7** |
| German | 23.5±2.2 | **23.3±2.1** | 23.9±2.2 |
| Heart | 16.1±3.1 | 16.0±3.4 | 16.0±3.4 |
| Image | **2.80±0.44** | **2.66±0.37** | 3.59±0.54 |
| Ringnorm | 2.64±0.35 | *4.08±0.58* | **1.90±0.23** |
| Splice | **10.8±0.71** | 11.3±0.51 | *12.3±0.83* |
| Thyroid | **4.05±2.3** | 4.84±2.5 | **4.00±2.2** |
| Titanic | 22.3±0.98 | 22.5±0.97 | 22.5±0.99 |
| Twonorm | 2.13±0.65 | **1.90±0.61** | *2.91±0.27* |
| Waveform | **9.88±0.47** | *14.9±0.98* | 12.2±0.82 |

Because the microarray problems have very small training samples but with large numbers of features, overfitting occurs quite easily. Thus, we used linear kernels and selected the values of $C$ from $\{0.0001, 0.001, 0.1, 1\}$ by fivefold cross-validation. If there were multiple values of $C$ with the same recognition rate for the validation data set, we selected the highest value.

Table 4.10 lists the values of $C$ and the numbers of support vectors for microarray problems. In the "SVs" column for the LP SVM, the numerals show the support vectors deleting zero support vectors and the numerals in parentheses are the numbers of support vectors including the zero support vectors. The numbers of support vectors for the L1 and L2 SVMs are almost the same and those for the LS SVM are equal to the numbers of training data. The numbers of support vectors for the LP SVM are the smallest among four classifiers. For the breast cancer (s), the number of nonzero support vectors is zero. This means that any data are classified into one class.

**Table 4.10** Values of $C$ and numbers of support vectors for microarray problems

| Data | L1 SVM | | LS SVM | | L2 SVM | | LP SVM | |
|------|--------|-----|--------|-----|--------|-----|--------|-----|
|      | $C$ | SVs | $C$ | SVs | $C$ | SVs | $C$ | SVs |
| C. cancer | 1 | 21 | 1 | 23 | 1 | 40 | 1 | 13 (19) |
| Leukemia | 1 | 30 | 1 | 30 | 1 | 38 | 1 | 23 (28) |
| B. cancer (1) | 1 | 13 | 1 | 13 | 0.001 | 14 | 1 | 9 (10) |
| B. cancer (2) | 1 | 14 | 1 | 14 | 1 | 14 | 1 | 13 (14) |
| B. cancer (s) | 1 | 12 | 0.1 | 12 | 1 | 14 | 0.001 | 0 (8) |
| H. carcinoma | 1 | 30 | 0.001 | 33 | 0.001 | 33 | 1 | 24 (26) |
| H. glioma | 1 | 19 | 1 | 19 | 1 | 21 | 1 | 16 (17) |
| P. cancer | 1 | 73 | 0.01 | 73 | 1 | 102 | 0.01 | 31 (51) |
| B. cancer (3) | 1 | 72 | 1 | 72 | 1 | 78 | 0.001 | 10 (69) |

Table 4.11 shows the recognition rates of the test data for the four classifiers. If the recognition rate of the training data is not 100%, it is shown in parentheses. The best recognition rate among the four classifiers is shown in boldface. Comparing the number of the best recognition rates, the L2 SVM is the best, the L1 SVM the second best, and the LP SVM the worst. For the b. cancer (s) problem, the LP SVM showed the best performance, but it classified any data into one class.

In most cases, the recognition rates of the training data were 100%, and for the b. cancer (s), h. glioma, and p. cancer problems, any classifier did not perform well. We checked the best recognition rate for these problems; for the b. cancer (s), 62.50% was the best; for the h. glioma problem, 65.52% by the L1 SVM with $C = 0.001$; and for the p. cancer problem, 88.24% by the L1 SVM with $C = 0.001$. Therefore, for the b. cancer (s) and h. glioma problems we will not be able to improve generalization ability only by improving model selection.

**Table 4.11** Recognition rates in percentage for microarray problems

| Data | L1 SVM | L2 SVM | LS SVM | LP SVM |
|------|--------|--------|--------|--------|
| C. cancer | **77.27** | **77.27** | **77.27** | 72.73 |
| Leukemia | 85.29 | 85.29 | 85.29 | **91.18** |
| B. cancer (1) | **87.50** | **87.50** | 62.50 (71.43) | **87.50** |
| B. cancer (2) | **87.50** | **87.50** | **87.50** | **87.50** |
| B. cancer (s) | 25 | 25 | 25 | **62.50** (71.43) |
| H. carcinoma | 51.85 | **81.48** (93.94) | **81.48** (93.94) | 51.85 |
| H. glioma | **55.17** | **55.17** | **55.17** | 48.28 |
| P. cancer | **52.94** | **52.94** | 50 | 26.47 (96.08) |
| B. cancer (3) | **89.47** | **89.47** | **89.47** | 78.95 |

For the multiclass problems we evaluated performance of the L1 SVM, L2 SVM, LS SVM, SLS SVM (sparse LS SVM), and LP SVM using RBF kernels. We determined the values of $\gamma$ and $C$ by fivefold cross-validation. We selected the $\gamma$ value from $\{0.1, 0.5, 1, 5, 10, 15, 20, 50, 100, 200\}$ and the $C$ value from $\{1, 10, 50, 100, 500, 1,000, 2,000\}$. For the SLS SVM, we set $\eta_L = 10^{-4}$ for all the data sets. Table 4.12 shows the determined parameter values. In the table, "1-A" and "Pair" denote one-against-all classification and pairwise classification, respectively. Two numerals in each column denote the $\gamma$ and $C$ values determined by fivefold cross-validation.

**Table 4.12** Values of $\gamma$ and $C$ determined by cross-validation

| Data | Mult. | L1 SVM | L2 SVM | LS SVM | SLS SVM | LP SVM |
|---|---|---|---|---|---|---|
| Iris | 1-A | 0.1, 1,000 | 0.1, 1,000 | 0.1, 2,000 | 5, 10 | 0.1, 10 |
| | Pair | 0.1, 100 | 0.1, 100 | 0.1, 2,000 | 0.1, 2,000 | 0.1, 500 |
| Numeral | 1-A | 5, 10 | 0.1, 500 | 1, 50 | 5, 50 | 0.1, 100 |
| | Pair | 5, 10 | 0.1, 1,000 | 0.1, 100 | 10, 1 | 1, 10 |
| Thyroid | 1-A | 20, 2,000 | 20, 2,000 | 200, 500 | 200, 1,000 | 10, 2,000 |
| | Pair | 10, 2,000 | 5, 2,000 | 50, 2,000 | 200, 2,000 | 5, 500 |
| Blood cell | 1-A | 5, 500 | 10, 100 | 15, 1,000 | 50, 2,000 | 1, 2,000 |
| | Pair | 5, 100 | 1, 500 | 5, 500 | 20, 1,000 | 10, 50 |
| Hiragana-50 | 1-A | 10, 50 | 10, 1,000 | 5, 2,000 | – | 20, 50 |
| | Pair | 5, 2,000 | 5, 100 | 10, 100 | 15, 2,000 | 10, 100 |
| Hiragana-13 | 1-A | 50, 2,000 | 15, 500 | 50, 2,000 | – | 15, 100 |
| | Pair | 15, 1,000 | 15, 2,000 | 15, 2,000 | 100, 2,000 | 5, 500 |
| Hiragana-105 | 1-A | 20, 2,000 | 50, 10 | 10, 2,000 | – | 10, 10 |
| | Pair | 15, 2,000 | 15, 2,000 | 1, 2,000 | 20, 100 | 10, 10 |
| Satimage | 1-A | 200, 1 | 200, 10 | 200, 10 | – | 20, 100 |
| | Pair | 200, 10 | 200, 10 | 200, 10 | 100, 500 | 100, 10 |
| USPS | 1-A | 10, 10 | 10, 50 | 10, 100 | – | 15, 10 |
| | Pair | 10, 100 | 10, 50 | 5, 500 | 10, 50 | 10, 10 |

Table 4.13 shows the recognition rates of the test data sets. For each row, the best recognition rate is shown in boldface. The number of best recognition rates of the LS SVM is 9 and those of the L1 SVM and L2 SVM are 6. But the LS SVM performed worse than the L1 or L2 SVM for the numeral and thyroid data sets. For the blood cell data set, the LS SVM by pairwise classification performed best. Therefore, performance of the LS SVM depends on data sets but performance of the L1 and L2 SVM performed well irrespective of data sets. The SLS SVM realizes performance comparable with or better than that of the LS SVM, but for some data sets, the recognition rates by one-against-all classification were not obtained because of slow training. Performance of

the LP SVM is somewhat worse than that of other classifiers except for the thyroid data set.

**Table 4.13** Comparison of recognition rates of the test data sets for different classifiers

| Data | Mult. | L1 SVM | L2 SVM | LS SVM | SLS SVM | LP SVM |
|------|-------|--------|--------|--------|---------|--------|
| Iris | 1-A | 94.67 | 96.00 | 94.67 | **97.33** | 93.33 |
|      | Pair | **97.33** | **97.33** | **97.33** | 96.00 | 92.00 |
| Numeral | 1-A | **99.39** | **99.39** | 98.66 | 98.66 | **99.39** |
|         | Pair | **99.76** | **99.76** | 99.15 | 99.27 | 99.39 |
| Thyroid | 1-A | 96.82 | 96.85 | 95.13 | 94.17 | **97.23** |
|         | Pair | 97.26 | 97.35 | 95.39 | 94.63 | **97.67** |
| Blood cell | 1-A | 93.61 | 93.58 | **93.90** | 93.87 | 93.29 |
|            | Pair | 93.19 | 92.55 | **94.23** | 93.97 | 92.52 |
| Hiragana-50 | 1-A | **99.31** | 99.28 | 99.20 | – | 98.39 |
|             | Pair | 99.05 | 99.05 | **99.48** | 99.37 | 98.11 |
| Hiragana-13 | 1-A | 99.77 | 99.74 | **99.90** | – | 99.33 |
|             | Pair | 99.78 | 99.80 | 99.87 | **99.89** | 99.38 |
| Hiragana-105 | 1-A | **100** | **100** | **100** | – | **100** |
|              | Pair | **100** | **100** | **100** | **100** | 99.99 |
| Satimage | 1-A | 91.60 | 91.70 | **91.85** | – | 90.05 |
|          | Pair | 91.90 | **92.10** | 91.95 | 91.85 | 91.15 |
| USPS | 1-A | 95.52 | 95.42 | **95.57** | – | 94.37 |
|      | Pair | 95.27 | 95.27 | 95.47 | **95.52** | 94.32 |

Table 4.14 shows the average number of support vectors per decision function. For the LS SVM, the number of support vectors is the number of training data for determining the decision function. For the LP SVM, we list the nonzero support vectors. From the table, the number of support vectors for LP SVM is the smallest and in general followed by the L1 SVM and the SLS SVM. And the L1 SVM shows smaller number of support vectors than the L2 SVM.

## 4.5 Robust Support Vector Machines

Compared to conventional classifiers, support vector machines are robust for outliers because of the margin parameter that controls the trade-off between the generalization ability and the training error. But there are several approaches to enhance robustness of support vector machines [43–49].

**Table 4.14** Comparison of numbers of support vectors

| Data | Mult. | L1 SVM | L2 SVM | LS SVM | SLS SVM | LP SVM |
|------|-------|--------|--------|--------|---------|--------|
| Iris | 1-A | 15 | 23 | 75 | 20 | 3 |
|      | Pair | 10 | 21 | 50 | 4 | 2 |
| Numeral | 1-A | 17 | 64 | 810 | 54 | 7 |
|         | Pair | 9 | 16 | 162 | 14 | 3 |
| Thyroid | 1-A | 219 | 357 | 3,772 | 364 | 85 |
|         | Pair | 90 | 161 | 2,515 | 254 | 37 |
| Blood cell | 1-A | 99 | 188 | 3,097 | 214 | 23 |
|            | Pair | 18 | 31 | 516 | 46 | 6 |
| Hiragana-50 | 1-A | 68 | 77 | 4,610 | – | 36 |
|             | Pair | 16 | 22 | 236 | 36 | 8 |
| Hiragana-13 | 1-A | 63 | 63 | 8,375 | – | 30 |
|             | Pair | 12 | 12 | 441 | 42 | 4 |
| Hiragana-105 | 1-A | 131 | 708 | 8,375 | – | 38 |
|              | Pair | 35 | 35 | 441 | 39 | 9 |
| Satimage | 1-A | 947 | 1,001 | 4,435 | – | 146 |
|          | Pair | 337 | 387 | 1,478 | 197 | 57 |
| USPS | 1-A | 586 | 608 | 7,291 | – | 300 |
|      | Pair | 214 | 220 | 1,458 | 143 | 64 |

Herbrich and Weston [43, 45, 46] introduced an adaptive margin for each training sample so that the margins for outliers become large and hence the classification becomes robust.

When outliers are included, the median of data is known to be more robust than the center of data [50]. Kou et al. [47] proposed median support vector machines (MSVM), an improved version of central support vector machines [44], to improve robustness of support vector machines. Instead of maximizing the separating margin of two classes, in MSVMs, the sum of distances of the class medians to the decision boundary is maximized.

# 4.6 Bayesian Support Vector Machines

Support vector machines can show high generalization ability for a wide range of applications. Incorporating prior knowledge into support vector machines is one way of further improving the generalization ability [51–53]. For example, Burges and Schölkopf [54] introduced virtual support vectors. First, train a support vector machine using a training data set. Then obtain virtual support vectors by applying, to the obtained support vectors, linear transformations, such as translation, scaling, and rotation for character recognition. Finally, train another support vector machine using the virtual support vectors. By this method, the generalization ability for the handwritten numeral classification was improved but the classification time was increased due to the increase in the number of support vectors. To shorten the classification time, Burges and Schölkopf [54] proposed approximating the decision function with a smaller set of data.

To improve class separability, Amari and Wu [55, 56] proposed enlarging the regions around support vectors by replacing $K(\mathbf{x}, \mathbf{x}')$ with

$$\tilde{K}(\mathbf{x}, \mathbf{x}') = c(\mathbf{x})\, c(\mathbf{x}')\, K(\mathbf{x}, \mathbf{x}'), \tag{4.100}$$

where $c(\mathbf{x})$ is a positive scalar function, and one example is

$$c(\mathbf{x}) = \sum_{i=1}^{M} \alpha_i \exp(-\|\mathbf{x} - \mathbf{x}_i\|^2 / 2\tau^2). \tag{4.101}$$

Here, $\tau$ is a parameter. It is shown that by (4.100) the volumes around support vectors are expanded in the feature space. Thus class separability is increased.

In LS support vector machines, after determination of the bias term of a hyperplane, it is further tuned to improve classification accuracy [57]. A similar technique is also discussed in [58, 59].

In training support vector machines, the minimum margin among training data is maximized. But a theoretical progress in machine learning suggests that the margin distribution affects the generalization ability and even maximizing the average of margins over the training samples may lead to good generalization ability. Aiolli et al. [60] proposed directly optimizing the weighted sum of margins and showed that their method outperformed the regular support vector machine for 6 problems out of 13 two-class problems that are also used in this book. But comparing the results with those in the book, the difference is not so prominent.

In the following, we discuss the generalization improvement by optimizing the bias term based on the Bayes' theory [61]. We call this support vector machine the *Bayesian support vector machine*.

In support vector machines, the optimal hyperplane is placed in the middle of the support vectors belonging to different classes. This is because the training data are assumed to be generated by an unknown probability distribution. Thus, if the distribution of each class is known, we can improve the generalization ability by the Bayes' theory. In the following we assume that the distribution of the class data in the direction perpendicular to the optimal hyperplane is normal. Under the assumption, the optimal separating hyperplane is no longer optimal; the boundary determined by the Bayes' theory becomes optimal. Thus, by the parallel displacement of the optimal hyperplane to the position determined by the Bayes' theory, we can improve the generalization ability.

In the following, we first discuss the one-dimensional Bayesian decision function. Then we move the optimal hyperplane in parallel so that the class boundary becomes the boundary given by the Bayesian decision function, and we discuss how to test whether the class data are normal.

## 4.6.1 One-Dimensional Bayesian Decision Functions

Let two classes be $C_1$, $C_2$, where $C_1$ is on the positive side of the hyperplane and $C_2$ on the negative side. The posterior probability that the observed $\mathbf{x}$ belongs to class $C_k$, $P(C_k \,|\, \mathbf{x})$, is given by

$$P(C_k \,|\, \mathbf{x}) = \frac{P(C_k)\, p(\mathbf{x} \,|\, C_k)}{p(\mathbf{x})}, \tag{4.102}$$

where $P(C_k)$ is the a priori probability of class $C_k$, $p(\mathbf{x} \,|\, C_k)$ is the conditional probability density function when $\mathbf{x}$ belonging to class $C_k$ is observed, and $p(\mathbf{x})$ is a probability density function given by

$$p(\mathbf{x}) = \sum_{k=1}^{2} P(C_k)\, p(\mathbf{x} \,|\, C_k), \quad \int p(\mathbf{x})\, d\mathbf{x} = 1. \tag{4.103}$$

Equation (4.102) is called *Bayes' rule.*

We assume that the one-dimensional data, which belong to class $C_k$, obey the normal distribution with the mean $\mu_k$ and the variance $\sigma_k^2$ given by

$$p(x \,|\, C_k) = \frac{1}{\sqrt{2\pi\,\sigma_k^2}} \exp\left(-\frac{1}{2\sigma_k^2}(x-\mu_k)^2\right). \tag{4.104}$$

According to Bayes' theory, the optimal classification is to classify a data sample to the class with the maximum posterior probability. Instead of comparing the posterior probabilities, we compare the logarithms of the posterior probabilities, deleting the term $p(x)$ common to the classes:

$$g_k(x) = \log P(C_k) - \frac{1}{2}\left(\log(2\pi\,\sigma_k^2) + \frac{1}{\sigma_k^2}(x-\mu_k)^2\right). \tag{4.105}$$

Thus the Bayesian decision function for the two classes is given by

$$\begin{aligned}
D_{\text{Bayes}}(x) &= g_1(x) - g_2(x) \\
&= \log \frac{P(C_1)}{P(C_2)} \\
&\quad -\frac{1}{2}\left(\log \frac{\sigma_1^2}{\sigma_2^2} + \frac{1}{\sigma_1^2}(x-\mu_1)^2 - \frac{1}{\sigma_2^2}(x-\mu_2)^2\right). \tag{4.106}
\end{aligned}$$

If $D_{\text{Bayes}}(x) > 0$, $x$ is classified into class $C_1$, and if $D_{\text{Bayes}}(x) < 0$, class $C_2$.

## 4.6.2 Parallel Displacement of a Hyperplane

We move the hyperplane obtained by training an L1 support vector machine, in parallel, by the Bayesian decision function.

Let the optimal hyperplane be

$$D(\mathbf{x}) = \mathbf{w}^\top \phi(\mathbf{x}) + b = 0, \tag{4.107}$$

where $\mathbf{w}$ is the $l$-dimensional weight vector, $\phi(\mathbf{x})$ is the mapping function from $\mathbf{x}$ to the $l$-dimensional feature space, and $b$ is the bias term.

First we calculate the component $(x_e)_j$ of data sample $\mathbf{x}_j$ in the direction orthogonal to the separating hyperplane:

$$(x_e)_j = \frac{\mathbf{w}^\top \phi(\mathbf{x}_j)}{\|\mathbf{w}\|} = \sum_{i=1}^{M} \delta \, \alpha_i \, y_i \, K(\mathbf{x}_i, \mathbf{x}_j), \tag{4.108}$$

where $\delta$ is the margin and $\delta \|\mathbf{w}\| = 1$, $y_i = 1$ if $\mathbf{x}_i$ belongs to class $C_1$ and $-1$ otherwise, and $\alpha_i$ is a dual variable associated with $\mathbf{x}_i$. Then we calculate the center

$$\mu_k = \frac{1}{n_k} \sum_{j=1}^{n_k} (x_e)_j$$

and the variance

$$v_k = \frac{1}{n_k} \sum_{j=1}^{n_k} ((x_e)_j - \mu_k)^2$$

for class $C_k$, where $n_k$ is the number of class $C_k$ data. Substituting these data into (4.106), we calculate the bias term of the Bayes' decision boundary, $b_{\text{Bayes}}$, namely, we solve

$$D_{\text{Bayes}}(x) = 0 \tag{4.109}$$

for $x$ and obtain $x_{b_1}$, $x_{b_2}$ ($x_{b_1} > x_{b_2}$). Then because $b_{\text{Bayes}} = -x_{b_1}$, we replace $b$ with $b_{\text{Bayes}}$. Here, we set $P(C_k)$ as follows:

$$P(C_k) = \frac{\text{Number of data belonging to } C_k}{\text{Total number of data}}. \tag{4.110}$$

If the recognition rate of the training data is decreased by the parallel displacement of the optimal hyperplane, we do not move the hyperplane so that the generalization ability is not worsened.

### 4.6.3 Normal Test

To move the optimal hyperplane, we first need to test if the distribution of training data in the direction orthogonal to the hyperplane is normal. The procedure is as follows.

First, using the number of data belonging to class $C_k$ $(k = 1, 2)$ and the center $\mu_k$, we determine the number of divisions and the width and generate a frequency distribution. Assume that we divided the interval into $p$ divisions. Let the probability that division $j$ occurs be $p_j$ and the probability assuming the normal distribution be $\pi_j$. We set the hypothesis:

$$K_0 : \quad p_j = \pi_j \quad \text{for } j = 1, \ldots, p,$$
$$K_1 : \quad p_j \neq \pi_j \quad \text{for } j \in \{1, \ldots, p\}.$$

If hypothesis $K_0$ holds, the normal distribution is assumed. Let among $n_k$ observations, $f_j$ observations belong to division $j$ and the number of expected observations under the normal distribution be $e_j = n_k \pi_j$. Then the deviation from the expected observation

$$x^2 = \sum_{j=1}^{p} \frac{(f_j - e_j)^2}{e_j} = \sum_{j=1}^{p} \frac{f_j^2}{e_j} - n_k \tag{4.111}$$

obeys the $\chi^2$ distribution with freedom $\nu = p - r - 1$ under the hypothesis $K_0$, where $r$ is the number of independent variables. Because we use the center and variances, $r = 2$. Therefore, we calculate the value of $x^2$ and if it is not within the critical region with the level of significance $s$:

$$R = \{x^2 \,|\, x^2 > \chi_{p-3}^2(s)\}, \tag{4.112}$$

we judge that the hypothesis that the distribution is normal is correct. If not, it is not normal.

## 4.7 Incremental Training

### 4.7.1 Overview

In this book, we tacitly assume that sufficient training data are available at the development stage of a classifier. But in some applications such as medical diagnosis, initially just a small number of training data are available and afterward they may be obtained sporadically as time passes. In such a situation, we need to update the classifier frequently so that it reflects the recently acquired data. One simple way of updating is to add the acquired data to the existing training data and retrain the classifier. *Incremental training* is considered to improve such inefficient retraining from both the training speed and memory size efficiency, namely, instead of keeping all the training data acquired so far, we improve memory size efficiency keeping important training data for classification and discarding training data that do not contribute in improving classification performance. And we speed up training using information of the current classifier and using the reduced training data.

In addition to the above, in incremental training, as opposed to *batch training*, we need to consider whether a given classification problem is time-dependent (nonstationary) or not. For instance, printed character recognition is considered to be time-independent. But face recognition may become time-dependent if the incremental training period is very long. To express such nonstationary conditions, *concept drift* [62], in which the concept that is going to grasp changes over time, is sometimes used. Under a concept drift environment, we need to discard old data that are harmful to produce class information. In addition, there may be cases where new classes need to be added during incremental training [63] such as in face recognition. But in the following we exclude this situation.

Support vector machines are suited for incremental training due to the fact that only support vectors are necessary for training. According to the purpose of incremental training, the methods for incremental training can be classified into three:

1. Exact and fast training without discarding training data [64–66]
2. Fast training with efficient memory utilization and without explicit treatment of concept drift [67–73]
3. Explicit treatment of concept drift [74]

In cross-validation, we repeatedly train a support vector machine changing parameter values. In such a situation, incremental training may speed up training. But because deleting training data may result in an approximate solution, it is not favorable to delete training data that seem to be redundant. Shilton et al. [65] consider initial value selection for exact incremental training using all the data previously used for training, namely, when new data are added, we "hot start" the training from the old solution plus the initial values for the added data with $\alpha_i = 0$ for $y_i D(\mathbf{x}_i) > 1$ and $\alpha_i = C$ for $y_i D(\mathbf{x}_i) \leq 1$. Cauwenberghs and Poggio [64] propose exact incremental or decremental training when one data sample is added or deleted. This method can be used to speed up leave-one-out method and extended to speeding up cross-validation [75]. This method is explained in Chapter 5.

When incremental training is used in an online environment, redundant training data may be deleted for efficient memory usage. When new incremental training data are obtained we estimate the candidates of support vectors, and using only these data we train the support vector machine. Xiao et al. [69] use the least-recently used (LRU) strategy, which is a page replacement algorithm for paged memory allocation of a computer, for discarding the least-recently used data. Here, exploiting the properties of support vector machines, we explain an incremental training method, which is slightly different from [69], for a one-against-all support vector machine. First we explain how to estimate the candidates of support vectors using Fig. 4.6. In the figure, assume that the data shown in filled circles and rectangles are newly obtained training data. The optimal hyperplane $D_i(\mathbf{x}) = 0$ that separates class $i$ from the remaining classes was determined using the data excluding those data. Then if we retrain the support vector machine, the data that satisfy $y(\mathbf{x}) D_i(\mathbf{x}) \leq 1$ are candidates for support vectors where $y(\mathbf{x}) = 1$ when $\mathbf{x}$ belongs to class $i$ and $y(\mathbf{x}) = -1$, otherwise. In addition, the data that satisfy $y(\mathbf{x}) D_i(\mathbf{x}) > 1$ but are near $y(\mathbf{x}) D_i(\mathbf{x}) = 1$ can be support vectors. Thus we determine that the data are candidates for support vectors if

$$y(\mathbf{x}) D_i(\mathbf{x}) \leq \beta + 1 \tag{4.113}$$

is satisfied, where $\beta \, (> 0)$ is a user-defined parameter. But if all the new data satisfy

$$y(\mathbf{x}) D_i(\mathbf{x}) \geq 1, \tag{4.114}$$

retraining of the support vector machine adding the new data will give the same support vector machine. Thus in this case, we only need to add the data that satisfy (4.113) to the training data for future training. (The same idea is discussed in "practical considerations" of [64].)

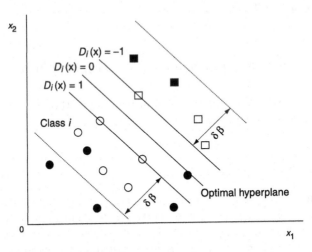

**Fig. 4.6** Estimating the candidates for support vectors

Assume that we have a training set $X_a$ in addition to the set $X_c$ that was used for training the current classifier. The general procedure for incremental training for an $n$-class problem is as follows:

1. Initialize the set $S$, i.e., $S = \phi$.
2. If for $\mathbf{x} \in X_a$, $y(\mathbf{x}) D_i(\mathbf{x}) < \beta + 1$ is satisfied for $i$ ($i \in \{1, \ldots, n\}$), add $\mathbf{x}$ to the set $S$, i.e., $S = S \cup \{\mathbf{x}\}$.
3. Add $S$ to $X_c$, i.e., $X_c = X_c \cup S$.
4. If all the data in $X_a$ satisfy (4.114), we do not retrain the support vector machine. Otherwise, go to Step 5.
5. Using $X_c$ we retrain the support vector machine. After the training, if for $\mathbf{x} \in X_c$, $y(\mathbf{x}) D_i(\mathbf{x}) > \beta + 1$ is satisfied for all $i$ ($i = 1, \ldots, n$), we delete $\mathbf{x}$ from the set $S$, i.e., $S = S - \{\mathbf{x}\}$.

Because $\delta \|\mathbf{w}_i\| = 1$, (4.113) is rewritten as follows:

$$\frac{y(\mathbf{x}) D_i(\mathbf{x})}{\|\mathbf{w}_i\|} \le \delta (\beta + 1), \tag{4.115}$$

where $\delta$ is the margin, namely, the distance of $\mathbf{x}$ from the hyperplane $y(\mathbf{x}) D_i(\mathbf{x}) = 1$ is $\delta \beta$. As the new training data are added to the training data, the value of the margin $\delta$ decreases or remains the same. Thus, the regions that extract support vector candidates are shrunk as the incremental training proceeds even for the fixed value of $\beta$. We call the above incremental training method *incremental training using hyperplanes*.

The problem of this method is that if the hyperplane rotates future support vectors may be lost and that it is difficult to set the optimal value to $\beta$. To solve these problems, in the following we discuss incremental training using hyperspheres [72, 73].

## 4.7.2 Incremental Training Using Hyperspheres

In this section, we discuss an incremental training method that is robust for the rotation of the separating hyperplane [72, 73].

### 4.7.2.1 Concept

The incremental training method is based on the assumption that candidates for support vectors exist near the separating hyperplane and are close to the surface of a region that includes training data of each class.

First, we explain how we can estimate the candidates for support vectors using the example shown in Fig. 4.7. In the figure, filled circles and squares show the initial data and open circles and squares show the added data. The separating hyperplane shown in the figure is determined by the initial data. Then after retraining, the new separating hyperplane may exist in the shaded region. Therefore we can predict that the candidates for support vectors exist near the boundary of the shaded region. The region, where the candidates for support vectors exist, can be approximated by the shaded regions in Fig. 4.8.

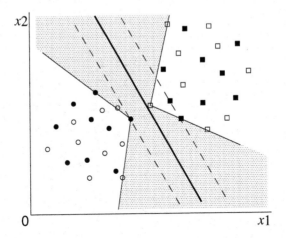

**Fig. 4.7** Estimation of the separating hyperplane after incremental training. Reprinted from [72, p. 1498] with permission from Elsevier

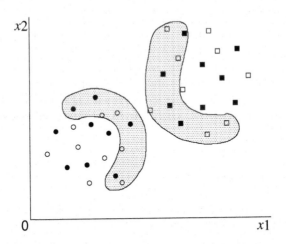

**Fig. 4.8** Estimation of candidates for support vectors. Reprinted from [72, p. 1498] with permission from Elsevier

We consider approximating the regions shown in Fig. 4.8, where candidates of support vectors exist, using hyperspheres. We explain the idea using the example shown in Fig. 4.9. In the figure, the data in the shaded regions for

Classes 1 and 2 are candidates for deletion; open circles and squares show
the deleted data after adding data and filled circles and squares show the
remained data. In the following, we explain how to approximate the shaded
region for each class.

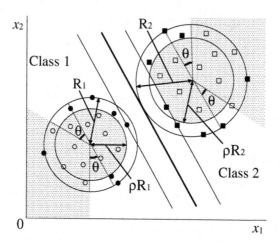

**Fig. 4.9** Deletion of the data using the hyperspheres. Reprinted from [72, p. 1498] with
permission from Elsevier

First, we generate the minimum-volume hypersphere in the feature space
that includes the training data of class $j$ $(j = 1, 2)$ with radius $R_j$. Next,
we define a concentric hypersphere with radius $\rho R_j$, where $\rho\,(0 < \rho < 1)$ is
the user-defined parameter. Then, we define the hypercone whose vertex is
at the center of the hyperspheres and which opens in the opposite direction
of the separating hyperplane. The user-defined parameter $\theta\,(-90 < \theta < 90)$
defines the angle between the separating hyperplane and the surface of the
hypercone.

### 4.7.2.2 Deletion of Data

If the added data are in the shaded regions in Fig. 4.9, we delete the data.
Now we explain how to delete data sample $\mathbf{x}$ using Fig. 4.10. If the distance
$r_j(\mathbf{x})$ between $\phi(\mathbf{x})$ and the center of the hypersphere, $\mathbf{a}_j$, is smaller than
$\rho R_j$, where $\phi(\mathbf{x})$ is the mapping function to the feature space:

$$r_j(\mathbf{x}) < \rho R_j, \tag{4.116}$$

we delete the data. Otherwise, if the angle between $\phi(\mathbf{x}) - \mathbf{a}_j$ and the separating hyperplane, $\psi_j(\mathbf{x})$, is larger than $\theta$:

$$\psi_j(\mathbf{x}) > \theta, \tag{4.117}$$

we judge that $\phi(\mathbf{x})$ exists inside of the hypercone and delete $\mathbf{x}$.

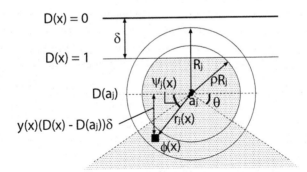

**Fig. 4.10** Judging whether the data are inside of the hypercone or not. Adapted from [72, p. 1498] with permission from Elsevier

But even if (4.116) or (4.117) is satisfied, if $\mathbf{x}$ satisfies

$$y(\mathbf{x})D(\mathbf{x}) \leq 1, \tag{4.118}$$

$\mathbf{x}$ is a candidate for support vectors, where $D(\mathbf{x}) = \mathbf{w}^\top \phi(\mathbf{x}) + b$ is the decision function. In such a case, we do not delete $\mathbf{x}$. In addition, we do not delete the data that are support vectors for hyperspheres because the support vectors for hyperspheres are candidates for the support vectors for hyperspheres at the next training.

The general procedure for incremental training is as follows:

1. Train the support vector machine using the initial data set $X_a$.
2. Add the additional data set $X_b$ to $X_a$: $X_a = X_a \cup X_b$.
3. If for $\mathbf{x} \in X_a$, (4.118) is not satisfied and $\mathbf{x}$ satisfies $r_j(\mathbf{x}) < \rho R_j$, where $j$ is the class label for $\mathbf{x}$ or $\psi_j(\mathbf{x}) > \theta$, delete $\mathbf{x}$ from $X_a$: $X_a = X_a - \{\mathbf{x}\}$.
4. If for $\mathbf{x} \in X_a$, (4.118) is satisfied, retrain the support vector machine.
5. Repeat (2), (3), and (4).

### 4.7.2.3 Generation of Hyperspheres

We approximate the region that includes training data of a class by a hypersphere, which will be discussed in Section 8.1. The procedure of generating the hypersphere for class $j$ ($j = 1, 2$) is as follows.

To generate the minimum-volume hypersphere that includes the data for class $j$, we need to solve the following optimization problem:

$$\text{minimize} \quad Q_{\mathrm{p}}(R_j, \mathbf{a}_j, \boldsymbol{\xi}^j) = R_j^2 + C_j \sum_{i \in X_j} \xi_i^j \tag{4.119}$$

$$\text{subject to} \quad ||\boldsymbol{\phi}(\mathbf{x}_i) - \mathbf{a}_j||^2 \le R_j^2 + \xi_i^j, \quad \xi_i^j \ge 0 \quad \text{for } i \in X_j, \tag{4.120}$$

where $X_j$ are sets of training data indices for class $j$ $(j = 1, 2)$, $X_1 \cup X_2 = \{1, 2, \ldots, M\}$, $X_1 \cap X_2 = \emptyset$, $\mathbf{a}_j$ is the class $j$ center, $R_j$ is the radius of the class $j$ hypersphere, $\xi_i^j$ are slack variables, and $C_j$ is the margin parameter that determines the trade-off between the volume of the hypersphere and outliers.

Introducing the Lagrange multipliers $\alpha_i^j$, $\mathbf{a}_j$ is given by

$$\mathbf{a}_j = \sum_{i \in X_j} \alpha_i^j \boldsymbol{\phi}(\mathbf{x}_i). \tag{4.121}$$

Then we obtain the following dual problem:

$$\text{maximize} \quad Q_{\mathrm{d}}(\boldsymbol{\alpha}^j) = \sum_{i \in X_j} \alpha_i^j K(\mathbf{x}_i, \mathbf{x}_i) - \sum_{i,k \in X_j} \alpha_i^j \alpha_k^j K(\mathbf{x}_i, \mathbf{x}_k) \tag{4.122}$$

$$\text{subject to} \quad \sum_{i \in X_j} \alpha_i^j = 1, \quad 0 \le \alpha_i^j \le C_j \quad \text{for } i \in X_j. \tag{4.123}$$

Using the unbounded support vector $\mathbf{x}$, which belongs to class $j$, $R_j$ is given by

$$R_j = ||\boldsymbol{\phi}(\mathbf{x}) - \mathbf{a}_j||. \tag{4.124}$$

The bounded support vectors with $\xi_i^j > 0$ are outside of the hypersphere and thus outliers. From (4.123), $\alpha_i^j$ do not exceed 1. Thus, if we set $C_j = 1$, all the support vectors are unbounded support vectors.

The mapping of an unknown data sample $\mathbf{x}$ is inside of the hypersphere if

$$K(\mathbf{x}, \mathbf{x}) - 2 \sum_{i \in S_j} \alpha_i^j K(\mathbf{x}, \mathbf{x}_i) + \sum_{i,k \in S_j} \alpha_i^j \alpha_j^j K(\mathbf{x}_i, \mathbf{x}_k) \le R_j^2, \tag{4.125}$$

where $S_j$ is the set of support vector indices of the class $j$ hypersphere.


#### 4.7.2.4 Evaluation of $\psi_j(\mathbf{x})$

We discuss how to evaluate $\psi_j(\mathbf{x})$. The distance $r_j(\mathbf{x})$ between $\boldsymbol{\phi}(\mathbf{x})$ and the center of the hypersphere is given by

$$
\begin{aligned}
r_j(\mathbf{x}) &= ||\boldsymbol{\phi}(\mathbf{x}) - \mathbf{a}_j|| \\
&= \sqrt{K(\mathbf{x}, \mathbf{x}) - 2 \sum_{i \in S_j} \alpha_i^j K(\mathbf{x}, \mathbf{x}_i) + \sum_{i,k \in S_j} \alpha_i^j \alpha_k^j K(\mathbf{x}_i, \mathbf{x}_k)}.
\end{aligned} \tag{4.126}
$$

From

$$\mathbf{w} = \sum_{i \in S} y(\mathbf{x}_i)\, \alpha_i\, \phi(\mathbf{x}_i) \tag{4.127}$$

and (4.121), where $\alpha_i$ and $S$ are the Lagrange multipliers and the set of support vector indices for the support vector machine, the value of the decision function at the center of the hypersphere is given by

$$D(\mathbf{a}_j) = \mathbf{w}^\top \mathbf{a}_j + b = \sum_{i \in S, k \in X_j} y(\mathbf{x}_i)\alpha_i \alpha_k^j K(\mathbf{x}_i, \mathbf{x}_k) + b. \tag{4.128}$$

The distance between $\phi(\mathbf{x})$ and the separating hyperplane is given by

$$\frac{|D(\mathbf{x})|}{||\mathbf{w}||} = \frac{y(\mathbf{x})D(\mathbf{x})}{||\mathbf{w}||} = y(\mathbf{x})D(\mathbf{x})\delta. \tag{4.129}$$

The distance between the separating hyperplane and the hyperplane, which is parallel to the separating hyperplane and which passes the center of the hypersphere, is given by

$$\frac{|D(\mathbf{a}_j)|}{||\mathbf{w}||} = \frac{y(\mathbf{x})D(\mathbf{a}_j)}{||\mathbf{w}||} = y(\mathbf{x})D(\mathbf{a}_j)\delta. \tag{4.130}$$

Thus the distance between $\phi(\mathbf{x})$ and the hyperplane which passes the center of the hypersphere is $|D(\mathbf{x}) - D(\mathbf{a}_j)|\delta$. Then if $y(\mathbf{x})(D(\mathbf{x}) - D(\mathbf{a}_j))\delta$ is negative, $\phi(\mathbf{x})$ lies between the separating hyperplane and the hyperplane, which is parallel to the separating hyperplane and which goes through $\mathbf{a}_j$. If it is positive, $\phi(\mathbf{x})$ and the separating hyperplane are on the opposite sides of the hyperplane which goes through $\mathbf{a}_j$. Therefore $\psi_j(\mathbf{x})$ $(-90 < \psi_j(\mathbf{x}) < 90)$ is given by

$$\psi_j(\mathbf{x}) = \sin^{-1}\left( \frac{y(\mathbf{x})(D(\mathbf{x}) - D(\mathbf{a}_j))\delta}{r_j(\mathbf{x})} \right). \tag{4.131}$$

If $\psi_j(\mathbf{x})$ is larger than $\theta$, we delete $\mathbf{x}$. The deleting region using the hypercone is the shaded region shown in Fig. 4.10.

We can readily extend the above method to multiclass problems. For example, in one-against-all support vector machines with $n$ classes, we apply the incremental training method for each two-class problem, namely, for the $i$th $(i = 1, \ldots, n)$ two-class problem, we approximate class $i$ and the remaining classes with hyperspheres. And we permanently delete data that are deleted by all the two-class problems.

## 4.7.2.5 Performance Evaluation

We compared the incremental training method using hyperspheres with that using hyperplanes using two-class problems listed in Table 1.1. As for the results for multiclass problems, see [72]. Hereafter, we call the incremental training method using hyperspheres the *hypersphere method* and that using hyperplanes the *hyperplane method*.

### Benchmark Data and Evaluation Conditions

Except for banana, ringnorm, and thyroid data sets, we normalized the input ranges into $[0, 1]$. In approximating hyperspheres, we used the same kernels as those of the classifiers and set $C_j = 1$ so that all the data were inside of the hyperspheres.

We randomly generated one incremental training data set for each of 100 (or 20) training data sets, dividing the training data set into the subsets whose size is about 5% of the total number. In incremental training, each subset was added to the classifier one at a time and the classifier was trained by the hyperplane and hypersphere methods. Thus, for each classification problem, we trained classifiers 100 (or 20) times and calculated the statistics.

The recognition rates of the test data, after incremental training was finished, were compared with those by batch training using all the training data. To evaluate the efficiency of data deletion, we evaluated the deletion ratio that is defined by the number of the deleted data divided by the number of training data. For batch training, the deletion ratio is the ratio of the deletable training data (i.e., training data minus support vectors) against training data. Thus assuming that all the support vectors are retained, the deletion ratio for the batch training is the upper bound for the deletion ratio for incremental training.

We also measured the computing time for batch training and incremental training using a personal computer (Xeon 2.8 GHz, 1 GB memory) with the Linux operating system. We used the primal–dual interior-point method combined with the decomposition technique to train support vector machines.

### Parameter Setting

In applying the hypersphere method to classification problems, we need to set the values to $\rho$ and $\theta$. But if we need to optimize the values for each problem, the applicability of the method is diminished. To solve this, we determined the values using one data set, namely, the twonorm data set. According to the experiment we found that the parameter values of $\rho = 0.5, \theta = 0$ are optimal to keep the recognition rates high while deleting sufficient training data. Therefore, in the following study we set $\rho = 0.5$ and $\theta = 0$. On the contrary, for the hyperplane method, we cannot set a value of $\beta$ that works well for most of the data set.

## Performance Comparison of Incremental Training Methods

We compared the hyperplane and hypersphere methods optimizing the kernel, its parameter value, and the $C$ value. We also compared the methods replacing the optimum $C$ value with $C = 10,000$. Here, we used a large value of $C$ so that comparison of incremental training with the optimum and non-optimum values of $C$ becomes clear. To optimize kernels and $C$, we use five-fold cross-validation. For polynomial kernels with $d = [2, 3, 4]$ and RBF kernels with $\gamma = [0.1, 1, 10]$, we performed fivefold cross-validation for the margin parameter $C = [1, 10, 50, 100, 500, 1,000, 2,000, 3,000, 5,000, 8,000, 10,000, 50,000, 100,000]$ and selected the kernel, its parameter, and the margin parameter with the highest average recognition rate in batch training.

Table 4.15 shows the parameter values. In the table, the column "Kernel" lists the kernels determined by cross-validation. For example, $\gamma 1$ and $d2$ denote RBF kernels with $\gamma = 1$ and polynomial kernels with $d = 2$, respectively. The same kernels were used for two cases: (1) optimal $C$ and (2) $C = 10,000$. The parameters for case (1) are listed in the columns "Batch 1," "Plane 1," and "Sphere 1" and those for case (2), "Batch 2," "Plane 2," and "Sphere 2." For each case, the same margin parameter value was used for batch training, the hyperplane, and hypersphere methods.

**Table 4.15** Parameters for two-class problems. Reprinted from [72, p. 1503] with permission from Elsevier

| Data | Kernel | Batch 1 | Plane 1 | Sphere 1 | | Batch 2 | Plane 2 | Sphere 2 | |
| | | $C$ | $\beta$ | $\rho$ | $\theta$ | $C$ | $\beta$ | $\rho$ | $\theta$ |
|---|---|---|---|---|---|---|---|---|---|
| Banana | $\gamma 1$ | 10 | 2 | 0.5 | 0 | 10,000 | 10 | 0.5 | 0 |
| B. cancer | $\gamma 1$ | 1 | 0.01 | 0.5 | 0 | 10,000 | 4 | 0.5 | 0 |
| Diabetes | $d2$ | 50 | 0.5 | 0.5 | 0 | 10,000 | 1 | 0.5 | 0 |
| Flare-solar | $d2$ | 10 | 0.5 | 0.5 | 0 | 10,000 | 2 | 0.5 | 0 |
| German | $\gamma 1$ | 10 | 0.1 | 0.5 | 0 | 10,000 | 2 | 0.5 | 0 |
| Heart | $\gamma 1$ | 50 | 0.5 | 0.5 | 0 | 10,000 | 4 | 0.5 | 0 |
| Image | $\gamma 1$ | 1,000 | 2 | 0.5 | 0 | 10,000 | 5 | 0.5 | 0 |
| Ringnorm | $\gamma 0.1$ | 1 | 0.1 | 0.5 | 0 | 10,000 | 0.1 | 0.5 | 0 |
| Splice | $\gamma 10$ | 1 | 0.1 | 0.5 | 0 | 10,000 | 0.1 | 0.5 | 0 |
| Thyroid | $d2$ | 1 | 1 | 0.5 | 0 | 10,000 | 3 | 0.5 | 0 |
| Twonorm | $d4$ | 50 | 2 | 0.5 | 0 | 10,000 | 2 | 0.5 | 0 |
| Waveform | $\gamma 1$ | 1 | 0.1 | 0.5 | 0 | 10,000 | 2 | 0.5 | 0 |

For the hypersphere method, we used $\rho = 0.5$ and $\theta = 0$ for all cases as listed in "Sphere 1" and "Sphere 2." As will be shown immediately, for the hyperplane method we could not select a value of $\beta$ common to all cases because the optimal value changed as the data set changed. Thus not to favor the hypersphere method, we set the optimal value to $\beta$, namely, we selected the value of $\beta$ so that the comparable recognition rate with that of batch training was obtained.

Table 4.16 shows the results. In each problem the maximum deletion ratios for Plane 1 and Sphere 1 and for Plane 2 and Sphere 2 are shown in boldface. The error rates for Batch 1 and Sphere 1, and Batch 2 and Sphere 2 are almost the same using the user-defined parameters ($\rho = 0.5, \theta = 0$). And about 30–60% of the training data are deleted. Because the error rates of the hypersphere method are almost the same with those of batch training, we can consider that almost all support vectors remained after incremental training. Consequently we can conclude that we delete many unnecessary data without deleting candidates for support vectors.

The deletion ratios of Sphere 1 are often larger than those of Plane 1 and the deletion ratios of Sphere 2 are often larger than those of Plane 2.

The optimal value of $\beta$ for the conventional method changes as the data set changes. Thus, it is difficult to set the optimal value of $\beta$ in advance. But for the hypersphere method, the choice of $\rho = 0.5, \theta = 0$ is almost always good.

Table 4.17 shows the computing time. We divided the total computing time "Total$_t$" into three parts: incremental training time "Trn$_t$," deletion time "Del$_t$," and total hypersphere generation time "Sph$_t$." The deletion time denotes the time to check whether the data can be deleted in incremental training. For Batch 1, at each incremental training step, batch training using all the data obtained so far is performed and Total$_t$ = Trn$_t$. For the hyperplane method, Total$_t$ = Trn$_t$ + Del$_t$ and for the hypersphere method, Total$_t$ = Trn$_t$ + Del$_t$ + Sph$_t$. For each problem, the shorter time between Plane 1 and Sphere 1 is shown in boldface.

We only show the results for five data sets. The tendency of the results for the remaining data sets is the same. From the table, the incremental training time, Trn$_t$, of the hypersphere method is shorter than that of batch training and comparable with that of the hyperplane method. The deletion time, Del$_t$, of the hypersphere method is longer than that of the hyperplane method because the deletion check is complicated. This tendency becomes prominent if the number of deleted data is large. In the hypersphere method, hypersphere generation time, Sph$_t$, is negligible except for the splice data set. Except for the image data set, the ratio of the deletion time in the total training time is small. For the image data set, because of the long deletion time, the total computing time is longer than that of batch training.

According to the computer experiments, the hypersphere method with $\rho = 0.5$ and $\theta = 0$ is shown to have generalization ability comparable to batch training while deleting 30–60% of the training data. This is especially favorable because it is difficult to tune parameters in incremental training.

**Table 4.16** Comparison between hyperplane and hypersphere methods (%). Reprinted from [72, p. 1503] with permission from Elsevier

| Data | Term | Batch 1 | Plane 1 | Sphere 1 | Batch 2 | Plane 2 | Sphere 2 |
|------|------|---------|---------|----------|---------|---------|----------|
| Banana | Test | 10.7±0.53 | 10.7±0.53 | 10.7±0.53 | 13.0±0.99 | 13.0±0.99 | 13.0±0.99 |
| | Trn | 8.05±1.30 | 8.07±1.30 | 8.07±1.30 | 5.33±1.33 | 5.36±1.33 | 5.33±1.33 |
| | Del | 73.6±2.3 | 64.3±3.7 | **69.8±2.7** | 76.9±3.1 | 47.1±7.3 | **53.2±6.6** |
| B. cancer | Test | 26.8±4.5 | 26.8±4.5 | 26.8±4.5 | 34.6±4.7 | 34.6±4.6 | 34.6±4.7 |
| | Trn | 17.2±1.7 | 17.2±1.7 | 17.2±1.7 | 1.67±0.62 | 1.86±0.75 | 1.67±0.62 |
| | Del | 34.6±2.7 | **0.07±0.27** | 0.06±0.22 | 50.1±3.4 | 26.3±8.4 | **37.4±5.5** |
| Diabetes | Test | 23.5±1.9 | 23.5±1.9 | 23.5±1.9 | 25.4±1.7 | 25.5±1.8 | 25.4±1.7 |
| | Trn | 21.5±1.2 | 21.5±1.2 | 21.5±1.2 | 18.2±1.2 | 18.3±1.1 | 18.2±1.2 |
| | Del | 45.5±1.7 | 15.7±8.3 | **27.8±13.4** | 49.3±2.3 | 36.6±5.1 | **44.1±8.1** |
| Flare-solar | Test | 31.7±1.9 | 31.7±1.9 | 31.7±1.9 | 33.7±2.1 | 33.8±2.0 | 33.8±2.0 |
| | Trn | 31.8±1.3 | 31.8±1.3 | 31.8±1.3 | 30.8±1.2 | 30.9±1.2 | 30.8±1.2 |
| | Del | 18.9±1.9 | 4.5±3.5 | **11.5±8.6** | 18.2±2.1 | 8.61±3.5 | **15.3±5.6** |
| German | Test | 23.4±2.1 | 23.3±2.3 | 23.4±2.1 | 30.1±2.5 | 30.2±2.5 | 30.1±2.5 |
| | Trn | 18.9±1.3 | 19.1±1.4 | 18.9±1.3 | 0.96±1.1 | 1.19±1.1 | 0.96±1.1 |
| | Del | 43.8±1.5 | 33.8±5.7 | **35.5±3.7** | 49.9±2.4 | 44.9±3.3 | **52.8±5.7** |
| Heart | Test | 16.3±3.4 | 16.3±3.4 | 16.3±3.4 | 20.3±3.4 | 20.3±3.5 | 20.3±3.5 |
| | Trn | 14.1±1.9 | 14.1±1.9 | 14.1±1.9 | 6.94±1.7 | 7.04±1.6 | 6.94±1.7 |
| | Del | 56.3±3.4 | 24.9±21.9 | **32.4±27.6** | 58.6±4.1 | 39.31±7.9 | **47.1±6.4** |
| Image | Test | 2.86±0.48 | 2.88±0.47 | 2.87±0.48 | 3.13±0.38 | 3.22±0.45 | 3.14±0.38 |
| | Trn | 1.40±0.18 | 1.40±0.17 | 1.40±0.18 | 0.49±0.17 | 0.58±0.21 | 0.5±0.17 |
| | Del | 88.3±0.7 | 60.2±3.8 | **61.7±4.7** | 91.2±0.58 | 59.1±3.8 | **61.6±6.8** |
| Ringnorm | Test | 1.59±0.10 | 1.59±0.10 | 1.59±0.10 | 1.66±0.12 | 1.66±0.12 | 1.66±0.12 |
| | Trn | 0.09±0.15 | 0.09±0.15 | 0.09±0.15 | 0.00±0.00 | 0.00±0.00 | 0.00±0.00 |
| | Del | 61.5±2.3 | **45.6±16.3** | 31.0±11.1 | 62.1±2.2 | **51.9±2.9** | 34.9±2.5 |
| Splice | Test | 11.3±0.71 | 11.3±0.72 | 11.3±0.71 | 10.8±0.71 | 89.20±0.71 | 10.8±0.71 |
| | Trn | 0.91±0.24 | 0.91±0.24 | 0.91±0.24 | 0.02±0.04 | 0.02±0.04 | 0.02±0.04 |
| | Del | 26.5±1.3 | 11.9±10.7 | **13.2±12.0** | 25.8±1.4 | 20.3±1.3 | **23.5±1.7** |
| Thyroid | Test | 3.69±1.90 | 3.69±1.90 | 3.69±1.90 | 4.49±2.23 | 4.49±2.23 | 4.49±2.23 |
| | Trn | 0.66±0.49 | 0.66±0.49 | 0.66±0.49 | 0.00±0.00 | 0.00±0.00 | 0.00±0.00 |
| | Del | 88.9±1.3 | **66.5±11.3** | 60.9±6.7 | 91.8±1.0 | 55.7±19.5 | **61.2±7.9** |
| Twonorm | Test | 2.43±0.12 | 2.43±0.12 | 2.43±0.12 | 3.65±0.45 | 3.65±0.45 | 3.65±0.45 |
| | Trn | 1.76±0.55 | 1.76±0.55 | 1.76±0.55 | 0.07±0.13 | 0.07±0.13 | 0.07±0.13 |
| | Del | 79.1±1.6 | **57.5±6.4** | 51.2±6.8 | 90.5±1.9 | **72.2±6.7** | 51.0±4.4 |
| Waveform | Test | 10.0±0.45 | 10.0±0.45 | 10.0±0.45 | 12.0±0.80 | 12.0±0.80 | 12.0±0.80 |
| | Trn | 6.49±1.37 | 6.49±1.37 | 6.49±1.37 | 0.00±0.00 | 0.00±0.00 | 0.00±0.00 |
| | Del | 61.6±2.2 | 36.8±28.0 | **38.2±25.7** | 73.0±3.1 | 45.6±4.1 | **61.8±3.2** |

# 4.8 Learning Using Privileged Information

Prior knowledge for a given classification problem can help improve the generalization ability. In this respect Vapnik proposed an LUPI (learning using

**Table 4.17** Training time and deleting time for two-class problems (s). Reprinted from [72, p. 1504] with permission from Elsevier

| Data | | Batch1 | Plane1 | Sphere1 |
|---|---|---|---|---|
| Diabetes | Total$_t$ | 24.2±3.6 | 19.7±8.4 | **14.5±5.8** |
| | Trn$_t$ | 24.2±3.6 | 19.4±8.3 | **11.9±4.7** |
| | Del$_t$ | — | **0.4±0.2** | 2.4±1.3 |
| | Sph$_t$ | — | — | 0.3±0.05 |
| Flare-solar | Total$_t$ | 108.2±11.5 | 75.1±41.4 | **73.8±41.2** |
| | Trn$_t$ | 108.2±11.5 | 74.2±40.7 | **65.7±35.4** |
| | Del$_t$ | — | **1.0±0.7** | 7.5±7.2 |
| | Sph$_t$ | — | — | 0.59±0.08 |
| German | Total$_t$ | 90.1±15.6 | **58.5±13.3** | 62.7±10.5 |
| | Trn$_t$ | 90.1±15.6 | 57.4±13.1 | **55.3±10.4** |
| | Del$_t$ | — | **1.2±0.2** | 6.6±1.2 |
| | Sph$_t$ | — | — | 0.8±0.08 |
| Image | Total$_t$ | 27.0±3.9 | **14.5±3.0** | 30.0±3.5 |
| | Trn$_t$ | 27.0±3.9 | 13.8±3.0 | **11.0±1.4** |
| | Del$_t$ | — | **0.64±0.07** | 18.3±2.5 |
| | Sph$_t$ | — | — | 0.71±0.07 |
| Splice | Total$_t$ | 514.3±27.1 | **314.3±161.7** | 314.4±146.3 |
| | Trn$_t$ | 514.3±27.1 | 309.1±158.2 | **256.0±123.8** |
| | Del$_t$ | — | **5.2±3.6** | 32.2±27.9 |
| | Sph$_t$ | — | — | 26.2±3.9 |

privileged information) paradigm [76, 53]. His idea is to use a teacher's hints and comments during training. But the teacher's hint is not provided at the testing stage. In the following we explain his idea based on [53].

For a given two-class problem, suppose we know the decision function that realizes the best generalization ability:

$$D(\mathbf{x}) = \mathbf{w}_0^\mathsf{T} \boldsymbol{\phi}(\mathbf{x}) + b_0, \qquad (4.132)$$

where $\boldsymbol{\phi}(\mathbf{x})$ is the mapping function that maps the input vector $\mathbf{x}$ into the feature space, $\mathbf{w}_0$ is the weight vector in the feature space, and $b_0$ is the bias term. Then for the training data pairs $(\mathbf{x}_i, y_i)$ $(i = 1, \ldots, M)$, we define the following oracle function:

$$\xi(\mathbf{x}_i) = \begin{cases} 1 - y_i\, D(\mathbf{x}_i) & \text{for} \quad y_i\, D(\mathbf{x}_i) \le 1, \\ 0 & \text{otherwise,} \end{cases} \qquad (4.133)$$

where $y_i = 1$ for Class 1 and $-1$ for Class 2. Here, if $1 > \xi(\mathbf{x}_i) \ge 0$, $\mathbf{x}_i$ is correctly classified by the decision function $D(\mathbf{x})$ and if $\xi(\mathbf{x}_i) \ge 1$, $\mathbf{x}_i$ is misclassified.

Suppose the values of the oracle function $\xi(\mathbf{x}_i)$ $(i = 1, \ldots, M)$ are given as hints of training. Then we define the following oracle support vector machine:

$$\text{minimize} \quad Q(\mathbf{w}, b) = \frac{1}{2} \|\mathbf{w}\|^2 \tag{4.134}$$

$$\text{subject to} \quad y_i \left(\mathbf{w}^\top \boldsymbol{\phi}(\mathbf{x}_i) + b\right) \geq r_i \quad \text{for } i = 1, \ldots, M, \tag{4.135}$$

where $r_i = 1 - \xi(\mathbf{x}_i)$. The oracle support vector machine is linearly separable and as $M$ goes to infinity, the solution converges to the best decision function given by (4.132). At least for the training data, the decision by the solution is the same with that by the best decision function. It is proved theoretically and experimentally that the oracle support vector machines show better generalization ability than the regular support vector machine, although the error rates of both machines converge to that of the Bayes decision function as $M$ goes to infinity.

In a real problem, the best decision function or the oracle function is not known. But we can obtain hints and comments. For instance, in a one-step-ahead prediction problem, an input–output pair consists of the input and the associated one-stead-ahead output. In the training stage, the $i$-stage $(i > 1)$ ahead output is available and can be used in training.

In the following, we will explain what is called $d$SVM+ model, which is defined in a one-dimensional $d$-space. Let the privileged information associated with $(\mathbf{x}_i, y_i)$ be $\mathbf{x}_i^*$ $(i = 1, \ldots, M)$. From the privileged information, we obtain the oracle function-like deviation variable, solving the following problem:

$$\text{minimize} \quad Q(\mathbf{w}^*, b^*, \boldsymbol{\xi}^*) = \frac{1}{2} \|\mathbf{w}^*\|^2 + C \sum_{i=1}^{M} \xi_i^* \tag{4.136}$$

$$\text{subject to} \quad y_i \left(\mathbf{w}^{*\top} \boldsymbol{\phi}^*(\mathbf{x}_i^*) + b^*\right) \geq 1 - \xi_i^*,$$
$$\xi_i^* \geq 0 \quad \text{for } i = 1, \ldots, M, \tag{4.137}$$

where $\boldsymbol{\phi}^*(\mathbf{x}^*)$ is the mapping function to the feature space, $\mathbf{w}^*$ is the weight vector in the feature space, $b^*$ is the bias term, and $C$ is the margin parameter. Using the solution of the above optimization problem, we define the values of the deviation variable:

$$d_i = 1 - y_i \left(\mathbf{w}^{*\top} \boldsymbol{\phi}^*(\mathbf{x}_i^*) + b^*\right) \quad \text{for } i = 1, \ldots, M. \tag{4.138}$$

Then using $d_i$ as privileged information, we train the so-called SVM+:

$$\text{minimize} \quad Q(\mathbf{w}, \mathbf{w}^*, b, b^*) = \frac{1}{2} \left(\|\mathbf{w}\|^2 + \gamma \|\mathbf{w}^*\|^2\right)$$
$$+ C \sum_{i=1}^{M} (\mathbf{w}^{*\top} \boldsymbol{\phi}^*(d_i) + b^*) \tag{4.139}$$

$$\text{subject to} \quad y_i \left(\mathbf{w}^\top \boldsymbol{\phi}(\mathbf{x}_i) + b\right) \geq 1 - (\mathbf{w}^{*\top} \boldsymbol{\phi}^*(d_i) + b^*) \quad \text{for } i = 1, \ldots, M,$$

$$(\mathbf{w}^{*\top}\phi^*(d_i) + b^*) \geq 0 \quad \text{for } i = 1, \dots, M. \tag{4.140}$$

Here, $\gamma$ is a trade-off parameter.

Introducing the Lagrange multipliers $\alpha_i$ and $\beta_i$ associated with $\mathbf{x}_i$ and $d_i$, respectively, we obtain the following optimization problem:

$$\text{maximize } Q(\boldsymbol{\alpha}, \boldsymbol{\beta}) = \sum_{i=1}^{M} \alpha_i - \frac{1}{2}\alpha_i\,\alpha_j\,y_i\,y_j\,K(\mathbf{x}_i, \mathbf{x}_j)$$

$$-\frac{1}{2\gamma}\sum_{i,j=1}^{M}(\alpha_i + \beta_i - C)\,(\alpha_j + \beta_j - C)\,K^*(d_i, d_j) \tag{4.141}$$

$$\text{subject to} \quad \sum_{i=1}^{M}(\alpha_i + \beta_i - C) = 0, \tag{4.142}$$

$$\sum_{i=1}^{M} y_i\,\alpha_i = 0, \tag{4.143}$$

$$\alpha_i \geq 0, \quad \beta_i \geq 0 \quad \text{for } i = 1, \dots, M, \tag{4.144}$$

where $K(\mathbf{x}_i, \mathbf{x}_j) = \phi^\top(\mathbf{x}_i)\,\phi(\mathbf{x}_j)$ and $K^*(d_i, d_j) = \phi^{*\top}(d_i)\,\phi^*(d_j)$. The decision function is given by

$$D(\mathbf{x}) = \sum_{i=1}^{M} y_i\,\alpha_i\,K(\mathbf{x}_i, \mathbf{x}) + b, \tag{4.145}$$

namely, no privileged information is used for classification.

In [53], the improvement of the generalization ability of LUPI over regular support vector machines is shown for some applications.

## 4.9 Semi-Supervised Learning

In training a support vector machine we need training inputs and their outputs, namely, class labels. These labels are assigned by humans. Training using input–output pairs is called *supervised learning*. And training using only training inputs is called *unsupervised learning*. Assigning labels requires time and cost especially when a huge number of data need to be labeled. To alleviate such labeling process, *semi-supervised learning* with labeled and un-labeled data has attracted much attention.

Performance of semi-supervised learning depends on the classification problems and the methods used and there are mixed evaluation results. Singh et al. [77] discussed the conditions for improving generalization ability but with rather severe assumptions.

As a variant of semi-supervised learning, *self-training* initially trains a classifier using only labeled data. Then unlabelled data are classified by the trained classifier and the newly labeled data with high confidence are added to the training data, and the classifier is retrained. This procedure is iterated several times. Adankon and Cheriet [78] use an LS support vector machine as a classifier and estimate posterior probabilities to select unlabeled data with high confidence.

In semi-supervised learning, unlabeled data are assumed to belong to one of the classes of labeled data. In a knowledge transfer environment, where knowledge obtained from a previously learned task is allowed to transfer to new and different kind of task learning, unlabeled data are not necessarily belong to the existing classes. This learning paradigm is called *self-taught learning* [79, 80].

*Transductive learning* is one type of semi-supervised learning. In transductive learning, the test data are known at the training stage and using test data as well as training data a support vector machine is trained. Using linear kernels, transduction training by a hard-margin support vector machine is formalized as follows [81]:

$$\text{minimize} \quad Q(\mathbf{y}^*, \mathbf{w}, b) = \frac{1}{2}\mathbf{w}^\top \mathbf{w} \tag{4.146}$$

$$\text{subject to} \quad y_i\left(\mathbf{w}^\top \mathbf{x}_i + b\right) \geq 1 \quad \text{for } i = 1, \ldots, M, \tag{4.147}$$

$$y_i^*\left(\mathbf{w}^\top \mathbf{x}_i + b\right) \geq 1 \quad \text{for } i = M+1, \ldots, M+N, \tag{4.148}$$

$$y_i^* \in \{-1, 1\} \quad \text{for } i = M+1, \ldots, M+N, \tag{4.149}$$

where $\mathbf{x}_i$ is the $i$th $(i = 1, \ldots, M)$ training input and the associated output is $y_i$ with label 1 for Class 1 and $-1$ for Class 2, $\mathbf{x}_i$ $(i = M+1, \ldots, M+N)$ are training inputs without training outputs and class labels $y_i^*$ are determined by training, and $\mathbf{w}$, $b$ are the coefficient vector and the bias term, respectively. Because the above optimization problem is not a quadratic programming problem, it is difficult to solve the problem and many training methods have been developed [82].

One application of transductive training is medical diagnosis where training data are scarce and test data sample is a patient to be diagnosed [83].

## 4.10 Multiple Classifier Systems

It is natural to consider realizing a higher generalization ability by combining several classifiers with different classification powers than that of a single classifier [84–86]. These classifiers are called *multiple classifier systems*, *ensembles of classifiers*, or *committee machines*.

The success of multiple classifier systems depends on how we can generate multiple classifiers with diverse classification powers. The well-known strategies for multiple classifier systems are *bagging* (bootstrap aggregating) and *boosting* or *AdaBoost* (adaptive boosting).

In bagging, classifiers are trained using different subsets of training data generated by sampling with replacement and the classification is carried out by voting the classifier outputs. In a narrow sense, the output of a committee machine is generated by voting.

In AdaBoost, weak classifiers whose classification powers are slightly higher than random guess are combined into a strong classifier. This is realized by repeatedly training classifiers with training data that are previously misclassified.

Because support vector machines are strong classifiers, we need to take special care in generating diverse classifiers especially for AdaBoost. There are several approaches in realizing multiple classifier systems. For example, Martinez and Millerioux [87] formulate a committee machine using support vector machines with different kernels. Schwaighofer and Tresp [88] alleviate the long training time caused by large training data by splitting the training data into approximately equal-size subsets, each of which constitutes a training data for each committee machine. Lima et al. [89] propose realizing AdaBoost using support vector machines with different values of the RBF kernel parameter.

A support vector machine is a global classifier, in that single parameter values are used for the input domain. In contrast to a global classifier, a local classifier consists of an ensemble of classifiers with each classifier covers a local region in the input domain. Classification is usually done by a classifier associated with the given input [90–92]. Yang and Kecman's adaptive local hyperplane (ALH) [91] generates a local classifier on the fly for a given input for classification, namely, the $k$-nearest neighbors of the input are selected based on the weighted Euclidean distance and for each class a local hyperplane is generated using the selected data belonging to the class and the input is classified into the class associated with the hyperplane with the minimum distance to the input. This method is an extension of [90], which selects the $k$-nearest neighbors for each class. It is interesting that the classification performance of ALH is comparable or better than that of the support vector machine but one of the problems is slow classification.

## 4.11 Multiple Kernel Learning

Because the weighted sum of kernels with positive weights is also a kernel, instead of a single kernel we can use the following multiple kernel:

$$K(\mathbf{x}, \mathbf{x}') = \sum_{i=1}^{p} \beta_i K_i(\mathbf{x}, \mathbf{x}'), \qquad (4.150)$$

where $p$ is the number of kernels, $K_i(\mathbf{x}, \mathbf{x}')$ are kernels, and $\beta_i$ are nonnegative parameters that satisfy

$$\sum_{i=1}^{p} \beta_i = 1. \tag{4.151}$$

Then the L1 support vector machine with the multiple kernel is defined by

$$\text{maximize} \quad Q(\boldsymbol{\alpha}, \boldsymbol{\beta}) = \sum_{i=1}^{M} \alpha_i - \frac{1}{2} \sum_{i,j=1}^{M} \alpha_i \alpha_j y_i y_j \sum_{k=1}^{p} \beta_k K_k(\mathbf{x}_i, \mathbf{x}_j) \tag{4.152}$$

$$\text{subject to} \quad \sum_{i=1}^{M} y_i \alpha_i = 0, \tag{4.153}$$

$$C \geq \alpha_i \geq 0 \quad \text{for } i = 1, \ldots, M, \tag{4.154}$$

$$\beta_i \geq 0 \quad \text{for } i = 1, \ldots, p, \tag{4.155}$$

$$\sum_{i=1}^{p} \beta_i = 1, \tag{4.156}$$

where $\mathbf{x}_i$ and $y_i$ $(i = 1, \ldots, M)$ are, respectively, the $i$th input vector and the associated class label with $y = 1$ for Class 1 and $y = -1$ for Class 2, $\alpha_i$ are Lagrange multipliers associated with $\mathbf{x}_i$, $\boldsymbol{\alpha} = (\alpha_1, \ldots, \alpha_M)^\top$, $\boldsymbol{\beta} = (\beta_1, \ldots, \beta_p)^\top$, and $C$ is a margin parameter.

The above optimization problem is reformulated and solved in several ways [93–95]. Dileep and Sekhar [95] solve the problem in two-stage procedure: in the first stage $\beta_i$ are fixed and $\alpha_i$ are obtained by regular support vector training; in the second stage, $\alpha_i$ are fixed and $\beta_i$ are obtained by linear programming; and the two-stage procedure is iterated until the solution is obtained.

Multiple kernels can be used to select important sets of features from heterogeneous sets of features. For instance, in image processing, there are many methods for extracting a set of features. And we need to select the set of features that is suited for the given classification problem from among many sets of features. By multiple kernel learning, we can select sets of features whose associated $\beta_i$'s are positive.

## 4.12 Confidence Level

Conventional support vector machines give a prediction result but without its confidence level. One way to give this is the introduction of fuzzy membership functions as discussed in Chapter 3.

Another approach is to obtain the confidence level assuming that the training pairs are generated by an independent and identically distributed (i.i.d.) process [96, 97]. Platt [98] proposed, in addition to support vector machine training, to train the sigmoid function that maps the support vector machine outputs into posterior probabilities. This method is extended to multiclass problems by Frasconi et al. [99].

Fumera and Roli [100] modified the objective function of the support vector machine so that the rejection region is obtained.

## 4.13 Visualization

Interpretability of the classifier or explanation of the decision process is especially important for decision support systems such as medical diagnosis. Without it, a classifier would not be used even if it has high generalization ability.[3] Support vector machines have high generalization ability compared to other classifiers but interpretability is relatively low, especially when nonlinear kernels are used. Thus improving interpretability is very important for the support vector machines to be used in such a field and there is much work in extracting rules from trained support vector machines [101–105].

## References

1. B. Baesens, S. Viaene, T. Van Gestel, J. A. K. Suykens, G. Dedene, B. De Moor, and J. Vanthienen. An empirical assessment of kernel type performance for least squares support vector machine classifiers. In *Proceedings of the Fourth International Conference on Knowledge-Based Intelligent Engineering Systems and Allied Technologies (KES 2000)*, volume 1, pages 313–316, Brighton, UK, 2000.
2. J. A. K. Suykens. Least squares support vector machines for classification and non-linear modelling. *Neural Network World*, 10(1–2):29–47, 2000.
3. J. A. K. Suykens, T. Van Gestel, J. De Brabanter, B. De Moor, and J. Vandewalle. *Least Squares Support Vector Machines*. World Scientific Publishing, Singapore, 2002.
4. J. A. K. Suykens and J. Vandewalle. Multiclass least squares support vector machines. In *Proceedings of International Joint Conference on Neural Networks (IJCNN '99)*, volume 2, pages 900–903, Washington, DC, 1999.
5. F. Masulli and G. Valentini. Comparing decomposition methods for classification. In *Proceedings of the Fourth International Conference on Knowledge-Based Intelligent Engineering Systems and Allied Technologies (KES 2000)*, volume 2, pages 788–791, Brighton, UK, 2000.
6. J. A. K. Suykens, L. Lukas, and J. Vandewalle. Sparse least squares support vector machine classifiers. In *Proceedings of the Eighth European Symposium on Artificial Neural Networks (ESANN 2000)*, pages 37–42, Bruges, Belgium, 2000.
7. G. C. Cawley and N. L. C. Talbot. A greedy training algorithm for sparse least-squares support vector machines. In J. R. Dorronsoro, editor, *Artificial Neural Networks (ICANN 2002)—Proceedings of International Conference, Madrid, Spain*, pages 681–686. Springer-Verlag, Berlin, Germany, 2002.

---

[3] Discussions with Prof. H. Motoda.

8. D. Tsujinishi, Y. Koshiba, and S. Abe. Why pairwise is better than one-against-all or all-at-once. In *Proceedings of International Joint Conference on Neural Networks (IJCNN 2004)*, volume 1, pages 693–698, Budapest, Hungary, 2004.

9. D. Tsujinishi and S. Abe. Fuzzy least squares support vector machines. In *Proceedings of International Joint Conference on Neural Networks (IJCNN 2003)*, volume 2, pages 1599–1604, Portland, OR, 2003.

10. D. Tsujinishi and S. Abe. Fuzzy least squares support vector machines for multiclass problems. *Neural Networks*, 16(5–6):785–792, 2003.

11. C.-W. Hsu and C.-J. Lin. A comparison of methods for multiclass support vector machines. *IEEE Transactions on Neural Networks*, 13(2):415–425, 2002.

12. J. A. K. Suykens and J. Vandewalle. Least squares support vector machine classifiers. *Neural Processing Letters*, 9(3):293–300, 1999.

13. K. P. Bennett. Combining support vector and mathematical programming methods for classification. In B. Schölkopf, C. J. C. Burges, and A. J. Smola, editors, *Advances in Kernel Methods: Support Vector Learning*, pages 307–326. MIT Press, Cambridge, MA, 1999.

14. A. Smola, B. Schölkopf, and G. Rätsch. Linear programs for automatic accuracy control in regression. In *Proceedings of the Ninth International Conference on Artificial Neural Networks (ICANN '99)*, volume 2, pages 575–580, Edinburgh, UK, 1999.

15. V. Kecman and I. Hadzic. Support vectors selection by linear programming. In *Proceedings of the IEEE-INNS-ENNS International Joint Conference on Neural Networks (IJCNN 2000)*, volume 5, pages 193–198, Como, Italy, 2000.

16. W. Zhou, L. Zhang, and L. Jiao. Linear programming support vector machines. *Pattern Recognition*, 35(12):2927–2936, 2002.

17. T. Graepel, R. Herbrich, B. Schölkopf, A. Smola, P. Bartlett, K.-R. Müller, K. Obermayer, and R. Williamson. Classification on proximity data with LP-machines. In *Proceedings of the Ninth International Conference on Artificial Neural Networks (ICANN '99)*, volume 1, pages 304–309, Edinburgh, UK, 1999.

18. B. Schölkopf and A. J. Smola. *Learning with Kernels: Support Vector Machines, Regularization, Optimization, and Beyond*. MIT Press, Cambridge, MA, 2002.

19. P. S. Bradley and O. L. Mangasarian. Massive data discrimination via linear support vector machines. *Optimization Methods and Software*, 13(1):1–10, 2000.

20. Y. Torii and S. Abe. Decomposition techniques for training linear programming support vector machines. *Neurocomputing*, 72(4-6):973–984, 2009.

21. S. Abe. Fuzzy LP-SVMs for multiclass problems. In *Proceedings of the Twelfth European Symposium on Artificial Neural Networks (ESANN 2004)*, pages 429–434, Bruges, Belgium, 2004.

22. C. J. C. Burges. Simplified support vector decision rules. In L. Saitta, editor, *Machine Learning, Proceedings of the Thirteenth International Conference (ICML '96), Bari, Italy*, pages 71–77. Morgan Kaufmann, San Francisco, 1996.

23. S. S. Keerthi, O. Chapelle, and D. DeCoste. Building support vector machines with reduced classifier complexity. *Journal of Machine Learning Research*, 7:1493–1515, 2006.

24. L. Wang, S. Sun, and K. Zhang. A fast approximate algorithm for training $L_1$-SVMs in primal space. *Neurocomputing*, 70(7–9):1554–1560, 2007.

25. T. Ban, Y. Kadobayashi, and S. Abe. Sparse kernel feature analysis using FastMap and its variants. In *Proceedings of the 2009 International Joint Conference on Neural Networks (IJCNN 2009)*, pages 256–263, Atlanta, GA, 2009.

26. M. Wu, B. Schölkopf, and G Bakir. A direct method for building sparse kernel learning algorithms. *Journal of Machine Learning Research*, 7:603–624, 2006.

27. S. Abe. Sparse least squares support vector training in the reduced empirical feature space. *Pattern Analysis and Applications*, 10(3):203–214, 2007.

28. S. Abe. Sparse least squares support vector machines by forward selection based on linear discriminant analysis. In L. Prevost, S. Marinai, and F. Schwenker, editors,

*Artificial Neural Networks in Pattern Recognition: Proceedings of Third IAPR Workshop, ANNPR 2008, Paris, France*, pages 54–65. Springer-Verlag, Berlin, Germany, 2008.

29. K. Iwamura and S. Abe. Sparse support vector machines trained in the reduced empirical feature space. In *Proceedings of the 2008 International Joint Conference on Neural Networks (IJCNN 2008)*, pages 2399–2405, Hong Kong, China, 2008.

30. K. Iwamura and S. Abe. Sparse support vector machines by kernel discriminant analysis. In *Proceedings of the Seventeenth European Symposium on Artificial Neural Networks (ESANN 2009)*, pages 367–372, Bruges, Belgium, 2009.

31. M. E. Tipping. Sparse Bayesian learning and the relevance vector machine. *Journal of Machine Learning Research*, 1:211–244, 2001.

32. S. Chen, X. Hong, C. J. Harris, and P. M. Sharkey. Sparse modelling using orthogonal forward regression with PRESS statistic and regularization. *IEEE Transactions on Systems, Man, and Cybernetics, Part B*, 34(2):898–911, 2004.

33. J. Zhu and T. Hastie. Kernel logistic regression and the import vector machine. *Journal of Computational and Graphical Statistics*, 14(1):185–205, 2005.

34. K. Tanaka, T. Kurita, and T. Kawabe. Selection of import vectors via binary particle swarm optimization and cross-validation for kernel logistic regression. In *Proceedings of the 2007 International Joint Conference on Neural Networks (IJCNN 2007)*, pages 1037–1042, Orlando, FL, 2007.

35. T. Hastie, R. Tibshirani, and J. Friedman. *The Elements of Statistical Learning: Data Mining, Inference, and Prediction*. Springer-Verlag, New York, 2001.

36. G. C. Cawley and N. L. C. Talbot. Improved sparse least-squares support vector machines. *Neurocomputing*, 48(1–4):1025–1031, 2002.

37. J. Valyon and G. Horváth. A sparse least squares support vector machine classifier. In *Proceedings of International Joint Conference on Neural Networks (IJCNN 2004)*, volume 1, pages 543–548, Budapest, Hungary, 2004.

38. L. Jiao, L. Bo, and L. Wang. Fast sparse approximation for least squares support vector machine. *IEEE Transactions on Neural Networks*, 18(3):685–697, 2007.

39. S. Abe. Comparison of sparse least squares support vector regressors trained in the primal and dual. In *Proceedings of the Sixteenth European Symposium on Artificial Neural Networks (ESANN 2008)*, pages 469–474, Bruges, Belgium, 2008.

40. Y. Xu, D. Zhang, Z. Jin, M. Li, and J.-Y. Yang. A fast kernel-based nonlinear discriminant analysis for multi-class problems. *Pattern Recognition*, 39(6):1026–1033, 2006.

41. K. Kaieda and S. Abe. KPCA-based training of a kernel fuzzy classifier with ellipsoidal regions. *International Journal of Approximate Reasoning*, 37(3):189–217, 2004.

42. M. Ashihara and S. Abe. Feature selection based on kernel discriminant analysis. In S. Kollias, A. Stafylopatis, W. Duch, and E. Oja, editors, *Artificial Neural Networks (ICANN 2006)—Proceedings of the Sixteenth International Conference, Athens, Greece, Part II*, pages 282–291. Springer-Verlag, Berlin, Germany, 2006.

43. J. Weston. Leave-one-out support vector machines. In *Proceedings of the Sixteenth International Joint Conference on Artificial Intelligence (IJCAI-99)*, volume 2, pages 727–733, Stockholm, Sweden, 1999.

44. X. Zhang. Using class-center vectors to build support vector machines. In *Neural Networks for Signal Processing IX—Proceedings of the 1999 IEEE Signal Processing Society Workshop*, pages 3–11, 1999.

45. R. Herbrich and J. Weston. Adaptive margin support vector machines for classification. In *Proceedings of the Ninth International Conference on Artificial Neural Networks (ICANN '99)*, volume 2, pages 880–885, Edinburgh, UK, 1999.

46. J. Weston and R. Herbrich. Adaptive margin support vector machines. In A. J. Smola, P. L. Bartlett, B. Schölkopf, and D. Schuurmans, editors, *Advances in Large Margin Classifiers*, pages 281–295. MIT Press, Cambridge, MA, 2000.

47. Z. Kou, J. Xu, X. Zhang, and L. Ji. An improved support vector machine using class-median vectors. In *Proceedings of the Eighth International Conference on Neural Information Processing (ICONIP-2001)*, Paper ID# 60, Shanghai, China, 2001.

48. H. Nakayama and T. Asada. Support vector machines using multi objective programming and goal programming. In *Proceedings of the Ninth International Conference on Neural Information Processing (ICONIP '02)*, volume 2, pages 1053–1057, Singapore, 2002.

49. J.-H. Chen. M-estimator based robust kernels for support vector machines. In *Proceedings of the Seventeenth International Conference on Pattern Recognition (ICPR 2004)*, volume 1, pages 168–171, Cambridge, UK, 2004.

50. S. Abe. *Pattern Classification: Neuro-Fuzzy Methods and Their Comparison.* Springer-Verlag, London, 2001.

51. D. Decoste and B. Schölkopf. Training invariant support vector machines. *Machine Learning*, 46(1–3):161–190, 2002.

52. Y. Bengio and Y. LeCun. Scaling learning algorithms toward AI. In L. Bottou, O. Chapelle, D. DeCoste, and J. Weston, editors, *Large-Scale Kernel Machines*, pages 321–359. MIT Press, Cambridge, MA, 2007.

53. V. Vapnik and A. Vashist. A new learning paradigm: Learning using privileged information. *Neural Networks*, 22(5–6):544–557, 2009.

54. C. J. C. Burges and B. Schölkopf. Improving the accuracy and speed of support vector machines. In M. C. Mozer, M. I. Jordan, and T. Petsche, editors, *Advances in Neural Information Processing Systems 9*, pages 375–381, 1997.

55. S. Amari and S. Wu. Improving support vector machine classifiers by modifying kernel functions. *Neural Networks*, 12(6):783–789, 1999.

56. S. Amari and S. Wu. An information-geometrical method for improving the performance of support vector machine classifiers. In *Proceedings of the Ninth International Conference on Artificial Neural Networks (ICANN '99)*, volume 1, pages 85–90, Edinburgh, UK, 1999.

57. L. Lukas, A. Devos, J. A. K. Suykens, L. Vanhamme, S. Van Huffel, A. R. Tate, C. Majós, and C. Arús. The use of LS-SVM in the classification of brain tumors based on magnetic resonance spectroscopy signals. In *Proceedings of the Tenth European Symposium on Artificial Neural Networks (ESANN 2002)*, pages 131–136, Bruges, Belgium, 2002.

58. J. Feng and P. Williams. The generalization error of the symmetric and scaled support vector machines. *IEEE Transactions on Neural Networks*, 12(5):1255–1260, 2001.

59. B. Goertzel and J. Venuto. Accurate SVM text classification for highly skewed data using threshold tuning and query-expansion-based feature selection. In *Proceedings of the 2006 International Joint Conference on Neural Networks (IJCNN 2006)*, pages 2199–2204, Vancouver, Canada, 2006.

60. F. Aiolli, G. Da San Martino and A. Sperduti. A kernel method for the optimization of the margin distribution. In V. Kůrková, R. Neruda, and J. Koutnik, editors, *Artificial Neural Networks (ICANN 2008)—Proceedings of the Eighteenth International Conference, Prague, Czech Republic, Part I*, pages 305–314. Springer-Verlag, Berlin, Germany, 2008.

61. T. Inoue and S. Abe. Improvement of generalization ability of multiclass support vector machines by introducing fuzzy logic and Bayes theory. *Transactions of the Institute of Systems, Control and Information Engineers*, 15(12):643–651, 2002 (in Japanese).

62. G. Widmer and M. Kubat. Learning in the presence of concept drift and hidden contexts. *Machine Learning*, 23(1):69–101, 1996.

63. Z. Erdem, R. Polikar, F. Gurgen, and N. Yumusak. Ensemble of SVMs for incremental learning. In N. C. Oza, R. Polikar, J. Kittler, and F. Roli, editors, *Multiple Classifier Systems—Proceedings of the Sixth International Workshop, MCS 2005, Seaside, CA*, pages 246–256. Springer-Verlag, Berlin, Germany, 2005.

64. G. Cauwenberghs and T. Poggio. Incremental and decremental support vector machine learning. In T. K. Leen, T. G. Dietterich, and V. Tresp, editors, *Advances in Neural Information Processing Systems 13*, pages 409–415. MIT Press, Cambridge, MA, 2001.

65. A. Shilton, M. Palaniswami, D. Ralph, and A. C. Tsoi. Incremental training of support vector machines. In *Proceedings of International Joint Conference on Neural Networks (IJCNN '01)*, Washington, DC, 2001.

66. P. Laskov, C. Gehl, S. Krüger, and K.-R. Müller. Incremental support vector learning: Analysis, implementation and applications. *Journal of Machine Learning Research*, 7:1909–1936, 2006.

67. P. Mitra, C. A. Murthy, and S. K. Pal. Data condensation in large databases by incremental learning with support vector machines. In *Proceedings of Fifteenth International Conference on Pattern Recognition (ICPR 2000)*, volume 2, pages 2708–2711, Barcelona, Spain, 2000.

68. J. P. Pedroso and N. Murata. Optimisation on support vector machines. In *Proceedings of the IEEE-INNS-ENNS International Joint Conference on Neural Networks (IJCNN 2000)*, volume 6, pages 399–404, Como, Italy, 2000.

69. R. Xiao, J. Wang, and F. Zhang. An approach to incremental SVM learning algorithm. In *Proceedings of the Twelfth IEEE International Conference on Tools with Artificial Intelligence (ICTAI 2000)*, pages 268–273, Vancouver, Canada, 2000.

70. C. Domeniconi and D. Gunopulos. Incremental support vector machine construction. In *Proceedings of the 2001 IEEE International Conference on Data Mining (ICDM 2001)*, pages 589–592, San Jose, CA, 2001.

71. L. Ralaivola and F. d'Alché-Buc. Incremental support vector machine learning: A local approach. In G. Dorffner, H. Bischof, and K. Hornik, editors, *Artificial Neural Networks (ICANN 2001)—Proceedings of International Conference, Vienna, Austria*, pages 322–330. Springer-Verlag, Berlin, Germany, 2001.

72. S. Katagiri and S. Abe. Incremental training of support vector machines using hyperspheres. *Pattern Recognition Letters*, 27(13):1495–1507, 2006.

73. S. Katagiri and S. Abe. Incremental training of support vector machines using truncated hypercones. In F. Schwenker and S. Marinai, editors, *Artificial Neural Networks in Pattern Recognition: Proceedings of Second IAPR Workshop, ANNPR 2006, Ulm, Germany*, pages 153–164. Springer-Verlag, Berlin, Germany, 2006.

74. A. Yalcin, Z. Erdem, and F. Gurgen. Ensemble based incremental SVM classifiers for changing environments. In *Proceedings of Twenty-Second International Symposium on Computer and Information Sciences (ISCIS 2007)*, pages 204–208, Ankara, Turkey, 2007.

75. T. Hastie, S. Rosset, R. Tibshirani, and J. Zhu. The entire regularization path for the support vector machine. *Journal of Machine Learning Research*, 5:1391–1415, 2004.

76. V. Vapnik. *Estimation of Dependences Based on Empirical Data, Second Edition*. Springer-Verlag, New York, 2006.

77. A. Singh, R. D. Nowak, and X. Zhu. Unlabeled data: Now it helps, now it doesn't. In D. Koller, D. Schuurmans, Y. Bengio, and L. Bottou, editors, *Advances in Neural Information Processing Systems 21*, pages 1513–1520. MIT Press, Cambridge, MA, 2009.

78. M. M. Adankon and M. Cheriet. Help-training semi-supervised LS-SVM. In *Proceedings of the 2009 International Joint Conference on Neural Networks (IJCNN 2009)*, pages 49–56, Atlanta, GA, 2009.

79. R. Raina, A. Battle, H. Lee, B. Packer, and A. Y. Ng. Self-taught learning: Transfer learning from unlabeled data. In *Proceedings of the Twenty-Fourth International Conference on Machine Learning (ICML 2007)*, pages 759–766, Corvallis, OR, 2007.

80. K. Huang, Z. Xu, I. King, M. R. Lyu, and C. Campbell. Supervised self-taught learning: Actively transferring knowledge from unlabeled data. In *Proceedings of*

the 2009 International Joint Conference on Neural Networks (IJCNN 2009), pages 1272–1277, Atlanta, GA, 2009.

81. T. Joachims. Transductive support vector machines. In O. Chapelle, B. Schölkopf, and A. Zien, editors, Semi-Supervised Learning, pages 105–117. MIT Press, Cambridge, MA, 2006.

82. V. Sindhwani and S. S. Keerthi. Newton methods for fast semisupervised linear SVMs. In L. Bottou, O. Chapelle, D. DeCoste, and J. Weston, editors, Large-Scale Kernel Machines, pages 155–174. MIT Press, Cambridge, MA, 2007.

83. S. Pang, T. Ban, Y. Kadobayashi, and N. Kasabov. Spanning SVM tree for personalized transductive learning. In C. Alippi, M. Polycarpou, C. Panayiotou, and G. Ellinas, editors, Artificial Neural Networks (ICANN 2009)—Proceedings of the Nineteenth International Conference, Limassol, Cyprus, Part I, pages 913–922. Springer-Verlag, Berlin, Germany, 2009.

84. S. Haykin. Neural Networks: A Comprehensive Foundation, Second Edition. Prentice Hall, Upper Saddle River, NJ, 1999.

85. T. Windeatt and F. Roli, editors. Multiple Classifier Systems—Proceedings of the fourth International Workshop, MCS 2003, Guildford, UK. Springer-Verlag, Berlin, Germany, 2003.

86. L. I. Kuncheva. Combining Pattern Classifiers: Methods and Algorithms. John Wiley & Sons, Hoboken, NJ, 2004.

87. D. Martinez and G. Millerioux. Support vector committee machines. In Proceedings of the Eighth European Symposium on Artificial Neural Networks (ESANN 2000), pages 43–48, Bruges, Belgium, 2000.

88. A. Schwaighofer and V. Tresp. The Bayesian committee support vector machine. In G. Dorffner, H. Bischof, and K. Hornik, editors, Artificial Neural Networks (ICANN 2001)—Proceedings of International Conference, Vienna, Austria, pages 411–417. Springer-Verlag, Berlin, Germany, 2001.

89. N. H. C. Lima, A. D. D. Neto, and J. D. de Melo. Creating an ensemble of diverse support vector machines using Adaboost. In Proceedings of the 2009 International Joint Conference on Neural Networks (IJCNN 2009), pages 1802–1806, Atlanta, GA, 2009.

90. P. Vincent and Y. Bengio. K-local hyperplane and convex distance nearest neighbor algorithms. In T. G. Dietterich, S. Becker, and Z. Ghahramani, editors, Advances in Neural Information Processing Systems 14, pages 985–992, Cambridge, MA, 2002. MIT Press.

91. T. Yang and V. Kecman. Adaptive local hyperplane classification. Neurocomputing, 71(13–15):3001–3004, 2008.

92. D. Martínez-Rego, O. Fontenla-Romero, I. Porto-Díaz, and A. Alonso-Betanzos. A new supervised local modelling classifier based on information theory. In Proceedings of the 2009 International Joint Conference on Neural Networks (IJCNN 2009), pages 2014–2020, Atlanta, GA, 2009.

93. G. R. G. Lanckriet, N. Cristianini, P. Bartlett, L. El Ghaoui, and M. I. Jordan. Learning the kernel matrix with semidefinite programming. Journal of Machine Learning Research, 5:27–72, 2004.

94. S. Sonnenburg, G. Rätsch, C. Schäfer, and B. Schölkopf. Large scale multiple kernel learning. Journal of Machine Learning Research, 7:1531–1565, 2006.

95. A. D. Dileep and C. C. Sekhar. Representation and feature selection using multiple kernel learning. In Proceedings of the 2009 International Joint Conference on Neural Networks (IJCNN 2009), pages 717–722, Atlanta, GA, 2009.

96. A. Gammerman, V. Vovk, and V. Vapnik. Learning by transduction. In Proceedings of the Fourteenth Conference on Uncertainty in Artificial Intelligence (UAI '98), pages 148–155, Madison, WI, 1998.

97. V. Vovk, A. Gammerman, and C. Saunders. Machine-learning applications of algorithmic randomness. In I. Bratko and S. Dzeroski, editors, Machine Learning,

*Proceedings of the Sixteenth International Conference (ICML '99), Bled, Slovenia,* pages 444–453. Morgan Kaufmann, San Francisco, 1999.

98.  J. C. Platt. Probabilities for SV machines. In A. J. Smola, P. L. Bartlett, B. Schölkopf, and D. Schuurmans, editors, *Advances in Large Margin Classifiers,* pages 61–73. MIT Press, Cambridge, MA, 2000.

99.  P. Frasconi, A. Passerini, and A. Vullo. A two-stage SVM architecture for predicting the disulfide bonding state of cysteines. In H. Bourlard, T. Adali, S. Bengio, J. Larsen, and S. Douglas, editors, *Neural Networks for Signal Processing XII—Proceedings of the 2002 IEEE Signal Processing Society Workshop,* pages 25–34, 2002.

100. G. Fumera and F. Roli. Support vector machines with embedded reject option. In S.-W. Lee and A. Verri, editors, *Pattern Recognition with Support Vector Machines: Proceedings of First International Workshop, SVM 2002, Niagara Falls, Canada,* pages 68–82. Springer-Verlag, Berlin, Germany, 2002.

101. H. Núñez, C. Angulo, and Català. Rule extraction from support vector machines. In *Proceedings of the Tenth European Symposium on Artificial Neural Networks (ESANN 2002),* pages 107–112, Bruges, Belgium, 2002.

102. D. Caragea, D. Cook, and V. Honavar. Towards simple, easy-to-understand, yet accurate classifiers. In *Proceedings of the Third IEEE International Conference on Data Mining (ICDM 2003),* pages 497–500, Melbourne, FL, 2003.

103. X. Fu, C.-J. Ong, S. S. Keerthi, G. G. Hung, and L. Goh. Extracting the knowledge embedded in support vector machines. In *Proceedings of International Joint Conference on Neural Networks (IJCNN 2004),* volume 1, pages 291–296, Budapest, Hungary, 2004.

104. L. Franke, E. Byvatov, O. Werz, D. Steinhilber, P. Schneider, and G. Schneider. Extraction and visualization of potential pharmacophore points using support vector machines: Application to ligand-based virtual screening for COX-2 inhibitors. *Journal of Medical Chemistry,* 48(22):6997–7004, 2005.

105. J. Diederich (Ed.). *Rule Extraction from Support Vector Machines.* Springer-Verlag, Berlin, 2008.

# Chapter 5
# Training Methods

In training an L1 or L2 support vector machine, we need to solve a quadratic programming problem with the number of variables equal to the number of training data. Computational complexity is of the order of $M^3$, where $M$ is the number of training data. Thus when $M$ is large, training takes long time. To speed up training, numerous methods have been proposed. One is to extract support vector candidates from the training data and then train the support vector machine using these data. Another method is to accelerate training by decomposing variables into a working set and a fixed set and by repeatedly solving the subproblem associated with the working set until convergence.

In this chapter, we discuss preselection of support vector candidates and training methods using decomposition techniques. We explain primal–dual interior-point methods, steepest ascent and Newton's methods, which are extensions of the sequential minimal optimization technique (SMO), and active set training methods, which retain the support vector candidates in the working set. We also discuss training methods for LP support vector machines.

## 5.1 Preselecting Support Vector Candidates

According to the architecture of the support vector machine, only the training data near the decision boundaries are necessary. In addition, because the training time becomes longer as the number of training data increases, the training time is shortened if the data far from the decision boundary are deleted. Therefore, if we can delete unnecessary data from the training data efficiently prior to training, we can speed up training. Several approaches have been developed to preselect support vector candidates [1–8]. In [2], all the training data are first clustered by the $k$-means clustering algorithm. Then if a generated cluster includes data that belong to the same class, the data are discarded. But if a generated cluster includes data with different

S. Abe, *Support Vector Machines for Pattern Classification,*
Advances in Pattern Recognition, DOI 10.1007/978-1-84996-098-4_5,
© Springer-Verlag London Limited 2010

classes, these data are retained for training, assuming that they may include support vectors. The $k$-means clustering algorithm is not a requisite. It may be the fuzzy $c$-means clustering algorithm or the Kohonen network. This method works when classes overlap, but if classes are well separated, it may not extract boundary data. In [3, 6], $k$ nearest neighbors are used to detect the boundary data.

In [1], a superset of support vectors called *guard vectors* is extracted by linear programming. Assume that the set of training input–output pairs $\{(\mathbf{x}_1, y_1), \ldots, (\mathbf{x}_M, y_M)\}$ is linearly separable, where $y_i$ are either 1 or $-1$. Then there exists a separating hyperplane that includes a data sample $\mathbf{x}_i$:

$$y_1 \left( \mathbf{w}^\top \mathbf{x}_1 + b \right) \geq 0,$$
$$\vdots$$
$$y_{i-1} \left( \mathbf{w}^\top \mathbf{x}_{i-1} + b \right) \geq 0,$$
$$y_i \left( \mathbf{w}^\top \mathbf{x}_i + b \right) = 0, \tag{5.1}$$
$$y_{i+1} \left( \mathbf{w}^\top \mathbf{x}_{i+1} + b \right) \geq 0,$$
$$\vdots$$
$$y_M \left( \mathbf{w}^\top \mathbf{x}_M + b \right) \geq 0.$$

We call the data sample $\mathbf{x}_i$ that satisfies (5.1) the *guard vector*. Clearly the support vectors are guard vectors but the reverse is not true. Thus the set of guard vectors is a superset of the set of support vectors. To obtain the set of guard vectors, we solve (5.1) by linear programming, changing $i$ from 1 to $M$. Unlike usual linear programming we need only to check whether $\mathbf{x}_i$ is a feasible solution of (5.1). If the training data set is not linearly separable, we need to map the original input space into the feature space. Instead, in [1], the original input space is extended to the $(m + M)$-dimensional space where $m$ is the dimension of $\mathbf{x}$ and the $(m + i)$th element is 1 for $\mathbf{x}_i$ and 0 otherwise.

In the following we discuss speedup of training by deleting unnecessary training data [4]. We estimate the data near the boundaries using the classifier based on the Mahalanobis distance and extracting the misclassified data and the data near the boundaries.

### 5.1.1 Approximation of Boundary Data

The decision boundaries of the classifier using the Mahalanobis distance are expressed by the polynomials of the input variables with degree 2. Therefore, the boundary data given by the classifier are supposed to approximate the boundary data for the support vector machine, especially with the polynomial kernels with degree 2.

For the class $i$ data $\mathbf{x}$, the Mahalanobis distance $d_i(\mathbf{x})$ is given by

$$d_i^2(\mathbf{x}) = (\mathbf{c}_i - \mathbf{x})^\top Q_i^{-1}(\mathbf{c}_i - \mathbf{x}), \tag{5.2}$$

where $\mathbf{c}_i$ and $Q_i$ are the center vector and the covariance matrix for the data belonging to class $i$, respectively,

$$\mathbf{c}_i = \frac{1}{|X_i|} \sum_{\mathbf{x} \in X_i} \mathbf{x}, \tag{5.3}$$

$$Q_i = \frac{1}{|X_i|} \sum_{\mathbf{x} \in X_i} (\mathbf{x} - \mathbf{c}_i)(\mathbf{x} - \mathbf{c}_i)^\top. \tag{5.4}$$

Here, $X_i$ denotes the set of data belonging to class $i$ and $|X_i|$ is the number of data in the set. The data $\mathbf{x}$ is classified into the class with the minimum Mahalanobis distance. The most important feature of the Mahalanobis distance is that it is invariant for linear transformation of input variables. Therefore, we do not worry about the scaling of each input variable.

For the data sample belonging to class $i$, we check whether

$$r(\mathbf{x}) = \frac{\min_{j \neq i, j=1,\dots,n} d_j(\mathbf{x}) - d_i(\mathbf{x})}{d_i(\mathbf{x})} \leq \eta \tag{5.5}$$

is satisfied, where $r(\mathbf{x})$ is the relative difference of distances, $\eta\,(>0)$ controls the nearness to the boundary, and $0 < \eta < 1$. If $r(\mathbf{x})$ is negative, the data sample is misclassified. We assume that the misclassified data are near the decision boundary. Equation (5.5) is satisfied when the second minimum Mahalanobis distance is shorter than or equal to $(1 + \eta)\, d_i(\mathbf{x})$ for the correctly classified $\mathbf{x}$.

In extracting boundary data, we set some appropriate value to $\eta$ and for each class we select the boundary data whose number is between $N_{\min}$ and $N_{\max}$. Here the minimum number is set so that the number of boundary data is not too small for some classes because the data that satisfy (5.5) are scarce. The maximum number is set not to allow too many data to be selected. A general procedure for extracting boundary data is as follows:

1. Calculate the centers and covariance matrices for all the classes using (5.3) and (5.4).
2. For the training sample $\mathbf{x}$ belonging to class $i$, calculate $r(\mathbf{x})$ and put the sample into the stack for class $i$, $S_i$, whose elements are sorted in increasing order of the value of $r(\mathbf{x})$ and whose maximum length is $N_{\max}$. Iterate this for all the training data.
3. If stack $S_i$ includes more than $N_{\min}$ data that satisfy (5.5), select these data as the boundary data for class $i$. Otherwise, select the first $N_{\min}$ data as the boundary data.

This procedure is refined according to the architecture of the multiclass support vector machine. If we use the pairwise classification, in determining the decision boundary for classes $i$ and $j$, we calculate $r(\mathbf{x})$ only for classes $i$

and $j$. If we use the one-against-all multiclass architecture, in determining the decision boundary for class $i$ and the remaining classes, we assume that the remaining classes consist of $n-1$ clusters and use (5.5) to extract boundary data.

## 5.1.2 Performance Evaluation

Although the performance varies as kernels vary, the polynomial kernels with degree 2 perform relatively well. Thus in the following study, we use the polynomials with degree 2 as the kernels. We evaluated the method using the iris data, blood cell data, and hiragana data listed in Table 1.3. Except for the iris data, we set $N_{max}$ as the half of the maximum number of class data, namely, 200. And we set $N_{min} = 50$ and evaluated the performance changing $\eta$.

We ran the software developed by Royal Holloway, University of London [9], on a SUN UltraSPARC-IIi (335 MHz) workstation. The software used pairwise classification. The training was done without decomposition. The data, which were originally scaled in [0, 1], were rescaled in [−1, 1].

Because the number of the iris data is small, we checked only the lowest rankings, in the relative difference of the Mahalanobis distances, of support vectors for the pairs of classes. Table 5.1 lists the results when the boundary data were extracted for each class. The numeral in the $i$th row and the $j$th column shows the lowest ranking of the support vectors, belonging to class $i$, for a pair of classes $i$ and $j$. The diagonal elements show the numbers of training data for the associated classes. The maximum value among the lowest rankings was 8, which was smaller than half the number of class data. Thus, the relative difference of the Mahalanobis distances well reflected the boundary data.

Table 5.1 The lowest rankings of support vectors for the iris data

| Class | 1 | 2 | 3 |
|-------|------|------|------|
| 1 | (25) | 1 | 2 |
| 2 | 8 | (25) | 3 |
| 3 | 2 | 3 | (25) |

Table 5.2 lists the results for the blood cell and hiragana data sets. The column "Selected" lists the number of data selected and the first row in each data set shows the results when all the training data were used. The column "Rate" lists the recognition rates of the test data and training data in parentheses. The numerals in parentheses in the "Time" column show the time for

extracting boundary data and the speedup ratios were calculated, including the time for extracting boundary data.

For $\eta = 2$–$4$, two to five times speedup was obtained without deteriorating the recognition rates very much.

**Table 5.2** Performance for the blood cell and hiragana data ($N_{\max} = 200$ and $N_{\min} = 50$)

| Data | $\eta$ | Selected | Rate (%) | Time (s) | Speedup |
|---|---|---|---|---|---|
| Blood cell | – | 3,097 | 92.13 (99.32) | 924 | 1 |
| | 2.0 | 2,102 | 92.13 (99.29) | 448 (2) | 2.1 |
| Hiragana-50 | – | 4,610 | 98.91 (100) | 2,862 | 1 |
| | 2.0 | 2,100 | 97.11 (98.68) | 644 (28) | 4.2 |
| | 4.0 | 3,611 | 98.79 (100) | 1,690 (28) | 1.7 |
| Hiragana-105 | – | 8,375 | 100 (100) | 11,656 | 1 |
| | 2.0 | 2,824 | 99.61 (99.65) | 1,908 (177) | 5.6 |
| | 3.0 | 5,231 | 99.99 (99.99) | 5,121 (187) | 2.2 |
| Hiragana-13 | – | 8,375 | 99.57 (100) | 9,183 | 1 |
| | 2.0 | 4,366 | 99.56 (100) | 2,219 (16) | 4.1 |

## 5.2 Decomposition Techniques

To reduce the number of variables in training, Osuna et al. [10] proposed dividing the set of variables into the working set and the fixed set, optimizing the subproblem associated with the working set, exchange variables between the working set and the fixed set, and repeating optimizing the subproblem until the solution is obtained.

Let sets $W$ and $N$ include indices in the working set and the fixed set, respectively, where $W \cup N = \{1, \ldots, M\}$ and $W \cap N = \phi$. Then decomposing $\{\alpha_i \mid i = 1, \ldots, M\}$ into $\boldsymbol{\alpha}_W = \{\alpha_i \mid i \in W\}$ and $\boldsymbol{\alpha}_N = \{\alpha_i \mid i \in N\}$, we define the following subproblem:

$$
\begin{aligned}
\text{maximize} \quad Q(\boldsymbol{\alpha}_W) = &\sum_{i \in W} \alpha_i - \frac{1}{2} \sum_{i,j \in W} \alpha_i \alpha_j y_i y_j K(\mathbf{x}_i, \mathbf{x}_j) \\
&- \sum_{\substack{i \in W, \\ j \in N}} \alpha_i \alpha_j y_i y_j K(\mathbf{x}_i, \mathbf{x}_j) \\
&- \frac{1}{2} \sum_{i,j \in N} \alpha_i \alpha_j y_i y_j K(\mathbf{x}_i, \mathbf{x}_j) + \sum_{i \in N} \alpha_i \quad (5.6)
\end{aligned}
$$

subject to $\displaystyle\sum_{i \in W} y_i \alpha_i = -\sum_{i \in N} y_i \alpha_i, \quad 0 \le \alpha_i \le C \quad \text{for } i \in W.$  (5.7)

Deleting the constant term in (5.6) we obtain

$$\text{maximize} \quad Q(\boldsymbol{\alpha}_W) = \sum_{i \in W} \alpha_i - \frac{1}{2} \sum_{i,j \in W} \alpha_i \alpha_j y_i y_j K(\mathbf{x}_i, \mathbf{x}_j)$$
$$- \sum_{\substack{i \in W, \\ j \in N}} \alpha_i \alpha_j y_i y_j K(\mathbf{x}_i, \mathbf{x}_j) \tag{5.8}$$

subject to $\displaystyle\sum_{i \in W} y_i \alpha_i = -\sum_{i \in N} y_i \alpha_i, \quad 0 \le \alpha_i \le C \quad \text{for } i \in W.$  (5.9)

Because the equality constraint in (5.9) is satisfied for $|W| \ge 2$, the minimum number of $|W|$ is 2. Optimizing (5.8) and (5.9) for a feasible solution, the obtained solution is also feasible. This is because the fixed variables satisfy the inequality constraints and the variables in the working set are optimized under the constraints. Therefore, if we start solving (5.8) and (5.9) with a feasible initial solution, the solution is also feasible and the objective function is non-decreasing. This makes the implementation of the decomposition technique successful.

After optimizing (5.8) and (5.9), we check whether the obtained solution satisfies the KKT conditions. If they are satisfied, the optimal solution is obtained. If not, we move the fixed variables that do not satisfy the KKT conditions into the working set. There are two ways in updating the working set: *variable-size chunking* and *fixed-size chunking* [9]. In the variable-size chunking, we keep support vector candidates in the working set and add the fixed variables that violate KKT conditions. A procedure for the variable-size chunking is as follows:

1. Set $F$ indices from those of the training data to $W$, where $F$ is a fixed positive integer.
2. Solve the subproblem for $\boldsymbol{\alpha}_W$.
3. Delete the variables except for the support vector candidates and add $F$ points that do not satisfy the KKT conditions and go to Step 2. Otherwise, terminate the algorithm.

By this algorithm the working set size increases as the iteration proceeds and when the algorithm terminates, the working set includes support vectors. Training based on variable-size chunking is sometimes called *active set training* because constraints associated with support vectors satisfy the equality constraints and these constraints are called *active* and the working set the *active set*.[1]

---

[1] In contrast to active set training, *active learning* tries to minimize a labeling task of unlabeled data by only labeling the data that are necessary for generating a classifier. During learning process, the learning machine asks labeling unlabeled data that are crucial

In the fixed-size chunking, we do not keep all the support vector candidates. The procedure of the fixed-size chunking is as follows [10]:

1. Select $|W|$ indices from those of the training data.
2. Solve the subproblem for $\alpha_W$.
3. If there exist such $\alpha_j$ ($\in \alpha_N$) that do not satisfy the KKT conditions, replace any $\alpha_i$ ($\in \alpha_W$) with $\alpha_j$ and go to Step 2. Otherwise, terminate the algorithm.

By both variable-size chunking and fixed-size chunking, the values of the objective function are non-decreasing. But this does not preclude the situation where the solution stays at some point and does not reach the optimal solution. The convergence characteristics of decomposition techniques have been investigated [12–14]. Especially for the working set size of 2, i.e., fixed-size chunking with $|W| = 2$, if the variable pair that violate the KKT conditions most are selected, the asymptotic convergence of the algorithm is guaranteed [13, 14]. Another working set selection strategy is to select the variable pair based on the second-order information [15].

By chunking computation complexity reduced from $O(M^3)$ to $O(N\,|W|^3)$, where $N$ is the number of iterations. Speedup of training by the decomposition technique depends on the initial working set and the strategy of working set update. In [16], to speed up the chunking algorithm, two methods for estimating the support vector candidates, which are used for the first chunk, are proposed. The procedure of the method that performed better for benchmark data sets is as follows:

1. For each data sample belonging to Class 1, find the data sample belonging to Class 2 with the minimum distance in the feature space (see Fig. 5.1a).
2. Among the selected pairs with the same Class 2 data sample, select the pair with the minimum distance as the support vector candidates (see Fig. 5.1b).
3. Iterate Steps 1 and 2 exchanging Class 1 and Class 2.
4. If the obtained data are not sufficient, delete the obtained data from the training data and iterate Steps 1–3 several times.

For the thyroid data, for the chunk sizes from 10 to 480, the training time was reduced by 12.3% on the average (50% in maximum) over the random chunking algorithm.

This algorithm can be used for preselection of training data [17]. Figure 5.2 shows the results when the algorithm was used for preselecting the data. In Step 1, the distance was measured in the input space, instead of the feature space,[2] and Step 2 was not executed. The figure shows the recognition rates of the thyroid test data when the training data were preselected

---

for generating a classifier, in a support vector machine environment, the data that change the optimal hyperplane most [11]. Active learning for the labeled data is considered to be one of the working set selection methods.

[2] See the discussions in Section 2.3.4.1 on p. 44.

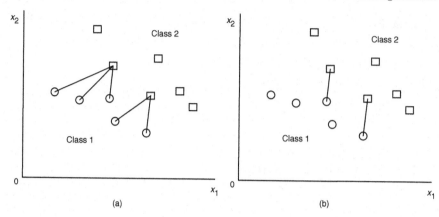

**Fig. 5.1** Selection of support vector candidates: **(a)** For each Class 1 data sample select a pair belonging to Class 2 with the minimum distance. **(b)** For the selected pairs with the same Class 2 data sample, select the pair with the minimum distance

and all the training data were used for training the one-against-all support vector machine. The rate of support vector selection shows the number of support vectors, which are included in the selected data, divided by the number of support vectors for all the training data. By one iteration of the algorithm, the recognition rate was 4% lower than using all the training data, and for two to three iterations, the recognition rate was 2% lower. The training time with three iterations was 5 s by a Pentium III 1G personal computer, compared to the 40 s using all the training data. But even if we increase the iterations, the recognition rate did not reach the rate by using all the training data. This is reflected to the rate of support vector selection.

For the one-against-all support vector machine, one class has a smaller number of training data than the other. Thus if the selected data are deleted in Step 4, the training data for one class become extinct before the support vectors for the other class are selected. Thus, to prevent this, we stop deleting the training data for one class. Figure 5.3 shows the result for the modified algorithm. After nine iterations, the recognition rates are almost the same, and the training time including the selection time was 29 s, and the selected data was one-third of the training data.

## 5.3 KKT Conditions Revisited

The KKT conditions for L1 support vector machines given by (2.63), (2.64), and (2.65) are satisfied for the optimal solution. But because the bias term

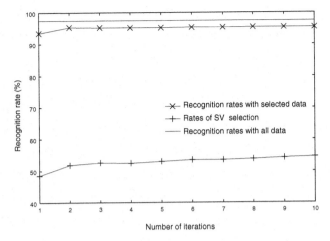

**Fig. 5.2** Recognition rate by preselecting the thyroid training data ($d = 3$, $C = 10{,}000$)

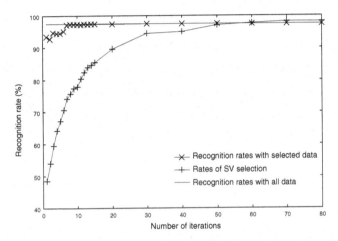

**Fig. 5.3** Recognition rate by preselecting the thyroid training data without deleting the data for one class ($d = 3$, $C = 10{,}000$)

$b$ is not included in the dual problem, if we detect the data that violate the KKT conditions using these equations, we need to estimate $b$. For variable-size chunking, all the non-zero variables are included in the working set. Thus, the estimate of $b$ using any non-zero variable will give the same value within the precision of a computer. However, by fixed-size chunking, some nonzero variables are in the fixed set. Therefore, the estimate of $b$ using the nonzero variable in the working set and that in the fixed set will be different. Thus, it will lead to incorrect estimation. We call the KKT conditions including the bias term *inexact KKT conditions*. To avoid this, we need to redefine the

KKT conditions for the dual problem that do not include $b$. We call them *exact KKT conditions*. In the following, we discuss the KKT conditions for the dual problem based on Keerthi and Gilbert [14].

We rewrite (2.63) as follows:

$$\alpha_i \left( y_i\, b - y_i\, F_i + \xi_i \right) = 0, \tag{5.10}$$

where

$$F_i = y_i - \sum_{j=1}^{M} y_j\, \alpha_j\, K(\mathbf{x}_i, \mathbf{x}_j). \tag{5.11}$$

Using (5.10), we can classify the KKT conditions into the following three cases:

1. $\alpha_i = 0$

$$y_i\, b \geq y_i\, F_i \quad \text{or} \quad b \geq F_i \;\; \text{if} \;\; y_i = 1; \quad b \leq F_i \;\; \text{if} \;\; y_i = -1$$

2. $0 < \alpha_i < C$

$$b = F_i,$$

3. $\alpha_i = C$

$$y_i\, b \leq y_i\, F_i \quad \text{or} \quad b \leq F_i \;\; \text{if} \;\; y_i = 1; \quad b \geq F_i \;\; \text{if} \;\; y_i = -1.$$

Then the KKT conditions are simplified as follows:

$$\bar{F}_i \geq b \geq \tilde{F}_i \quad \text{for } i = 1, \ldots, M, \tag{5.12}$$

where

$$\begin{aligned}
\tilde{F}_i = F_i \quad &\text{if} \quad (y_i = 1, \alpha_i = 0), \quad 0 < \alpha_i < C \\
&\text{or} \quad (y_i = -1, \alpha_i = C), \tag{5.13} \\
\bar{F}_i = F_i \quad &\text{if} \quad (y_i = -1, \alpha_i = 0), \quad 0 < \alpha_i < C \\
&\text{or} \quad (y_i = 1, \alpha_i = C). \tag{5.14}
\end{aligned}$$

Depending on the value of $\alpha_i$, $\bar{F}_i$ or $\tilde{F}_i$ is not defined. For instance, if $y_i = 1$ and $\alpha_i = C$, $\tilde{F}_i$ is not defined.

To detect the violating variables, we define $b_{\text{low}}$ and $b_{\text{up}}$ as follows:

$$\begin{aligned}
b_{\text{low}} &= \max_i \tilde{F}_i, \\
b_{\text{up}} &= \min_i \bar{F}_i. \tag{5.15}
\end{aligned}$$

If the KKT conditions are satisfied,

$$b_{\text{up}} \geq b_{\text{low}}. \tag{5.16}$$

Figure 5.4a illustrates the KKT conditions that are satisfied. The filled rectangles show $\bar{F}_i$ and the filled circles show $\tilde{F}_i$.

We can use (5.16) as a stopping condition for training. To lighten a computational burden, we loosen (5.16) as follows:

$$b_{\text{up}} \geq b_{\text{low}} - \tau, \tag{5.17}$$

where $\tau$ is a positive tolerance parameter.

Figure 5.4b shows the case where the KKT conditions are not satisfied. The data between the two dotted lines are the data that violate the KKT conditions. We define the $\tau$-violating set $V_{\text{KKT}}$:

$$V_{\text{KKT}} = \{\mathbf{x}_i \,|\, b_{\text{up}} < \tilde{F}_i - \tau \quad \text{or} \quad b_{\text{low}} > \bar{F}_i + \tau$$
$$\text{for } i \in \{1, \ldots, M\}\}. \tag{5.18}$$

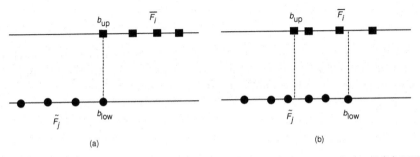

**Fig. 5.4** KKT conditions for a dual problem: **(a)** KKT conditions satisfied and **(b)** KKT conditions violated

For $\boldsymbol{\alpha} = \mathbf{0}$, $b_{\text{low}} = \tilde{F}_i = 1$ and $b_{\text{up}} = \bar{F}_i = -1$. Thus, if we set $\boldsymbol{\alpha} = \mathbf{0}$ for the initial vector, all the data violate the KKT conditions.

*Example 5.1.* We examine if the solution $(\alpha_1, \alpha_2, \alpha_3, \alpha_4) = (1, 0, 1, 0)$ in Example 2.8 satisfies the KKT conditions. Because

$$\bar{F}_1 = \bar{F}_3 = \bar{F}_4 = \tilde{F}_1 = \tilde{F}_2 = \tilde{F}_3 = 0,$$

we obtain

$$b_{\text{up}} = b_{\text{low}}.$$

Thus the KKT conditions are satisfied.

Let $(\alpha_1, \alpha_2, \alpha_3, \alpha_4) = (0.8, 0, 1, 0)$. Then because

$$\bar{F}_1 = \tilde{F}_1 = 0.2, \quad \tilde{F}_2 = \bar{F}_3 = \tilde{F}_3 = 0, \quad \tilde{F}_4 = -0.2,$$

we obtain

$$b_{\text{up}} = -0.2, \quad b_{\text{low}} = 0.2.$$

Thus the $\tau$-violating set with $\tau = 0$ is

$$V_{\text{KKT}} = \{1, 2, 3\}.$$

The following example shows that $b_{\text{up}} > b_{\text{low}}$ when the primal solution is not unique.

*Example 5.2.* Reconsider Example 2.9. When $C < 1/2$, $\alpha_1 = \alpha_2 = C$. Thus,

$$\tilde{F}_1 = -1 + 2C,$$
$$\bar{F}_2 = 1 - 2C.$$

Thus the KKT conditions are satisfied and

$$1 - 2C \geq b \geq -1 + 2C.$$

Similar to L1 soft-margin support vector machines, we can derive the KKT conditions for L2 soft-margin support vector machines. For the L2 support vector machine, the KKT conditions are as follows:

1. $\alpha_i = 0$

$$y_i\, b \geq y_i\, F_i,$$

2. $\alpha_i > 0$

$$b = F_i - \frac{y_i\, \alpha_i}{C}.$$

The KKT conditions are simplified as follows:

$$\bar{F}_i \geq b \geq \tilde{F}_i \quad \text{for } i = 1, \ldots, M, \tag{5.19}$$

where

$$\tilde{F}_i = F_i - \frac{y_i\, \alpha_i}{C} \quad \text{if} \quad (y_i = 1, \alpha_i = 0) \quad \text{or} \quad \alpha_i > 0, \tag{5.20}$$

$$\bar{F}_i = F_i - \frac{y_i\, \alpha_i}{C} \quad \text{if} \quad (y_i = -1, \alpha_i = 0) \quad \text{or} \quad \alpha_i > 0. \tag{5.21}$$

The remaining procedure is the same as that of the L1 support vector machine.

## 5.4 Overview of Training Methods

Training of a support vector machine results in solving a quadratic programming (QP) program. But commercially available QP solvers are not suited for training a support vector machine with a large number of training data due to long training time and large memory consumption. To overcome this problem, QP solvers have been combined with the decomposition techniques or other training methods have been developed. There are now many downloadable programs that train support vector machines.

Conventional learning algorithms developed for perceptrons are adapted to be used in the feature space [18–22]. The kernel-Adatron algorithm [18, 19] is extended from the Adatron algorithm [23]. Because the kernel-Adatron algorithm does not consider the equality constraint, the optimality for the bias term is not guaranteed. To compensate for this, in [24], a method to augment an additional input is discussed. Mangasarian and Musicant [25] added $b^2/2$ to the primal objective function of the L1 support vector machine. Then the dual problem becomes as follows:

$$\text{maximize} \quad Q(\boldsymbol{\alpha}) = \sum_{i=1}^{M} \alpha_i - \frac{1}{2} \sum_{i,j=1}^{M} \alpha_i \, \alpha_j \, y_i \, y_j \, (K(\mathbf{x}_i, \mathbf{x}_j) + 1) \quad (5.22)$$

$$\text{subject to} \quad 0 \le \alpha_i \le C \quad \text{for } i = 1, \ldots, M. \quad (5.23)$$

By this formulation, the linear constraint is not included, and they applied the successive overrelaxation method for training.

The most well-known training method using fixed-size chunking is sequential minimal optimization (SMO) developed by Platt [26]. It optimizes two data at a time. Thus it is equivalent to the fixed-size chunking with $|W| = 2$ in Section 5.2. Because it is inefficient to solve a large-size problem modifying only two variables at a time, many sophisticated techniques such as for working set selection and limiting the search space (called *shrinking*) are used to speed up training [27, 28]. Abe et al. [29] extended SMO so that more than two variables are optimized at a time by Newton's method. Vogt [30] extended SMO when the bias term is not used, i.e., there is no equality constraint in the dual problem. Instead of two variables, one variable is modified at a time. Kecman et al. [31] showed that this algorithm is equivalent to the kernel-Adatron algorithm. Sentelle et al. [32] extended SMO to optimizing four data at a time and Hernandez et al. [33] multiple pairs at a time.

Cauwenberghs and Poggio's incremental and decremental training [34], which keeps track of the status change among unbounded support vectors, bounded support vectors, and non-support vectors, is extended to batch training [35–39].

Usually training methods using either fixed-size chunking or variable-size chunking correct variables under the constraints to guarantee monotonic convergence of the objective function. Chapelle [40] proposed solving primal problems allowing the solution to be infeasible during training. He showed that in some cases his method is faster than SMO. This method is extended to dual problems [41].

Comparing fixed-size chunking and variable-size chunking, the former can handle large-sized problems where the latter cannot store the support vectors in the memory but the accuracy of the latter is usually better because the entire solution is solved or updated at the same time.

Some training methods are based on the geometrical properties of support vectors [42–45]. Roobaert [42] proposed DirectSVM in which the optimal separating hyperplane is determined in a geometrical way. Initially, we find two nearest points of opposite classes and determine the hyperplane so that it contains the center of the two points and is orthogonal to the line connecting these points (see Fig. 5.5a). Then we find the point that violates separation the most with respect to this hyperplane and rotates the hyperplane so that it goes through the center of the point and the initial point with the opposite class (see Fig. 5.5b). According to the algorithm, without any proper explanation, the weight vector is updated so that the correction is orthogonal to the previous updates. The algorithm is extended to the feature space, again without an elaborate explanation.

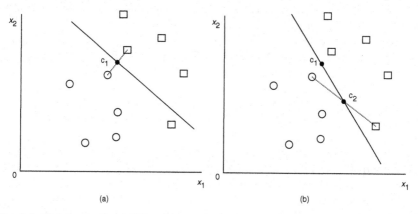

(a)                                          (b)

**Fig. 5.5** Training by DirectSVM: (a) Initialization of a hyperplane and (b) modification of the hyperplane by the point that violates separation the most

Assuming that the problem is linearly separable in the feature space, Keerthi et al. [43] showed the equivalence of training a support vector machine with finding the minimum distance between the two convex hulls for the two classes and derived an improved algorithm for finding the minimum distance. For the non-separable case, the separating hyperplane is proposed determining by reducing the convex hulls. [46–50].

Navia-Vázquez et al. [51] proposed an iterative least-squares training algorithm. They first transform the objective function for the input space given by (2.41) into

$$Q = \frac{1}{2} \|\mathbf{w}\|^2 - \sum_{i=1}^{M} a_i^2 \left( y_i - (\mathbf{w}^\top \mathbf{x} + b)^2 \right) + \sum_{i=1}^{M} \xi_i \left( C - \beta_i - \alpha_i \right), \quad (5.24)$$

where $a_i$ are weights given by

$$
\begin{aligned}
a_i &= \left( \frac{\alpha_i y_i}{y_i - (\mathbf{w}^\top \mathbf{x}_i + b)} \right)^{1/2} \\
&= \left( \frac{\alpha_i}{1 - y_i (\mathbf{w}^\top \mathbf{x}_i + b)} \right)^{1/2} \\
&= \left( \frac{\alpha_i}{\xi_i} \right)^{1/2}.
\end{aligned}
\quad (5.25)
$$

Assuming $\alpha_i$, $\beta_i$, and $\xi_i$ are constant, $Q$ given by (5.24) is a square function of $\mathbf{w}$ and $b$. Thus by the least-squares method we obtain $\mathbf{w}$ and $b$. Then using these $\mathbf{w}$ and $b$, we renew $\alpha_i$, $\beta_i$, and $\xi_i$ considering the inequality constraints. This procedure is iterated until the solution converges. Because this method is based on the primal problem, the extension to the feature space with an infinite dimension is not straightforward. Navia-Vázquez et al. proposed reducing the dimension by the kernel PCA and applying the preceding method.

To speeding up training support vector machines for huge-size problems parallel processing [52–54] is useful. For instance, Ferreira et al. [53] converted the optimization problem into a gradient system (a set of differential equations) and solve it by a second-order Runge–Kutta method using a shared memory multiprocessor system. The computer experiment showed the training speedup by parallel processing. Another approach for speedup is to develop new classifiers or modify classifiers suited for huge-size problems [55–58].

In the rest of the chapter, we discuss training methods based on primal–dual interior-point methods, steepest ascent methods and Newton's methods, which are extensions of SMO, batch training methods based on exact incremental training, active set training in primal and dual, and training methods for LP support vector machines.

## 5.5 Primal–Dual Interior-Point Methods

In the following, first we briefly summarize the primal–dual interior-point method for linear programming based on [59, 60] and then we summarize the primal–dual interior-point method for quadratic programming. Last, we evaluate characteristics of support vector machine training by the primal–dual interior-point method combined with the decomposition technique.

### 5.5.1 Primal–Dual Interior-Point Methods for Linear Programming

Consider the following linear programming problem:

$$\text{minimize} \quad \mathbf{c}^\top \mathbf{x} \tag{5.26}$$

$$\text{subject to} \quad A\mathbf{x} \geq \mathbf{b}, \quad \mathbf{x} \geq \mathbf{0}, \tag{5.27}$$

where $A$ is an $n \times m$ matrix, $\mathbf{x}$ and $\mathbf{c}$ are $m$-dimensional vectors, and $\mathbf{b}$ is an $n$-dimensional vector. Here, $\mathbf{x} \geq \mathbf{0}$ means that all the elements of $\mathbf{x}$ are nonnegative.

Consider that this problem is a primal problem. Then the dual problem of (5.26) and (5.27) is given as follows [61, 59]:

$$\text{maximize} \quad \mathbf{b}^\top \mathbf{y} \tag{5.28}$$

$$\text{subject to} \quad A^\top \mathbf{y} \leq \mathbf{c}, \quad \mathbf{y} \geq \mathbf{0}, \tag{5.29}$$

where $\mathbf{y}$ is an $n$-dimensional dual variable vector.

Now consider the following linear programming problem with equality constraints:

$$\text{minimize} \quad \mathbf{c}^\top \mathbf{x} \tag{5.30}$$

$$\text{subject to} \quad A\mathbf{x} = \mathbf{b}, \quad \mathbf{x} \geq \mathbf{0}, \tag{5.31}$$

where $A$ is an $n \times m$ matrix, $\mathbf{x}$ and $\mathbf{c}$ are $m$-dimensional vectors, and $\mathbf{b}$ is an $n$-dimensional vector.

Because the first equation in (5.31) is equivalent to

$$\begin{pmatrix} A \\ -A \end{pmatrix} \mathbf{x} \geq \begin{pmatrix} \mathbf{b} \\ -\mathbf{b} \end{pmatrix}, \tag{5.32}$$

the dual problem of (5.30) and (5.31) is given as follows:

$$\text{maximize} \quad \left( \mathbf{b}^\top - \mathbf{b}^\top \right) \mathbf{Y} \tag{5.33}$$

$$\text{subject to} \quad \left( A^\top - A^\top \right) \mathbf{Y} \leq \mathbf{c}, \quad \mathbf{Y} \geq \mathbf{0}, \tag{5.34}$$

where $\mathbf{Y}$ is the $(2\,n)$-dimensional dual variable vector. Now we define

$$\mathbf{Y} = \begin{pmatrix} \mathbf{y}^+ \\ \mathbf{y}^- \end{pmatrix}, \quad \mathbf{y}^+ \geq \mathbf{0}, \quad \mathbf{y}^- \geq \mathbf{0}, \quad \mathbf{y} = \mathbf{y}^+ - \mathbf{y}^-. \tag{5.35}$$

Then the dual problem of (5.30) and (5.31) is given as follows:

$$\text{maximize} \quad \mathbf{b}^\top \mathbf{y} \tag{5.36}$$
$$\text{subject to} \quad A^\top \mathbf{y} + \mathbf{z} = \mathbf{c}, \quad \mathbf{z} \geq \mathbf{0}, \tag{5.37}$$

where $\mathbf{z}$ is the $m$-dimensional slack variable vector. Here we must notice that the elements of $\mathbf{y}$ need not be nonnegative.

The solution that satisfies the constraints is called the *feasible solution*. It is known that for feasible solutions $\mathbf{x}$ for (5.30) and (5.31) and $(\mathbf{y}, \mathbf{z})$ for (5.36) and (5.37),

$$\mathbf{c}^\top \mathbf{x} \geq \mathbf{b}^\top \mathbf{y} \tag{5.38}$$

is satisfied. The difference of the objective functions $\mathbf{c}^\top \mathbf{x} - \mathbf{b}^\top \mathbf{y}$ is called the *duality gap*. It can be proved that if the primal solution has the optimal solution, the dual problem also has the optimal solution and the duality gap is zero, namely, the strict equality holds in (5.38).

Then if the primal problem has the optimal solution,

$$\begin{aligned} \mathbf{x}^\top \mathbf{z} &= \mathbf{x}^\top (\mathbf{c} - A^\top \mathbf{y}) \\ &= \mathbf{x}^\top \mathbf{c} - \mathbf{b}^\top \mathbf{y} \\ &= 0 \end{aligned} \tag{5.39}$$

is satisfied. Because $x_i \geq 0$ and $z_i \geq 0$, (5.39) is equivalent to

$$x_i z_i = 0 \quad \text{for } i = 1, \dots, m. \tag{5.40}$$

This is called the *complementarity condition*.

Then if $\mathbf{x}$ and $(\mathbf{y}, \mathbf{z})$ satisfy the equality constraints for the primal and dual problems, respectively, and (5.40) is satisfied, $\mathbf{x}$ and $(\mathbf{y}, \mathbf{z})$ are the optimal solutions, namely, solving the primal or dual problem is equivalent to solving

$$\begin{aligned} A\mathbf{x} &= \mathbf{b}, \\ A^\top \mathbf{y} + \mathbf{z} &= \mathbf{c}, \\ x_i z_i &= 0, \quad x_i \geq 0, z_i \geq 0 \quad \text{for } i = 1, \dots, m. \end{aligned} \tag{5.41}$$

This problem is called the *primal–dual problem*. In solving the primal–dual problem by the interior-point methods, the complementarity condition is changed to

$$x_i z_i = \mu \quad \text{for } i = 1, \dots, m, \tag{5.42}$$

where $\mu \geq 0$, and the value of $\mu$ is decreased to zero in solving the problem. The locus of the solutions when $\mu$ is decreased to zero is called the *central path*.

The primal–dual problem can be obtained by introducing the barrier (objective) function [59, pp. 277–285]. Consider the primal problem given by (5.30) and (5.31). We subtract the sum of $\log x_i$ from the objective function

$$\mathbf{c}^\top \mathbf{x} - \mu \sum_{i=1}^m \log x_i, \tag{5.43}$$

where $\mu \geq 0$. Because $-\log x_i$ is finite for positive $x_i$ and approaches positive infinity as $x_i$ approaches 0, we can eliminate the inequality constraints. We call this the *barrier function* because $\log x_i$ works as a barrier and the solution cannot go into the infeasible region: $x_i < 0$. We then introduce the Lagrange multipliers $\mathbf{y} = (y_1, \ldots, y_n)^\top$:

$$L(\mathbf{x}, \mathbf{y}) = \mathbf{c}^\top \mathbf{x} - \mu \sum_{i=1}^m \log x_i - \mathbf{y}^\top (A\mathbf{x} - \mathbf{b}). \tag{5.44}$$

The optimality conditions satisfy

$$\frac{\partial L(\mathbf{x}, \mathbf{y})}{\partial \mathbf{x}} = \mathbf{c} - \mu \begin{pmatrix} 1/x_1 & & 0 \\ & \ddots & \\ 0 & & 1/x_n \end{pmatrix} - A^\top \mathbf{y} = \mathbf{0}; \tag{5.45}$$

$$\frac{\partial L(\mathbf{x}, \mathbf{y})}{\partial \mathbf{y}} = A\mathbf{x} - \mathbf{b} = \mathbf{0}. \tag{5.46}$$

By defining $x_i z_i = \mu$ for $j = 1, \ldots, m$, (5.45) becomes

$$A^\top \mathbf{y} + \mathbf{z} = \mathbf{c}. \tag{5.47}$$

Thus, the primal–dual problem is obtained.

Primal–dual interior-point methods are classified according to whether a solution satisfies the constraints $x_i \geq 0, z_i \geq 0$ at every iteration step, into feasible and infeasible methods. The solution at each iteration step is generated by potential reduction methods, path-following methods, and predictor–corrector methods [60]. In the following we will discuss the path-following method [59].

We write the primal–dual interior-point method in a matrix form:

$$\begin{aligned} A\mathbf{x} &= \mathbf{b}, \\ A^\top \mathbf{y} + \mathbf{z} &= \mathbf{c}, \\ X Z \mathbf{e} &= \mu \mathbf{e}, \end{aligned} \tag{5.48}$$

where $\mathbf{x} > \mathbf{0}$, $\mathbf{z} > \mathbf{0}$, $X = \text{diag}(x_1, \ldots, x_m)$, $Z = \text{diag}(z_1, \ldots, z_m)$, $\mathbf{e}$ is an $m$-dimensional vector, and $\mathbf{e} = (1, \ldots, 1)^\top$. Here $\mathbf{x}$ and $\mathbf{z}$ are positive vectors because of positive $\mu$.

In the path-following method, starting from the feasible solution $(\mathbf{x}, \mathbf{y}, \mathbf{z})$ and positive $\mu$, we alternately calculate the corrections of $(\mathbf{x}, \mathbf{y}, \mathbf{z})$ and estimation of $\mu$, and we stop calculations when the solution reaches the optimal solution.

For the given $(\mathbf{x}, \mathbf{y}, \mathbf{z})$ and $\mu$, we calculate the corrections of $(\mathbf{x}, \mathbf{y}, \mathbf{z})$, $(\Delta\mathbf{x}, \Delta\mathbf{y}, \Delta\mathbf{z})$, by Newton's method:

$$
\begin{aligned}
A\,\Delta\mathbf{x} &= \mathbf{b} - A\,\mathbf{x}, \\
A^\top \Delta\mathbf{y} + \Delta\mathbf{z} &= \mathbf{c} - A^\top\mathbf{y} - \mathbf{z}, \\
Z\,\Delta\mathbf{x} + X\,\Delta\mathbf{z} &= \mu\mathbf{e} - X\,Z\,\mathbf{e}.
\end{aligned}
\tag{5.49}
$$

Assuming that $A$ is nonsingular, we can solve (5.49) for $(\Delta\mathbf{x}, \Delta\mathbf{y}, \Delta\mathbf{z})$. However, it is not guaranteed that the calculated corrections will satisfy the positive constraints on $\mathbf{x}$ and $\mathbf{z}$. Thus, introducing the positive parameter $\theta\,(\leq 1)$, we determine the maximum value of $\theta$, which satisfies

$$
\mathbf{x} + \theta\Delta\mathbf{x} \geq \mathbf{0}, \tag{5.50}
$$

$$
\mathbf{z} + \theta\Delta\mathbf{z} \geq \mathbf{0}. \tag{5.51}
$$

Then we set

$$
\begin{aligned}
\theta &\leftarrow \min(r\,\theta, 1), \tag{5.52} \\
\mathbf{x} &\leftarrow \mathbf{x} + \theta\,\Delta\mathbf{x}, \tag{5.53} \\
\mathbf{y} &\leftarrow \mathbf{y} + \theta\,\Delta\mathbf{y}, \tag{5.54} \\
\mathbf{z} &\leftarrow \mathbf{z} + \theta\,\Delta\mathbf{z}, \tag{5.55}
\end{aligned}
$$

where $r$ is smaller than but close to 1 to guarantee that $\mathbf{x}$ and $\mathbf{z}$ are positive vectors.

Because current $\mu$ is estimated by $\mu = x_i z_i$ for $i = 1, \ldots, m$, we update the value of $\mu$ by

$$
\mu = \frac{\mathbf{z}^\top\mathbf{x}}{m}\delta, \tag{5.56}
$$

where $1 > \delta > 0$ and $\mathbf{z}^\top\mathbf{x}/m$ is the average estimated value of $\mu$.

We stop calculation if elements of $\mathbf{x}$ or $\mathbf{y}$ increase indefinitely or if the complementarity conditions are satisfied, namely, $\mathbf{z}^\top\mathbf{x}$ is within a specified value.

## 5.5.2 Primal–Dual Interior-Point Methods for Quadratic Programming

Consider the following quadratic programming problem:

$$\text{minimize} \quad \mathbf{c}^{\top}\mathbf{x} + \frac{1}{2}\mathbf{x}^{\top}K\mathbf{x} \tag{5.57}$$

$$\text{subject to} \quad A\mathbf{x} = \mathbf{b}, \qquad x_i \geq 0 \quad \text{for } i = 1, \ldots, m, \tag{5.58}$$

where $K$ is an $m \times m$ symmetric, positive semidefinite matrix, $A$ is an $n \times m$ matrix with rank $n$, $\mathbf{x}$ and $\mathbf{c}$ are $m$-dimensional vectors, and $\mathbf{b}$ is an $n$-dimensional vector. Let this be a primal problem.

The dual problem of (5.57) and (5.58) is given as follows[3]:

$$\text{maximize} \quad \mathbf{b}^{\top}\mathbf{y} - \frac{1}{2}\mathbf{x}^{\top}K\mathbf{x} \tag{5.59}$$

$$\text{subject to} \quad A^{\top}\mathbf{y} - K\mathbf{x} + \mathbf{z} = \mathbf{c},$$

$$x_i \geq 0, \quad z_i \geq 0 \quad \text{for } i = 1, \ldots, m, \tag{5.60}$$

where $\mathbf{y}$ is the $n$-dimensional dual variable vector and $\mathbf{z}$ is the $n$-dimensional slack variable vector.

If $\mathbf{x}$ is the optimal solution for the primal problem and $(\mathbf{x}, \mathbf{y}, \mathbf{z})$ is the optimal solution for the dual problem, the following conditions are satisfied:

$$A\mathbf{x} = \mathbf{b},$$
$$A^{\top}\mathbf{y} - K\mathbf{x} + \mathbf{z} = \mathbf{c}, \tag{5.61}$$
$$x_i z_i = 0, \quad x_i \geq 0, \quad z_i \geq 0 \quad \text{for } i = 1, \ldots, m.$$

This is called the *primal–dual problem*. If $(\mathbf{x}, \mathbf{y}, \mathbf{z})$ is the solution, $\mathbf{x}$ is the optimal solution of the primal problem and $(\mathbf{y}, \mathbf{z})$ is the optimal solution of the dual problem.

Because a primal–dual problem of quadratic programming is similar to a primal–dual problem of linear programming, there is not much difference in solving linear programming and quadratic programming problems by primal–dual interior-point methods.

Now derive the optimality conditions of the L1 support vector machine. Training of the L1 support vector machine is given as follows:

$$\text{minimize} \quad Q(\boldsymbol{\alpha}) = -\sum_{i=1}^{M}\alpha_i + \frac{1}{2}\sum_{i,j=1}^{M}\alpha_i\,\alpha_j\,y_i\,y_j\,K(\mathbf{x}_i, \mathbf{x}_j) \tag{5.62}$$

$$\text{subject to} \quad \sum_{i=1}^{M}y_i\,\alpha_i = 0, \tag{5.63}$$

---

[3] If the inequality constraint $A\mathbf{x} \geq \mathbf{b}$ is used, $\mathbf{y}$ needs to be a nonnegative vector.

$$C \geq \alpha_i \geq 0 \quad \text{for } i = 1, \dots, M. \tag{5.64}$$

Here we multiplied (2.61) by the minus sign to make the problem a minimization problem. Introducing slack variables $\beta_i$ ($i = 1, \dots, M$), we convert the inequality constraints (5.64) into equality constraints:

$$\alpha_i + \beta_i = C, \quad \alpha_i \geq 0, \quad \beta_i \geq 0 \qquad \text{for } i = 1, \dots, M. \tag{5.65}$$

Then the dual problem of the problem given by (5.62), (5.63), and (5.65) is given as follows:

$$\text{maximize} \quad C \sum_{i=1}^{M} \delta_i - \frac{1}{2} \sum_{i,j=1}^{M} \alpha_i \alpha_j \, y_i \, y_j \, K(\mathbf{x}_i, \mathbf{x}_j) \tag{5.66}$$

$$\text{subject to} \quad \delta_i + y_i \, \delta_{M+1} - \sum_{j=1}^{M} y_i \, y_j \, \alpha_j K(\mathbf{x}_i, \mathbf{x}_j) + z_i = -1$$
$$\text{for } i = 1, \dots, M, \tag{5.67}$$

$$\delta_i + z_{M+i} = 0 \quad \text{for } i = 1, \dots, M, \tag{5.68}$$

where $\boldsymbol{\delta} = (\delta_1, \dots, \delta_{M+1})^{\mathsf{T}}$, $\boldsymbol{\delta}$ corresponds to $\mathbf{y}$ in (5.60), and $\mathbf{z} = (z_1, \dots, z_{2M})^{\mathsf{T}}$.

Substituting $\delta_i = -z_{M+i}$ obtained from (5.68) into (5.67), we obtain

$$y_i \, \delta_{M+1} - \sum_{j=1}^{M} y_i \, y_j \, \alpha_j K(\mathbf{x}_i, \mathbf{x}_j) + z_i - z_{M+i} = -1$$
$$\text{for } i = 1, \dots, M. \tag{5.69}$$

Then the optimality conditions for the L1 support vector machine are given as follows:

$$\sum_{i=1}^{M} y_i \, \alpha_i = 0, \tag{5.70}$$

$$y_i \, \delta_{M+1} - \sum_{j=1}^{M} y_i \, y_j \, \alpha_j K(\mathbf{x}_i, \mathbf{x}_j) + z_i - z_{M+i} = -1$$
$$\text{for } i = 1, \dots, M, \tag{5.71}$$

$$\alpha_i \, z_i = 0 \quad \text{for } i = 1, \dots, M, \tag{5.72}$$

$$(C - \alpha_i) \, z_{M+i} = 0 \quad \text{for } i = 1, \dots, M, \tag{5.73}$$

$$C \geq \alpha_i \geq 0, \quad z_i \geq 0, \quad z_{M+i} \geq 0 \quad \text{for } i = 1, \dots, M. \tag{5.74}$$

Similarly, we can obtain the optimality conditions for the L2 support vector machine as follows:

$$\sum_{i=1}^{M} y_i \, \alpha_i = 0, \tag{5.75}$$

$$y_i \, \delta - \sum_{j=1}^{M} y_i \, y_j \, \alpha_j \left( K(\mathbf{x}_i, \mathbf{x}_j) + \frac{\delta_{ij}}{C} \right) + z_i = -1 \quad \text{for } i = 1, \ldots, M, \tag{5.76}$$

$$\alpha_i \, z_i = 0 \quad \text{for } i = 1, \ldots, M, \tag{5.77}$$

$$\alpha_i \geq 0, \quad z_i \geq 0 \quad \text{for } i = 1, \ldots, M, \tag{5.78}$$

where $\delta$ is a scalar variable associated with the linear constraint (5.75).

### 5.5.3 Performance Evaluation

Using the primal–dual interior-point method [62] combined with the variable-size chunking technique, we compared L1 and L2 support vector machines from the standpoint of training time and the generalization ability [63] using the data sets listed in Table 1.3. We used one-against-all fuzzy SVMs to resolve unclassifiable regions. We used linear, polynomial, and RBF kernels and set some appropriate value to $C$. We ran the C program on an Athlon MP 2000+ personal computer.

We used the inexact and exact KKT conditions for working set selection. For the inexact KKT conditions, we randomly selected $F$ points that violated the KKT conditions. For the exact KKT conditions, we set $\tau = 0.01$. Initially, we randomly selected the working set variables. Then we sorted $\tilde{F}_i$ and $\bar{F}_i$ in descending order of KKT violations. Then we alternately selected $F$ points from the top of $\tilde{F}_i$ and $\bar{F}_i$. We set the initial working set size of 50 and added 50 variables at a time. Because the initial working set was randomly selected, for each training condition, we trained the SVM 100 times and calculated the average values of the recognition rates and training time.

Table 5.3 shows the results for L1 and L2 SVMs using the inexact KKT conditions. Numerals in parentheses show the recognition rates of the training data when they were not 100%. In each row of the results, the higher recognition rate and the smaller training time are shown in boldface. The recognition rates of the test data by the L2 SVM are higher than those by the L1 SVM for 16 cases out of 30. But those by the L1 SVM are higher for seven cases. Thus the L2 SVM performed better than the L1 SVM, but the difference in recognition rates is small.

For linear kernels, the maximum rank of the Hessian matrix for the L1 SVM is the number of input variables plus 1 (see Theorem 2.11). Thus, if the working set size exceeds this value, the Hessian matrix for the L1 SVM is positive semidefinite and the Hessian matrix for L2 SVMs is always positive definite. But this fact is not reflected in the training time. From the table, excluding the iris data with nonlinear kernels, training of the L1 SVM was faster than that of the L2 SVM.

**Table 5.3** Recognition rates and training time of L1 SVM and L2 SVM using inexact KKT conditions

| Data | Kernel | L1 SVM (%) | Time (s) | L2 SVM (%) | Time (s) |
|---|---|---|---|---|---|
| Iris | Linear | 96.00 (97.33) | **0.02** | **97.33 (98.67)** | 0.04 |
| ($C$ = 5,000) | $d2$ | 94.67 | 0.01 | 94.67 | 0.01 |
| | $d3$ | 94.67 | 0.01 | 94.67 | 0.01 |
| Numeral | Linear | 99.63 | **0.43** | 99.63 | 0.65 |
| ($C$ = 50) | $d2$ | 99.39 | **0.40** | **99.63** | 0.62 |
| | $d3$ | 99.51 | **0.40** | **99.63** | 0.65 |
| | $d4$ | 99.51 | **0.45** | **99.63** | 0.65 |
| Thyroid | Linear | **96.70 (97.56)** | 39 | 94.22 (94.67) | 4,008 |
| ($C$ = 10,000) | $d2$ | **97.14 (98.75)** | 60 | 96.47 (98.38) | 328 |
| | $d3$ | **97.49 (99.31)** | 27 | 97.26 (99.10) | 151 |
| | $d4$ | **97.43 (99.34)** | 19 | 97.35 (99.23) | 73 |
| | $\gamma 1$ | **96.79 (99.02)** | 83 | 96.50 (99.02) | 57 |
| | $\gamma 2$ | 96.53 (99.36) | 55 | 96.53 (99.34) | 212 |
| Blood cell | Linear | 87.23 (91.02) | 214 | **87.87 (90.64)** | 872 |
| ($C$ = 2,000) | $d2$ | 92.97 (96.67) | 27 | **93.52 (97.06)** | 44 |
| | $d3$ | 93.19 (98.22) | 24 | **93.71 (98.55)** | 32 |
| | $d4$ | 92.68 (98.93) | 24 | **93.42 (99.00)** | 30 |
| | $\gamma 1$ | 93.35 (97.74) | 27 | **93.74 (98.06)** | 42 |
| | $\gamma 2$ | 93.42 (98.84) | 26 | **93.58 (98.90)** | 38 |
| Hiragana-50 | Linear | 92.60 (98.07) | 129 | **93.28 (98.79)** | 253 |
| ($C$ = 5,000) | $d2$ | 99.24 | 109 | 99.24 | 137 |
| | $d3$ | **99.31** | 112 | 99.26 | 137 |
| | $d4$ | **99.33** | 112 | 99.28 | 147 |
| Hiragana-105 | Linear | 97.03 (97.50) | 806 | **97.45 (98.08)** | 1,652 |
| ($C$ = 2,000) | $d2$ | 100 | 430 | 100 | 531 |
| | $d3$ | 100 | 434 | 100 | 532 |
| Hiragana-13 | Linear | 96.37 (97.92) | 360 | **96.43 (97.89)** | 705 |
| ($C$ = 5,000) | $d2$ | 99.27 (99.59) | 295 | **99.35 (99.67)** | 333 |
| | $d3$ | 99.37 (99.64) | 289 | **99.39 (99.69)** | 342 |
| | $d4$ | 99.34 (99.62) | 290 | **99.37 (99.68)** | 368 |

Table 5.4 shows the results when the exact KKT conditions were used. In each row of the results, the higher recognition rate and the smaller training time are shown in boldface. Due to the difference in the convergence tests, in some cases, the recognition rates were slightly different from those in Table 5.3. Similar to the inexact KKT conditions, in most cases, training time of the L1 SVM was shorter than that of the L2 SVM.

Comparing Tables 5.3 and 5.4, training time of the L1 SVM by the inexact KKT conditions was shorter than by the exact KKT conditions except for the iris data. The tendency was the same for the L2 SVM.

To investigate why the exact KKT conditions were not always better than the inexact KKT conditions, we studied the two-class problem that separates

**Table 5.4** Recognition rates and training time of L1 SVM and L2 SVM using exact KKT conditions

| Data | Kernel | L1 SVM (%) | Time (s) | L2 SVM (%) | Time (s) |
|---|---|---|---|---|---|
| Iris | Linear | 96.00 (97.33) | 0.03 | **97.33** (98.67) | 0.03 |
| ($C = 5,000$) | $d2$ | 94.67 | 0.01 | 94.67 | 0.01 |
| | $d3$ | 94.67 | 0.01 | 94.67 | 0.01 |
| Numeral | Linear | 99.63 | **0.53** | 99.63 | 0.69 |
| ($C = 50$) | $d2$ | 99.39 | **0.52** | **99.63** | 0.69 |
| | $d3$ | 99.51 | **0.56** | **99.63** | 0.75 |
| | $d4$ | 99.51 | **0.59** | **99.63** | 0.79 |
| Thyroid | Linear | **96.32** (97.40) | **98** | 94.22 (94.70) | 69,840 |
| ($C = 10,000$) | $d2$ | **97.14** (98.81) | **162** | 96.44 (98.33) | 1867 |
| | $d3$ | **97.52** (99.26) | **69** | 97.08 (99.07) | 363 |
| | $d4$ | **97.46** (99.34) | **54** | 97.32 (99.23) | 150 |
| | $\gamma1$ | **96.82** (99.02) | **218** | 96.50 (99.02) | 1362 |
| | $\gamma2$ | 96.53 (99.36) | **170** | 96.53 (99.34) | 410 |
| Blood cell | Linear | 87.77 (91.41) | **267** | **88.45** (91.12) | 4,180 |
| ($C = 2,000$) | $d2$ | 93.00 (96.74) | 51 | **93.48** (97.06) | **49** |
| | $d3$ | 93.26 (98.22) | 39 | **93.71** (98.55) | **37** |
| | $d4$ | 92.65 (98.93) | 34 | **93.42** (99.00) | 34 |
| | $\gamma1$ | 93.35 (97.74) | **42** | **93.74** (98.06) | 52 |
| | $\gamma2$ | 93.42 (98.84) | **39** | **93.58** (98.90) | 41 |
| Hiragana-50 | Linear | 92.58 (98.09) | **187** | **93.34** (98.79) | 366 |
| ($C = 5,000$) | $d2$ | 99.24 | **124** | 99.24 | 135 |
| | $d3$ | **99.31** | 129 | 99.26 | 135 |
| | $d4$ | **99.33** | 129 | 99.28 | 146 |
| Hiragana-105 | Linear | 97.04 (97.55) | **1323** | **97.47** (98.10) | 2,063 |
| ($C = 2,000$) | $d2$ | 100 | **508** | 100 | 527 |
| | $d3$ | 100 | **524** | 100 | 530 |
| Hiragana-13 | Linear | 96.42 (97.93) | **563** | **96.47** (97.89) | 1,121 |
| ($C = 5,000$) | $d2$ | 99.26 (99.59) | **356** | **99.37** (99.67) | 364 |
| | $d3$ | 99.39 (99.64) | **318** | 99.39 (99.69) | 344 |
| | $d4$ | 99.34 (99.62) | **335** | **99.39** (99.69) | 367 |

Class 2 from the remaining classes for the blood cell data. We trained the L1 SVM with $d = 3$ and $C = 2,000$.

Figure 5.6 shows the working set sizes of the inexact and exact KKT conditions against the number of iterations. From the figure, the number of iterations for the exact KKT conditions is smaller than that of the inexact KKT conditions, but the working set size of the exact KKT conditions is larger after the second iteration. This means that the inexact KKT conditions estimate the violating variables conservatively. Thus, with the smaller working set size, training by the inexact KKT conditions was faster.

Figure 5.7 shows the training time of the inexact and exact KKT conditions for the change of $F$, namely the number of variables added to the working set $W$. For the inexact KKT conditions, training was fastest when 100 variables were added, but for the exact KKT conditions, training was fastest when 20 variables were added. Because for the exact KKT conditions the training time was not monotonic for the increase of the added variables, the exact KKT conditions are less suitable as a variable selection strategy.

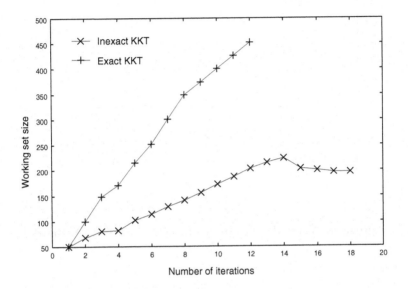

**Fig. 5.6** Relationship between the working set size and number of iterations ($d = 3$, $C = 2,000$)

According to the computer experiments, we have found that

1. training of L2 SVMs was not always faster than that of L1 SVMs,
2. the difference of the generalization abilities between the L1 and L2 SVMs was small, and
3. training by the exact KKT conditions was not always faster than by the inexact KKT conditions. This was due to the conservative estimation of violating variables by the inexact KKT conditions.

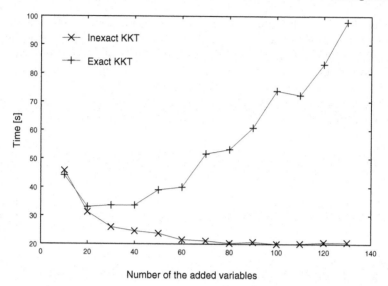

Number of the added variables

**Fig. 5.7** Relationship between the number of variables added and training time ($d = 3$, $C = 2,000$)

## 5.6 Steepest Ascent Methods and Newton's Methods

In this section, we discuss training methods based on the steepest ascent method and Newton's method [29]. The methods reduce to SMO when two data are optimized simultaneously.[4]

### 5.6.1 Solving Quadratic Programming Problems Without Constraints

First we consider solving the following general quadratic programming problem without constraints:

$$\text{maximize} \quad Q(\mathbf{x}) = \mathbf{c}^\top \mathbf{x} - \frac{1}{2}\mathbf{x}^\top H\,\mathbf{x}, \qquad (5.79)$$

where $\mathbf{x}$ is an $m$-dimensional variable vector, $\mathbf{c}$ is an $m$-dimensional constant vector, and $H$ is an $m \times m$ symmetric positive semidefinite matrix.

We consider solving the above problem by the steepest ascent method and Newton's method combined with the decomposition technique. We decom-

---

[4] We have changed steepest ascent methods used in the first edition to Newton's methods [64] to follow the common usage.

pose the set of variables $\{x_i \,|\, i, \ldots, m\}$ into a working set $\mathbf{x}_W = \{x_i \,|\, i \in W\}$ and a fixed set $\mathbf{x}_N = \{x_i \,|\, i \in N\}$, where $W$ and $N$ are the index sets, $W \cup N = \{1, \ldots, m\}$, and $W \cap N = \phi$. Then deleting constant terms associated with only $\mathbf{x}_N$, (5.79) becomes

$$\text{maximize} \quad Q(\mathbf{x}_W) = \mathbf{c}_W^\top \mathbf{x}_W - \frac{1}{2}\mathbf{x}_W^\top H_{WW}\,\mathbf{x}_W - \mathbf{x}_W^\top H_{WN}\,\mathbf{x}_N, \quad (5.80)$$

where $\mathbf{c}_W$ is the sub-vector of $\mathbf{c}$ associated with the elements of $\mathbf{x}_W$, and $H_{WW}$ and $H_{WN}$ are sub-matrices of $H$ associated with the elements of $\mathbf{x}_W$ and the elements of $\mathbf{x}_W$ and $\mathbf{x}_N$, respectively.

In the first approach we consider solving the problem by Newton's method. For the current $\mathbf{x}_W^{\text{old}}$ the new $\mathbf{x}_W^{\text{new}}$ is obtained by

$$\mathbf{x}_W^{\text{new}} = \mathbf{x}_W^{\text{old}} + \Delta\mathbf{x}_W^{\text{old}}, \quad (5.81)$$

where for $\mathbf{x}_W^{\text{new}}$ to be optimal the following relation must be satisfied:

$$\frac{\partial Q(\mathbf{x}_W^{\text{new}})}{\partial \Delta\mathbf{x}_W^{\text{old}}} = \mathbf{0}, \quad (5.82)$$

where $\mathbf{0}$ is the $m$-dimensional zero vector. Thus, from (5.80) and (5.82), $\Delta\mathbf{x}_W^{\text{old}}$ is obtained by

$$\Delta\mathbf{x}_W^{\text{old}} = H_{WW}^{-1}(\mathbf{c}_W - H_{WW}\,\mathbf{x}_W^{\text{old}} - H_{WN}\,\mathbf{x}_N). \quad (5.83)$$

Here we assume that $H_{WW}$ is regular. If it is singular we add a small value to the diagonal elements of $H_{WW}$. Actually, without constraints, we can obtain the solution of the subproblem without iterations:

$$\mathbf{x}_W^{\text{new}} = \mathbf{x}_W^{\text{old}} + \Delta\mathbf{x}_W^{\text{old}} = H_{WW}^{-1}(\mathbf{c}_W - H_{WN}\,\mathbf{x}_N). \quad (5.84)$$

The second method is to solve (5.80) by the steepest ascent method with the optimized step size:

$$\begin{aligned}
\mathbf{x}_W^{\text{new}} &= \mathbf{x}_W^{\text{old}} + r\,\frac{\partial Q(\mathbf{x}_W)}{\partial \mathbf{x}_W}\bigg|_{\mathbf{x}_W = \mathbf{x}_W^{\text{old}}} \\
&= \mathbf{x}_W^{\text{old}} + r(\mathbf{c}_W - H_{WW}\,\mathbf{x}_W^{\text{old}} - H_{WN}\,\mathbf{x}_N), \quad (5.85)
\end{aligned}$$

where $r$ is a small positive parameter.

Now the value of $r$ is determined by maximizing the objective function with respect to $r$ [65]:

$$\arg\max_r Q\left(\mathbf{x}_W + r\,\frac{\partial Q(\mathbf{x}_W)}{\partial \mathbf{x}_W}\right), \quad (5.86)$$

where

$$Q\left(\mathbf{x}_W + r\,\frac{\partial Q(\mathbf{x}_W)}{\partial \mathbf{x}_W}\right) = \left(\mathbf{x}_W + r\,\frac{\partial Q(\mathbf{x}_W)}{\partial \mathbf{x}_W}\right)^{\top}(\mathbf{c}_W - H_{WN}\,\mathbf{x}_N)$$
$$-\frac{1}{2}\left(\mathbf{x}_W + \frac{\partial Q(\mathbf{x}_W)}{\partial \mathbf{x}_W}\right)^{\top} H_{WW}\left(\mathbf{x}_W + r\,\frac{\partial Q(\mathbf{x}_W)}{\partial \mathbf{x}_W}\right).$$
$$(5.87)$$

Taking the derivative of (5.87) with respect to $r$ and equating it to zero, we obtain

$$r = \frac{\dfrac{\partial Q(\mathbf{x}_W)}{\partial \mathbf{x}_W}^{\top}(\mathbf{c}_W - H_{WN}\,\mathbf{x}_N - H_{WW}\,\mathbf{x}_W)}{\dfrac{\partial Q(\mathbf{x}_W)}{\partial \mathbf{x}_W}^{\top} H_{WW}\,\dfrac{\partial Q(\mathbf{x}_W)}{\partial \mathbf{x}_W}}$$

$$= \frac{(\mathbf{c}_W - H_{WN}\,\mathbf{x}_N - H_{WW}\,\mathbf{x}_W)^{\top}(\mathbf{c}_W - H_{WN}\,\mathbf{x}_N - H_{WW}\,\mathbf{x}_W)}{(\mathbf{c}_W - H_{WN}\,\mathbf{x}_N - H_{WW}\,\mathbf{x}_W)^{\top} H_{WW}\,(\mathbf{c}_W - H_{WN}\,\mathbf{x}_N - H_{WW}\,\mathbf{x}_W)}(5.88)$$

Because $H_{WW}$ is positive semidefinite, the denominator in the right-hand side of (5.88) is nonnegative and the numerator is also nonnegative. Thus if the denominator is not zero, $r$ is nonnegative. If the denominator is zero or $r$ is zero, we set a small value to $r$.

In an extreme case where only one variable is selected as a working set, (5.84) and (5.85) with $r$ given by (5.88) reduce to the same formula:

$$\mathbf{x}_W^{\text{new}} = \mathbf{x}_W^{\text{old}} + \frac{\mathbf{c}_W - H_{WN}\,\mathbf{x}_N - H_{WW}\,\mathbf{x}_W^{\text{old}}}{H_{WW}}, \qquad (5.89)$$

where $\mathbf{c}_W$ and $H_{WW}$ become scalars.

As discussed in the next section, this corresponds to sequential minimal optimization that optimizes two variables at a time in support vector machine training.

If $\mathbf{x}$ is upper or lower bounded, we start from a feasible solution and if the solution becomes infeasible by the correction given by (5.85), we reduce $r$ so that the solution remains feasible.

## 5.6.2 Training of L1 Soft-Margin Support Vector Machines

Based on Newton's method in the previous section, we discuss training an L1 soft-margin support vector machine. An extension to training by the steepest ascent method is straightforward.

We decompose the set of variables $\{\alpha_i \mid i \ldots, M\}$ into a working set $\boldsymbol{\alpha}_W = \{\alpha_i \mid i \in W\}$ and a fixed set $\boldsymbol{\alpha}_N = \{\alpha_i \mid i \in N\}$, where $W$ and $N$ are the index sets, $W \cup N = \{1, \ldots, M\}$, and $W \cap N = \phi$. Then fixing $\boldsymbol{\alpha}_N$, we consider the following optimization problem:

$$\text{maximize} \quad Q(\boldsymbol{\alpha}_W) = \sum_{i \in W} \alpha_i - \frac{1}{2} \sum_{i,j \in W} \alpha_i \, \alpha_j \, y_i \, y_j \, K(\mathbf{x}_i, \mathbf{x}_j)$$

$$- \sum_{\substack{i \in W, \\ j \in N}} \alpha_i \, \alpha_j \, y_i \, y_j \, K(\mathbf{x}_i, \mathbf{x}_j)$$

$$- \frac{1}{2} \sum_{i,j \in N} \alpha_i \, \alpha_j \, y_i \, y_j \, K(\mathbf{x}_i, \mathbf{x}_j) + \sum_{i \in N} \alpha_i \quad (5.90)$$

$$\text{subject to} \quad \sum_{i \in W} y_i \, \alpha_i = - \sum_{i \in N} y_i \, \alpha_i, \quad 0 \le \alpha_i \le C \quad \text{for } i \in W. (5.91)$$

Now consider solving the subprogram for $\boldsymbol{\alpha}_W$. Solving the equality in (5.91) for $\alpha_s \in \boldsymbol{\alpha}_W$, we obtain

$$\alpha_s = - \sum_{\substack{i \ne s, \\ i=1}}^{M} y_s \, y_i \, \alpha_i. \quad (5.92)$$

Substituting (5.92) into (5.90), we can eliminate the equality constraint. Let $\boldsymbol{\alpha}_{W'} = \{\alpha_i \,|\, i \ne s, i \in W\}$. Now because $Q(\boldsymbol{\alpha}_{W'})$ is quadratic, we can express the change of $Q(\boldsymbol{\alpha}_{W'})$, $\Delta Q(\boldsymbol{\alpha}_{W'})$, as a function of the change of $\boldsymbol{\alpha}_{W'}$, $\Delta \boldsymbol{\alpha}_{W'}$, by

$$\Delta Q(\boldsymbol{\alpha}_{W'}) = \frac{\partial Q(\boldsymbol{\alpha}_{W'})}{\partial \boldsymbol{\alpha}_{W'}} \Delta \boldsymbol{\alpha}_{W'} + \frac{1}{2} \Delta \boldsymbol{\alpha}_{W'}^\top \frac{\partial^2 Q(\boldsymbol{\alpha}_{W'})}{\partial \boldsymbol{\alpha}_{W'}^2} \Delta \boldsymbol{\alpha}_{W'}. \quad (5.93)$$

Considering that

$$\frac{\partial \alpha_s}{\partial \alpha_i} = -y_s \, y_i, \quad (5.94)$$

we derive the partial derivatives of $Q(\boldsymbol{\alpha}_{W'})$ with respect to $\alpha_i$ $(i \ne s, i \in W')$:

$$\frac{\partial Q(\boldsymbol{\alpha}_{W'})}{\partial \alpha_i} = 1 + \frac{\partial \alpha_s}{\partial \alpha_i} - \sum_{j=1}^{M} \alpha_j \, y_i \, y_j \, K(\mathbf{x}_i, \mathbf{x}_j)$$

$$- \sum_{j=1}^{M} \alpha_j \, y_s \, y_j \, \frac{\partial \alpha_s}{\partial \alpha_i} \, K(\mathbf{x}_s, \mathbf{x}_j)$$

$$= 1 - y_s \, y_i - \sum_{j=1}^{M} \alpha_j \, y_i \, y_j \, K(\mathbf{x}_i, \mathbf{x}_j)$$

$$+ \sum_{j=1}^{M} \alpha_j \, y_j \, y_i \, K(\mathbf{x}_s, \mathbf{x}_j). \quad (5.95)$$

Using (5.95), the second partial derivatives of $Q(\boldsymbol{\alpha}_{W'})$ with respect to $\alpha_i$ and $\alpha_j$ $(i, j \neq s, i, j \in W')$ are

$$\frac{\partial^2 Q(\boldsymbol{\alpha}_{W'})}{\partial \alpha_i^2} = -K(\mathbf{x}_i, \mathbf{x}_i) + 2\,K(\mathbf{x}_i, \mathbf{x}_s) - K(\mathbf{x}_s, \mathbf{x}_s), \qquad (5.96)$$

$$\frac{\partial^2 Q(\boldsymbol{\alpha}_{W'})}{\partial \alpha_i \partial \alpha_j} = y_i\, y_j\, \left(-K(\mathbf{x}_i, \mathbf{x}_j) + K(\mathbf{x}_i, \mathbf{x}_s) + K(\mathbf{x}_s, \mathbf{x}_j)\right.$$
$$\left. -K(\mathbf{x}_s, \mathbf{x}_s)\right). \qquad (5.97)$$

From (2.59),

$$\frac{\partial^2 Q(\boldsymbol{\alpha}_{W'})}{\partial \alpha_{W'}^2} = -\left(\cdots y_i(\boldsymbol{\phi}(\mathbf{x}_i) - \boldsymbol{\phi}(\mathbf{x}_s)) \cdots\right)^{\top}$$
$$\times \left(\cdots y_j(\boldsymbol{\phi}(\mathbf{x}_j) - \boldsymbol{\phi}(\mathbf{x}_s)) \cdots\right). \qquad (5.98)$$

Thus, if $\mathbf{x}_i = \mathbf{x}_s$ $(i, s \in W)$, $\partial^2 Q(\boldsymbol{\alpha})/\partial \alpha_{W'}^2$ is singular. Therefore, we need to avoid choosing $\mathbf{x}_i$ that is the same as $\mathbf{x}_s$, if included. Because $-\partial^2 Q(\boldsymbol{\alpha})/\partial \alpha_{W'}^2$ is expressed by the product of a transposed matrix and the matrix, it is positive semidefinite. Assuming that it is positive definite and neglecting the bounds, $\Delta Q(\boldsymbol{\alpha}_{W'})$ has the maximum at

$$\Delta \boldsymbol{\alpha}_{W'} = -\left(\frac{\partial^2 Q(\boldsymbol{\alpha}_{W'})}{\partial \alpha_{W'}^2}\right)^{-1} \frac{\partial Q(\boldsymbol{\alpha}_{W'})}{\partial \boldsymbol{\alpha}_{W'}}. \qquad (5.99)$$

Assume that $\alpha_i$ $(i = 1, \ldots, M)$ satisfy (5.91). Then from (5.91) and (5.99), we obtain the correction of $\alpha_s$:

$$\Delta \alpha_s = -\sum_{i \in W'} y_s\, y_i\, \Delta \alpha_i. \qquad (5.100)$$

Now the correction of $\boldsymbol{\alpha}_W$, $\Delta \boldsymbol{\alpha}_W$, is obtained by (5.99) and (5.100). For $\alpha_i$ $(i \in W)$, if

$$\alpha_i = 0, \quad \Delta \alpha_i < 0 \quad \text{or} \qquad (5.101)$$
$$\alpha_i = C, \quad \Delta \alpha_i > 0, \qquad (5.102)$$

we cannot modify $\boldsymbol{\alpha}_W$, because this will violate the lower or upper bound. If this happens, we delete these variables from the working set and repeat the procedure for the reduced working set.

If (5.101) or (5.102) does not hold for any variable in the working set, we can modify the variables. Let $\Delta \alpha_i'$ be the maximum or minimum correction of $\alpha_i$ that is within the bounds. Then if $\alpha_i + \Delta \alpha_i < 0$, $\Delta \alpha_i' = -\alpha_i$. And if $\alpha_i + \Delta \alpha_i > C$, $\Delta \alpha_i' = C - \alpha_i$. Otherwise $\Delta \alpha_i' = \Delta \alpha_i$. Now we calculate

$$r = \min_{i \in W} \frac{\Delta \alpha_i'}{\Delta \alpha_i}, \qquad (5.103)$$

where $0 < r \leq 1$.

We modify $\boldsymbol{\alpha}_W$ by

$$\boldsymbol{\alpha}_W^{\text{new}} = \boldsymbol{\alpha}_W^{\text{old}} + r\,\Delta\boldsymbol{\alpha}_W. \tag{5.104}$$

It is clear that the obtained $\boldsymbol{\alpha}_W^{\text{new}}$ satisfies the equality constraint. In addition, $Q(\boldsymbol{\alpha}_W^{\text{new}}) \geq Q(\boldsymbol{\alpha}_W^{\text{old}})$ is satisfied.

If $\partial^2 Q(\boldsymbol{\alpha}_{W'})/\partial\boldsymbol{\alpha}_{W'}^2$ is singular, we avoid singularity adding a small positive value to the diagonal elements. Or we delete the variables that cause singularity from the working set, and if the number of reduced variables is more than 1, do the procedure for the reduced working set.

Consider deleting the variables using the symmetric Cholesky factorization in calculating (5.99). In factorizing the $i$th column of $\partial^2 Q(\boldsymbol{\alpha}_{W'})/\partial\boldsymbol{\alpha}_{W'}^2$, if the value of the $i$th diagonal element of the lower triangular matrix is smaller than the prescribed small value, we discard the $i$th column and row of $\partial^2 Q(\boldsymbol{\alpha}_{W'})/\partial\boldsymbol{\alpha}_{W'}^2$ and proceed to the $(i+1)$th column.

In the following we consider fixed-size chunking, but an extension to variable-size chunking is straightforward. Let $V$ be the index set of support vector candidates. At the start of training, $V$ includes all the indices, i.e., $V = \{1, \ldots, M\}$. Assume that the maximum working set size $|W|_{\max}$ is given. We change the working set size according to the change of $|V|$ as follows:

$$|W| = \begin{cases} |W|_{\max} & \text{for} \quad |V| \geq |W|_{\max}, \\ \max\{2, |V|\} & \text{for} \quad |V| < |W|_{\max}. \end{cases} \tag{5.105}$$

We call the corrections of all the variables $\alpha_i$ $(i \in V)$ *one epoch of training*. At the beginning of a training epoch, we select $|W|$ variables $\alpha_i$ $(i \in V)$, where $|W|$ is given by (5.105), and delete and add the associated indices from $V$ to $W$, respectively. And we select one variable $\alpha_i$ $(i \in W)$ as $\alpha_s$. We calculate the variable vector $\boldsymbol{\alpha}_W^{\text{new}}$ and iterate the procedure until $V$ is empty.

For each calculation of $\boldsymbol{\alpha}_W^{\text{new}}$, we check if

$$|r\,\Delta\alpha_i| < \varepsilon_{\text{var}} \tag{5.106}$$

holds for all $i$ $(i \in W)$, where $\varepsilon_{\text{var}}$ is a tolerance of convergence for the variables. If we use inexact KKT conditions, at the end of a training epoch, we calculate the bias term $b_i$ for $\alpha_i$ $(i \in W)$ that is within the bounds, i.e., $0 < \alpha_i < C$ by

$$b_i = y_i - \sum_{j \in W} \alpha_j\, y_j\, K(\mathbf{x}_j, \mathbf{x}_i), \tag{5.107}$$

and their average $b_{\text{ave}}$ by

$$b_{\text{ave}} = \frac{1}{N_{\text{b}}} \sum_{\substack{i \in W, \\ 0 < \alpha_i < C}} b_i, \tag{5.108}$$

where $N_b$ is the number of $\alpha_i$ $(\in V)$ that satisfies $0 < \alpha_i < C$.

To accelerate training, for $N_{ite}$ consecutive iterations, where $N_{ite}$ is the positive integer, we confine calculations within the support vector candidates [26], namely, at the end of the training epoch we delete the index $i$ with $\alpha_i = 0$ from $V$ and proceed with training.

If this does not happen and (5.106) holds for all $\alpha_W$'s in the training epoch and if

$$\frac{|b_{ave} - b_i|}{|b_{ave}|} \leq \varepsilon_b \qquad (5.109)$$

holds for all $i$ $(i \in V, 0 < \alpha_i < C)$, we check whether there is a new candidate $\mathbf{x}_i$ $(i \notin V)$ of support vectors that satisfy

$$y_j \left( \sum_{i \in V} \alpha_i y_i K(\mathbf{x}_i, \mathbf{x}_j) + b_{ave} \right) < 1. \qquad (5.110)$$

If there is none, we stop training. Otherwise we add the indices to $V$ and proceed with training. At the end of every $(I_{ite} + 1)$th training epoch, we check (5.110) and add the support vector candidates if there are some.

To further speed training, we impose the following condition:

$$\frac{Q^{(n+1)}(\boldsymbol{\alpha}) - Q^{(n)}(\boldsymbol{\alpha})}{Q^{(n)}(\boldsymbol{\alpha})} \leq \varepsilon_q, \qquad (5.111)$$

where $\varepsilon_q$ is a small positive parameter. If this criterion is applied at the early stage of convergence, the calculation is terminated before the solution reaches the optimal solution. Therefore, we apply the criterion after $N_q$ $(> 1)$ epoch of training.

To reduce the number of calculations of $K(\mathbf{x}_i, \mathbf{x}_j)$, we prepare an $N_M \times N_M$ matrix $K$ for $K(\mathbf{x}_i, \mathbf{x}_j)$. When we calculate $K(\mathbf{x}_i, \mathbf{x}_j)$, where $\alpha_i$ and $\alpha_j$ are support vector candidates, i.e., $i, j \in V$, we store it into $K$ in a way similar to how the cache memory does, namely, if $i = j$, $K(\mathbf{x}_i, \mathbf{x}_i)$ is not stored in $K$, and the $k$th row and column of $K$ are empty, we assign $i$ to the $k$th column (row) and store $K(\mathbf{x}_i, \mathbf{x}_i)$ in $K_{kk}$. If $i \neq j$ and $i$ and $j$ are not assigned, we assign $i$ and $j$ to the empty columns (rows) $k_1$ and $k_2$, respectively, if they exist and store $K(\mathbf{x}_i, \mathbf{x}_j)$ to $K_{k_1 k_2}$. If either of $i$ and $j$ is assigned already, we assign the one that is not and store $K(\mathbf{x}_i, \mathbf{x}_j)$ in $K$. When $\alpha_i$ is deleted from $V$ and $i$ is assigned to a column (row) of $K$, we clear the column (row). We define the hit ratio of $K$ by

$$\frac{\text{Number of successful accesses}}{\text{Total number of accesses}} \times 100(\%), \qquad (5.112)$$

where a successful access means that $K(\mathbf{x}_i, \mathbf{x}_j)$ is stored in $K$ and its calculation is not necessary.

## 5.6.3 Sequential Minimal Optimization

In the following, we discuss correction of variables for SMO, which is Newton's method or the steepest ascent method with $|W| = 2$. We restate the optimization problem:

$$\text{maximize} \quad Q(\boldsymbol{\alpha}) = \sum_{i=1}^{M} \alpha_i - \frac{1}{2} \sum_{i,j=1}^{M} \alpha_i \, \alpha_j \, y_i \, y_j \, K(\mathbf{x}_i, \mathbf{x}_j) \quad (5.113)$$

$$\text{subject to} \quad \sum_{i=1}^{M} y_i \, \alpha_i = 0, \quad 0 \le \alpha_i \le C \quad \text{for} \quad i = 1, \dots, M. \quad (5.114)$$

Solving the equality in (5.114) for $\alpha_s$, we get

$$\alpha_s = - \sum_{\substack{i \ne s, \\ i = 1}}^{M} y_s \, y_i \, \alpha_i. \quad (5.115)$$

Substituting (5.115) into (5.114), we can eliminate the equality constraint. Now because $Q(\boldsymbol{\alpha})$ is quadratic, we can express the change of $Q(\boldsymbol{\alpha})$, $\Delta Q(\boldsymbol{\alpha})$, for the change of $\alpha_i$, $\Delta \alpha_i$, by

$$\Delta Q(\boldsymbol{\alpha}) = \frac{\partial Q(\boldsymbol{\alpha})}{\partial \alpha_i} \Delta \alpha_i + \frac{1}{2} \frac{\partial^2 Q(\boldsymbol{\alpha})}{\partial \alpha_i^2} (\Delta \alpha_i)^2. \quad (5.116)$$

Considering that

$$\frac{\partial \alpha_s}{\partial \alpha_i} = -y_s \, y_i, \quad (5.117)$$

we derive the partial derivatives of $Q(\boldsymbol{\alpha})$ with respect to $\alpha_i$ ($i \ne s, i = 1, \dots, M$),

$$\frac{\partial Q(\boldsymbol{\alpha})}{\partial \alpha_i} = 1 + \frac{\partial \alpha_s}{\partial \alpha_i} - \sum_{j=1}^{M} \alpha_j \, y_i \, y_j \, K(\mathbf{x}_i, \mathbf{x}_j)$$

$$- \sum_{j=1}^{M} \alpha_j \, y_s \, y_j \, \frac{\partial \alpha_s}{\partial \alpha_i} \, K(\mathbf{x}_i, \mathbf{x}_j)$$

$$= 1 - y_s \, y_i - \sum_{j=1}^{M} \alpha_j \, y_i \, y_j \, K(\mathbf{x}_i, \mathbf{x}_j)$$

$$+ \sum_{j=1}^{M} \alpha_j \, y_j \, y_i \, K(\mathbf{x}_s, \mathbf{x}_j). \quad (5.118)$$

Using (5.118), the second partial derivatives of $Q(\alpha)$ with respect to $\alpha_i$ ($i \neq s, i = 1, \ldots, M$) are

$$\frac{\partial^2 Q(\alpha)}{\partial \alpha_i^2} = -K(\mathbf{x}_i, \mathbf{x}_i) + 2\,K(\mathbf{x}_i, \mathbf{x}_s) - K(\mathbf{x}_s, \mathbf{x}_s). \qquad (5.119)$$

From (2.59),

$$\begin{aligned}
\frac{\partial^2 Q(\alpha)}{\partial \alpha_i^2} &= -K(\mathbf{x}_i, \mathbf{x}_i) + 2\,K(\mathbf{x}_i, \mathbf{x}_s) - K(\mathbf{x}_s, \mathbf{x}_s) \\
&= -\phi(\mathbf{x}_i)^\top \phi(\mathbf{x}_s) + 2\,\phi(\mathbf{x}_i)^\top \phi(\mathbf{x}_i) - \phi(\mathbf{x}_s)^\top \phi(\mathbf{x}_s) \\
&= -\left(\phi(\mathbf{x}_i) - \phi(\mathbf{x}_s)\right)^\top \left(\phi(\mathbf{x}_i) - \phi(\mathbf{x}_s)\right) \leq 0. \qquad (5.120)
\end{aligned}$$

Thus, if $\partial^2 Q(\alpha)/\partial \alpha_i^2 < 0$, $\Delta Q(\alpha)$ has the maximum at

$$\Delta \alpha_i = -\frac{\partial Q(\alpha)/\partial \alpha_i}{\partial^2 Q(\alpha)/\partial \alpha_i^2}. \qquad (5.121)$$

As initial values, we set $\alpha_i = 0$ so that they satisfy (5.114). For $i \neq s$, $i = 1, \ldots, M$, if $\partial^2 Q(\alpha)/\partial \alpha_i^2 > 0$, we modify $\alpha_i$ by

$$\alpha_i^{\text{new}} = \alpha_i^{\text{old}} + r\,\Delta \alpha_i, \qquad (5.122)$$

where $0 < r \leq 1$ and is determined so that $\alpha_i^{\text{new}}$ and $\alpha_s^{\text{new}}$ satisfy the upper and lower bounds (see Fig. 5.8).

Then $\alpha_s$ is given as follows:

1. if $y_i \neq y_s$,
$$\alpha_s^{\text{new}} = \alpha_s^{\text{old}} + r\,\Delta \alpha_i; \qquad (5.123)$$

2. if $y_i = y_s$,
$$\alpha_s^{\text{new}} = \alpha_s^{\text{old}} - r\,\Delta \alpha_i. \qquad (5.124)$$

Because $Q(\alpha)$ is monotonic for the change of $\alpha_i + r\,\Delta \alpha_i$ ($r \in (0, 1]$), by this modification the resultant $Q(\alpha)$ is nondecreasing. Thus the steepest ascent is guaranteed.

## 5.6.4 Training of L2 Soft-Margin Support Vector Machines

Newton's method obtained for the L1 soft-margin support vector machine can be extended to the L2 soft-margin support vector machine by replacing $K(\mathbf{x}_i, \mathbf{x}_j)$ with $K(\mathbf{x}_i, \mathbf{x}_j) + \delta_{ij}/C$ except for calculating the decision function

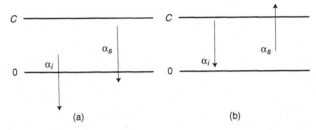

**Fig. 5.8** Satisfying the upper and lower bounds: **(a)** $y_i \neq y_s$. Because $\alpha_i$ violates the constraint more than $\alpha_s$, $r$ is determined so that $\alpha_i^{\text{new}} = 0$, i.e., $r \Delta\alpha_i = -\alpha_i^{\text{old}}$. **(b)** $y_i = y_s$. Because $\alpha_s$ violates the constraint, $r$ is adjusted so that $\alpha_s^{\text{new}} = C$, i.e., $r \Delta\alpha_i = \alpha_i^{\text{old}} - C$

(see Section 2.4) and by replacing the box constraints with positive constraints. In the following performance evaluation, we compare training time of L1 and L2 support vector machines.

## 5.6.5 Performance Evaluation

We used a Sun UltraSPARC-IIi (335 MHz) workstation to train fuzzy pairwise support vector machines for the thyroid, blood cell, and hiragana data sets listed in Table 1.3. The input ranges were scaled into $[-1, 1]$. We used polynomial kernels and RBF kernels. The parameters used for computer experiment for L1 and L2 SVMs are as follows: $C = 5,000$, $\varepsilon_{\text{b}} = 10^{-2}$, $\varepsilon_{\text{var}} = 10^{-6}$, $\varepsilon_{\text{q}} = 10^{-7}$, $N_{\text{ite}} = 10$, and $N_{\text{q}} = 50$. Because for the thyroid data and blood cell data, $N_{\text{q}} = 50$ was too small, we set as follows: for the thyroid data except for $d = 4$ and the blood cell data, we set $N_{\text{q}} = 100$; for the thyroid data with $d = 4$, we set $N_{\text{q}} = 600$.

To compare the training time with the quadratic programming technique, we used the software developed by London University[5] [9]. We used the primal–dual interior-point method (PDIP) as the optimization technique. The working set size was set to be 50.

Table 5.5 shows the results for the different training methods. In the table, columns PDIP, SMO, NM (Newton's method for L1 SVM), L2 SMO (for L2 SVM), and L2 NM (for L2 SVM) show the training time and in parentheses the speedup ratios to SMO. In each row of the results the shortest training time (largest speedup ratio) is shown in boldface. For the thyroid data and blood cell data with linear kernels, training by the L2 NM is much faster than that by the NM. But except for this, there is not much difference between NM and the L2 NM. Because the implementation of SMO is different from that in [26], only comparison of the training time with that of NM is meaningful.

---

[5] http://svm.cs.rhbnc.ac.uk/

In [26], SMO was shown to be faster than the projected conjugate gradient method with the working set size of 500. But, comparing PDIP and SMO, PDIP outperformed SMO for all the data sets evaluated. For PDIP also the working set size affected the training time considerably. For instance, without the decomposition technique, the training time by PDIP for blood cell data with $d = 4$ was 936 s. Thus the optimal selection of the working set size for the primal–dual interior-point method is important.

Comparing SMO and NM, there is not much difference for the hiragana data but for the thyroid and blood cell data training by NM is much faster for most cases. NM is comparable to PDIP for hiragana data, but for thyroid data and blood cell data with $d = 1$, PDIP is much faster. L2 NM is comparable to PDIP for the blood cell data, but for the thyroid data with $d = 1$, PDIP is faster.

**Table 5.5** Performance comparison of training methods

| Data | Parm | PDIP (s) | SMO (s) | NM (s) | L2 SMO (s) | L2 NM (s) |
|------|------|----------|---------|--------|-----------|----------|
| Thyroid | Linear | **38 (40)** | 1531 (1) | 931 (1.6) | 3650 (0.42) | 154 (9.9) |
| | $d4$ | **14 (145)** | 2032 (1) | 109 (19) | 2362 (0.86) | 134 (15) |
| | $\gamma 10$ | **108 (37)** | 4044 (1) | 424 (9.5) | 4309 (0.94) | 390 (10) |
| Blood cell | Linear | **20 (23)** | 463 (1) | 150 (3.1) | 932 (0.50) | 28 (17) |
| | $d4$ | **19 (18)** | 338 (1) | 31 (11) | 353 (0.96) | 31 (11) |
| | $\gamma 10$ | 185 (1.3) | 238 (1) | **165 (1.4)** | 235 (1.0) | **165 (1.4)** |
| Hiragana-50 | $d2$ | 143 (1.4) | 202 (1) | **117 (1.7)** | 215 (0.94) | 124 (1.6) |
| | $\gamma 0.1$ | 288 (1.1) | 321 (1) | **194 (1.6)** | 333 (0.96) | 199 (1.6) |
| Hiragana-105 | $d2$ | **318 (2.1)** | 671 (1) | 394 (1.7) | 698 (0.96) | 414 (1.6) |
| | $\gamma 0.1$ | 1889 (1.2) | 2262 (1) | **1479 (1.5)** | 2316 (0.98) | 1525 (1.5) |
| Hiragana-13 | $d2$ | 99 (1.4) | 136 (1) | **93 (1.5)** | 147 (0.93) | 104 (1.3) |
| | $\gamma 1$ | **181 (1.8)** | 325 (1) | 190 (1.7) | 333 (0.98) | 195 (1.7) |

According to the simulations, there is not much difference between L1 SVMs and L2 SVMs. The difference occurred for the thyroid data and blood cell data with $d = 1$. For larger $|W|$, training by L2 SVMs was faster. This is because the thyroid data and blood cell data were difficult to classify with $d = 1$ and L2 SVMs exploited the positive definiteness of the Hessian matrix.

## 5.7 Batch Training by Exact Incremental Training

Although the support vector machine is formulated as the quadratic programming problem, support vectors satisfy a set of linear equations derived from KKT conditions. Cauwenberghs and Poggio's idea in incremental training [34] is to incrementally update the set of support vectors solving the set of linear equation. This method is further analyzed in [66] and is extended to batch training [67, 37, 38]. Along this line Hastie et al. [68] proposed the entire path method to obtain solutions of different values of the margin parameter from the initial solution.

In this section, first we discuss exact incremental training of L1 and L2 support vector machines based on [34]. Then we extend incremental training to batch training [37]. Because the solution is updated using all the current support vectors, the training methods are one type of active set training.

### 5.7.1 KKT Conditions

In this section, we summarize KKT conditions. In training a support vector machine, we solve the following optimization problem:

$$\text{minimize} \quad Q(\mathbf{w}, \boldsymbol{\xi}) = \frac{1}{2}||\mathbf{w}||^2 + \frac{C}{p}\sum_{i=1}^{M}\xi_i^p \tag{5.125}$$

$$\text{subject to} \quad y_i\left(\mathbf{w}^\top\boldsymbol{\phi}(\mathbf{x}_i) + b\right) \geq 1 - \xi_i \quad \text{for } i = 1, ..., M, \tag{5.126}$$

where $\mathbf{w}$ is the weight vector, $\boldsymbol{\phi}(\mathbf{x})$ is the mapping function that maps $m$-dimensional input vector $\mathbf{x}$ into the feature space, $b$ is the bias term, $(\mathbf{x}_i, y_i)$ $(i = 1, ..., M)$ are $M$ training input–output pairs, with $y_i = 1$ if $\mathbf{x}_i$ belongs to Class 1, and $y_i = -1$ if Class 2, $C$ is the margin parameter that determines the trade-off between the maximization of the margin and minimization of the classification error, $\xi_i$ is the nonnegative slack variable for $\mathbf{x}_i$, and $p = 1$ for an L1 support vector machine and $p = 2$ for an L2 support vector machine.

Introducing the Lagrange multipliers $\alpha_i$, we obtain the following dual problem for the L1 support vector machine:

$$\text{maximize} \quad Q(\boldsymbol{\alpha}) = \sum_{i=1}^{M}\alpha_i - \frac{1}{2}\sum_{i,j=1}^{M}\alpha_i\alpha_j\, y_i\, y_j K(\mathbf{x}_i, \mathbf{x}_j) \tag{5.127}$$

$$\text{subject to} \quad \sum_{i=1}^{M} y_i\,\alpha_i = 0, \quad 0 \leq \alpha_i \leq C \quad \text{for } i = 1, ..., M, \tag{5.128}$$

and for the L2 support vector machine:

$$\text{maximize} \quad Q(\boldsymbol{\alpha}) = \sum_{i=1}^{M}\alpha_i - \frac{1}{2}\sum_{i,j=1}^{M}\alpha_i\alpha_j\, y_i\, y_j \left(K(\mathbf{x}_i, \mathbf{x}_j) + \frac{\delta_{ij}}{C}\right) \tag{5.129}$$

$$\text{subject to} \quad \sum_{i=1}^{M} y_i \, \alpha_i = 0, \quad \alpha_i \geq 0 \quad \text{for } i = 1, ..., M, \tag{5.130}$$

where $K(\mathbf{x}, \mathbf{x}')$ is a kernel function that is given by $K(\mathbf{x}, \mathbf{x}') = \boldsymbol{\phi}^\top(\mathbf{x}) \, \boldsymbol{\phi}(\mathbf{x}')$. Now we define

$$Q_{ij} = y_i \, y_j \, K(\mathbf{x}_i, \mathbf{x}_j) \tag{5.131}$$

for the L1 support vector machine and

$$Q_{ij} = y_i \, y_j \, K(\mathbf{x}_i, \mathbf{x}_j) + \delta_{ij}/C \tag{5.132}$$

for the L2 support vector machine. Then the KKT complementarity conditions for the L1 support vector machine are given by

$$\alpha_i \left( \sum_{j=1}^{M} \alpha_j \, Q_{ij} + y_i \, b - 1 + \xi_i \right) = 0 \quad \text{for } i = 1, \ldots, M, \tag{5.133}$$

$$(C - \alpha_i) \, \xi_i = 0 \quad \alpha_i \geq 0, \quad \xi_i \geq 0 \quad \text{for } i = 1, \ldots, M. \tag{5.134}$$

For the solution of (5.127) and (5.128), if $\alpha_i > 0$, $\mathbf{x}_i$ are called support vectors, especially if $\alpha_i = C$, bounded support vectors and if $0 < \alpha_i < C$, unbounded support vectors.

The KKT complementarity conditions for the L2 support vector machine are given by

$$\alpha_i \left( \sum_{j=1}^{M} \alpha_j \, Q_{ij} + y_i \, b - 1 + \xi_i \right) = 0, \quad \alpha_i \geq 0 \quad \text{for } i = 1, \ldots, M. \tag{5.135}$$

## 5.7.2 Training by Solving a Set of Linear Equations

### 5.7.2.1 Incremental Training

In the following we discuss incremental training for the L1 support vector machine. Because we use the notation of $Q_{ij}$, which is common to L1 and L2 support vector machines, the only difference of L1 and L2 support vector machines is that there is no upper bound in L2 support vector machines. Thus, in the following, if we exclude the discussions on the upper bound and consider that the bounded support vectors do not exist for L2 support vector machines, we obtain the training method for L2 support vector machines.

From the KKT conditions and the optimality of the bias term, the optimal solution of (5.127) and (5.128) must satisfy the following set of linear equations:

$$g_i = \sum_{j \in S} Q_{ij}\alpha_j + y_i b - 1 = 0 \quad \text{for } i \in S_\text{U}, \tag{5.136}$$

$$\sum_{i \in S} y_i\,\alpha_i = 0, \tag{5.137}$$

where $g_i$ is a margin for $\mathbf{x}_i$[6] and $S_\text{U}$ is the index set of unbounded support vectors.

In incremental training if an added data sample satisfies the KKT conditions, we do nothing. But if not, we modify the decision hyperplane under the constraints of (5.136) and (5.137). Let $\mathbf{x}_c$ be the added data sample. Then if the associated $\alpha_c$ is increased from 0, from (5.136) and (5.137), the following equations for the small increment of $\alpha_c$, $\Delta\alpha_c$, must be satisfied:

$$\sum_{j \in S_\text{U}} Q_{ij}\Delta\alpha_j + Q_{ic}\Delta\alpha_c + y_i\,\Delta b = 0 \quad \text{for } i \in S_\text{U}, \tag{5.138}$$

$$\sum_{i \in S_\text{U}} y_i\,\Delta\alpha_i + y_c\,\Delta\alpha_c = 0, \tag{5.139}$$

The above set of equations is valid so long as bounded support vectors remain bounded, unbounded support vectors remain unbounded, and non-support vectors remain non-support vectors. If any of the above conditions is violated, we call it the *status change*.

From (5.138) and (5.139),

$$\Delta b = \beta\Delta\alpha_c, \tag{5.140}$$

$$\Delta\alpha_j = \beta_j\Delta\alpha_c \quad \text{for } j \in S_\text{U}, \tag{5.141}$$

where

$$\begin{bmatrix} \beta \\ \beta_{s1} \\ \vdots \\ \beta_{sl} \end{bmatrix} = - \begin{bmatrix} 0 & y_{s1} & \cdots & y_{sl} \\ y_{s1} & Q_{s1\,s1} & \cdots & Q_{s1\,sl} \\ \vdots & \vdots & \ddots & \vdots \\ y_{sl} & Q_{sl\,s1} & \cdots & Q_{sl\,sl} \end{bmatrix}^{-1} \begin{bmatrix} y_c \\ Q_{s1\,c} \\ \vdots \\ Q_{sl\,c} \end{bmatrix} \Delta\alpha_c \tag{5.142}$$

and $\{s1, \ldots, sl\} = S_\text{U}$.

For a bounded support vector or a non-support vector, $\mathbf{x}_i$, the margin change due to $\Delta\alpha_c$, $\Delta g_i$, is calculated by

$$\Delta g_i = \gamma_i\Delta\alpha_c, \tag{5.143}$$

where

---

[6] Here, the margin is not measured from the separating hyperplane. Thus, to measure the margin from it we need to add 1.

$$\gamma_i = Q_{ic} + \sum_{j \in S_u} Q_{ij}\,\beta_j + y_i\,\beta. \tag{5.144}$$

The value of $\Delta\alpha_c$ is determined as the minimum value of $\Delta\alpha_c$ that causes status change.

### 5.7.2.2 Batch Training

Initial Solution

In batch training, we need to start from a solution that satisfy (5.136) and (5.137). The simplest one is the solution with one data sample for each class, namely, $\mathbf{x}_{s1}$ for Class 1 and $\mathbf{x}_{s2}$ for Class 2:

$$\alpha_{s1} = \alpha_{s2} = \frac{2}{Q_{s1\,s1} + 2Q_{s1\,s2} + Q_{s2\,s2}}, \tag{5.145}$$

$$b = -\frac{Q_{s1\,s1} - Q_{s2\,s2}}{Q_{s1\,s1} + 2Q_{s1\,s2} + Q_{s2\,s2}}. \tag{5.146}$$

We choose two data with the minimum distance. If $\alpha_{s1} = \alpha_{s2} \le C$, $\alpha_{s1} = \alpha_{s2}$ is the solution of the L1 SVM for the training data set consisting of $\mathbf{x}_{s1}$ and $\mathbf{x}_{s2}$. If the solution does not satisfy the upper bound, we select two data with the maximum distance. By this selection of data, however, we cannot obtain the solution for $C \le 1$ if we use RBF kernels. This is because

$$\alpha_{s1} = \alpha_{s2} = \frac{1}{1 - \exp(-\gamma \|\mathbf{x}_{s1} - \mathbf{x}_{s2}\|^2)} > 1. \tag{5.147}$$

To obtain a feasible solution, let

$$\alpha_{s1} = \alpha_{s2} = C, \tag{5.148}$$

and $\mathbf{x}_{s1}$ satisfy (5.136):

$$(Q_{s1s1} + Q_{s1s2})\,C + y_{s1}\,b - 1 = 0. \tag{5.149}$$

Using (5.149), $b$ is determined by

$$b = y_{s1}\,(1 - C\,(Q_{s1s1} + Q_{s1s2})). \tag{5.150}$$

Then,

$$\begin{aligned}
g_{s2} &= (Q_{s2s2} + Q_{s2s1})\,C + y_{s2}\,b - 1 \\
&= C\,(Q_{s1\,s1} + 2Q_{s1\,s2} + Q_{s2\,s2}) - 2 < 0,
\end{aligned} \tag{5.151}$$

namely, $\mathbf{x}_{s2}$ is a bounded support vector. Thus, if (5.147) is satisfied, we set $\alpha_{s1}$ and $\alpha_{s2}$ by (5.148), $b$ by (5.150). Then $\mathbf{x}_{s1}$ is a bounded support vector with $g_{s1} = 0$ and $\mathbf{x}_{s2}$ is a bounded support vector with $g_{s2} < 0$. Here, we need to set $\mathbf{x}_c$ with the same class as that of $\mathbf{x}_{s1}$ so that $\alpha_{s1}$ is decreased when $\alpha_c$ is increased.

For L2 support vector machines we set (5.145) and (5.146) as the initial solution.

## Training by the Decomposition Technique

If we apply incremental training to batch training, after one data sample is processed, we need to search a violating data sample in the remaining training data. But this is time-consuming. Thus to speed up training, we divide the training data into $o$ chunk data sets. Let the $T_i$ be the index set of the $i$th chunk data set and an active set be $A$, which includes current and previous indices of support vectors.

Initially $A = \{s1, s2\}$. One iteration consists of $o$ sub-iterations, in which at the $i$th sub-iteration the support vector machine is trained using the training data associated with the combined set of $T_i$ and $A$: $T_i \cup A$. At the beginning of a sub-iteration we check the KKT conditions using the hyperplane. We divide the combined set $T_i \cup A$ into $T_s$ and $T_u$, where training data associated with $T_s$ satisfy the KKT conditions and those with $T_u$ do not. We choose $\mathbf{x}_c$ ($c \in T_u$) that maximally violates KKT conditions.

We modify the optimal hyperplane so that $\mathbf{x}_i$ ($i \in T_s$) and $\mathbf{x}_c$ satisfy the KKT conditions. Then we move indices from $T_u$ to $T_s$ whose associated data satisfy the KKT conditions and add to $A$ the indices whose associated data become support vectors. If there is no modification of the hyperplane for the $j$th iteration, namely, if there is no modification of the hyperplane in any of sub-iteration of the $j$th iteration, we stop training.

Because previous support vectors may be near the hyperplane, they may resume being support vectors afterward. To prevent an additional iteration we include in the active set previous support vectors in addition to current support vectors.

### 5.7.2.3 Determining the Minimum Correction

In the following we consider the four cases where the status changes. For each case we calculate the correction of $\alpha_c$ to reach the status change [34]. For the L2 support vector machine, we need not consider (5.153) and (5.154), which will appear later.

1. Correction for $\mathbf{x}_c$.
   Because $\mathbf{x}_c$ violates the KKT conditions, $g_c < 0$. Then if $\gamma_c \geq \varepsilon$, where $\varepsilon$ takes a small positive value, and $g_c + \Delta g_c = 0$, $\mathbf{x}_c$ becomes an unbounded

support vector. From (5.143), correction $\Delta\alpha_c$ is calculated by

$$\Delta\alpha_c = -\frac{g_c}{\gamma_c}. \tag{5.152}$$

But because $\alpha_c$ is upper bounded by $C$, we need to check this bound; if $\alpha_c + \Delta\alpha_c < C$, $\mathbf{x}_c$ becomes an unbounded support vector. If $\alpha_c + \Delta\alpha_c \geq C$, we bound the correction by

$$\Delta\alpha_c = C - \alpha_c. \tag{5.153}$$

Then $\mathbf{x}_c$ becomes a bounded support vector.

2. Correction for $\mathbf{x}_{si}$.
   If $|\beta_{si}| \geq \varepsilon$, $\mathbf{x}_{si}$ becomes a non-support vector or a bounded support vector. From (5.141), if $\beta_{si} \geq \varepsilon$, $\mathbf{x}_{si}$ becomes a bounded support vectors for

$$\Delta\alpha_c = \frac{C - \alpha_{si}}{\beta_{si}}. \tag{5.154}$$

If $\beta_{si} \leq -\varepsilon$, $\mathbf{x}_{si}$ becomes a non-support vectors for

$$\Delta\alpha_c = -\frac{\alpha_{si}}{\beta_{si}}. \tag{5.155}$$

3. Correction for a bounded support vector $\mathbf{x}_i$ ($g_i < 0$).
   If $\gamma_i \geq \varepsilon$, for

$$\Delta\alpha_c = -\frac{g_i}{\gamma_i}, \tag{5.156}$$

the margin becomes zero and $\mathbf{x}_i$ becomes an unbounded support vector.

4. Correction for a non-support vector $\mathbf{x}_i$ ($g_i > 0$).
   The margin becomes zero when $\gamma_i \leq -\varepsilon$ and $g_i + \Delta g_i = 0$. Thus from (5.152),

$$\Delta\alpha_c = -\frac{g_i}{\gamma_i}. \tag{5.157}$$

We calculated the smallest collection among Cases 1–4. If the correction calculated by Case 1 is not the smallest, this means that the status changes before $\mathbf{x}_c$ becomes a support vector.

In any case, we set up a new set of support vectors and recalculate $\beta, \beta_j, g_i,$ and $\gamma_i$.

### 5.7.2.4 Inverse Matrix Update and Numerical Instability

If $\mathbf{x}_c$ is added as an unbounded support vector, the $(l+1) \times (l+1)$ matrix $Q^{-1}$ is updated to the $(l+2) \times (l+2)$ matrix $Q^{-1}$ as follows [34]:

$$Q^{-1} \leftarrow \begin{bmatrix} & & & 0 \\ & Q^{-1} & & \vdots \\ & & & 0 \\ 0 & \cdots & 0\,0 \end{bmatrix} + \frac{1}{\gamma_c} \begin{bmatrix} \beta \\ \beta_{s1} \\ \vdots \\ \beta_{sl} \\ 1 \end{bmatrix} \begin{bmatrix} \beta & \beta_{s1} & \cdots & \beta_{sl} & 1 \end{bmatrix}. \qquad (5.158)$$

If the $k$th support vector is deleted from the set of support vectors, $\{Q^{-1}\}_{ij}$ is updated as follows [34]:

$$\{Q^{-1}\}_{ij} \leftarrow \{Q^{-1}\}_{ij} - \frac{1}{\{Q^{-1}\}_{kk}} \{Q^{-1}\}_{ik} \{Q^{-1}\}_{kj}. \qquad (5.159)$$

Because $Q$ is positive semidefinite, numerical stability in calculating (5.158) and (5.159) is not guaranteed; namely, in some conditions, because of the round-off errors in calculating (5.158) and (5.159) the equalities in (5.138) and (5.139) are violated. The occurrence of round-off errors can be detected during or after finishing training by checking the margins of support vectors, the optimality of the bias term, and the duality gap.

### 5.7.2.5 Maintenance of Unbounded Support Vectors

If the initial solution is bounded, there is only one support vector index $s1$ in $S_U$ with $\alpha_{s1} = \alpha_{s2} = C$. In this case because the support vector $\mathbf{x}_{s1}$ is bounded, we need to select $\mathbf{x}_c$ with the same class as that of $\mathbf{x}_{s1}$.

During training also, there is a case where the number of elements of $S_U$ is 1. In such a case $\alpha_{s1}$ is either 0 or $C$, namely, non-support vector becomes a support vector or a bounded support vector becomes a non-bounded support vector for an L1 support vector machine. In such a case $\mathbf{x}_c$ and $\mathbf{x}_{s1}$ need to belong to different classes for $\alpha_{s1} = 0$ and the same class for $\alpha_{s1} = C$. If there is no such $\mathbf{x}_c$, we resolve the conflict adjusting the bias term. Figure 5.9 shows one such example. In the figure, $\alpha_{s1}$ associated with $\mathbf{x}_{s1}$ is zero and $\mathbf{x}_c$ and $\mathbf{x}_{s1}$ belong to the same class, namely, Class 1. Therefore, we cannot increase both $\alpha_{s1}$ and $\alpha_c$. Thus, in this condition, we cannot correct $\alpha_{s1}$ and $\alpha_c$. Now consider moving the separating hyperplane in parallel in the direction of Class 2. We assume that by this movement the margins $g_i$ for Class 1 increase and those for Class 2 decrease by the change of the bias term. As we move the hyperplane, $\mathbf{x}_{s1}$ becomes a non-support vector. If we move the hyperplane until $g_c = 0$ and there is no bounded support vector belonging to Class 1 whose margin is larger than $g_c$ before moving the hyperplane and there is no non-support vector belonging to Class 2 whose margin is smaller than $|g_c|$ before moving the hyperplane, we can safely move the hyperplane and make $\mathbf{x}_c$ be an unbounded support vector.

In the following we discuss the procedure more in detail:

1. For $\alpha_{s1} = 0$ with $\mathbf{x}_c$ and $\mathbf{x}_{s1}$ belonging to the same class

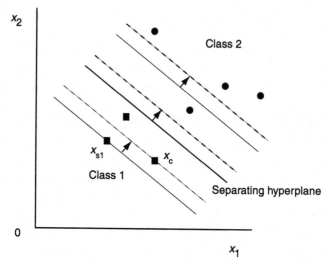

**Fig. 5.9** Adjusting the bias term to resolve conflict

We can move the separating hyperplane in parallel in the direction of the class associated with $-y_{s1}$, because the margin of $\mathbf{x}_{s1}$, $g_{s1}$, becomes positive and $\mathbf{x}_{s1}$ becomes a non-support vector. The nearest bounded support vector $\mathbf{x}_{i_{\text{bsv}}}$ belonging to the class associated with $y_{s1}$ is

$$i_{\text{bsv}} = \arg \max_{\substack{i \in S - S_{\text{U}}, \\ y_i = y_{s1}}} g_i. \tag{5.160}$$

And the non-support vector belonging to the class associated with $-y_{s1}$ with the smallest positive margin is

$$i_{\text{nsv}} = \arg \min_{\substack{i \notin S, \\ y_i = -y_{s1}}} g_i. \tag{5.161}$$

Now we move the separating hyperplane according to the following conditions:

a. $|g_{i_{\text{bsv}}}| \geq |g_c|$ and $g_{i_{\text{nsv}}} \geq |g_c|$

The separating hyperplane can be moved in parallel until $\mathbf{x}_c$ becomes a support vector. Thus we change the bias term as follows:

$$b \leftarrow b - y_c \, g_c. \tag{5.162}$$

Support vector $\mathbf{x}_{s1}$ becomes a non-support vector and the remaining data do not change status.

b. $|g_{i_{\text{bsv}}}| \geq g_{i_{\text{nsv}}}$ and $|g_c| > g_{i_{\text{nsv}}}$

When the separating hyperplane is moved in parallel in the direction of the class associated with $-y_{s1}$, the margin of $\mathbf{x}_{i_{\mathrm{nsv}}}$ becomes zero and it becomes a support vector. Thus, we modify the bias term as follows:

$$b \leftarrow b - y_{i_{\mathrm{nsv}}} \, g_{i_{\mathrm{nsv}}}. \tag{5.163}$$

c. $|g_c| > |g_{i_{\mathrm{bsv}}}|$ and $g_{i_{\mathrm{nsv}}} \geq |g_{i_{\mathrm{bsv}}}|$

When the separating hyperplane is moved in parallel in the direction of the class associated with $-y_{s1}$, the margin of $\mathbf{x}_{i_{\mathrm{bsv}}}$ becomes zero and it becomes an unbounded support vector with $\alpha_{i_{\mathrm{bsv}}} = C$. Thus, we modify the bias term as follows:

$$b \leftarrow b - y_{i_{\mathrm{bsv}}} \, g_{i_{\mathrm{bsv}}}. \tag{5.164}$$

2. For $\alpha_{s1} = C$ with $\mathbf{x}_c$ and $\mathbf{x}_{s1}$ belonging to the different classes

Similarly if we move the separating hyperplane in parallel, in the direction of the class associated with $y_{s1}$, $\mathbf{x}_{s1}$ becomes a bounded support vector. The nearest bounded support vector $\mathbf{x}_{i_{\mathrm{bsv}}}$ belonging to class $-y_{s1}$ is

$$i_{\mathrm{bsv}} = \arg \max_{\substack{i \in S - S_{\mathrm{U}}, \\ y_i = -y_{s1}}} g_i. \tag{5.165}$$

And the non-support vector belonging to the class associated with $y_{s1}$ with the smallest positive margin is

$$i_{\mathrm{nsv}} = \arg \min_{\substack{i \notin S, \\ y_i = y_{s1}}} g_i. \tag{5.166}$$

Now the hyperplane is moved in parallel as follows:

a. $|g_{i_{\mathrm{bsv}}}| \geq |g_c|$ and $g_{i_{\mathrm{nsv}}} \geq |g_c|$

The separating hyperplane can be moved in parallel until $\mathbf{x}_c$ becomes a support vector. Thus we change the bias term as follows:

$$b \leftarrow b - y_c \, g_c. \tag{5.167}$$

b. $|g_{i_{\mathrm{bsv}}}| \geq g_{i_{\mathrm{nsv}}}$ and $|g_c| > g_{i_{\mathrm{nsv}}}$

When the separating hyperplane is moved in parallel in the direction of the class associated with $y_{s1}$, $\mathbf{x}_{i_{\mathrm{nsv}}}$ becomes a support vector by

$$b \leftarrow b - y_{i_{\mathrm{nsv}}} \, g_{i_{\mathrm{nsv}}}. \tag{5.168}$$

c. $|g_c| > |g_{i_{\mathrm{bsv}}}|$ and $g_{i_{\mathrm{nsv}}} \geq |g_{i_{\mathrm{bsv}}}|$

The separating hyperplane, moving in parallel in the direction of the class associated with $-y_{s1}$, $\mathbf{x}_{i_{\mathrm{bsv}}}$ becomes a support vector by

$$b \leftarrow b - y_{i_{\mathrm{bsv}}} \, g_{i_{\mathrm{bsv}}}. \tag{5.169}$$

## 5.7.3 Performance Evaluation

We evaluated performance of the batch training method using the benchmark data sets in Table 1.3. Table 5.6 shows the parameter values for L1 and L2 SVMs determined by fivefold cross-validation.[7] For instance, $d4$ and $\gamma 10$ mean that the kernels are polynomial kernel with degree 4 and RBF kernels with $\gamma = 10$, respectively. We used fuzzy one-against-all SVMs in Section 3.1.2 and measured the training time using a workstation (3.6 GHz, 2 GB memory, Linux operating system). For the obtained solution, we checked the duality gap, the margins of support vectors, and optimality of the bias term to conform that the optimal solution was obtained.

**Table 5.6** Benchmark data sets and parameters for L1 and L2 support vector machines

| Data | L1 parameters | | L2 parameters | |
|------|--------|--------|--------|--------|
|      | Kernel | $C$ | Kernel | $C$ |
| Thyroid | $d4$ | $10^5$ | $d4$ | $10^5$ |
| Blood cell | $d2$ | 3,000 | $\gamma 10$ | 100 |
| Hiragana-50 | $\gamma 10$ | 5,000 | $\gamma 10$ | 1,000 |
| Hiragana-13 | $\gamma 10$ | 3,000 | $\gamma 10$ | 500 |
| Hiragana-105 | $\gamma 10$ | $10^4$ | $\gamma 10$ | $10^4$ |
| Satimage | $\gamma 200$ | 1 | $\gamma 200$ | 10 |
| USPS | $\gamma 10$ | 10 | $\gamma 10$ | 100 |

Table 5.7 lists the training results using the batch training method and the primal–dual interior-point method with the decomposition technique. The chunk size for the batch training method was 100 and that for the interior-point method was 50. "SVs," "Active Set," "Time," and "Rec." denote, respectively, the number of support vectors, the average number of elements in the active set, the time for training, and recognition rate for the test data set and that for the training data set in parentheses. The asterisk denotes that we used full matrix inversion instead of (5.158) and (5.159) because of numerical instability and the bold letters in the Time columns denote the shorter training time. As for the training time the interior-point method is faster for seven data sets and the batch training method for four data sets but except for the hiragana-105 data set both methods are comparable. The number of support vectors by the batch training method is usually smaller. This is because in the batch training method training proceeds by status changes of training data and if the status change occurs simultaneously, we only consider one data sample.

---

[7] The satimage and USPS data sets will be evaluated in the following section.

**Table 5.7** Comparison of the batch training method and the interior-point method with the decomposition technique using L1 and L2 support vector machines. Reprinted from [37, p. 303] with kind permission of Springer Science+Business Media

| SVM | Data | Batch training | | | | Interior-point method | | |
|-----|------|------|--------|------|------------|------|------|------------|
| | | SVs | A. Set | Time | Rec. | SVs | Time | Rec. |
| | Thyroid | 87 | 268 | 16 | 97.93 (99.97) | 141 | **8.5** | 97.32 (99.76) |
| | Blood cell | 78 | 197 | 35 | 93.13 (98.16) | 100 | **12** | 93.03 (96.77) |
| L1 | H-50 | 67* | 182 | 160 | 99.26 (100) | 70 | **71** | 99.26 (100) |
| | H-13 | 39 | 122 | **64** | 99.64 (100) | 40 | 134 | 99.64 (100) |
| | H-105 | 91 | 285 | 806 | 100 (100) | 91 | **256** | 100 (100) |
| | Thyroid | 99 | 279 | **17** | 97.81 (100) | 217 | 40 | 97.20 (99.79) |
| | Blood cell | 186 | 310 | 101 | 93.71 (97.13) | 136 | 51 | 93.26 (99.55) |
| L2 | H-50 | 78 | 201 | 184 | 99.28 (100) | 77 | **101** | 99.26 (100) |
| | H-13 | 71 | 157 | **118** | 99.70 (99.95) | 73 | 174 | 99.70 (99.96) |
| | H-105 | 91 | 285 | 811 | 100 (100) | 91 | **295** | 100 (100) |

# 5.8 Active Set Training in Primal and Dual

In batch training based on exact incremental training discussed in Section 5.7, a support vector machine is trained by repeatedly solving a set of equations based on the idea that the solution is piecewise linear for the change of a variable. The working set is updated by keeping track of the status change. In this section we discuss another approach of training L2 support vector machines by solving a set of primal or dual equations based on [40, 41].

## 5.8.1 Training Support Vector Machines in the Primal

In this section, we consider whether we can train L1 and L2 support vector machines in the primal.

In training a support vector machine, we solve the following optimization problem:

$$\text{minimize} \quad Q(\mathbf{w}, \boldsymbol{\xi}) = \frac{1}{2}||\mathbf{w}||^2 + \frac{C}{p} \sum_{i=1}^{M} \xi_i^p \quad (5.170)$$

$$\text{subject to} \quad y_i \left( \mathbf{w}^\top \boldsymbol{\phi}(\mathbf{x}_i) + b \right) \geq 1 - \xi_i \quad \text{for } i = 1, ..., M, \quad (5.171)$$

where $\mathbf{w}$ is the weight vector, $\boldsymbol{\phi}(\mathbf{x})$ is the mapping function that maps an $m$-dimensional input vector $\mathbf{x}$ into the feature space, $b$ is the bias term, $(\mathbf{x}_i, y_i)$ $(i = 1, ..., M)$ are $M$ training input–output pairs, with $y_i = 1$ if $\mathbf{x}_i$ belongs to

Class 1, and $y_i = -1$ if Class 2, $C$ is the margin parameter that determines the trade-off between the maximization of the margin and minimization of the classification error, $\xi_i$ are the nonnegative slack variables for $\mathbf{x}_i$, and $p = 1$ for an L1 SVM and $p = 2$ for an L2 SVM. We call the value of $y_i (\mathbf{w}^\top \phi(\mathbf{x}_i) + b)$ the margin for $\mathbf{x}_i$.

Introducing the Lagrange multipliers $\alpha_i$, we obtain the following dual problem for the L1 SVM:

$$\text{maximize} \quad Q(\boldsymbol{\alpha}) = \sum_{i=1}^{M} \alpha_i - \frac{1}{2} \sum_{i,j=1}^{M} \alpha_i \alpha_j \, y_i \, y_j K(\mathbf{x}_i, \mathbf{x}_j) \quad (5.172)$$

$$\text{subject to} \quad \sum_{i=1}^{M} y_i \, \alpha_i = 0, \quad 0 \le \alpha_i \le C \quad \text{for } i = 1, ..., M, \quad (5.173)$$

and for the L2 SVM:

$$\text{maximize} \quad Q(\boldsymbol{\alpha}) = \sum_{i=1}^{M} \alpha_i - \frac{1}{2} \sum_{i,j=1}^{M} \alpha_i \alpha_j \, y_i \, y_j \left( K(\mathbf{x}_i, \mathbf{x}_j) + \frac{\delta_{ij}}{C} \right) \quad (5.174)$$

$$\text{subject to} \quad \sum_{i=1}^{M} y_i \, \alpha_i = 0, \quad \alpha_i \ge 0 \quad \text{for } i = 1, ..., M, \quad (5.175)$$

where $K(\mathbf{x}, \mathbf{x}')$ is a kernel function that is given by $K(\mathbf{x}, \mathbf{x}') = \phi^\top(\mathbf{x}) \phi(\mathbf{x}')$ and $\delta_{ij} = 1$ for $i = j$ and 0 for $i \ne j$.

The KKT complementarity conditions for the L1 SVM are given by

$$\alpha_i \left( \sum_{j=1}^{M} y_i \, y_j \, \alpha_j \, K(\mathbf{x}_i, \mathbf{x}_j) + y_i \, b - 1 + \xi_i \right) = 0 \quad \text{for } i = 1, \ldots, M, \quad (5.176)$$

$$(C - \alpha_i) \, \xi_i = 0 \qquad \alpha_i \ge 0, \quad \xi_i \ge 0 \qquad \text{for } i = 1, \ldots, M. \quad (5.177)$$

For the solution of (5.172) and (5.173), if $\alpha_i > 0$, $\mathbf{x}_i$ are called support vectors, especially if $\alpha_i = C$, bounded support vectors and if $0 < \alpha_i < C$, unbounded support vectors.

The KKT complementarity conditions for the L2 SVM are given by

$$\alpha_i \left( \sum_{j=1}^{M} y_i \, y_j \, \alpha_j \, K(\mathbf{x}_i, \mathbf{x}_j) + y_i \, b - 1 + \frac{\alpha_i}{C} \right) = 0 \quad \text{for } i = 1, \ldots, M, \quad (5.178)$$

$$\alpha_i \ge 0 \qquad \text{for } i = 1, \ldots, M. \quad (5.179)$$

Here, $\alpha_i = C \, \xi_i$.

The optimization problem given by (5.170) and (5.171) is converted to the following optimization problem without constraints:

$$\text{minimize} \quad \frac{1}{2}\mathbf{w}^\top \mathbf{w} + \frac{C}{p}\sum_{i=1}^{M}\max(0, 1 - y_i\,(\mathbf{w}^\top \boldsymbol{\phi}(\mathbf{x}_i) + b))^p \quad (5.180)$$

Assume that $\mathbf{w}$ is expressed by

$$\mathbf{w} = \sum_{i=1}^{M}\beta_i\,\boldsymbol{\phi}(\mathbf{x}_i), \tag{5.181}$$

where $\beta_i$ $(i = 1,\dots,M)$ are constants. Substituting (5.181) into (5.180), we obtain

$$\text{minimize} \quad Q(\boldsymbol{\beta}, b) = \frac{1}{2}\sum_{i,j=1}^{M}K(\mathbf{x}_i, \mathbf{x}_j)\,\beta_i\,\beta_j$$

$$+\frac{C}{p}\sum_{i=1}^{M}\max\left(0, 1 - y_i\left(\sum_{j=1}^{M}\beta_j\,K(\mathbf{x}_j, \mathbf{x}_i) + b\right)\right)^p \tag{5.182}$$

Let define a set of indices associated with the data that give the optimal solution for the L1 SVM:

$$S = \{i \mid y_i\,D(\mathbf{x}_i) \le 1 \text{ for } i \in \{1,\dots,M\}\}, \tag{5.183}$$

where $D(\mathbf{x})$ is the decision function and is given by $D(\mathbf{x}) = \mathbf{w}^\top \boldsymbol{\phi}(\mathbf{x}) + b$ and for the L2 SVM

$$S = \{i \mid y_i\,D(\mathbf{x}_i) < 1 \text{ for } i \in \{1,\dots,M\}\}. \tag{5.184}$$

Here equality does not hold because of $\alpha_i = C\,\xi_i$.

We consider whether we can obtain the solution restricting the training data to the data associated with $S$. Then (5.182) reduces to

$$\text{minimize} \quad Q(\boldsymbol{\beta}, b) = \frac{1}{2}\sum_{i,j\in S}K(\mathbf{x}_i, \mathbf{x}_j)\,\beta_i\,\beta_j$$

$$+\frac{C}{p}\sum_{i\in S}\left(1 - y_i\left(\sum_{j\in S}\beta_j\,K(\mathbf{x}_j, \mathbf{x}_i) + b\right)\right)^p \tag{5.185}$$

For the L1 support vector machine ($p = 1$), from the KKT conditions (5.176) and (5.177), the slack variables $\xi_i$ associated with unbounded support vectors $\mathbf{x}_i$ $(i \in S)$ are zero. But in (5.185), the sum of the slack variables is minimized. Thus, each constraint is not necessarily enforced to zero. In addition, because there is no quadratic term for $b$, $b$ cannot be determined by this formulation. Therefore, we cannot obtain the solution by solving (5.185). To solve this problem, Chappell used a Huber loss function, in which a linear loss is combined with a quadratic loss. Because this method gives the approximate solution for the L1 support vector machine, we do not consider solving L1 support vector machines in the primal form.

For the L2 support vector machine, from (5.184) $\xi_i$ associated with support vectors $\mathbf{x}_i$ $(i \in S)$ are positive. Thus (5.182) with the restriction of $i \in S$ is equivalent to (5.185). Let $\boldsymbol{\beta}'_S = (\boldsymbol{\beta}_S^\top, b)^\top$, where $\boldsymbol{\beta}_S = \{\beta_i \mid i \in S\}$ and (5.185) be

$$\text{minimize} \quad Q(\boldsymbol{\beta}'_S) = \mathbf{c}_S^\top \boldsymbol{\beta}'_S + \frac{1}{2} {\boldsymbol{\beta}'_S}^\top K_S \boldsymbol{\beta}'_S, \tag{5.186}$$

where $K_S$ is the $(|S|+1) \times (|S|+1)$ matrix, $\mathbf{c}_S$ is the $(|S|+1)$-dimensional vector, and

$$\mathbf{c}_{S_i} = -C \sum_{k \in S} y_k K(\mathbf{x}_k, \mathbf{x}_i) \quad \text{for } i \in S, \tag{5.187}$$

$$\mathbf{c}_{S_b} = -C \sum_{i \in S} y_i, \tag{5.188}$$

$$K_{S_{ij}} = K(\mathbf{x}_i, \mathbf{x}_j) + C \sum_{k \in S} K(\mathbf{x}_i, \mathbf{x}_k) K(\mathbf{x}_k, \mathbf{x}_j) \quad \text{for } i, j \in S, \tag{5.189}$$

$$K_{S_{ib}} = C \sum_{j \in S} K(\mathbf{x}_i, \mathbf{x}_j) \quad \text{for } i \in S, \tag{5.190}$$

$$K_{S_{bb}} = C |S|. \tag{5.191}$$

Solving $\partial Q / \partial \boldsymbol{\beta}'_S = \mathbf{0}$, the optimal solution is given by

$$\boldsymbol{\beta}'_S = -K_S^{-1} \mathbf{c}_S. \tag{5.192}$$

Here, notice that $K_S$ is positive semidefinite. If $K_S$ is singular, usually a small value is added to the diagonal elements [40]. But because this will increase the number of support vectors, we delete data that cause singularity of the matrix from the working set.

## 5.8.2 Comparison of Training Support Vector Machines in the Primal and the Dual

### 5.8.2.1 Training in the Primal

The differences of the training method of the primal support vector machine and Chapelle's method are that we start training with a small number of chunking data and we avoid singularity of $K_S$ by deleting the associated row and column in decomposing $K_S$ by the Cholesky factorization, instead of adding a small value to the diagonal elements of $K_S$. We use the variable chunking algorithm, namely, we solve (5.192) for the initial working set, delete, from the working set, the data with zero slack variables (the associated margins larger than or equal to 1), add to the working set the data with the positive slack variables (the associated margins smaller than 1), and solve (5.192), and repeat the above procedure until the same working set is obtained. Let the chunk size be $h$, where $h$ is a positive integer. Then the procedure is as follows:

1. Set $h$ training data to the working set and go to Step 2.
2. Solve (5.192) for the working set by the Cholesky factorization. If the diagonal element is smaller than the prescribed value, delete the associated row and column, overwrite the column and row using the next data sample, and resume factorization and obtain $\beta'_S$.
3. Delete from the working set the data with zero slack variables, namely $\mathbf{x}_i$ that satisfy $y_i D(\mathbf{x}_i) \geq 1$. And add to the working set at most $h$ most-violating data, namely $\mathbf{x}_i$ that satisfy $y_i D(\mathbf{x}_i) < 1$ from the smallest $y_i D(\mathbf{x}_i)$ in order. If the obtained working set is the same with the previous iteration, stop training. Otherwise, go to Step 2.

### 5.8.2.2 Training in the Dual

Similar to the primal support vector machine, we train the dual support vector machine. The idea is to eliminate the equality constraint (5.175) by solving it for one variable and substitute it into (5.174). Then the problem is reduced to the maximization problem with the positive constraints. We solve the subproblem without considering the positive constraints and delete negative variables from the working set. Other procedure is the same with that of the primal support vector machine.

Consider solving (5.174) and (5.175) for the index set $S$. Solving the equality constraint in (5.175) for $\alpha_s$ $(s \in S)$, we obtain

$$\alpha_s = -\sum_{\substack{i \neq s, \\ i \in S}} y_s\, y_i\, \alpha_i. \tag{5.193}$$

Substituting (5.193) into (5.174), we obtain the following optimization problem:

$$\text{maximize} \quad Q(\boldsymbol{\alpha}_S) = \mathbf{c}_S^\top\, \boldsymbol{\alpha}'_S - \frac{1}{2}\boldsymbol{\alpha}'^{\top}_S\, K_S\, \boldsymbol{\alpha}'_S \tag{5.194}$$

$$\text{subject to} \quad \boldsymbol{\alpha}_S \geq \mathbf{0}, \tag{5.195}$$

where $\boldsymbol{\alpha}_S = \{\alpha_i | i \in S\}$, $\boldsymbol{\alpha}'_S = \{\alpha_i | i \neq s, i \in S\}$, $\mathbf{c}_S$ is the $(|S| - 1)$-dimensional vector, $K_S$ is the $(|S| - 1) \times (|S| - 1)$ positive definite matrix, and

$$\mathbf{c}_{S_i} = 1 - y_s\, y_i \qquad \text{for } i \neq s, \quad i \in S, \tag{5.196}$$

$$K_{S_{ij}} = y_i\, y_j\, \big(K(\mathbf{x}_i, \mathbf{x}_j) - K(\mathbf{x}_i, \mathbf{x}_s) - K(\mathbf{x}_s, \mathbf{x}_j)$$

$$+ K(\mathbf{x}_s, \mathbf{x}_s) + \frac{1 + \delta_{ij}}{C}\big) \qquad \text{for } i, j \neq s, \quad i, j \in S. \tag{5.197}$$

Now the procedure for training the dual support vector machine is as follows:

1. Set $h$ training data to the working set and go to Step 2.
2. Solve $K_S\, \boldsymbol{\alpha}'_S = \mathbf{c}_S$ for $\boldsymbol{\alpha}_S$ and using (5.193) obtain $\alpha_s$. Determine $b$ by

$$b = y_i - \sum_{j \in S} \alpha_j\, y_j \left( K(\mathbf{x}_i, \mathbf{x}_j) + \frac{\delta_{ij}}{C} \right) \qquad \text{for } i \in S. \tag{5.198}$$

3. Delete from the working set the data with negative variables, as well as the data $\mathbf{x}_i$ that satisfy $y_i\, D(\mathbf{x}_i) > 1$. And add to the working set at most $h$ most-violating data, namely $\mathbf{x}_i$ that satisfy $y_i\, D(\mathbf{x}_i) < 1$ from the smallest $y_i\, D(\mathbf{x}_i)$ in order. If the obtained working set is the same with the previous iteration, stop training. Otherwise, go to Step 2.

Although there may be negative $\alpha_i$ in Step 2, the first equation in (5.178) is satisfied because $\boldsymbol{\alpha}_S$ is obtained by solving the set of linear equations. Thus any $i\, (\in S)$ will give the same value. Because we ignore the positive constraints in solving $K_S\, \boldsymbol{\alpha}'_S = \mathbf{c}_S$ and delete negative variables afterward, the convergence of the above algorithm is not guaranteed but in almost all cases training is finished within 10–20 iterations.

### 5.8.2.3 Comparison

The differences of the dual support vector machine and the primal support vector machine are summarized as follows.

1. Matrix $K_S$ for the dual support vector machine is positive definite while that for the primal is positive semidefinite. Comparing (5.187), (5.188), (5.189), (5.190), and (5.191) with (5.196) and (5.197), $K_s$ and $\mathbf{c}_S$ for the dual support vector machine require less kernel evaluations than the primal. Thus, the dual support vector machine will give more stable solution with less computation time.
2. Mapped support vectors for the primal support vector machine are interpreted as the independent data that span the empirical feature space. Thus for the linear kernels the number of support vectors for the primal support vector machine is at most the number of the input variables. And any data can be support vectors so long as they span the empirical feature space.
3. Unlike the conventional training methods based on the decomposition technique such as SMO, training for primal and dual SVMs does not guarantee

monotonic convergence or may not converge. This is because there is no guarantee that the objective function is monotonic as the iteration proceeds. Thus, to avoid divergence, we may need to incorporate acceleration mechanism such as developed for linear programming support vector machines discussed in Section 5.9.

## 5.8.3 Performance Evaluation

We evaluated performance of the dual SVM with that of the primal SVM using the benchmark data sets shown in Table 5.6 on p. 272, which lists the parameter values determined by fivefold cross-validation. We used fuzzy one-against-all SVMs and measured the training time using a personal computer (3 GHz, 2 GB memory, Windows XP operating system). As in [40], we prepared a cache memory with the size equal to the kernel matrix.

Table 5.8 shows the effect of the chunk size on the performance of primal and dual SVMs for the USPS data set. In the table "Chunk," "SVs," "Iterations," "Kernels," "Rec.," and "Time" denote, respectively, the chunk size, the average number of support vectors per decision function, the average number of iterations, the total number of kernel accesses, the recognition rate of the test (training) data set, and training time. "Kernels" means that kernel values are provided through the cache memory if they are in it, and after evaluation if they are not. For "SVs," "Iterations," "Kernels," and "Time" columns, the better value is shown in boldface between dual and primal SVMs.

From the table, the numbers of support vectors for the dual SVM are the same for the five cases, but for the primal SVM, the number increases as the chunk size increases. Although the number of iterations for the primal SVM is smaller, the number of kernel accesses is smaller and training time is shorter for the dual SVM. This means that the computation burden per iteration for the primal SVM is larger as discussed previously.

We examined performance of the case where singularity of the matrix occurs. Table 5.9 shows the result for the blood cell data set for the linear kernel and $C = 100$. For the primal SVM we also include the results when the diagonal elements were added with 0.00001 denoted as primal (added). The numeral in parentheses in the SVs column shows the working set size after training. Thus, for example, for the chunk size of 50, among 497 data only 15 data are support vectors and the remaining data are deleted because of singularity of the matrix. For the same chunk sizes among the three methods, the smallest kernel accesses and shortest training time are shown in boldface.

From the table, training of the dual SVM was the fastest for all the chunk sizes. Comparing the results for the primal SVM with the primal SVM (added), training is faster for the primal SVM (added) but the solutions are different and the number of support vectors increased as the number of the chunk size was increased. The results clearly show that the addition of a small

**Table 5.8** Effect of the chunk size for primal and dual support vector machines for the USPS data set. Reprinted from [41, p. 861] with kind permission of Springer Science+Business Media

| Method | Chunk | SVs | Iterations | Kernels | Rec. | Time (s) |
|--------|-------|-----|------------|---------|------|----------|
| Dual | 10 | 597 | 68 | **1,889,358,476** | 95.47 (99.99) | **253** |
|  | 50 | **597** | 19 | **593,529,638** | 95.47 (100) | **114** |
|  | 100 | **597** | 15 | **585,495,205** | 95.47 (99.99) | **129** |
|  | 500 | **597** | 26 | **1,926,210,878** | 95.47 (99.99) | **650** |
|  | 1,000 | **597** | 26 | **2,627,126,816** | 95.47 (99.99) | **1,527** |
| Primal | 10 | 597 | **66** | 28,647,597,296 | 95.47 (99.99) | 1,437 |
|  | 50 | 604 | **16** | 8,116,273,966 | 95.47 (99.99) | 412 |
|  | 100 | 604 | **11** | 6,928,602,363 | 95.42 (99.99) | 385 |
|  | 500 | 724 | **7** | 33,628,904,338 | 95.47 (99.99) | 1,826 |
|  | 1,000 | 724 | **7** | 103,084,629,720 | 95.47 (99.99) | 5,074 |

positive value to the diagonal elements is not a good strategy for avoiding the singularity of the matrix.

**Table 5.9** Comparison of primal and dual support vector machines for the blood cell data set with $d = 1$ and $C = 100$. Reprinted from [41, p. 861] with kind permission of Springer Science+Business Media

| Method | Chunk | SVs | Iterations | Kernels | Rec. | Time (s) |
|--------|-------|-----|------------|---------|------|----------|
| Dual | 10 | 497 | 55 | **1,043,872,194** | 88.97 (91.99) | **212** |
|  | 50 | 497 | 17 | **326,186,052** | 88.97 (91.99) | **71** |
|  | 100 | 497 | 13 | **274,730,520** | 88.97 (91.99) | **63** |
|  | 500 | 497 | 13 | **561,872,500** | 88.97 (91.99) | **252** |
| Primal | 10 | **15** (497) | 78 | 110,167,255,737 | 88.97 (91.99) | 4,735 |
|  | 50 | **15** (497) | 20 | 27,464,672,230 | 88.97 (91.99) | 1,203 |
|  | 100 | **16** (497) | 14 | 19,333,796,617 | 88.97 (91.99) | 848 |
|  | 500 | **16** (497) | 9 | 36,858,682,541 | 88.97 (91.99) | 1753 |
| Primal (added) | 10 | 498 | 53 | 53,685,575,039 | 88.97 (91.99) | 2,387 |
|  | 50 | 517 | 14 | 14,051,320,440 | 88.97 (91.99) | 644 |
|  | 100 | 545 | 9 | 10,042,016,165 | 88.90 (91.73) | 475 |
|  | 500 | 1,073 | 5 | 27,336,218,781 | 87.45 (89.60) | 1,287 |

Table 5.10 lists the results for the primal and dual SVMs using the benchmark data sets. For each problem, the smaller number of support vectors, the smaller number of kernel accesses, and the shorter training time are shown in boldface. We set the chunk size of 50 for all the cases. For the thyroid data set, training of the dual SVM was very slow. And for the primal SVM the working set size fluctuated considerably and the training did not converge within 10,000 iterations. Except for hiragana-105 data set, the dual SVM was faster and except for the hiragana-50 data set the number of support vectors for the dual SVM was smaller.

From our experiments it is clear that the dual SVM is better than the primal SVM from the standpoints of stable convergence and fast training.

**Table 5.10** Comparison of primal and dual support vector machines for the benchmark data sets. Reprinted from [41, p. 862] with kind permission of Springer Science+Business Media

| SVM | Data | SVs | Iterations | Kernels | Rec. | Time (s) |
|---|---|---|---|---|---|---|
| Dual | Thyroid | 98 | 1,439 | 2,660,742,540 | 97.81 (100) | **220** |
| | Blood cell | **188** | 13 | **86,324,885** | 93.58 (97.19) | **10** |
| | H-50 | 77 | 20 | **386,391,968** | 99.28 (100) | **50** |
| | H-13 | **71** | 17 | **477,373,912** | 99.69 (99.96) | **46** |
| | H-105 | **91** | 22 | **812,259,183** | 100 (100) | 147 |
| | Satimage | **1,001** | 28 | **600,106,924** | 91.70 (99.71) | **157** |
| | USPS | **597** | 19 | **593,529,638** | 95.47 (100) | **114** |
| Primal | Thyroid | | | No convergence | | |
| | Blood cell | 203 | **10** | 445,319,582 | 93.61 (97.19) | 21 |
| | H-50 | **70** (78) | **15** | 605,637,116 | 99.28 (100) | 52 |
| | H-13 | 99 | **12** | 749,712,635 | 99.70 (99.96) | 93 |
| | H-105 | 111 | **13** | 907,360,383 | 100 (100) | **140** |
| | Satiamge | 1006 | **25** | 26,125,955,619 | 91.70 (99.71) | 1,258 |
| | USPS | 604 | **16** | 8,116,273,966 | 95.47 (99.99) | 412 |

We compared the training time for different training methods using the benchmark data sets. Table 5.11 lists the recognition rate of the training data and training time. The shortest training time among four methods is shown in boldface. For the steepest ascent method (SAM) and the Newton's method (NM) we set the working set size of 20 and for the batch training based on incremental training (Batch) and active set training of the dual SVM (Dual) 50. The recognition rates of SAM and NM are inferior to those of Batch and Dual for the thyroid and satimage data sets. For the thyroid data set, SAM did not converge within 5,000 iterations and the SAM is the slowest. Training time of dual is shortest except for the thyroid data set and batch is slower than NM for the blood cell, satimage, and USPS data sets. Therefore, if the convergence is stable, dual is the fastest among the four training methods.

# 5.9 Training of Linear Programming Support Vector Machines

Training an LP support vector machine becomes difficult as the problem size becomes large. To cope with this problem, in this section, based on [69] we discuss three decomposition techniques for linear programming problems.

**Table 5.11** Comparison of different training methods for L2 support vector machines

| Data | Recognition rate (%) | | | | Training time (s) | | | |
|------|------|------|-------|------|------|------|-------|------|
|      | SAM | NM | Batch | Dual | SAM | NM | Batch | Dual |
| Thyroid | 98.36 | 99.42 | 100 | 100 | 10,412 | 939 | **12** | 220 |
| Blood cell | 97.19 | 97.22 | 97.19 | 97.19 | 296 | 52 | 56 | **10** |
| H-50 | 100 | 100 | 100 | 100 | 1,881 | 492 | 79 | **50** |
| H-13 | 99.96 | 99.96 | 99.96 | 99.96 | 10,885 | 1,588 | 113 | **46** |
| H-105 | 100 | 100 | 100 | 100 | 3,862 | 1,565 | 259 | **147** |
| Satimage | 99.66 | 99.64 | 99.71 | 99.71 | 851 | 208 | 1,974 | **157** |
| USPS | 99.99 | 99.99 | 99.99 | 100 | 1,532 | 510 | 2,330 | **114** |

Then, we apply the methods to LP support vector machines and show the effectiveness of the methods by computer experiment.

## 5.9.1 Decomposition Techniques

In training LP support vector machines, we need to solve linear programming problems with the number of variables more than the number of training data. Thus training an LP support vector machine for a large-size problem, decomposition techniques are essential. Bradley and Mangasarian [70] proposed a decomposition technique, in which only a part of linear constraints are used for linear support vector machines. This method confirms monotonic convergence of the objective function and is useful for the problems with a large number of constraints but a small number of variables. Torii and Abe [71, 69] proposed three decomposition methods for LP programs: Method 1, in which variables are divided into working variables and fixed variables but constraints are all used; Method 2, in which constraints are divided into two but variables are all used; and Method 3, in which both variables and constraints are divided into two.

In the following, based on [69], we discuss three decomposition methods for LP problems and their application to LP support vector machines.

### 5.9.1.1 Formulation

We consider the following problem, which is a generalized version of an LP support vector machine:

$$\text{minimize} \quad \mathbf{c}^\top \mathbf{x} + \mathbf{d}^\top \boldsymbol{\xi} \tag{5.199}$$

$$\text{subject to} \quad A\mathbf{x} \geq \mathbf{b} - \boldsymbol{\xi}, \quad \mathbf{x} \geq \mathbf{0}, \quad \boldsymbol{\xi} \geq \mathbf{0}, \tag{5.200}$$

where $\mathbf{c}$ is an $m$-dimensional constant vector, $\mathbf{d}$ is an $M$-dimensional vector and $\mathbf{d} > 0$, which means that all the elements of $\mathbf{d}$ are positive, $A$ is an $M \times m$ constant matrix, $\mathbf{b}$ is an $M$-dimensional positive constant vector, and $\boldsymbol{\xi}$ is a slack variable vector to make $\mathbf{x} = \mathbf{0}$ and $\boldsymbol{\xi} = \mathbf{b}$ be a feasible solution. Therefore, the optimal solution always exists.

Introducing an $M$-dimensional slack variable vector $\mathbf{u}$, (5.200) becomes

$$\text{subject to} \quad A\mathbf{x} = \mathbf{b} + \mathbf{u} - \boldsymbol{\xi}, \quad \mathbf{x} \geq \mathbf{0}, \quad \mathbf{u} \geq \mathbf{0}, \quad \boldsymbol{\xi} \geq \mathbf{0}. \tag{5.201}$$

The dual problem of (5.199) and (5.201) is as follows:

$$\text{maximize} \quad \mathbf{b}^\top \mathbf{z} \tag{5.202}$$

$$\text{subject to} \quad A^\top \mathbf{z} + \mathbf{v} = \mathbf{c}, \quad \mathbf{z} + \mathbf{w} = \mathbf{d},$$

$$\mathbf{v} \geq \mathbf{0}, \quad \mathbf{z} \geq \mathbf{0}, \quad \mathbf{w} \geq \mathbf{0}, \tag{5.203}$$

where $\mathbf{z}$ is an $M$-dimensional vector, $\mathbf{v}$ is an $m$-dimensional slack variable vector, and $\mathbf{w}$ is an $M$-dimensional slack variable vector.

The optimal solution $(\mathbf{x}^*, \boldsymbol{\xi}^*, \mathbf{u}^*, \mathbf{z}^*, \mathbf{v}^*, \mathbf{w}^*)$ must satisfy the following complementarity conditions:

$$x_i^* v_i^* = 0 \quad \text{for } i = 1, \ldots, m, \tag{5.204}$$

$$\xi_i^* w_i^* = 0, \quad z_i^* u_i^* = 0 \quad \text{for } i = 1, \ldots, M. \tag{5.205}$$

Now solving the primal or dual problem is equivalent to solving

$$A\mathbf{x} = \mathbf{b} + \mathbf{u} - \boldsymbol{\xi}, \quad \mathbf{x} \geq \mathbf{0}, \quad \mathbf{u} \geq \mathbf{0}, \quad \boldsymbol{\xi} \geq \mathbf{0},$$

$$A^\top \mathbf{z} + \mathbf{v} = \mathbf{c}, \quad \mathbf{z} + \mathbf{w} = \mathbf{d},$$

$$\mathbf{z} \geq \mathbf{0}, \quad \mathbf{w} \geq \mathbf{0}, \quad \mathbf{v} \geq \mathbf{0},$$

$$x_i v_i = 0 \quad \text{for } i = 1, \ldots, m,$$

$$\xi_i w_i = 0, \quad z_i u_i = 0 \quad \text{for } i = 1, \ldots, M.$$

Here, we call $x_i$ active if $x_i > 0$ and inactive if $x_i = 0$. Likewise, the $i$th constraint is active if $u_i = 0$ and inactive if $u_i > 0$. Notice that even if we delete inactive variables and constraints, we can obtain the same solution as that of the original problem.

By the primal–dual interior-point method, the above set of equations is solved. By the simplex method, if we solve the primal or dual problem, the primal and dual solutions are obtained simultaneously [61]. Therefore, either by the primal–dual interior-point method or the simplex method, we obtain the primal and dual solutions.

**5.9.1.2 Three Decomposition Techniques**

Now we consider the following three decomposition methods to solve (5.199) and (5.201).

**Method 1**, in which a subset of the variables in $\mathbf{x}$ is optimized using all the constraints, while fixing the remaining variables. Let the set of indices of the subset be $W_v$ and the remaining subset be $F_v$, where $W_v \cap F_v = \emptyset$ and $W_v \cup F_v = \{1, \ldots, m\}$. Assuming $x_i = 0 \, (i \in F_v)$, the original problem given by (5.199) and (5.201) reduces as follows:

$$\text{minimize} \quad \sum_{i \in W_v} c_i x_i + \mathbf{d}^\top \boldsymbol{\xi} \tag{5.206}$$

$$\text{subject to} \quad \sum_{j \in W_v} A_{ij} x_j = b_i + u_i - \xi_i \quad \text{for } i = 1, \ldots, M,$$

$$x_i \geq 0 \quad \text{for } i \in W_v, \quad \mathbf{u} \geq \mathbf{0}, \quad \boldsymbol{\xi} \geq \mathbf{0}. \tag{5.207}$$

The dual problem of (5.206) and (5.207) is as follows:

$$\text{maximize} \quad \mathbf{b}^\top \mathbf{z} \tag{5.208}$$

$$\text{subject to} \quad \sum_{j=1}^{M} A_{ji} z_j + v_i = c_i, \quad v_i \geq 0 \quad \text{for } i \in W_v,$$

$$\mathbf{z} + \mathbf{w} = \mathbf{d}, \quad \mathbf{z} \geq \mathbf{0}, \quad \mathbf{w} \geq \mathbf{0}. \tag{5.209}$$

Therefore from (5.208) and (5.209), if we solve (5.206) and (5.207), in addition to the solution of the primal problem, we obtain the solution of the dual problem except for $v_i \, (i \in F_v)$, namely, except for $x_i v_i = 0 \, (i \in F_v)$, the complementarity conditions given by (5.204) and (5.205) are satisfied. Using the first equation in (5.203) for $i \in F_v$, we can calculate $v_i \, (i \in F_v)$. Because we assume that $x_i = 0 \, (i \in F_v)$, if $v_i \geq 0 \, (i \in F_v)$, $v_i$ satisfy the constraints and the obtained primal solution is optimal. But if some of $v_i$ are negative, the obtained solution is not optimal.

If the obtained solution is not optimal, we move the indices associated with inactive variables from $W_v$ to $F_v$, move, from $F_v$ to $W_v$, the indices associated with the violating variables, and iterate the previous procedure.

By this method, the optimal solution at each iteration step is obtained by restricting the original space

$$\{\mathbf{x} \mid A\mathbf{x} \geq \mathbf{b} - \boldsymbol{\xi}, \ \mathbf{x} \geq \mathbf{0}, \ \boldsymbol{\xi} \geq \mathbf{0}\} \tag{5.210}$$

to

$$\{\mathbf{x} \mid A\mathbf{x} \geq \mathbf{b} - \boldsymbol{\xi}, \ \boldsymbol{\xi} \geq \mathbf{0}, \ x_i \geq 0 \text{ for } i \in W_v, \quad x_i = 0 \text{ for } i \in F_v\}. \tag{5.211}$$

If the solution is not optimal, we repeat solving the subproblem with the non-zero $x_i$ ($i \in W_v$) and with the violating variables $x_i$ ($i \in F_v$). Because of the added violating variables, the objective function of the minimization problem for the newly obtained solution does not increase at least, namely, the values of the objective function are monotonically non-increasing during the iteration process. Thus, the following theorem holds.

**Theorem 5.1.** *For Method 1 the sequence of the objective function values is non-increasing and is bounded below by the global minimum of (5.206).*

**Method 2**, in which we optimize $\mathbf{x}$ using a subset of the constraints. Let the set of indices for the subset be $W_c$ and the set of the remaining indices be $F_c$. Then we consider the following optimization problem:

$$\text{minimize} \quad \mathbf{c}^\top \mathbf{x} + \sum_{i \in W_c} d_i \, \xi_i \qquad (5.212)$$

$$\text{subject to} \quad A_i \mathbf{x} = b_i + u_i - \xi_i, \quad u_i \geq 0, \quad \xi_i \geq 0$$
$$\text{for } i \in W_c, \quad \mathbf{x} \geq \mathbf{0}, \qquad (5.213)$$

where $A_i$ is the $i$th row vector of $A$. The dual problem is given as follows:

$$\text{maximize} \quad \sum_{i \in W_c} b_i \, z_i \qquad (5.214)$$

$$\text{subject to} \quad \sum_{j \in W_c} A_{ji} z_j + v_i = c_i \quad \text{for } i = 1, \ldots, M, \quad \mathbf{v} \geq \mathbf{0},$$
$$z_i + w_i = d_i, \quad z_i \geq 0, \quad w_i \geq 0, \quad \text{for } i \in W_c. \qquad (5.215)$$

Using the above primal and dual solutions, we can generate the solution of (5.199) and (5.201) as follows. From (5.213), for $i \in F_c$

1. if $A_i \mathbf{x} - b_i > 0$, $\xi_i = 0$ and $u_i = A_i \mathbf{x} - b_i$,
2. otherwise, $\xi_i = b_i - A_i \mathbf{x}$ and $u_i = 0$.

From (5.215), the first equation of (5.203) is satisfied if $z_i = 0$ for $i \in F_c$. Thus from the second equation of (5.203), $w_i = d_i$ ($i \in F_c$). Now the optimal solution $\mathbf{x}$ obtained from (5.212) and (5.213) is also the optimal solution of (5.199) and (5.201) if

$$\xi_i \, w_i = 0 \quad \text{for } i \in F_c. \qquad (5.216)$$

If (5.216) is not satisfied for some $i$, we move the indices for inactive constraints from $W_c$ to $F_c$, move some indices for violating constraints from $F_c$ to $W_c$, and iterate the preceding procedure.

The optimal solution at each iteration step is obtained by restricting the original space

$$\{ \mathbf{x} \mid A \mathbf{x} \geq \mathbf{b} - \boldsymbol{\xi}, \ \mathbf{x} \geq \mathbf{0}, \ \boldsymbol{\xi} \geq \mathbf{0} \} \qquad (5.217)$$

to

$$\{\mathbf{x} \,|\, A_i\,\mathbf{x} \geq b_i - \xi_i, \quad \xi_i \geq 0 \quad \text{for } i \in W_\mathrm{c}, \quad \mathbf{x} \geq \mathbf{0}\}. \tag{5.218}$$

For the non-optimal solution, we repeat solving the subproblem with the active constraints in the working set and with the violating constraints in the fixed set. Therefore, because new constraints are added, the value of the objective function for the newly obtained solution does not decrease at least [72], namely, the objective function is monotonically non-decreasing during the iteration process. Thus the following theorem holds.

**Theorem 5.2.** *For Method 2 the sequence of the objective function values is non-decreasing and is bounded above by the global minimum of (5.212).*

Unlike Theorem 3.2 in [70], we do not claim the finite convergence of Method 2, because according to the implementation of LP infinite loops may occur even if decomposition techniques are not used [61].

**Method 3**, in which we optimize a subset of variables using a subset of the constraints:

$$\text{minimize} \quad \sum_{i \in W_\mathrm{v}} c_i\,x_i + \sum_{i \in W_\mathrm{c}} d_i\,\xi_i \tag{5.219}$$

$$\text{subject to} \quad \sum_{j \in W_\mathrm{v}} A_{ij}\,x_j = b_i + u_i - \xi_i, \quad u_i \geq 0, \quad \xi_i \geq 0$$

$$\text{for } i \in W_\mathrm{c}, \quad x_j \geq 0 \quad \text{for } j \in W_\mathrm{v}. \tag{5.220}$$

The dual problem of (5.219) and (5.220) is as follows:

$$\text{maximize} \quad \sum_{i \in W_\mathrm{c}} b_i\,z_i \tag{5.221}$$

$$\text{subject to} \quad \sum_{j \in W_\mathrm{c}} A_{ji}\,z_j + v_i = c_i, \quad v_i \geq 0 \quad \text{for } i \in W_\mathrm{v},$$

$$z_j + w_j = d_j, \quad z_j \geq 0, \quad w_j \geq 0 \quad \text{for } j \in W_\mathrm{c}. \tag{5.222}$$

Now we construct, from the solution of (5.219) and (5.220), the solution of $\mathbf{x}$ in (5.199) and (5.201). Assuming $x_i = 0\,(i \in F_\mathrm{v})$, $\mathbf{x}$ satisfies

$$A_i\,\mathbf{x} = b_i + u_i - \xi_i \quad \text{for } i \in W_\mathrm{c}. \tag{5.223}$$

We generate $\xi_i$ and $u_i$ $(i \in F_\mathrm{c})$ as follows:

1. if $A_i\,\mathbf{x} - b_i > 0$, $\xi_i = 0$ and $u_i = A_i\,\mathbf{x} - b_i$,
2. otherwise, $\xi_i = b_i - A_i\,\mathbf{x}$ and $u_i = 0$.

Now assuming $z_i = 0$ for $i \in F_\mathrm{c}$,

$$(A^\top)_i\,\mathbf{z} + v_i = c_i \quad \text{for } i \in W_\mathrm{v}, \tag{5.224}$$

and $w_i = d_i$ for $i \in W_\mathrm{c}$. Further,

$$v_i = c_i - \sum_{j \in W_c} A_{ji} z_i \quad \text{for } i \in F_v. \tag{5.225}$$

Now, if $v_i \geq 0 \, (i \in F_v)$ and $\xi_i \, w_i = 0 \, (i \in F_c)$, the generated solution is optimal. If the solution is not optimal, we move the indices for inactive variables from $W_v$ to $F_v$, the indices for violating variables from $F_v$ to $W_v$, the indices for inactive constraints from $W_c$ to $F_c$, and some indices for violating constraints from $F_c$ to $W_c$, and iterate the preceding procedure.

Because Method 3 is a combination of Methods 1 and 2, whose objective functions are non-increasing and non-decreasing, respectively, monotonicity of the objective function of Method 3 is not guaranteed, namely, the following corollary holds:

**Corollary 5.1.** *For Method 3 the sequence of the objective function values is not guaranteed to be monotonic.*

The problem with a non-monotonic objective function is that the solution may not be obtained because of an infinite loop. Because the combinations of the working sets are finite, in an infinite loop, the same working set selection occurs infinitely. Let the working set sequence be

$$\ldots, \; W_k, \; W_{k+1}, \ldots, \; W_{k+t}, \; W_{k+t+1}, \; W_{k+t+2}, \ldots, W_{k+2t+1} \cdots,$$

where $W_k$ is the working set indices at the $k$th iteration and $W_k = W_{v,k} \cup W_{c,k}$. If

$$W_k = W_{k+t+1}, \quad W_{k+1} = W_{k+t+2}, \quad \cdots \quad, W_{k+t} = W_{k+2t+1} \tag{5.226}$$

are satisfied, the same sequence of working set selection occurs infinitely. Equation (5.226) means that all the variables and constraints that are moved out of the working set are moved back afterward. Thus the infinite loop can be avoided if we keep all the variables and constraints that are fed into the working set even after they become inactive, namely, for the initial $W_c$ and $W_v$, we solve (5.219) and (5.220) and delete the indices for the inactive variables and constraints from $W_v$ and $W_c$, respectively. Then we repeat solving (5.219) and (5.220) adding some violating indices to $W_v$ and $W_c$. But we do not delete the indices for inactive variables and constraints from $W_v$ and $W_c$, respectively. By this method, the working set size monotonically increases and the method terminates when there is no violating variables and constraints. Evidently the method terminates in finite steps, but the memory usage is inefficient.

To improve memory efficiency, we consider detecting and resolving infinite loops. If an infinite loop given by (5.226) is detected at the $(k + 2t + 1)$th step, we set

$$W_{k+2t+2} = W_k \cup W_{k+1} \cup \cdots \cup W_{k+t}. \tag{5.227}$$

We do not remove the indices included in $W_{k+2t+2}$ for the subsequent iterations. We call this procedure infinite loop resolution. This guarantees the convergence of the method in finite steps as the following theorem shows.

**Theorem 5.3.** *If infinite loop resolution is adopted, Method 3 terminates in finite steps.*

*Proof.* If an infinite loop is detected and infinite lop resolution is done, the same infinite loop does not occur in the subsequent iterations. Because the numbers of variables and constraints are finite, the number of infinite loops that will occur is also finite. Thus, the infinite loops are eventually resolved in finite steps. Thus Method 3 with infinite loop resolution terminates in finite steps.

### 5.9.1.3 Comparison of the Three Methods

Method 1 is useful for problems with a large number of variables but with a small number of constraints. For instance, microarray data sets have usually a large or sometimes huge number of variables but a small number of training data, namely constraints. In addition, they are usually linearly separable. Thus, we can use a linear LP support vector machine applying Method 1 to (5.229) and (5.230) discussed later.

Method 2 is suited for problems with a small number of variables but a large number of constraints. But similar to Method 1, Method 2 is only applicable to linear LP support vector machine expressed by (5.229) and (5.230). Method 3 is useful for problems with large numbers of variables and constraints.

### 5.9.1.4 Working Set Selection and Stopping Conditions

Bradley and Mangasarian [70] discussed a decomposition technique for linear LP support vector machines, which is similar to Method 2. They divide the constraints into several sets of constraints and solve the problem with the first set of constraints. Then they solve the problem with the active constraints in the first set and the constraints in the second set. In this way they solve the problem with the active constraints and the next set of constraints and terminate calculations if the solution does not change after several additions of the full set of constraints. (They state that four times of addition are enough.) In this method, we need not use complementarity conditions either for the addition of constraints or stopping calculations but with the expense of an additional computation.

We discuss selection of $q$ variables for Method 1 extending the above method. To simplify discussions, we do not discuss deletion of variables in the working set. Let $p$ be the pointer to the set of indices $\{1, \ldots, m\}$. Initially, we use the first $q$ variables as working variables. Thus, $W_v = \{1, \ldots, q\}$ and $p = q + 1$. After solving the subproblem, we check if $x_p$ satisfies the complementarity conditions. If not, we add index $p$ to $W_v$. And incrementing $p$ we iterate the above procedure until $q$ indices are added to $W_v$. If $p$ exceeds $m$ we set $p = 1$ and repeat the above procedure. Or if $p$ returns back to the point where the search started, we terminate working set selection. We can use similar methods for Method 2. But because by Method 3 the objective function values are not monotonic during iteration, it is difficult to apply the above method. In the computer experiments in Section 5.9.3, we randomly selected $q$ indices. Taking the similar selection strategies as in support vector machines [73–76], we may be able to improve convergence of the decomposition technique further.

We can stop training using the decomposition techniques when the complementarity conditions are satisfied. But in some cases the conditions are too strict and it may increase iterations. One way to alleviate the conditions is to slacken the conditions by introducing a threshold and assume that the conditions are satisfied if the conditions are within the threshold. Or we can stop training if the change of the objective function values is within a threshold.

## 5.9.2 Decomposition Techniques for Linear Programming Support Vector Machines

In this section, first we define and then discuss LP support vector machines with Methods 1 and 3.

### 5.9.2.1 Formulation of Linear Programming Support Vector Machines

Let $M$ $m$-dimensional input vector $\mathbf{x}_i$ $(i = 1, \ldots, M)$ belong to Class 1 or 2, and the class label be $y_i = 1$ for Class 1 and $y_i = -1$ for Class 2. We map the input space into the high-dimensional feature space by the mapping function $\phi(\mathbf{x}) = (\phi_1(\mathbf{x}), \ldots, \phi_l(\mathbf{x}))^\top$, where $l$ is the dimension of the feature space, and determine the following decision function:

$$D(\mathbf{x}) = \mathbf{w}^\top \phi(\mathbf{x}) + b \tag{5.228}$$

so that the margin is maximized, where $\mathbf{w}$ is an $l$-dimensional vector and $b$ is a bias term.

The LP support vector machine [77–79] is given by

$$\text{minimize} \quad Q(\mathbf{w}, b, \boldsymbol{\xi}) = \sum_{i=1}^{l} |w_i| + C \sum_{i=1}^{M} \xi_i \tag{5.229}$$

subject to    $y_i(\mathbf{w}^\top \phi(\mathbf{x}_i) + b) \geq 1 - \xi_i, \quad \xi_i \geq 0 \quad \text{for } i = 1, \ldots, M, (5.230)$

where $\xi_i$ are slack variables and $C$ is a margin parameter to control the trade-off between the classification error of the training data and the generalization ability.

For linear kernels, where $\phi(\mathbf{x}) = \mathbf{x}$, we can solve (5.229) and (5.230) by linear programming, but for nonlinear kernels we need to treat feature space variables explicitly. To avoid this, we redefine the decision function by [80]

$$D(\mathbf{x}) = \sum_{i=1}^{M} \alpha_i K(\mathbf{x}, \mathbf{x}_i) + b, \tag{5.231}$$

where $\alpha_i$ and $b$ take real values and $K(\mathbf{x}, \mathbf{x}')$ is a kernel function:

$$K(\mathbf{x}, \mathbf{x}') = \phi^\top(\mathbf{x})\, \phi(\mathbf{x}'). \tag{5.232}$$

We define the LP support vector machine by

minimize    $$Q(\boldsymbol{\alpha}, b, \boldsymbol{\xi}) = \sum_{i=1}^{M}(|\alpha_i| + C\xi_i) \tag{5.233}$$

subject to    $$y_j \left( \sum_{i=1}^{M} \alpha_i K(\mathbf{x}_j, \mathbf{x}_i) + b \right) \geq 1 - \xi_j, \quad \xi_j \geq 0$$
$$\text{for} \quad j = 1, \ldots, M. \tag{5.234}$$

To solve the problem by the simplex method or the primal–dual interior-point method, we need to change variables into nonnegative variables. Then using $\alpha_i^+ \geq 0$, $\alpha_i^- \geq 0$, $b^+ \geq 0$, $b^- \geq 0$, we define $\alpha_i = \alpha_i^+ - \alpha_i^-$, $b = b^+ - b^-$ and convert (5.233) and (5.234) into the following linear programming problem:

minimize    $$Q(\boldsymbol{\alpha}^+, \boldsymbol{\alpha}^-, b^+, b^-, \boldsymbol{\xi}) = \sum_{i=1}^{M}(\alpha_i^+ + \alpha_i^- + C\xi_i) \tag{5.235}$$

subject to    $$y_j \left( \sum_{i=1}^{M}(\alpha_i^+ - \alpha_i^-)K(\mathbf{x}_j, \mathbf{x}_i) + b^+ - b^- \right) + \xi_j \geq 1$$
$$\text{for} \quad j = 1, \ldots, M \tag{5.236}$$

which has $(3M + 2)$ variables and $M$ constraints.

By introducing slack variables $u_i$ ($i = 1, \ldots, M$) into (5.236), (5.235) and (5.236) become

minimize    $$Q(\boldsymbol{\alpha}^+, \boldsymbol{\alpha}^-, b^+, b^-, \boldsymbol{\xi}, \mathbf{u}) = \sum_{i=1}^{M}(\alpha_i^+ + \alpha_i^- + C\xi_i) \tag{5.237}$$

$$\text{subject to} \quad y_j \left( \sum_{i=1}^{M} (\alpha_i^+ - \alpha_i^-) K(\mathbf{x}_j, \mathbf{x}_i) + b^+ - b^- \right) + \xi_j = 1 + u_j$$

$$\text{for} \quad j = 1, \ldots, M, \tag{5.238}$$

respectively, which have $(4M + 2)$ variables and $M$ constraints.

Assuming the problem given by (5.235) and (5.236) primal, the dual problem is as follows:

$$\text{maximize} \quad Q(\mathbf{z}) = \sum_{i=1}^{M} z_i \tag{5.239}$$

$$\text{subject to} \quad \sum_{i=1}^{M} y_i K(\mathbf{x}_i, \mathbf{x}_j) z_i \leq 1 \quad \text{for} \quad j = 1, \ldots, M, \tag{5.240}$$

$$\sum_{i=1}^{M} y_i K(\mathbf{x}_i, \mathbf{x}_j) z_i \geq -1 \quad \text{for} \quad j = 1, \ldots, M, \tag{5.241}$$

$$z_j \leq C \qquad\qquad \text{for} \quad j = 1, \ldots, M, \tag{5.242}$$

$$\sum_{i=1}^{M} y_i z_i = 0, \tag{5.243}$$

where $z_i$ $(i = 1, \ldots, M)$ are dual variables and the number of constraints is $(3M + 1)$. Introducing nonnegative slack variables $v_i^+$, $v_i^-$, $w_i$ $(i = 1, \ldots, M)$, (5.239), (5.240), (5.241), (5.242), and (5.243) become as follows:

$$\text{maximize} \quad Q(\mathbf{z}, \mathbf{v}^+, \mathbf{v}^-, \mathbf{w}) = \sum_{i=1}^{M} z_i \tag{5.244}$$

$$\text{subject to} \quad \sum_{i=1}^{M} y_i K(\mathbf{x}_i, \mathbf{x}_j) z_i + v_j^+ = 1 \quad \text{for} \ j = 1, \ldots, M, \tag{5.245}$$

$$\sum_{i=1}^{M} y_i K(\mathbf{x}_i, \mathbf{x}_j) z_i = v_i^- - 1 \quad \text{for} \ j = 1, \ldots, M, \tag{5.246}$$

$$z_j + w_i = C \qquad\qquad \text{for} \ j = 1, \ldots, M, \tag{5.247}$$

$$\sum_{i=1}^{M} y_i z_i = 0. \tag{5.248}$$

The linear programming problem given by (5.244), (5.245), (5.246), (5.247), and (5.248) has $4M$ variables and $(3M + 1)$ constraints.

Let the optimal solution of the primal problem given by (5.237) and (5.238) be $(\boldsymbol{\alpha}^{+*}, \boldsymbol{\alpha}^{-*}, b^{+*}, b^{-*}, \boldsymbol{\xi}^*, \mathbf{u}^*)$ and that of the dual problem given by (5.244), (5.245), (5.246), (5.247), and (5.248) be $(\mathbf{z}^*, \mathbf{v}^{+*}, \mathbf{v}^{-*}, \mathbf{w}^*)$. Then the following complementarity conditions are satisfied:

$$\alpha_i^{+*} v_i^{+*} = 0 \qquad \text{for} \quad i = 1, \ldots, M, \qquad (5.249)$$

$$\alpha_i^{-*} v_i^{-*} = 0 \qquad \text{for} \quad i = 1, \ldots, M, \qquad (5.250)$$

$$\xi_i^* w_i^* = 0 \qquad \text{for} \quad i = 1, \ldots, M, \qquad (5.251)$$

$$u_i^* z_i^* = 0 \qquad \text{for} \quad i = 1, \ldots, M. \qquad (5.252)$$

Even if we delete $(\mathbf{x}_i, y_i)$ that satisfies

$$\alpha_i^{+*} = 0, \qquad (5.253)$$

$$\alpha_i^{-*} = 0, \qquad (5.254)$$

$$\xi_i^* = 0, \qquad (5.255)$$

$$z_i^* = 0, \qquad (5.256)$$

the optimal solution does not change, namely, training data that do not satisfy either of (5.253), (5.254), (5.255), and (5.256) are support vectors. Therefore, unlike support vector machines, $\mathbf{x}_i$ is a support vector even if $\alpha_i = 0$ so long as either of $\xi_i$ and $z_i$ is nonzero. In classification, however, only nonzero $\alpha_i$ are necessary and the small number of nonzero $\alpha_i$ is important to speed up classification.

### 5.9.2.2 Linear Programming Support Vector Machines Using Method 1

For linear LP support vector machines defined by (5.229) and (5.230), i.e., $\phi(\mathbf{x}) = \mathbf{x}$, the number of variable is $m$ and the number of constraints is $M$. Because the definition of linear LP support vector machines is similar to nonlinear LP support vector machines defined by (5.233) and (5.234), we only discuss the latter, in which $\alpha_i$ is associated with the $i$th training sample and for each training sample a constraint is defined. Therefore, $m = M$.

We divide the index set of training data, $T = \{1, \ldots, M\}$, into $W$ and $F$. Because there is no confusion we do not append the subscript v to the working sets. Then we divide variables in the primal problem given by (5.237) and (5.238) and those by the dual problem given by (5.244) (5.245), (5.246), (5.247), and (5.248), namely, $\boldsymbol{\alpha}^+ = \{\alpha_i^+ | i = 1, \ldots, M\}$ into $\boldsymbol{\alpha}_W^+ = \{\alpha_i^+ | i \in W\}$ and $\boldsymbol{\alpha}_F^+ = \{\alpha_i^+ | i \in F\}$; $\boldsymbol{\alpha}^-$ into $\boldsymbol{\alpha}_W^-$ and $\boldsymbol{\alpha}_F^-$; $\mathbf{v}^+$ into $\mathbf{v}_W^+$ and $\mathbf{v}_F^+$; $\mathbf{v}^-$ into $\mathbf{v}_W^-$ and $\mathbf{v}_F^-$; $\mathbf{w}$ into $\mathbf{w}_W$ and $\mathbf{w}_F$; $\mathbf{z}$ into $\mathbf{z}_W$ and $\mathbf{z}_F$. Here, we do not divide $\boldsymbol{\xi}$ and $\mathbf{u}$ because they are slack variables.

Fixing $\boldsymbol{\alpha}_F^+$ and $\boldsymbol{\alpha}_F^-$ we optimize the following subproblem:

$$\text{maximize} \quad Q(\boldsymbol{\alpha}_W^+, \boldsymbol{\alpha}_W^-, \boldsymbol{\xi}, b^+, b^-, \mathbf{u}) = \sum_{i \in W} (\alpha_i^+ + \alpha_i^-) + \sum_{i=1}^{M} C\xi_i \quad (5.257)$$

$$\text{subject to} \quad y_j \left( \sum_{i \in W} (\alpha_i^+ - \alpha_i^-) K(\mathbf{x}_i, \mathbf{x}_j) + b^+ - b^- \right.$$

$$+ \sum_{i \in F} (\alpha_i^+ - \alpha_i^-) K(\mathbf{x}_i, \mathbf{x}_j) \Bigg) + \xi_j = 1 + u_j$$

$$\text{for} \quad j = 1, \ldots, M. \tag{5.258}$$

After solving the subproblem fixing $\alpha_i = 0$ $(i \in F)$, we check whether the solution is the optimal solution of the entire problem using the complementarity conditions.

Because $\boldsymbol{\alpha}_F^+$ and $\boldsymbol{\alpha}_F^-$ are fixed to zero and $\boldsymbol{\xi}_F$ and $\mathbf{u}_F$ are determined when (5.257) and (5.258) are solved, all the primal variables are determined. Because $\mathbf{w}_F^+$ and $\mathbf{z}_F^+$ are dual variables associated with $\boldsymbol{\xi}_F$ and $\mathbf{u}_F$, respectively, the values of their variables are determined when solved by the simplex method or the primal–dual interior-point method.

But dual variables $\mathbf{v}_F^+$ and $\mathbf{v}_F^-$ are not determined yet. They can be determined so that the following constraints are satisfied:

$$v_j^+ = 1 - \sum_{i=1}^{M} y_i K(\mathbf{x}_i, \mathbf{x}_j) z_i \quad \text{for} \ j \in F, \tag{5.259}$$

$$v_j^- = 1 + \sum_{i=1}^{M} y_i K(\mathbf{x}_i, \mathbf{x}_j) z_i \quad \text{for} \ j \in F. \tag{5.260}$$

If some of them are negative, the entire solution does not satisfy the complementarity conditions. Thus, we need to solve the subproblem again adding the indices associated with violating variables into $W$ and deleting the indices associated with zero variables in $W$.

The algorithm of training an LP support vector machine by Method 1 is as follows:

1. Set $\alpha_i^+ = 0$ and $\alpha_i^- = 0$ for $i = 1, \ldots, M$. And initialize the iteration count: $k = 1$. Go to Step 2.
2. The initial working set be $W_1$. Select $q$ elements from $T$ and set the remaining elements to $F_1$. Go to Step 3.
3. Setting $W = W_k$ and $F = F_k$, optimize (5.257) and (5.258).
4. Calculate $\mathbf{v}_F^+$ and $\mathbf{v}_F^-$. Go to Step 5.
5. For data corresponding to $F_k$, check whether $v_i^+ \geq 0$ or $v_i^- \geq 0$ is satisfied. If there are violating variables, go to Step 6. Otherwise, stop training. For $k \geq 2$ if $Q_k - Q_{k-1} < \varepsilon$ is satisfied, stop training, where $Q_k$ is the value of the objective function for $W_k$ and $F_k$ and $\varepsilon$ is a small positive value.
6. Move the indices associated with zero variables in $W_k$ to the fixed set and add at most $q$ indices associated with the violating variables to the working set. Let the working set and the fixed set determined be $W_{k+1}$ and $F_{k+1}$, respectively, and add $k$ to 1 and go to Step 3.

### 5.9.2.3 Linear Programming Support Vector Machines Using Method 3

In LP support vector machines, $\alpha_i$ is associated with the $i$th training sample and for each training sample a constraint is defined, namely, $M = m$. It is possible to treat the variables and the constraints separately, but to make the definition of the LP support vector machine simpler, we set $W_v = W_c$ and $F_v = F_c$. Therefore, to simplify notations, we denote the working and fixed sets by $W$ and $F$, respectively.

In the following, we define a subproblem for Method 3 and then discuss infinite loop resolution of Method 3 based on Theorem 5.3. Finally, we discuss working set selection for training speedup [71].

Definition of Subproblems

We divide the index set $T = \{1, \ldots, M\}$ into the working set $W$ and the fixed set $F$. Then in (5.237) and (5.238) we fix $\alpha_F^+$, $\alpha_F^-$, $\boldsymbol{\xi}_F$, and $\mathbf{u}_F$, delete constraints associated with the fixed set $F$, and obtain the following subproblem:

$$\text{minimize} \quad Q(\boldsymbol{\alpha}_W^+, \boldsymbol{\alpha}_W^-, \boldsymbol{\xi}_W, \mathbf{u}_W, b^+, b^-) = \sum_{i \in W} (\alpha_i^+ + \alpha_i^- + C\xi_i) \quad (5.261)$$

$$\text{subject to} \quad y_j \left( \sum_{i \in W} (\alpha_i^+ - \alpha_i^-) K(\mathbf{x}_i, \mathbf{x}_j) + b^+ - b^- \right.$$

$$\left. + \sum_{i \in F} (\alpha_i^+ - \alpha_i^-) K(\mathbf{x}_i, \mathbf{x}_j) \right) + \xi_j = 1 + u_j \text{ for } j \in W. \quad (5.262)$$

We solve (5.261) and (5.262) for $\boldsymbol{\alpha}_W^+$, $\boldsymbol{\alpha}_W^-$, $\boldsymbol{\xi}_W$, $\mathbf{u}_W$, $\mathbf{v}_W^+$, $\mathbf{v}_W^-$, $\mathbf{w}_W$, and $\mathbf{z}_W$. To check whether the obtained solution satisfies the entire solution of (5.235) and (5.236), we generate solutions for the fixed set and check the complementarity conditions and constraints for the data associated with the fixed set.

For the primal variables, assuming $\alpha_i^+ = 0$ and $\alpha_i^- = 0$ ($i \in F$) we generate $\boldsymbol{\xi}_F$ and $\mathbf{u}_F$ using

$$y_j \left( \sum_{i \in W, F} (\alpha_i^+ - \alpha_i^-) K(\mathbf{x}_i, \mathbf{x}_j) + b^+ - b^- \right) + \xi_j = 1 + u_j \text{ for } j \in F. (5.263)$$

Here, (5.263) is included in (5.238). Using (5.231), (5.263) reduces to

$$y_j D(\mathbf{x}_j) + \xi_j = 1 + u_j \quad \text{for} \quad j \in F. \quad (5.264)$$

Thus, we generate $\boldsymbol{\xi}_F$ and $\mathbf{u}_F$ as follows:

1. If $y_j D(\mathbf{x}_j) > 1$, $\xi_j = 0$. Thus, from (5.264), $u_j = y_j D(\mathbf{x}_j) - 1$.
2. If $y_j D(\mathbf{x}_j) \leq 1$, $\xi_j = 1 - y_j D(\mathbf{x}_j)$. Thus, from (5.264), $u_j = 0$.

Then we generate the dual variables $\mathbf{v}_F^+$, $\mathbf{v}_F^-$, $\mathbf{w}_F$, and $\mathbf{z}_F$. Fixing $z_i = 0$ ($i \in F$), we generate $\mathbf{v}_F^+$, $\mathbf{v}_F^-$, and $\mathbf{w}_F$ by

$$v_j^+ = 1 - \sum_{i=1}^{M} y_i K(\mathbf{x}_i, \mathbf{x}_j) z_i \quad \text{for} \quad j \in F, \tag{5.265}$$

$$v_j^- = 1 + \sum_{i=1}^{M} y_i K(\mathbf{x}_i, \mathbf{x}_j) z_i \quad \text{for} \quad j \in F, \tag{5.266}$$

$$w_j = C \qquad \qquad \text{for} \quad j \in F. \tag{5.267}$$

Equations (5.265), (5.266), and (5.267) are included in (5.245), (5.246), (5.247), and (5.248). The values of $v_j^+$, $v_j^-$ ($j \in F$) obtained from (5.265) and (5.266) may be negative, which violate the constraints.

Instead of (5.262), if we solve the subproblem using (5.238), the subproblem is optimized by Method 1. Unlike support vector machines, deleting the constraints associated with the fixed set in (5.238), the optimal solution changes. This is because the constraints associated with the fixed set include $\alpha_i^+$ and $\alpha_i^-$ ($i \in W$). And because Method 3 is the combination of Methods 1 and 2, which have opposite convergence characteristics, the values of the objective function for Method 3 are not monotonic during training. However, from Theorem 5.3, if we apply infinite loop resolution to Method 3 the optimal solution is obtained in finite steps.

Infinite Loop Resolution (Method 3-1)

We call Method 3 with the infinite loop resolution discussed in Section 5.9.1.2 Method 3-1. This method guarantees the finite steps of convergence. In the following we show the algorithm.

1. Set $\alpha_i^+ = 0$, $\alpha_i^- = 0$ $z_i = 0$ for $i = 1, \ldots, M$ and $k = 1$. Go to Step 2.
2. Let the initial working set be $W_1$ and set $q$ elements from $T$ to $W_1$ and the remaining elements to $F_1$. Go to Step 3.
3. Setting $W = W_k$, $F = F_k$, solve (5.261) and (5.262), and go to Step 4.
4. Calculate the values of variables in $F_k$ and go to Step 5.
5. Check whether the training data associated with $F_k$ satisfy complementarity conditions (5.249), (5.250), (5.251), and (5.252) and $v_i^+ \geq 0$, $v_i^- \geq 0$. If not go to Step 6, otherwise stop training.
6. If an infinite loop occurs, set $W_{k+1}$ so that it includes all the indices of the working set in the loop and set the remaining indices to $F_{k+1}$. The indices

that are newly added to $W_{k+1}$ are kept until the algorithm terminates. Increment $k$ by 1 and go to Step 3. If there is no infinite loop go to Step 7.

7. Select at most $q$ indices in $F_k$ that correspond to violating variables and/or constraints and add them to the working set. Delete the indices in $W_k$ that are not support vectors and move them to the fixed set. Let the working and fixed sets thus generated be $W_{k+1}$ and $F_{k+1}$, respectively. Increment $k$ by 1 and go to Step 3.

## Working Set Selection Checking Violating Variables (Method 3-2)

By Method 3-1 we avoid infinite loops, but because the values of the objective function fluctuate during training, the number of iterations increases tremendously in some cases. To accelerate training, we further improve the working set selection strategy.

Let the number of variables that violate the complementarity conditions or constraints at step $k$ be $V_k$. In general, initially the value of $V_k$ is large but as training proceeds, it decreases and at the final stage $V_k = 0$ or near zero and training is terminated. However, according to our experiments in training of an LP support vector machine by Method 3-1, it frequently occurred that $V_k$ increased at some iteration step, i.e., $V_k > V_{k-1}$. This slowed down training.

Then if $V_k \geq V_{k-1}$, we consider that important data are deleted from $W_{k-1}$ and moved into the fixed set. Thus among the data that were moved away from $W_{k-1}$ we return back the data that violate the complementarity conditions or the constraints. In addition we keep the data in $W_k$ from deleting even if they satisfy the complementarity conditions.

We call the above method including infinite loop resolution Method 3-2. The detailed flow is as follows.

1. Set $\alpha_i^+ = 0$, $\alpha_i^- = 0$, and $z_i = 0$ for $i = 1, \ldots, M$ and $k = 1$. Go to Step 2.
2. Set the initial working set $W_1$ by $q$ elements from the index set $T$ and $F_1$ by the remaining elements. Go to Step 3.
3. Setting $W = W_k$, $F = F_k$, solve (5.261) and (5.262) and go to Step 4.
4. Generate values of the variables associated with $F_k$ and go to Step 5.
5. If some training data associated with $F_k$ violate (5.249), (5.250), (5.251), and (5.252) or $v_i^+ \geq 0$, $v_i^- \geq 0$, go to Step 6. Otherwise terminate training.
6. If an infinite loop exists, select all the indices associated with the working sets that form infinite loops as $W_{k+1}$. The indices that are newly added to $W_{k+1}$ are kept until the algorithm terminates. Increment $k$ by 1 and go to Step 3. Otherwise, go to Step 7.
7. If $V_k < V_{k-1}$, go to Step 8. Otherwise, add the indices of the data whose indices moved away from $W_{k-1}$ and which violate the complementarity

conditions or the constraints at the $k$th step to the working set. Let the obtained working set be $W_{k+1}$ and the set of the indices of the remaining data be $F_{k+1}$. Increment $k$ by 1 and go to Step 3. If there are not such data go to Step 8.

8. Check whether the data associated with $F_k$ violate the complementarity conditions or constraints. Add at most $q$ indices of the violating variables to the working set. Delete the indices from the working set whose associated data are not support vectors and move them to the fixed set. Let the working set and the fixed set thus determined be $W_{k+1}$ and $F_{k+1}$, respectively. Increment $k$ by 1 and go to Step 3.

### *5.9.3 Computer Experiments*

In this section we investigate speedup by Methods 1, 3-1, and 3-2 for the microarray data sets and the multiclass data sets in Tables 1.2 and 1.3, respectively.

#### 5.9.3.1 Evaluation Conditions

The microarray data sets are characterized by a large number of input variables but a small number of training/test data. Thus the classification problems are linearly separable and overfitting occurs quite easily. Therefore, usually, feature selection or extraction is performed to improve generalization ability. But because we were interested in the speedup by the decomposition technique, we used the original data sets for classification. We applied Method 1 to the linear LP SVM given by (5.229) and (5.230) and evaluated training speedup of the linear LP SVM for microarray data sets.

We evaluated the speedup by Methods 3-1 and 3-2 using the multiclass problems. We used one-against-all classification, in which one class is separated from the remaining classes. Thus, all the training data were used in training each decision function. To speed up training, in Methods 3-1 and 3-2 we freed the non-support vectors after infinite loop resolution. But for all the cases that we tested the optimal solutions were obtained in finite steps.

We used the revised simplex method [61] to solve linear programming problems. In measuring training time we used a workstation (3.6 GHz, 2 GB memory, Linux operating system).

#### 5.9.3.2 Speedup by Decomposition Techniques

Using the microarray data sets, we investigated the speedup by Method 1 for the linear LP SVM. We set $C = 1,000$ and $q = 50$. Table 5.12 lists the results. The "Rates" column lists the recognition rates for the test data and the training data in parentheses. The numeral in parentheses in the "SVs" column shows the number of nonzero $\alpha_i$. Only the nonzero $\alpha_i$ constitute the solution of the LP SVM. The "Non" column lists the training time without decomposition. The "M. 1" column lists the training time using Method 1. From the table, the speedup by Method 1 is 1.3 to 4. The low speedup ratio is because the number of constraints is too small and without decomposition, training is not so slow.

**Table 5.12** Speedup by linear LP SVMs combined with Method 1 for the microarray problems. Reprinted from [69, p. 983] with permission from Elsevier

| Data | Rates | SVs | Non [s] | M. 1 [s] | Speedup |
|------|-------|-----|---------|----------|---------|
| C. cancer | 77.27 (100) | 19 (18) | 0.50 | 0.37 | 1.4 |
| Leukemia | 61.76 (100) | 34 (33) | 2.2 | 1.0 | 2.2 |
| B. cancer (1) | 87.50 (100) | 12 (11) | 0.19 | 0.15 | 1.3 |
| B. cancer (2) | 87.50 (100) | 12 (11) | 0.19 | 0.15 | 1.3 |
| B. cancer (s) | 37.50 (100) | 14 (13) | 0.20 | 0.15 | 1.3 |
| Carcinoma | 81.48 (100) | 32 (31) | 2.0 | 1.0 | 2.0 |
| Glioma | 48.28 (100) | 20 (10) | 1.7 | 1.3 | 1.3 |
| P. cancer | 26.47 (100) | 97 (96) | 51 | 15 | 3.4 |
| B. cancer (4) | 52.63 (100) | 76 (75) | 37 | 9.2 | 4.0 |

We investigated the speedup of Methods 3-1 and 3-2 for the nonlinear LP SVM using the multiclass problems. We considered two cases: (1) RBF kernels with $\gamma = 1$ and $C = 10$ and (2) RBF kernels with $\gamma = 10$ and $C = 10,000$. We set $q = 50$ and measured the training time. Table 5.13 shows the result. In the table, in the "Cond." column, for example, "Numeral 1" and "Numeral 2" denote that RBF kernels with $\gamma = 1$ and $C = 10$ and RBF kernels with $\gamma = 10$ and $C = 10,000$ were used for the numeral training data set, respectively. The "Rates" column lists the recognition rates of the test data and the training data in parentheses. And the "SVs" column lists the number of support vectors and the numeral in parentheses shows the number of nonzero $\alpha_i$. The "Speedup" column lists the speedup of Method 3-2 over LP SVM training without using the decomposition techniques for the numeral, blood cell, and thyroid data sets but for the other data sets we list the speedup of Method 3-2 over Method 3-1 because the training time was too long if decomposition techniques were not used. Method 3-2 obtained the speedup of three to four orders of magnitudes over the training without the decomposition technique and speedup from 31 to 1.3 over Method 3-1.

**Table 5.13** Speedup by the decomposition techniques for multiclass problems. Reprinted from [69, p. 982] with permission from Elsevier

| Cond.    | Rates        | SVs      | Non [s]   | M. 3-1 [s] | M. 3-2 [s] | Speedup |
|----------|--------------|----------|-----------|------------|------------|---------|
| Numeral 1 | 99.27 (99.63) | 18 (6)   | 1,827     | 2.54       | 1.88       | 972     |
| Numeral 2 | 99.51 (100)   | 18 (9)   | 1,606     | 14.7       | 2.18       | 737     |
| Blood 1   | 88.77 (91.19) | 200 (9)  | 787,073   | 1141       | 681        | 1,156   |
| Blood 2   | 91.52 (99.77) | 120 (62) | 1,164,302 | 27,089     | 4,831      | 241     |
| Thyroid 1 | 94.22 (94.43) | 315 (11) | 541,896   | 2,605      | 2,485      | 218     |
| Thyroid 2 | 97.23 (99.81) | 179 (83) | 1,481,248 | 8,517      | 2,443      | 606     |
| H50 1     | 89.61 (91.32) | 107 (14) | –         | 2,710      | 1,387      | 2.0     |
| H50 2     | 98.11 (100)   | 57 (31)  | –         | 4,838      | 478        | 10      |
| H13 1     | 91.49 (91.47) | 183 (9)  | –         | 4,434      | 3,014      | 1.5     |
| H13 2     | 99.23 (100)   | 57 (29)  | –         | 46,663     | 1,503      | 31      |
| H105 1    | 96.69 (96.97) | 134 (22) | –         | 3,627      | 2,792      | 1.3     |
| H105 2    | 100 (100)     | 70 (38)  | –         | 26,506     | 1,645      | 16      |

# References

1. M.-H. Yang and N. Ahuja. A geometric approach to train support vector machines. In *Proceedings of IEEE Conference on Computer Vision and Pattern Recognition*, volume 1, pages 430–437, Hilton Head Island, SC, 2000.
2. M. B. de Almeida, A. de Pádua Braga, and J. P. Braga. SVM-KM: Speeding SVMs learning with a priori cluster selection and k-means. In *Proceedings of the Sixth Brazilian Symposium on Neural Networks (SBRN 2000)*, pages 162–167, Rio de Janeiro, Brazil, 2000.
3. S. Sohn and C. H. Dagli. Advantages of using fuzzy class memberships in self-organizing map and support vector machines. In *Proceedings of International Joint Conference on Neural Networks (IJCNN '01)*, volume 3, pages 1886–1890, Washington, DC, 2001.
4. S. Abe and T. Inoue. Fast training of support vector machines by extracting boundary data. In G. Dorffner, H. Bischof, and K. Hornik, editors, *Artificial Neural Networks (ICANN 2001)—Proceedings of International Conference, Vienna, Austria*, pages 308–313. Springer-Verlag, Berlin, Germany, 2001.
5. W. Zhang and I. King. Locating support vectors via $\beta$-skeleton technique. In *Proceedings of the Ninth International Conference on Neural Information Processing (ICONIP '02)*, volume 3, pages 1423–1427, Singapore, 2002.
6. H. Shin and S. Cho. How many neighbors to consider in pattern pre-selection for support vector classifiers? In *Proceedings of International Joint Conference on Neural Networks (IJCNN 2003)*, volume 1, pages 565–570, Portland, OR, 2003.
7. S.-Y. Sun, C. L. Tseng, Y. H. Chen, S. C. Chuang, and H. C. Fu. Cluster-based support vector machines in text-independent speaker identification. In *Proceedings of International Joint Conference on Neural Networks (IJCNN 2004)*, volume 1, pages 729–734, Budapest, Hungary, 2004.
8. B. Li, Q. Wang, and J. Hu. A fast SVM training method for very large datasets. In *Proceedings of the 2009 International Joint Conference on Neural Networks (IJCNN 2009)*, pages 1784–1789, Atlanta, GA, 2009.
9. C. Saunders, M. O. Stitson, J. Weston, L. Bottou, B. Schölkopf, and A. Smola. Support vector machine: Reference manual. Technical Report CSD-TR-98-03, Royal Holloway, University of London, London, UK, 1998.

10. E. Osuna, R. Freund, and F. Girosi. An improved training algorithm for support vector machines. In *Neural Networks for Signal Processing VII—Proceedings of the 1997 IEEE Signal Processing Society Workshop*, pages 276–285, 1997.

11. G. Schohn and D. Cohn. Less is more: Active learning with support vector machines. In *Proceedings of the Seventeenth International Conference on Machine Learning (ICML-2000)*, pages 839–846, Stanford, CA, 2000.

12. C.-J. Lin. On the convergence of the decomposition method for support vector machines. *IEEE Transactions on Neural Networks*, 12(6):1288–1298, 2001.

13. C.-J. Lin. Asymptotic convergence of an SMO algorithm without any assumptions. *IEEE Transactions on Neural Networks*, 13(1):248–250, 2002.

14. S. S. Keerthi and E. G. Gilbert. Convergence of a generalized SMO algorithm for SVM classifier design. *Machine Learning*, 46(1–3):351–360, 2002.

15. R.-E. Fan, P.-H. Chen, and C.-J. Lin. Working set selection using second order information for training support vector machines. *Journal of Machine Learning Research*, 6:1889–1918, 2005.

16. M. Rychetsky, S. Ortmann, M. Ullmann, and M. Glesner. Accelerated training of support vector machines. In *Proceedings of International Joint Conference on Neural Networks (IJCNN '99)*, volume 2, pages 998–1003, Washington, DC, 1999.

17. Y. Koshiba. Acceleration of training of support vector machines. Master's thesis (in Japanese), Graduate School of Science and Technology, Kobe University, Japan, 2004.

18. C. Campbell, T.-T. Frieß, and N. Cristianini. Maximal margin classification using the KA algorithm. In *Proceedings of the First International Symposium on Intelligent Data Engineering and Learning (IDEAL '98)*, pages 355–362, Hong Kong, China, 1998.

19. T.-T. Frieß, N. Cristianini, and C. Campbell. The Kernel-Adatron algorithm: A fast and simple learning procedure for support vector machines. In *Proceedings of the Fifteenth International Conference on Machine Learning (ICML '98)*, pages 188–196, Madison, WI, 1998.

20. Y. Freund and R. E. Schapire. Large margin classification using the perceptron algorithm. *Machine Learning*, 37(3):277–296, 1999.

21. I. Guyon and D. G. Stork. Linear discriminant and support vector classifiers. In A. J. Smola, P. L. Bartlett, B. Schölkopf, and D. Schuurmans, editors, *Advances in Large Margin Classifiers*, pages 147–169. MIT Press, Cambridge, MA, 2000.

22. J. Xu, X. Zhang, and Y. Li. Large margin kernel pocket algorithm. In *Proceedings of International Joint Conference on Neural Networks (IJCNN '01)*, volume 2, pages 1480–1485, Washington, DC, 2001.

23. J. K. Anlauf and M. Biehl. The Adatron: An adaptive perceptron algorithm. *Europhysics Letters*, 10:687–692, 1989.

24. N. Cristianini and J. Shawe-Taylor. *An Introduction to Support Vector Machines and Other Kernel-Based Learning Methods*. Cambridge University Press, Cambridge, UK, 2000.

25. O. L. Mangasarian and D. R. Musicant. Successive overrelaxation for support vector machines. *IEEE Transactions on Neural Networks*, 10(5):1032–1037, 1999.

26. J. C. Platt. Fast training of support vector machines using sequential minimal optimization. In B. Schölkopf, C. J. C. Burges, and A. J. Smola, editors, *Advances in Kernel Methods: Support Vector Learning*, pages 185–208. MIT Press, Cambridge, MA, 1999.

27. J.-X. Dong, A. Krzyżak, and C. Y. Suen. Fast SVM training algorithm with decomposition on very large data sets. *IEEE Transactions on Pattern Analysis and Machine Intelligence*, 27(4):603–618, 2005.

28. L. Bottou and C.-J. Lin. Support vector machine solvers. In L. Bottou, O. Chapelle, D. DeCoste, and J. Weston, editors, *Large-Scale Kernel Machines*, pages 1–27. MIT Press, Cambridge, MA, 2007.

29. S. Abe, Y. Hirokawa, and S. Ozawa. Steepest ascent training of support vector machines. In E. Damiani, L. C. Jain, R. J. Howlett, and N. Ichalkaranje, editors, *Knowledge-Based Intelligent Engineering Systems and Allied Technologies (KES 2002)*, volume Part 2, pages 1301–1305, IOS Press, Amsterdam, The Netherlands, 2002.

30. M. Vogt. SMO algorithms for support vector machines without bias term. Technical report, Institute of Automatic Control, TU Darmstadt, Germany, 2002.

31. V. Kecman, M. Vogt, and T. M. Huang. On the equality of kernel AdaTron and Sequential Minimal Optimization in classification and regression tasks and alike algorithms for kernel machines. In *Proceedings of the Eleventh European Symposium on Artificial Neural Networks (ESANN 2003)*, pages 215–222, Bruges, Belgium, 2003.

32. C. Sentelle, M. Georgiopoulos, G. C. Anagnostopoulos, and C. Young. On extending the SMO algorithm sub-problem. In *Proceedings of the 2007 International Joint Conference on Neural Networks (IJCNN 2007)*, pages 886–891, Orlando, FL, 2007.

33. R. A. Hernandez, M. Strum, J. C. Wang, and J. A. Q. Gonzalez. The multiple pairs SMO: A modified SMO algorithm for the acceleration of the SVM training. In *Proceedings of the 2009 International Joint Conference on Neural Networks (IJCNN 2009)*, pages 1221–1228, Atlanta, GA, 2009.

34. G. Cauwenberghs and T. Poggio. Incremental and decremental support vector machine learning. In T. K. Leen, T. G. Dietterich, and V. Tresp, editors, *Advances in Neural Information Processing Systems 13*, pages 409–415. MIT Press, Cambridge, MA, 2001.

35. A. Shilton, M. Palaniswami, D. Ralph, and A. C. Tsoi. Incremental training of support vector machines. *IEEE Transactions on Neural Networks*, 16(1):114–131, 2005.

36. K. Scheinberg. An efficient implementation of an active set method for SVMs. *Journal of Machine Learning Research*, 7:2237–2257, 2006.

37. S. Abe. Batch support vector training based on exact incremental training. In V. Kůrková, R. Neruda, and J. Koutnik, editors, *Artificial Neural Networks (ICANN 2008)—Proceedings of the Eighteenth International Conference, Prague, Czech Republic, Part I*, pages 527–536. Springer-Verlag, Berlin, Germany, 2008.

38. H. Gâlmeanu and R. Andonie. Implementation issues of an incremental and decremental SVM. In V. Kůrková, R. Neruda, and J. Koutnik, editors, *Artificial Neural Networks (ICANN 2008)—Proceedings of the Eighteenth International Conference, Prague, Czech Republic, Part I*, pages 325–335. Springer-Verlag, Berlin, Germany, 2008.

39. C. Sentelle, G. C. Anagnostopoulos, and M. Georgiopoulos. An efficient active set method for SVM training without singular inner problems. In *Proceedings of the 2009 International Joint Conference on Neural Networks (IJCNN 2009)*, pages 2875–2882, Atlanta, GA, 2009.

40. O. Chapelle. Training a support vector machine in the primal. In L. Bottou, O. Chapelle, D. DeCoste, and J. Weston, editors, *Large-Scale Kernel Machines*, pages 29–50. MIT Press, Cambridge, MA, 2007.

41. S. Abe. Is primal better than dual. In C. Alippi, M. Polycarpou, C. Panayiotou, and G. Ellinas, editors, *Artificial Neural Networks (ICANN 2009)—Proceedings of the Nineteenth International Conference, Limassol, Cyprus, Part I*, pages 854–863. Springer-Verlag, Berlin, Germany, 2009.

42. D. Roobaert. DirectSVM: A fast and simple support vector machine perceptron. In *Neural Networks for Signal Processing X—Proceedings of the 2000 IEEE Signal Processing Society Workshop*, volume 1, pages 356–365, 2000.

43. S. S. Keerthi, S. K. Shevade, C. Bhattacharyya, and K. R. K. Murthy. A fast iterative nearest point algorithm for support vector machine classifier design. *IEEE Transactions on Neural Networks*, 11(1):124–136, 2000.

44. T. Raicharoen and C. Lursinsap. Critical support vector machine without kernel function. In *Proceedings of the Ninth International Conference on Neural Information Processing (ICONIP '02)*, volume 5, pages 2532–2536, Singapore, 2002.

45. S. V. N. Vishwanathan and M. N. Murty. SSVM: A simple SVM algorithm. In *Proceedings of the 2002 International Joint Conference on Neural Networks (IJCNN '02)*, volume 3, pages 2393–2398, Honolulu, Hawaii, 2002.

46. K. P. Bennett and E. J. Bredensteiner. Geometry in learning. In C. A. Gorini, editor, *Geometry at Work*, pages 132–145. Mathematical Association of America, Washington, DC 2000.

47. D. J. Crisp and C. J. C. Burges. A geometric interpretation of ν-SVM classifiers. In S. A. Solla, T. K. Leen, and K.-R. Müller, editors, *Advances in Neural Information Processing Systems 12*, pages 244–250. MIT Press, Cambridge, MA, 2000.

48. Q. Tao, G. Wu, and J. Wang. A generalized S-K algorithm for learning ν-SVM classifiers. *Pattern Recognition Letters*, 25(10):1165–1171, 2004.

49. M. E. Mavroforakis and S. Theodoridis. A geometric approach to support vector machine (SVM) classification. *IEEE Transactions on Neural Networks*, 17(3):671–682, 2006.

50. M. E. Mavroforakis, M. Sdralis, and S. Theodoridis. A geometric nearest point algorithm for the efficient solution of the SVM classification task. *IEEE Transactions on Neural Networks*, 18(5):1545–1549, 2007.

51. A. Navia-Vázquez, F. Pérez-Cruz, A. Artés-Rodríguez, and A. R. Figueiras-Vidal. Weighted least squares training of support vector classifiers leading to compact and adaptive schemes. *IEEE Transactions on Neural Networks*, 12(5):1047–1059, 2001.

52. G. Zanghirati and L. Zanni. A parallel solver for large quadratic programs in training support vector machines. *Parallel Computing*, 29(4):535–551, 2003.

53. L. Ferreira, E. Kaszkurewicz, and A. Bhaya. Parallel implementation of gradient-based neural networks for SVM training. In *Proceedings of the 2006 International Joint Conference on Neural Networks (IJCNN 2006)*, pages 731–738, Vancouver, Canada, 2006.

54. I. Durdanovic, E. Cosatto, and H.-P. Graf. Large-scale parallel SVM implementation. In L. Bottou, O. Chapelle, D. DeCoste, and J. Weston, editors, *Large-Scale Kernel Machines*, pages 105–138. MIT Press, Cambridge, MA, 2007.

55. I. W. Tsang, J. T. Kwok, and P.-M. Cheung. Core vector machines: Fast SVM training on very large data sets. *Journal of Machine Learning Research*, 6:363–392, 2005.

56. I. W.-H. Tsang, J. T.-Y. Kwok, and J. M. Zurada. Generalized core vector machines. *IEEE Transactions on Neural Networks*, 17(5):1126–1140, 2006.

57. G. Loosli and S. Canu. Comments on the "Core vector machines: Fast SVM training on very large data sets". *Journal of Machine Learning Research*, 8:291–301, 2007.

58. L. Bo, L. Wang, and L. Jiao. Training hard-margin support vector machines using greedy stepwise algorithm. *IEEE Transactions on Neural Networks*, 19(8):1446–1455, 2008.

59. R. J. Vanderbei. *Linear Programming: Foundations and Extensions, Second Edition*. Kluwer Academic Publishers, Norwell, MA, 2001.

60. S. J. Wright. *Primal-Dual Interior-Point Methods*. Society for Industrial and Applied Mathematics, Philadelphia, PA, 1997.

61. V. Chvátal. *Linear Programming*. W. H. Freeman and Company, New York, NY, 1983.

62. R. J. Vanderbei. LOQO: An interior point code for quadratic programming. Technical Report SOR-94-15, Princeton University, 1998.

63. Y. Koshiba and S. Abe. Comparison of L1 and L2 support vector machines. In *Proceedings of International Joint Conference on Neural Networks (IJCNN 2003)*, volume 3, pages 2054–2059, Portland, OR, 2003.

64. D. P. Bertsekas. *Nonlinear Programming, Second Edition*. Athena Scientific, Belmont, MA, 1999.

65. S. Boyd and L. Vandenberghe. *Convex Optimization*. Cambridge University Press, Cambridge, 2004.

66. P. Laskov, C. Gehl, S. Krüger, and K.-R. Müller. Incremental support vector learning: Analysis, implementation and applications. *Journal of Machine Learning Research*, 7:1909–1936, 2006.

67. C. P. Diehl and G. Cauwenberghs. SVM incremental learning, adaptation and optimization. In *Proceedings of International Joint Conference on Neural Networks (IJCNN 2003)*, volume 4, pages 2685–2690, Portland, OR, 2003.

68. T. Hastie, S. Rosset, R. Tibshirani, and J. Zhu. The entire regularization path for the support vector machine. *Journal of Machine Learning Research*, 5:1391–1415, 2004.

69. Y. Torii and S. Abe. Decomposition techniques for training linear programming support vector machines. *Neurocomputing*, 72(4–6):973–984, 2009.

70. P. S. Bradley and O. L. Mangasarian. Massive data discrimination via linear support vector machines. *Optimization Methods and Software*, 13(1):1–10, 2000.

71. Y. Torii and S. Abe. Fast training of linear programming support vector machines using decomposition techniques. In F. Schwenker and S. Marinai, editors, *Artificial Neural Networks in Pattern Recognition: Proceedings of Second IAPR Workshop, ANNPR 2006, Ulm, Germany*, pages 165–176. Springer-Verlag, Berlin, Germany, 2006.

72. P. S. Bradley and O. L. Mangasarian. Feature selection via concave minimization and support vector machines. In *Proceedings of the Fifteenth International Conference on Machine Learning (ICML '98)*, pages 82–90, Madison, WI, 1998.

73. T. Joachims. Making large-scale support vector machine learning practical. In B. Schölkopf, C. J. C. Burges, and A. J. Smola, editors, *Advances in Kernel Methods: Support Vector Learning*, pages 169–184. MIT Press, Cambridge, MA, 1999.

74. C.-W. Hsu and C.-J. Lin. A simple decomposition method for support vector machines. *Machine Learning*, 46(1–3):291–314, 2002.

75. P. Laskov. Feasible direction decomposition algorithms for training support vector machines. *Machine Learning*, 46(1–3):315–349, 2002.

76. D. Hush and C. Scovel. Polynomial-time decomposition algorithms for support vector machines. *Machine Learning*, 51(1):51–71, 2003.

77. B. Schölkopf and A. J. Smola. *Learning with Kernels: Support Vector Machines, Regularization, Optimization, and Beyond*. MIT Press, Cambridge, MA, 2002.

78. V. Kecman and I. Hadzic. Support vectors selection by linear programming. In *Proceedings of the IEEE-INNS-ENNS International Joint Conference on Neural Networks (IJCNN 2000)*, volume 5, pages 193–198, Como, Italy, 2000.

79. W. Zhou, L. Zhang, and L. Jiao. Linear programming support vector machines. *Pattern Recognition*, 35(12):2927–2936, 2002.

80. B. Schölkopf, P. Simard, A. Smola, and V. Vapnik. Prior knowledge in support vector kernels. In M. I. Jordan, M. J. Kearns, and S. A. Solla, editors, *Advances in Neural Information Processing Systems 10*, pages 640–646. MIT Press, Cambridge, MA, 1998.

# Chapter 6
# Kernel-Based Methods

Inspired by the success of support vector machines, to improve generalization and classification abilities, conventional pattern classification techniques have been extended to incorporate maximizing margins and mapping to a feature space. For example, perceptron algorithms [1–4], neural networks (Chapter 9), and fuzzy systems (Chapter 10) have incorporated maximizing margins and/or mapping to a feature space.

There are numerous conventional techniques that are extended to be used in the high-dimensional feature space, e.g., kernel least squares [5, 6], kernel principal component analysis [7–9], kernel discriminant analysis [9, pp. 457–468], the kernel Mahalanobis distance [6], kernel $k$-means clustering algorithms [10, 11], kernel fuzzy $c$-means clustering algorithms [12, 13], the kernel self-organizing feature map [14], kernel subspace methods [15–18], and other kernel-based methods [19, 20].

In this chapter, we discuss some of the kernel-based methods: kernel least squares, kernel principal component analysis, the kernel Mahalanobis distance, and kernel discriminant analysis.

## 6.1 Kernel Least Squares

### 6.1.1 Algorithm

Least-squares methods in the input space can be readily extended to the feature space using kernel techniques [5, 6].

Assume that we have training input–output pairs $\{\mathbf{x}_i, y_i\}$ $(i = 1, \ldots, M)$. We approximate the output $y$ by

$$y = \mathbf{a}^\top \phi(\mathbf{x}), \tag{6.1}$$

S. Abe, *Support Vector Machines for Pattern Classification*,
Advances in Pattern Recognition, DOI 10.1007/978-1-84996-098-4_6,
© Springer-Verlag London Limited 2010

where $\phi(\mathbf{x})$ is the mapping function that maps $\mathbf{x}$ into the $l$-dimensional feature space and $\mathbf{a}$ is the $l$-dimensional vector. Without loss of generality, we can assume that the last element of $\phi(\mathbf{x})$ is 1. By this assumption, we need not add a constant term in (6.1).

We determine vector $\mathbf{a}$ so that

$$J = \sum_{i=1}^{M} \left( y_i - \mathbf{a}^{\top} \phi(\mathbf{x}_i) \right)^2 \tag{6.2}$$

is minimized.

Without loss of generality we can assume that the set of $M'$ vectors, $\{\phi(\mathbf{x}_1), \ldots, \phi(\mathbf{x}_{M'})\}$ $(M' \leq M)$, spans the space generated by $\{\phi(\mathbf{x}_1), \ldots, \phi(\mathbf{x}_M)\}$. Then, because $\mathbf{a}$ is in the space spanned by $\{\phi(\mathbf{x}_1), \ldots, \phi(\mathbf{x}_{M'})\}$, $\mathbf{a}$ is expressed by

$$\mathbf{a} = \sum_{i=1}^{M'} \alpha_i \, \phi(\mathbf{x}_i), \tag{6.3}$$

where $\alpha_i$ are parameters. Then (6.1) becomes

$$\begin{aligned} y &= \sum_{i=1}^{M'} \alpha_i \, \phi^{\top}(\mathbf{x}_i) \, \phi(\mathbf{x}) \\ &= \sum_{i=1}^{M'} \alpha_i \, K(\mathbf{x}, \mathbf{x}_i), \end{aligned} \tag{6.4}$$

where $K(\mathbf{x}, \mathbf{x}_i) = \phi^{\top}(\mathbf{x}) \, \phi(\mathbf{x}_i) = \phi^{\top}(\mathbf{x}_i) \, \phi(\mathbf{x})$.

When RBF kernels are used, (6.4) is equivalent to radial basis function neural networks with $\mathbf{x}_i$ $(i = 1, \ldots, M')$ being the centers of radial bases [5].

Substituting (6.3) into (6.2), we obtain

$$\begin{aligned} J &= \frac{1}{2} \sum_{i=1}^{M} \left( y_i - \sum_{j=1}^{M'} \alpha_j K(\mathbf{x}_i, \mathbf{x}_j) \right)^2 \\ &= \frac{1}{2} (\mathbf{y} - K\boldsymbol{\alpha})^{\top} (\mathbf{y} - K\boldsymbol{\alpha}), \end{aligned} \tag{6.5}$$

where $\mathbf{y} = (y_1, \ldots, y_M)^{\top}$, $\boldsymbol{\alpha} = (\alpha_1, \ldots, \alpha_{M'})^{\top}$, $K$ is an $M \times M'$ matrix, and $K = \{K(\mathbf{x}_i, \mathbf{x}_j)\}$ $(i = 1, \ldots, M, j = 1, \ldots, M')$.

Taking the partial derivative of $J$ with respect to $\boldsymbol{\alpha}$ and setting it to zero, we obtain

$$\frac{\partial J}{\partial \boldsymbol{\alpha}} = -K^{\top} (\mathbf{y} - K\boldsymbol{\alpha}) = \mathbf{0}. \tag{6.6}$$

Because $\{\phi(\mathbf{x}_1), \ldots, \phi(\mathbf{x}_{M'})\}$ are linearly independent, the rank of $K$ is $M'$. Thus, $K^\top K$ is positive definite. Then, from (6.6),

$$\boldsymbol{\alpha} = (K^\top K)^{-1} K^\top \mathbf{y}. \tag{6.7}$$

If $\{\phi(\mathbf{x}_1), \ldots, \phi(\mathbf{x}_M)\}$ are linearly independent, $K$ is a square matrix and positive definite. Then $\boldsymbol{\alpha}$ is given by

$$\boldsymbol{\alpha} = K^{-1} \mathbf{y}. \tag{6.8}$$

But in general $M' < M$. Thus, to use (6.7) we need to select linearly independent $\{\phi(\mathbf{x}_1), \ldots, \phi(\mathbf{x}_{M'})\}$. Otherwise, we need to solve

$$\boldsymbol{\alpha} = K^+ \mathbf{y}, \tag{6.9}$$

where $K$ is an $M \times M$ matrix and $K^+$ is the pseudo-inverse of $K$. If $K$ is singular, $K^+$ is calculated by singular value decomposition (see Section B.2). This is a time-consuming task, and in estimating $\mathbf{y}$ we need to use all $M$ training data. Thus it is favorable to select linearly independent vectors in advance.

Baudat and Anouar [5] proposed selecting linearly independent vectors by sequential forward selection, in which starting from an empty set, a vector that maximizes the objective function is selected. Cawley and Talbot [21] proposed speeding up basis selection by using the matrix inversion lemma, deleting data that make the kernel matrix singular when added to the kernel matrix, sampling of the candidate basis vectors, and stochastic approximation of the objective function for basis vector selection.

Here, we consider using the Cholesky factorization to select linearly independent vectors [22]. Let $K$ be positive definite. Then $K$ is decomposed by the Cholesky factorization into

$$K = L L^\top, \tag{6.10}$$

where $L$ is the regular lower triangular matrix and each element $L_{ij}$ is given by

$$L_{op} = \frac{K_{op} - \sum_{n=1}^{p-1} L_{pn} L_{on}}{L_{pp}} \quad \text{for } o = 1, \ldots, M, \quad p = 1, \ldots, o-1, \tag{6.11}$$

$$L_{aa} = \sqrt{K_{aa} - \sum_{n=1}^{a-1} L_{an}^2} \quad \text{for } a = 1, 2, \ldots, M. \tag{6.12}$$

Here, $K_{ij} = K(\mathbf{x}_i, \mathbf{x}_j)$.

Then during the Cholesky factorization, if the argument of the root in (6.12) is smaller than the prescribed value $\eta \, (> 0)$:

$$K_{aa} - \sum_{n=1}^{a-1} L_{an}^2 \leq \eta, \tag{6.13}$$

we delete the associated row and column and continue decomposing the matrix. The training data that are not deleted in the Cholesky factorization are linearly independent. If no training data are deleted, the training data are all linearly independent in the feature space.

After factorization, $\boldsymbol{\alpha}$ is obtained by solving

$$L\mathbf{c} = \mathbf{y}, \tag{6.14}$$
$$L^{\top}\boldsymbol{\alpha} = \mathbf{c}, \tag{6.15}$$

where $\mathbf{c}$ is an $M'$-dimensional vector.

Selection of the value of $\eta$ influences the generalization ability. For instance, for RBF kernels, a small value of $\eta$ may result in selecting all the training data. Thus, there is the optimal value of $\eta$, which is determined by model selection.

If we use linear kernels, kernel least squares result in conventional least squares. The advantage of using kernel least squares is that we avoid using singular value decomposition (see Section B.2), when the dimension of the space spanned by the training data is lower than that of the input space.

In the kernel least squares, the regularization term is not included. Thus overfitting may occur. To avoid overfitting, in [23] the regularized least-squares methods were proposed, in which the solution is obtained by solving

$$(K + D)\boldsymbol{\alpha} = \mathbf{y}, \tag{6.16}$$

where $K$ is the kernel matrix and $D$ is a diagonal matrix associated with the regularization term. Because $K + D$ is positive definite, (6.16) is solved without singular value decomposition. According to [24], the generalization abilities of support vector machines and regularized least-squares methods are shown to be comparable for several benchmark data sets. But, unlike support vector machines, the solution is not sparse. This is the same situation with the least-squares support vector machines.

### 6.1.2 Performance Evaluation

We evaluate the kernel least-squares method using pattern classification and function approximation problems listed in Tables 1.3 and 1.4 [25].

We measured the training time of our method using a personal computer (Xeon 2.4 GHz dual, memory 4 GB) with the Linux operating system.

For an $n$-class classification problem, we determined $n$ hyperplanes, with each hyperplane separating one class from the others, setting the target value of 1 for the class and $-1$ for the remaining classes. In classification, we classified unknown data sample to the class with the maximum output. This is a one-against-all classification strategy. When RBF kernels are used, the kernel least squares are equivalent to radial basis function neural networks.

Because the generalization ability depends on the value of $\eta$, we determined the value by fivefold cross-validation for linear kernels, polynomial kernels with $d = [2, 3, 4]$, and RBF kernels with $\gamma = [1, 5, 10]$ with $\eta = [10^{-8}, 10^{-7}, \ldots, 10^{-2}]$, and we selected the value with the maximum recognition rate.

If the same recognition rate was obtained for the validation data set, we broke the tie by selecting the simplest structure as follows:

1. Select the kernel and parameters with the highest recognition rate for the training data.
2. Select polynomial kernels from polynomial and RBF kernels.
3. Select the polynomial kernel with the smallest degree from polynomial kernels.
4. Select the RBF kernel with the smallest value of $\gamma$ from RBF kernels.
5. Select the largest value of $\eta$.

Table 6.1 shows the cross-validation results for the blood cell data. In the table, the "Test," "Train.," and "Valid." columns list the recognition rates of the test data, training data, and cross-validation data, respectively. The "Num." and "Time" columns list the number of selected linearly independent data and training and classification time, respectively. By selecting the value of $\eta$ by cross-validation, the number of linearly independent data was reduced to less than one-fifteenth of that of the training data. In addition, training and classification time was short.

Because the recognition rate of the cross-validation data with RBF kernels with $\gamma = 5$ was the highest, we selected RBF kernels with $\gamma = 5$ and $\eta = 10^{-4}$. In this case the recognition rate of the test data was the second highest.

**Table 6.1** Recognition rates for the blood cell data

| Kernel | $\eta$ | Test (%) | Train. (%) | Valid. (%) | Num. | Time (s) |
|--------|--------|----------|------------|------------|------|----------|
| Linear | $10^{-3}$ | 67.71 | 69.78 | 74.15 | 10 | 25 |
| $d2$ | $10^{-5}$ | 90.29 | 91.12 | 91.44 | 49 | 28 |
| $d3$ | $10^{-5}$ | 91.16 | 93.51 | 92.37 | 79 | 30 |
| $d4$ | $10^{-5}$ | 90.74 | 92.64 | 91.15 | 70 | 30 |
| $\gamma1$ | $10^{-5}$ | 82.90 | 84.99 | 92.37 | 119 | 37 |
| $\gamma5$ | $10^{-4}$ | 93.03 | 95.96 | **93.86** | 253 | 58 |
| $\gamma10$ | $10^{-3}$ | 93.32 | 95.87 | 93.63 | 227 | 52 |

Table 6.2 lists the cross-validation results for the Mackey-Glass data. Under the same conditions as pattern classification, we performed fivefold

**Table 6.2** Approximation errors of the Mackey–Glass data

| Kernel | $\eta$ | Test (NRMSE) | Train. (NRMSE) | Valid. (NRMSE) | Num. | Time (s) |
|--------|--------|--------------|----------------|----------------|------|----------|
| Linear | $10^{-2}$ | 0.0586 | 0.0584 | 0.0603 | 1 | 0 |
| d2 | $10^{-7}$ | 0.0096 | 0.0098 | 0.0138 | 15 | 12 |
| d3 | $10^{-5}$ | 0.0062 | 0.0063 | 0.0093 | 15 | 0 |
| d4 | $10^{-5}$ | 0.0063 | 0.0064 | 0.0064 | 15 | 0 |
| $\gamma1$ | $10^{-5}$ | 0.0026 | 0.0026 | 0.0022 | 51 | 1 |
| $\gamma5$ | $10^{-4}$ | 0.0021 | 0.0021 | 0.0011 | 122 | 1 |
| $\gamma10$ | $10^{-4}$ | 0.0015 | 0.0015 | **0.0008** | 190 | 1 |

cross-validation for each kernel. Approximation performance was measured by the normalized root-mean-square error (NRMSE). Because the Mackey–Glass data did not include noise, NRMSEs for the training and test data did not vary very much. Because the NRMSE of the cross-validation data with RBF kernels with $\gamma = 10$ was the smallest, we selected RBF kernels with $\gamma = 10$ and $\eta = 10^{-4}$. For this case, the linearly independent data were reduced to less than half of the training data and the training time was very short. If the training time was shorter than 0.5 s, it is listed as 0 in the "Time" column.

Table 6.3 lists the results for the benchmark data sets. For comparison, for pattern classification problems we also list the recognition rates of the one-against-all fuzzy L1 SVM in Table 3.2. For function approximation we used the performance of the SVM listed in Tables 11.6 and 11.8. Higher or equal recognition rates (lower or equal approximation errors) are shown in boldface. For the water purification data set, the "Test," "Train.," and "SVM" columns list the average errors in mg/l. And for this data set, the cross-validation for the RBF kernels, we used $\gamma = [0.1, 0.5, 1.0]$.

**Table 6.3** Recognition performance

| Data | Kernel | $\eta$ | Test (%) | Train. (%) | Num. | Time (s) | SVM (%) |
|------|--------|--------|----------|------------|------|----------|---------|
| Iris | d2 | $10^{-5}$ | **96.00** | 100 | 15 | 0 | 94.67 |
| Numeral | d3 | $10^{-5}$ | 99.15 | 100 | 199 | 2 | **99.27** |
| Thyroid | $\gamma5$ | $10^{-4}$ | 93.99 | 95.97 | 471 | 163 | **97.93** |
| Blood cell | $\gamma5$ | $10^{-4}$ | 93.03 | 95.96 | 253 | 57 | **93.16** |
| Hiragana-50 | $\gamma10$ | $10^{-4}$ | 99.13 | 100 | 3,511 | 1,641 | **99.26** |
| Hiragana-13 | $\gamma10$ | $10^{-4}$ | 99.44 | 99.61 | 860 | 1,715 | **99.63** |
| Hiragana-105 | $\gamma10$ | $10^{-3}$ | **100** | 100 | 7,197 | 92,495 | **100** |
| Mackey–Glass | $\gamma10$ | $10^{-4}$ | **0.002**[1] | 0.002[1] | 190 | 1 | 0.003[1] |
| Water purif. | $\gamma0.1$ | $10^{-4}$ | 1.08[2] | 0.72[2] | 62 | 0 | **1.03**[2] |

[1] NRMSE
[2] mg/l

Except for the hiragana-50 and hiragana-105 data sets, by fivefold cross-validation, small numbers of data were selected and training time was relatively short.

Except for the thyroid data, the generalization abilities of the kernel least squares and the SVM are comparable. For the thyroid test data, the recognition rate of the kernel least squares is much lower than that of the L1 SVM. This is the same tendency as that of the least-squares SVM.

## 6.2 Kernel Principal Component Analysis

Principal component analysis (PCA) is a well-known feature extraction method, in which the principal components of the input vector relative to the mean vector are extracted by orthogonal transformation. Similarly kernel PCA extracts principal components in the feature space [9, 7, 8]. We call them *kernel principal components*.

Unlike the approach discussed in [7], here we consider that the mean of the data is nonzero from the beginning. Consider extracting kernel principal components of a set of data $\{\mathbf{x}_1, \ldots, \mathbf{x}_M\}$. The covariance matrix $Q$ of the data in the feature space is calculated by

$$Q = \frac{1}{M} \sum_{i=1}^{M} (\boldsymbol{\phi}(\mathbf{x}_i) - \mathbf{c})(\boldsymbol{\phi}(\mathbf{x}_i) - \mathbf{c})^{\top}$$

$$= \frac{1}{M} \sum_{i=1}^{M} \boldsymbol{\phi}(\mathbf{x}_i)\,\boldsymbol{\phi}^{\top}(\mathbf{x}_i) - \mathbf{c}\,\mathbf{c}^{\top}, \tag{6.17}$$

where $\boldsymbol{\phi}(\mathbf{x})$ is the mapping function that maps $\mathbf{x}$ into the $l$-dimensional feature space and $\mathbf{c}$ is the mean vector of the mapped training data and is calculated by

$$\mathbf{c} = \frac{1}{M} \sum_{i=1}^{M} \boldsymbol{\phi}(\mathbf{x}_i)$$

$$= (\boldsymbol{\phi}(\mathbf{x}_1), \ldots, \boldsymbol{\phi}(\mathbf{x}_M)) \begin{pmatrix} \frac{1}{M} \\ \vdots \\ \frac{1}{M} \end{pmatrix}. \tag{6.18}$$

Substituting (6.18) into (6.17) and rewriting it in a matrix form, we obtain

$$Q = \frac{1}{M}(\boldsymbol{\phi}(\mathbf{x}_1), \ldots, \boldsymbol{\phi}(\mathbf{x}_M))\,(I_M - \mathbf{1}_M) \begin{pmatrix} \boldsymbol{\phi}^{\top}(\mathbf{x}_1) \\ \vdots \\ \boldsymbol{\phi}^{\top}(\mathbf{x}_M) \end{pmatrix}, \tag{6.19}$$

where $I_M$ is the $M \times M$ unit matrix and $\mathbf{1}_M$ is the $M \times M$ matrix with all elements being $1/M$.

Let $\lambda$ and $\mathbf{z}$ be the eigenvalue and the associated eigenvector of $Q$:

$$Q\mathbf{z} = \lambda\mathbf{z}. \tag{6.20}$$

Substituting (6.19) into (6.20),

$$\frac{1}{M}(\phi(\mathbf{x}_1), \ldots, \phi(\mathbf{x}_M))(I_M - \mathbf{1}_M) \begin{pmatrix} \phi^\top(\mathbf{x}_1)\mathbf{z} \\ \vdots \\ \phi^\top(\mathbf{x}_M)\mathbf{z} \end{pmatrix} = \lambda\mathbf{z}. \tag{6.21}$$

Thus $\mathbf{z}$ is expressed by a linear sum of $\{\phi(\mathbf{x}_1), \ldots, \phi(\mathbf{x}_M)\}$.

According to the formulation of [7], all the training data are used to represent $\mathbf{z}$. But by this method we need to retain all the training data after training. To overcome this problem, in [9, 8], sparsity is introduced into KPCA by imposing some restriction to the parameter range. Here, instead of restricting the parameter range, we select a set of linearly independent data from the training data as discussed in Section 6.1. By this method, we can retain only the selected data after training.

Without loss of generality, we can assume that a set of vectors, $\{\phi(\mathbf{x}_1), \ldots, \phi(\mathbf{x}_{M'})\}$ ($M' \leq M$), spans the space generated by $\{\phi(\mathbf{x}_1), \ldots, \phi(\mathbf{x}_M)\}$. Then $\mathbf{z}$ is expressed by

$$\mathbf{z} = (\phi(\mathbf{x}_1), \ldots, \phi(\mathbf{x}_{M'})) \begin{pmatrix} \rho_1 \\ \vdots \\ \rho_{M'} \end{pmatrix}, \tag{6.22}$$

where $\rho_1, \ldots, \rho_{M'}$ are scalars.

In (6.21) we are only interested in the components for $\phi(\mathbf{x}_i)$ ($i = 1, \ldots, M'$). Multiplying both terms of (6.21) by $\phi^\top(\mathbf{x}_i)$ from the left and substituting (6.22) into (6.21), we obtain

$$\frac{1}{M}(K_{i1}, \ldots, K_{iM})(I_M - \mathbf{1}_M)K\boldsymbol{\rho}' = \lambda(K_{i1}, \ldots, K_{iM'})\boldsymbol{\rho}', \tag{6.23}$$

where $K_{ij} = K(\mathbf{x}_i, \mathbf{x}_j) = \phi^\top(\mathbf{x}_i)\phi(\mathbf{x}_j)$, $\boldsymbol{\rho}' = (\rho_1, \ldots, \rho_{M'})^\top$, $K = \{K_{ij}\}$ ($i = 1, \ldots, M$, $j = 1, \ldots, M'$). Thus, combining (6.23) for $i = 1, \ldots, M'$,

$$\frac{1}{M}K^\top(I_M - \mathbf{1}_M)K\boldsymbol{\rho}' = \lambda K^s\boldsymbol{\rho}', \tag{6.24}$$

where $K^s = \{K_{ij}\}$ ($i = 1, \ldots, M'$, $j = 1, \ldots, M'$).[1]

---

[1] If $M' = M$, (6.24) is the same as (20.13) in [7].

Equation (6.24) has the form $A\mathbf{x} = \lambda B\mathbf{x}$, where $A$ and $B$ are $n \times n$ matrices. Solving the equation for $\lambda$ and $\mathbf{x}$ is called the *generalized eigenvalue problem* [26, pp. 375–376]. And if $B$ is nonsingular, there are $n$ eigenvalues. Because $\{\phi(\mathbf{x}_1), \ldots, \phi(\mathbf{x}_{M'})\}$ are linearly independent, $K^s$ is positive definite. Thus by the Cholesky factorization, $K^s = LL^\top$, where $L$ is a lower triangular matrix. Multiplying both terms of (6.24) by $L^{-1}$ from the left [27, pp. 462–463], we obtain

$$\frac{1}{M} L^{-1} K^\top (I_M - \mathbf{1}_M) K (L^{-1})^\top (L^\top \boldsymbol{\rho}') = \lambda (L^\top \boldsymbol{\rho}'). \qquad (6.25)$$

Here, $\lambda$ is the eigenvalue and the $L^\top \boldsymbol{\rho}'$ is the eigenvector. Because the coefficient matrix is a symmetric, positive semidefinite matrix, the eigenvalues are nonnegative. Thus solving the eigenvalue $\lambda$ and the eigenvector $L^\top \boldsymbol{\rho}'$ of (6.25), we obtain the generalized eigenvalue $\lambda$ and the eigenvector $\boldsymbol{\rho}'$ of (6.24).

Let $\lambda_1 \geq \lambda_2 \geq \cdots \geq \lambda_{M'}$ be the eigenvalues and $\mathbf{z}_1, \ldots, \mathbf{z}_{M'}$ be the associated eigenvectors for (6.20) and

$$\mathbf{z}_i = (\phi(\mathbf{x}_1), \ldots, \phi(\mathbf{x}_{M'})) \begin{pmatrix} \rho_{i1} \\ \vdots \\ \rho_{iM'} \end{pmatrix} \quad \text{for } i = 1, \ldots, M', \qquad (6.26)$$

where $\rho_{i1}, \ldots, \rho_{iM'}$ are scalars. We normalize $\mathbf{z}_i$ for $i = 1, \ldots, M'$, namely,

$$\mathbf{z}_i^\top \mathbf{z}_i = 1. \qquad (6.27)$$

Let the adjusted $\rho_{ij}$ be $\rho'_{ij}$.

For $\mathbf{x}$ we call

$$\mathbf{z}_i^\top (\phi(\mathbf{x}) - \mathbf{c}) = \sum_{j=1}^{M'} \rho'_{ij} K(\mathbf{x}_j, \mathbf{x}) - \mathbf{z}_i^\top \mathbf{c} \qquad (6.28)$$

the $i$th *kernel principal component* of $\mathbf{x}$. Here, $\mathbf{z}_i^\top \mathbf{c}$ is calculated by (6.18), (6.26), and (6.24). Thus, after training we can discard $(\mathbf{x}_{M'+1}, \ldots, \mathbf{x}_M)$.

The eigenvalue $\lambda_i$ is the variance in the $\mathbf{z}_i$ direction. The trace of $Q$ is defined as the sum of the diagonal elements of $Q$:

$$\text{tr}(Q) = \sum_{i=1}^{M'} Q_{ii}. \qquad (6.29)$$

Then $\text{tr}(Q) = \lambda_1 + \cdots + \lambda_{M'}$ [26, p. 310]. Thus the sum of the variances of $\mathbf{x}$ is the same as the sum of the variances of $\mathbf{z}$. Suppose we select the first $d$ principal components. We define the accumulation of $d$ eigenvalues as follows:

$$A_c(d) = \frac{\sum\limits_{i=1}^{d} \lambda_i}{\sum\limits_{i=1}^{M'} \lambda_i} \times 100\,(\%). \tag{6.30}$$

The accumulation of eigenvalues shows how well the reduced feature vector reflects the characteristics of the original feature vector in the feature space.

Kernel PCA can be used for feature extraction for pattern classification and noise filtering of image data called *denoising*. In denoising by kernel PCA we choose the principal components and discard the remaining components. Then we restore the image called the *preimage*. Because usually, inverse mapping does not exist, this is done by minimizing

$$\|\Phi - \phi(\mathbf{x})\|, \tag{6.31}$$

where $\Phi$ is the reduced image in the feature space and $\mathbf{x}$ is the preimage in the input space. Mika et al. [28, 29] proposed an iterative method for RBF kernels. To suppress the effect of outliers, Takahashi and Kurita [30] proposed modifying the principal components during iterations applying robust statistical techniques. Kwok and Tsang [31] derived a non-iterative algorithm to calculate the preimage based on the idea that the distances between the preimage and its neighbors are similar to those between the image and its neighbors in the feature space. This method is also applicable to preimages such as those of the kernel $k$-means clustering algorithm.

## 6.3 Kernel Mahalanobis Distance

The Mahalanobis distance for a class is a distance, from the center of the class, normalized by the inverse of the covariance matrix for that class. The Mahalanobis distance is linear transformation invariant and is widely used for pattern classification. In this section, we discuss two methods to calculate the kernel version of the Mahalanobis distance: (1) the kernel Mahalanobis distance calculated by singular value decomposition (SVD) [6] and (2) the kernel Mahalanobis distance calculated by KPCA. In the former method, all the training data are used for calculating the pseudo-inverse. But in the latter method, linearly independent vectors in the feature space are selected by Cholesky factorization. Thus, usually, the latter method takes less time in calculation.

## 6.3.1 SVD-Based Kernel Mahalanobis Distance

The kernel Mahalanobis distance $d_{\phi_i}(\mathbf{x})$ for class $i$ is given by

$$d^2_{\phi_i}(\mathbf{x}) = (\phi(\mathbf{x}) - \mathbf{c}_i)^\top Q_{\phi_i}^{-1}(\phi(\mathbf{x}) - \mathbf{c}_i), \qquad (6.32)$$

where $\phi(\mathbf{x})$ is the mapping function from the input space to the $l$-dimensional feature space and $\mathbf{c}_i$ is the center of class $i$ in the feature space:

$$\mathbf{c}_i = \frac{1}{M_i} \sum_{j=1}^{M_i} \phi(\mathbf{x}_{ij}). \qquad (6.33)$$

Here $\mathbf{x}_{ij} = (x_{ij1} \cdots x_{ijm})^\top$ is the $j$th training data sample for class $i$ and $M_i$ is the number of training data for class $i$.

The covariance matrix for class $i$ in the feature space, $Q_{\phi_i}$, is expressed in a matrix form as follows:

$$\begin{aligned}
Q_{\phi_i} &= \frac{1}{M_i} \sum_{j=1}^{M_i} (\phi(\mathbf{x}_{ij}) - \mathbf{c}_i)(\phi(\mathbf{x}_{ij}) - \mathbf{c}_i)^\top \\
&= \phi^\top(X_i)\left(\frac{1}{M_i}(I_{M_i} - \mathbf{1}_{M_i})\right)\phi(X_i), \qquad (6.34)
\end{aligned}$$

where the second equation is derived by substituting (6.33) into the right-hand side of the first equation, $I_{M_i}$ is the $M_i \times M_i$ unit matrix, $\mathbf{1}_{M_i}$ is the $M_i \times M_i$ matrix with all the components equal to $1/M_i$, and $\phi(X_i)$ is an $M_i \times l$ matrix:

$$\phi(X_i) = \begin{pmatrix} \phi^\top(\mathbf{x}_{i1}) \\ \vdots \\ \phi^\top(\mathbf{x}_{iM_i}) \end{pmatrix}. \qquad (6.35)$$

To calculate the kernel Mahalanobis distance given by (6.32) without explicitly treating the variables in the feature space, we need to use the kernel trick, namely, we transform (6.32) so that only the dot products $\phi^\top(\mathbf{x})\,\phi(\mathbf{x})$ appear in (6.32) [6].

Because $I_{M_i} - \mathbf{1}_{M_i}$ is a symmetric, positive semidefinite matrix, we can define the square root of the matrix, $Z_i$, by

$$Z_i = \left(\frac{1}{M_i}(I_{M_i} - \mathbf{1}_{M_i})\right)^{\frac{1}{2}}. \qquad (6.36)$$

Substituting (6.36) into (6.34), we obtain

$$Q_{\phi_i} = \phi^\top(X_i)\, Z_i^2\, \phi(X_i). \qquad (6.37)$$

Substituting (6.37) into (6.32) gives

$$d^2_{\phi_i}(\mathbf{x}) = (\phi(\mathbf{x}) - \mathbf{c}_i)^\top \left( \phi^\top(X_i) Z_i^2 \, \phi(X_i) \right)^{-1} (\phi(\mathbf{x}) - \mathbf{c}_i). \qquad (6.38)$$

Now the following equation is valid for any integer $n$, symmetric, positive semidefinite matrix $A$, and any vectors $\mathbf{t}$ and $\mathbf{u}$ [6]:

$$\mathbf{t}^\top (X^\top A X)^n \, \mathbf{u} = \mathbf{t}^\top X^\top (A^{\frac{1}{2}} (A^{\frac{1}{2}} K A^{\frac{1}{2}})^{n-1} A^{\frac{1}{2}}) X \, \mathbf{u}, \qquad (6.39)$$

where $K = X X^\top$. If $n$ is negative, it means pseudo-inverse. We calculate the pseudo-inverse using the singular value decomposition.

Here because $Z_i^2$ in (6.38) is a symmetric, positive semidefinite matrix, we can apply (6.39) to (6.38):

$$d^2_{\phi_i}(\mathbf{x}) = (\phi(X_i)\,\phi(\mathbf{x}) - \phi(X_i)\,\mathbf{c}_i)^\top \left( Z_i \left( Z_i \, \phi(X_i)\,\phi^\top(X_i)\,Z_i \right)^{-2} Z_i \right)$$
$$\times (\phi(X_i)\,\phi(\mathbf{x}) - \phi(X_i)\,\mathbf{c}_i). \qquad (6.40)$$

Because (6.40) consists of only dot products in the feature space, we can replace them with kernels:

$$K(\mathbf{x}, \mathbf{y}) = \phi^\top(\mathbf{x})\,\phi(\mathbf{y}). \qquad (6.41)$$

Using (6.41), we can rewrite the dot products in (6.40) as follows:

$$\phi(X_i)\,\phi(\mathbf{x}) = \begin{pmatrix} \phi^\top(\mathbf{x}_{i1})\,\phi(\mathbf{x}) \\ \vdots \\ \phi^\top(\mathbf{x}_{iM_i})\,\phi(\mathbf{x}) \end{pmatrix}$$

$$= \begin{pmatrix} K(\mathbf{x}_{i1}, \mathbf{x}) \\ \vdots \\ K(\mathbf{x}_{iM_i}, \mathbf{x}) \end{pmatrix} = K(X_i, \mathbf{x}), \qquad (6.42)$$

$$\phi(X_i)\,\mathbf{c}_i = \frac{1}{M_i}\,\phi(X_i) \sum_{j=1}^{M_i} \phi(\mathbf{x}_{ij})$$

$$= \frac{1}{M_i} \sum_{j=1}^{M_i} K(X_i, \mathbf{x}_{ij}), \qquad (6.43)$$

$$\phi(X_i)\,\phi^\top(X_i) = \begin{pmatrix} \phi^\top(\mathbf{x}_{i1}) \\ \vdots \\ \phi^\top(\mathbf{x}_{iM_i}) \end{pmatrix} \left( \phi(\mathbf{x}_{i1}) \cdots \phi(\mathbf{x}_{iM_i}) \right)$$

$$= \begin{pmatrix} K(\mathbf{x}_{i1}, \mathbf{x}_{i1}) & \cdots & K(\mathbf{x}_{i1}, \mathbf{x}_{iM_i}) \\ \vdots & \ddots & \vdots \\ K(\mathbf{x}_{iM_i}, \mathbf{x}_{i1}) & \cdots & K(\mathbf{x}_{iM_i}, \mathbf{x}_{iM_i}) \end{pmatrix}$$

$$= K(X_i, X_i^\top). \tag{6.44}$$

Substituting (6.42), (6.43), and (6.44) into (6.40), we obtain

$$d_{\phi_i}^2(\mathbf{x}) = \left( K(X_i, \mathbf{x}) - \frac{1}{M_i} \sum_{j=1}^{M_i} K(X_i, \mathbf{x}_{ij}) \right)^\top \left( Z_i \left( Z_i \, K(X_i, X_i^\top) \, Z_i \right)^{-2} Z_i \right)$$

$$\times \left( K(X_i, \mathbf{x}) - \frac{1}{M_i} \sum_{j=1}^{M_i} K(X_i, \mathbf{x}_{ij}) \right). \tag{6.45}$$

Using (6.45) we can calculate the kernel Mahalanobis distance without treating variables in the feature space. But $(Z_i \, K(X_i, X_i^\top) \, Z_i)^{-2}$ needs to be calculated by singular value decomposition. In the following, we discuss the singular value decomposition and its variant to improve generalization ability when the number of data is small.

Any matrix $A$ is decomposed into $A = S \Lambda U^\top$ by singular value decomposition, where $S$ and $U$ are orthogonal matrices ($S \, S^\top = I, U \, U^\top = I$, where $I$ is a unit matrix) and $\Lambda$ is a diagonal matrix. If $A$ is an $m \times m$ positive semidefinite matrix, the singular value decomposition is equivalent to the diagonalization of the matrix, namely, $S$, $\Lambda$, $U$ are $m \times m$ square matrices and $S = U$. Because $Z_i \, K(X_i, X_i^\top) \, Z_i$ is a symmetric, positive semidefinite matrix, in the following discussion we assume that $A$ is symmetric and positive semidefinite.

If $A$ is positive definite, the inverse of $A$ is expressed as follows:

$$A^{-1} = (U \, \Lambda U^\top)^{-1} = (U^\top)^{-1} \Lambda^{-1} \, U^{-1}$$
$$= U \Lambda^{-1} U^\top. \tag{6.46}$$

Assume that $A$ is positive semidefinite with rank $r$ ($m > r$). In this case the pseudo-inverse of $A$, $A^+$, is used. In the following we discuss two methods to calculate the pseudo-inverse: the conventional and improved methods.

### 6.3.1.1 Conventional Method

Because $m > r$ holds, the $(r + 1)$th to $m$th diagonal elements of $\Lambda$ are zero. In the conventional pseudo-inverse, if a diagonal element $\lambda_i$ is larger than or equal to $\sigma$, where $\sigma$ is a predefined threshold, we set $1/\lambda_i$ to the $i$th element of $\Lambda^+$. But if it is smaller, we set 0.

### 6.3.1.2 Improved Method

In the conventional method, if a diagonal element is smaller than $\sigma$, the associated diagonal element of the pseudo-inverse is set to 0. This means that all the components with small singular values are neglected, namely, the subspace corresponding to zero diagonal elements is neglected. This leads to decreasing the approximation ability of the subspace. To avoid this we set $1/\sigma$ instead of 0, namely, we calculate the pseudo-inverse as follows:

$$
A^+ = U\,\Lambda^+\,U^\top
$$

$$
= U \begin{pmatrix} \lambda_1^{-1} & & & & & \\ & \ddots & & & & \\ & & \lambda_r^{-1} & & & \\ & & & \dfrac{1}{\sigma} & & \\ & & & & \ddots & \\ & & & & & \dfrac{1}{\sigma} \end{pmatrix} U^\top. \tag{6.47}
$$

## 6.3.2 KPCA-Based Mahalanobis Distance

In this section we discuss a KPCA-based method to calculate a Mahalanobis distance in the feature space. This method is equivalent to the kernel Mahalanobis distance discussed in Section 6.3.1. But in this method because we can discard the redundant input vectors, calculation time can be shortened compared with that of the conventional kernel Mahalanobis distance given by (6.45).

Let $\lambda_j$ and $\mathbf{z}_j$ be the $j$th eigenvalue and the associated eigenvector of the covariance matrix $Q_{\phi_i}$ given by (6.34). Then

$$
Q_{\phi_i}\mathbf{z}_j = \lambda_j\,\mathbf{z}_j \quad \text{for } j = 1,\ldots,l, \tag{6.48}
$$

where $l$ is the dimension of the feature space. Then for the pseudo-inverse of $Q_{\phi_i}$, $Q_{\phi_i}^+$, the following equation holds:

$$
Q_{\phi_i}^+\,\mathbf{z}_j = \begin{cases} \lambda_j^{-1}\,\mathbf{z}_j & \lambda_j > 0, \\ 0 & \lambda_j = 0. \end{cases} \tag{6.49}
$$

Then the $j$th kernel principal component of input $\mathbf{x}$ is defined by

$$
y_j = \begin{cases} \mathbf{z}_j^\top\,(\phi(\mathbf{x}) - \mathbf{c}_i) & \lambda_j > 0, \\ 0 & \lambda_j = 0. \end{cases} \tag{6.50}
$$

From (6.50), the Mahalanobis distance in that space can be calculated as follows:

$$d_{\phi_i}^2(\mathbf{x}) = \frac{y_1^2}{\lambda_1} + \cdots + \frac{y_{M_i'}^2}{\lambda_{M_i'}}, \tag{6.51}$$

where $M_i'$ is the number of nonzero eigenvalues.

To calculate the component given by (6.50) efficiently, we use the KPCA discussed in Section 6.2.

# 6.4 Principal Component Analysis in the Empirical Feature Space

If we use the empirical feature space instead of the feature space, we can perform principal component analysis in the similar way as in the input space.

For the $M$ $m$-dimensional data $\mathbf{x}_i$, the $M \times M$ kernel matrix be $K = \{K_{ij}\}$ ($i, j = 1, \ldots M$), where $K_{ij} = K(\mathbf{x}_i, \mathbf{x}_j)$. Let the rank of $K$ be $N (\leq M)$. Then as discussed in Section 2.3.6.1 the mapping function that maps the input space to the $N$-dimensional empirical feature space is given by

$$\mathbf{h}(\mathbf{x}) = \Lambda^{-1/2} P^\top \left( K(\mathbf{x}_1, \mathbf{x}), \ldots, K(\mathbf{x}_M, \mathbf{x}) \right)^\top, \tag{6.52}$$

where $\Lambda$ is the $N \times N$ diagonal matrix whose diagonal elements are the nonzero, i.e., positive eigenvalues of $K$ and $P$ is the $M \times N$ matrix whose $i$th column is the eigenvector associated with the $i$th diagonal element of $\Lambda$.

As discussed in Section 4.3.2 to speed up generating the empirical feature space we may select linearly independent training data that span the empirical feature space. Let the $N (\leq M)$ training data $\mathbf{x}_{i_1}, \ldots, \mathbf{x}_{i_N}$ be linearly independent in the empirical feature space, where $\mathbf{x}_{i_j} \in \{\mathbf{x}_1, \ldots, \mathbf{x}_M\}$ and $j = 1, \ldots, N$. Then, we use the following mapping function:

$$\mathbf{h}(\mathbf{x}) = \left( K(\mathbf{x}_{i_1}, \mathbf{x}), \ldots, K(\mathbf{x}_{i_N}, \mathbf{x}) \right)^\top. \tag{6.53}$$

The above mapping function maps the input space into the empirical feature space with different coordinates.

The covariance matrix $Q_e$ of the data in the empirical feature space is calculated by

$$Q_e = \frac{1}{M} \sum_{i=1}^{M} (\mathbf{h}(\mathbf{x}_i) - \mathbf{c})(\mathbf{h}(\mathbf{x}_i) - \mathbf{c})^\top$$

$$= \frac{1}{M} \sum_{i=1}^{M} \mathbf{h}(\mathbf{x}_i)\, \mathbf{h}^\top(\mathbf{x}_i) - \mathbf{c}\,\mathbf{c}^\top. \tag{6.54}$$

where $\mathbf{c}$ is the mean vector of the mapped training data and is calculated by

$$\mathbf{c} = \frac{1}{M} \sum_{i=1}^{M} \mathbf{h}(\mathbf{x}_i). \tag{6.55}$$

Let $\lambda$ and $\mathbf{z}$ be the eigenvalue and the associated eigenvector of $Q_e$:

$$Q_e \, \mathbf{z} = \lambda \, \mathbf{z}. \tag{6.56}$$

Then calculating the eigenvalues and eigenvectors of (6.56), we carry our principal component analysis.

Likewise, the kernel Mahalanobis distance $d_{\mathbf{h}_i}(\mathbf{x})$ for class $i$ in the empirical feature space is given by

$$d_{\mathbf{h}_i}^2(\mathbf{x}) = (\mathbf{h}(\mathbf{x}) - \mathbf{c}_i)^\top Q_{\mathbf{h}_i}^{-1}(\mathbf{h}(\mathbf{x}) - \mathbf{c}_i), \tag{6.57}$$

where $\mathbf{c}_i$ is the center of class $i$ in the empirical feature space:

$$\mathbf{c}_i = \frac{1}{M_i} \sum_{j=1}^{M_i} \mathbf{h}(\mathbf{x}_{ij}). \tag{6.58}$$

Here $\mathbf{x}_{ij} = (x_{ij1} \cdots x_{ijm})^\top$ is the $j$th training sample for class $i$ and $M_i$ is the number of training data for class $i$.

The covariance matrix for class $i$ in the empirical feature space, $Q_{\mathbf{h}_i}$, is expressed in a matrix form as follows:

$$Q_{\mathbf{h}_i} = \frac{1}{M_i} \sum_{j=1}^{M_i} (\mathbf{h}(\mathbf{x}_{ij}) - \mathbf{c}_i)(\mathbf{h}(\mathbf{x}_{ij}) - \mathbf{c}_i)^\top. \tag{6.59}$$

We must notice that we use the same mapping function $\mathbf{h}(\mathbf{x})$ for different classes to avoid degeneracy of the Mahalanobis distance which will be discussed in Section 10.1.3. If the covariance matrix $Q_{\mathbf{h}_i}$ is singular, we add a small positive value to the diagonal elements to avoid degeneracy.

Principal component analysis and the Mahalanobis distance in the empirical feature space is much easier to calculate than those in the feature space.

## 6.5 Kernel Discriminant Analysis

Principal component analysis does not use class information. Thus, the first principal component is not necessarily useful for class separation. On the other hand, linear discriminant analysis, defined for a two-class problem, finds the line that maximally separates two classes in the sense that the training data of different classes projected on the line are maximally separated [32, pp. 118–121]. Figure 6.1 shows an example of linear discriminant analysis in the two-dimensional input space. As seen from the figure, this method works well when each class consists of one cluster and clusters are separated.

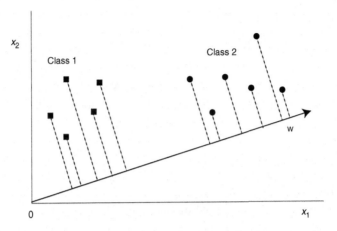

**Fig. 6.1** Linear discriminant analysis in a two-dimensional space

Kernel discriminant analysis is extended in the case where each class consists of more than one cluster or clusters overlap to find the line that maximally separates the training data of different classes projected on the line in the feature space [33, 34], [9, pp. 457–468]. It is extended to multiclass problems, and there are also variants [19, 35–37].

In the following, we discuss kernel discriminant analysis for two-class problems and multiclass problems.

### 6.5.1 Kernel Discriminant Analysis for Two-Class Problems

Let the sets of $m$-dimensional data belonging to Class $i$ ($i = 1, 2$) be $\{\mathbf{x}_{i1}, \ldots, \mathbf{x}_{iM_i}\}$, where $M_i$ is the number of data belonging to Class $i$, and data $\mathbf{x}$ be mapped into the $l$-dimensional feature space by the mapping function $\phi(\mathbf{x})$. Now we find the $l$-dimensional vector $\mathbf{w}$ in which the training data of two classes projected on $\mathbf{w}$ are separated maximally in the feature space. We call $\mathbf{w}$ the *projection axis*.

The projection of $\phi(\mathbf{x})$ on $\mathbf{w}$ is $\mathbf{w}^\top \phi(\mathbf{x})/\|\mathbf{w}\|$. In the following we assume that $\|\mathbf{w}\| = 1$, but this is not necessary. We find such $\mathbf{w}$ that maximizes the difference of the centers and minimizes the variances of the projected data.

The square difference of the centers of the projected data, $d^2$, is

$$d^2 = (\mathbf{w}^\top (\mathbf{c}_1 - \mathbf{c}_2))^2$$
$$= \mathbf{w}^\top (\mathbf{c}_1 - \mathbf{c}_2)(\mathbf{c}_1 - \mathbf{c}_2)^\top \mathbf{w}, \tag{6.60}$$

where $\mathbf{c}_i$ are the centers of class $i$ data:

$$\mathbf{c}_i = \frac{1}{M_i} \sum_{j=1}^{M_i} \phi(\mathbf{x}_{ij})$$

$$= (\phi(\mathbf{x}_{i1}), \dots, \phi(\mathbf{x}_{iM_i})) \begin{pmatrix} \frac{1}{M_i} \\ \vdots \\ \frac{1}{M_i} \end{pmatrix} \quad \text{for } i = 1, 2. \tag{6.61}$$

We define

$$Q_{\mathrm{B}} = (\mathbf{c}_1 - \mathbf{c}_2)(\mathbf{c}_1 - \mathbf{c}_2)^\top \tag{6.62}$$

and call $Q_{\mathrm{B}}$ the *between-class scatter matrix*.

The variances of the projected data, $s_i^2$, are

$$s_i^2 = \mathbf{w}^\top Q_i \mathbf{w} \quad \text{for } i = 1, 2, \tag{6.63}$$

where

$$Q_i = \frac{1}{M_i} \sum_{j=1}^{M_i} (\phi(\mathbf{x}_{ij}) - \mathbf{c}_i)(\phi(\mathbf{x}_{ij}) - \mathbf{c}_i)^\top$$

$$= \frac{1}{M_i} \sum_{j=1}^{M} \phi(\mathbf{x}_{ij})\phi(\mathbf{x}_{ij})^\top - \mathbf{c}_i \mathbf{c}_i^\top$$

$$= \frac{1}{M_i} (\phi(\mathbf{x}_{i1}), \dots, \phi(\mathbf{x}_{iM_i}))(I_{M_i} - \mathbf{1}_{M_i}) \begin{pmatrix} \phi^\top(\mathbf{x}_{i1}) \\ \vdots \\ \phi^\top(\mathbf{x}_{iM_i}) \end{pmatrix}$$

$$\text{for } i = 1, 2. \tag{6.64}$$

Here, $I_{M_i}$ is the $M_i \times M_i$ unit matrix and $\mathbf{1}_{M_i}$ is the $M_i \times M_i$ matrix with all elements being $1/M_i$. We define

$$Q_{\mathrm{W}} = Q_1 + Q_2 \tag{6.65}$$

and call $Q_{\mathrm{W}}$ the *within-class scatter matrix*.

Now, we want to maximize

$$J(\mathbf{w}) = \frac{d^2}{s_1^2 + s_2^2}$$

$$= \frac{\mathbf{w}^\top Q_{\mathrm{B}}\,\mathbf{w}}{\mathbf{w}^\top Q_{\mathrm{W}}\,\mathbf{w}}, \tag{6.66}$$

but because $\mathbf{w}$, $Q_{\mathrm{B}}$, and $Q_{\mathrm{W}}$ are defined in the feature space, we need to use kernel tricks. Assume that a set of $M'$ vectors $\{\phi(\mathbf{y}_1), \ldots, \phi(\mathbf{y}_{M'})\}$ spans the space generated by $\{\phi(\mathbf{x}_{11}), \ldots, \phi(\mathbf{x}_{1M_1}), \phi(\mathbf{x}_{21}), \ldots, \phi(\mathbf{x}_{2M_2})\}$, where $\{\mathbf{y}_1, \ldots, \mathbf{y}_{M'}\} \subset \{\mathbf{x}_{11}, \ldots, \mathbf{x}_{1M_1}, \mathbf{x}_{21}, \ldots, \mathbf{x}_{2M_2}\}$ and $M' \leq M_1 + M_2$. Then $\mathbf{w}$ is expressed as

$$\mathbf{w} = (\phi(\mathbf{y}_1), \ldots, \phi(\mathbf{y}_{M'}))\,\boldsymbol{\alpha}, \tag{6.67}$$

where $\boldsymbol{\alpha} = (\alpha_1, \ldots, \alpha_{M'})^\top$ and $\alpha_1, \ldots, \alpha_{M'}$ are scalars. Substituting (6.67) into (6.66), we obtain

$$J(\boldsymbol{\alpha}) = \frac{\boldsymbol{\alpha}^\top K_{\mathrm{B}}\,\boldsymbol{\alpha}}{\boldsymbol{\alpha}^\top K_{\mathrm{W}}\,\boldsymbol{\alpha}}, \tag{6.68}$$

where

$$K_{\mathrm{B}} = (\mathbf{k}_{\mathrm{B}_1} - \mathbf{k}_{\mathrm{B}_2})\,(\mathbf{k}_{\mathrm{B}_1} - \mathbf{k}_{\mathrm{B}_2})^\top, \tag{6.69}$$

$$\mathbf{k}_{Bi} = \begin{pmatrix} \dfrac{1}{M_i} \sum_{j=1}^{M_i} K(\mathbf{y}_1, \mathbf{x}_{ij}) \\ \cdots \\ \dfrac{1}{M_i} \sum_{j=1}^{M_i} K(\mathbf{y}_{M'}, \mathbf{x}_{ij}) \end{pmatrix} \quad \text{for } i = 1, 2, \tag{6.70}$$

$$K_{\mathrm{W}} = K_{\mathrm{W}_1} + K_{\mathrm{W}_2}, \tag{6.71}$$

$$K_{\mathrm{W}_i} = \frac{1}{M_i} \begin{pmatrix} K(\mathbf{y}_1, \mathbf{x}_{i1}) \cdots K(\mathbf{y}_1, \mathbf{x}_{iM_i}) \\ \cdots \\ K(\mathbf{y}_{M'}, \mathbf{x}_{i1}) \cdots K(\mathbf{y}_{M'}, \mathbf{x}_{iM_i}) \end{pmatrix} (I_{M_i} - \mathbf{1}_{M_i})$$

$$\times \begin{pmatrix} K(\mathbf{y}_1, \mathbf{x}_{i1}) \cdots K(\mathbf{y}_1, \mathbf{x}_{iM_i}) \\ \cdots \\ K(\mathbf{y}_{M'}, \mathbf{x}_{i1}) \cdots K(\mathbf{y}_{M'}, \mathbf{x}_{iM_i}) \end{pmatrix}^\top \quad \text{for } i = 1, 2. \tag{6.72}$$

Taking the partial derivative of (6.68) with respect to $\mathbf{w}$ and equating the resulting equation to zero, we obtain the following generalized eigenvalue problem:

$$K_{\mathrm{B}}\,\boldsymbol{\alpha} = \lambda\,K_{\mathrm{W}}\,\boldsymbol{\alpha}, \tag{6.73}$$

where $\lambda$ is a generalized eigenvalue.

Substituting

$$K_{\mathrm{W}}\,\boldsymbol{\alpha} = \mathbf{k}_{\mathrm{B}_1} - \mathbf{k}_{\mathrm{B}_2} \tag{6.74}$$

into the left-hand side of (6.73), we obtain

$$(\boldsymbol{\alpha}^\top K_{\mathrm{W}}\,\boldsymbol{\alpha})\,K_{\mathrm{W}}\,\boldsymbol{\alpha}. \tag{6.75}$$

Thus, by letting $\lambda = \alpha^\top K_W \alpha$, (6.74) is a solution of (6.73).

Because $K_{W_1}$ and $K_{W_2}$ are positive semidefinite, $K_W$ is positive semidefinite. If $K_W$ is positive definite, $\alpha$ is given by

$$\alpha = K_W^{-1}(k_{B_1} - k_{B_2}). \tag{6.76}$$

Even if we choose linearly independent vectors $y_1, \ldots, y_{M'}$, for nonlinear kernels, $K_W$ may be positive semidefinite, i.e., singular. One way to overcome singularity is to add positive values to the diagonal elements [33]:

$$\alpha = (K_W + \varepsilon I)^{-1}(k_{B_1} - k_{B_2}), \tag{6.77}$$

where $\varepsilon$ is a small positive parameter. Another way is to use the pseudo-inverse:

$$\alpha = K_W^+(k_{B_1} - k_{B_2}), \tag{6.78}$$

which gives the solution in a least-squares sense.

In (6.66), instead of the within-class scatter matrix $Q_W$, we can use the *total-scatter matrix* $Q_T$:

$$Q_T = \frac{1}{M} \sum_{j=1}^M (\phi(x_j) - c)(\phi(x_j) - c)^\top, \tag{6.79}$$

where $M = M_1 + M_2$, $\{x_1, \ldots, x_M\} = \{x_{11}, \ldots, x_{1M_1}, x_{21}, \ldots, x_{2M_2}\}$, and

$$c = \frac{1}{M} \sum_{i=1}^M \phi(x_i) = \frac{M_1 c_1 + M_2 c_2}{M_1 + M_2}. \tag{6.80}$$

## 6.5.2 Linear Discriminant Analysis for Two-Class Problems in the Empirical Feature Space

Using the idea of empirical feature space, we can define linear discriminant analysis in the empirical feature space, which is equivalent to kernel discriminant analysis. In the following we discuss linear discriminant analysis using the total scatter matrix.

Let $h(x)$ be the mapping function that maps the input space into the $N$-dimensional empirical feature space and is given either by (6.52) or by (6.53).

The objective function of linear discriminant analysis in the empirical feature space is given by

$$J(w) = \frac{w^\top Q_B w}{w^\top Q_T w}, \tag{6.81}$$

where

$$Q_B = (c_1 - c_2)(c_1 - c_2)^\top \tag{6.82}$$

$$c_i = \frac{1}{M_i} \sum_{j=1}^{M_i} h(x_{ij}) \quad \text{for } i = 1, 2, \tag{6.83}$$

$$Q_T = \frac{1}{M} \sum_{j=1}^{M} h(x_j) h^\top(x_j) - c\, c^\top, \tag{6.84}$$

$$c = \frac{1}{M} \sum_{j=1}^{M} h(x_j) = \frac{M_1 c_1 + M_2 c_2}{M_1 + M_2}. \tag{6.85}$$

Taking the partial derivative of (6.81) with respect to $w$ and equating the resulting equation to zero, we obtain the following generalized eigenvalue problem:

$$Q_B\, w = \lambda\, Q_W\, w, \tag{6.86}$$

where $\lambda$ is a generalized eigenvalue.

Substituting

$$Q_T\, w = c_1 - c_2 \tag{6.87}$$

into the left-hand side of (6.86), we obtain $(w^\top Q_T\, w) Q_T\, w = \lambda\, Q_T\, w$. Thus, by letting $\lambda = w^\top Q_T\, w$, (6.87) is a solution of (6.86).

Because $Q_T$ is positive definite, the optimum $w$, $w_{\text{opt}}$, is given by

$$w_{\text{opt}} = Q_T^{-1}(c_1 - c_2). \tag{6.88}$$

Substituting (6.88) into (6.81), we obtain

$$J(w_{\text{opt}}) = (c_1 - c_2)^\top w_{\text{opt}}. \tag{6.89}$$

Linear discriminant analysis in the empirical feature space discussed above is equivalent to kernel discriminant analysis in the feature space if (6.52) is used. Because we can explicitly treat the variables in the empirical feature space, the calculation is much simpler.

### 6.5.3 Kernel Discriminant Analysis for Multiclass Problems

In this section, we explain kernel discriminant analysis for multiclass problems based on [34].

We assume that the center of training data in the feature space is zero. Then the total scatter matrix $Q_T$ and the between-class scatter matrix $Q_B$ are given, respectively, by

$$Q_{\mathrm{T}} = \frac{1}{M} \sum_{k=1}^{n} \sum_{j=1}^{M_k} \phi(\mathbf{x}_{kj}) \phi^{\top}(\mathbf{x}_{kj}), \tag{6.90}$$

$$Q_{\mathrm{B}} = \frac{1}{M} \sum_{k=1}^{n} M_k \mathbf{c}_k \mathbf{c}_k^{\top}, \tag{6.91}$$

where $n$ is the number of classes and notations in Section 6.5.1 are extended to $n$ classes, e.g., $M = M_1 + \cdots + M_n$.

For $n$ class problems, we obtain $n - 1$ projection axes. Let them be $\mathbf{w}_i$ $(i = 1, \ldots, n - 1)$. Then the total scatter and the between-class scatter on this axis are given, respectively, by

$$\frac{1}{M} \sum_{k=1}^{n} \sum_{j=1}^{M_k} (\mathbf{w}_i^{\top} \phi(\mathbf{x}_{kj}))^2 = \mathbf{w}_i^{\top} Q_{\mathrm{T}} \mathbf{w}_i, \tag{6.92}$$

$$\frac{1}{M} \sum_{k=1}^{n} M_k (\mathbf{w}_i^{\top} \mathbf{c}_k)^2 = \mathbf{w}_i^{\top} Q_{\mathrm{B}} \mathbf{w}_i. \tag{6.93}$$

We seek the projection axis $\mathbf{w}_i$ that maximizes the between-class scatter and minimizes the total scatter, namely,

$$\text{maximize} \quad J(\mathbf{w}_i) = \frac{\mathbf{w}_i^{\top} Q_{\mathrm{B}} \mathbf{w}_i}{\mathbf{w}_i^{\top} Q_{\mathrm{T}} \mathbf{w}_i}. \tag{6.94}$$

Here, $\mathbf{w}_i$ can be expressed by the linear combination of the mapped training data:

$$\mathbf{w}_i = \sum_{k=1}^{n} \sum_{j=1}^{M_k} a_i^{kj} \phi(\mathbf{x}_{kj}), \tag{6.95}$$

where $a_i^{kj}$ are constants.

Substituting (6.95) into (6.94), we obtain

$$J(\mathbf{a}_i) = \frac{\mathbf{a}_i^{\top} K W K \mathbf{a}_i}{\mathbf{a}_i^{\top} K K \mathbf{a}_i}, \tag{6.96}$$

where $\mathbf{a}_i = \{a_i^{kj}\}$ $(i = 1, ..., n - 1, k = 1, ..., n, j = 1, ..., M_k)$, $K$ is the kernel matrix, and $W = \{W_{ij}\}$ is a block diagonal matrix given by

$$W_{ij} = \begin{cases} \dfrac{1}{M_k} & \mathbf{x}_i, \mathbf{x}_j \in \text{class } k, \\[2mm] 0 & \text{otherwise.} \end{cases} \tag{6.97}$$

Taking the partial derivative of (6.96) with respect to $\mathbf{w}_i$, and the resulting equation to 0, we obtain the following generalized eigenvalue problem:

$$\lambda_i KK\mathbf{a}_i = KWK\mathbf{a}_i, \qquad (6.98)$$

where $\lambda_i$ are eigenvalues.

Let singular value decomposition of $K$ be $K = P\Gamma P^\top$, where $\Gamma$ is the diagonal matrix with nonzero eigenvalues and $P^\top P = I$. Substituting $K = P\Gamma P^\top$ into (6.96) and replacing $\Gamma P^\top \mathbf{a}_i$ with $\beta_i$, we obtain

$$J(\beta_i) = \frac{\beta_i^\top P^\top WP\beta_i}{\beta_i^\top P^\top P\beta_i} = \frac{\beta_i^\top P^\top WP\beta_i}{\beta_i^\top \beta_i}. \qquad (6.99)$$

Therefore, the resulting eigenvalue problem is

$$P^\top WP\beta_i = \lambda_i \beta_i. \qquad (6.100)$$

Solving (6.100) for $\beta_i$ we obtain $\mathbf{a}_i$ from $\mathbf{a}_i = P\Gamma^{-1}\beta_i$.

If we use the empirical feature space, kernel discriminant analysis for multiclass problems can be replaced by linear discriminant analysis in the empirical feature space.

# References

1. Y. Freund and R. E. Schapire. Large margin classification using the perceptron algorithm. *Machine Learning*, 37(3):277–296, 1999.
2. C. Gentile. A new approximate maximal margin classification algorithm. *Journal of Machine Learning Research*, 2:213–242, 2001.
3. Y. Li and P. M. Long. The relaxed online maximum margin algorithm. *Machine Learning*, 46(1–3):361–387, 2002.
4. K. Crammer and Y. Singer. Ultraconservative online algorithms for multiclass problems. *Journal of Machine Learning Research*, 3:951–991, 2003.
5. G. Baudat and F. Anouar. Kernel-based methods and function approximation. In *Proceedings of International Joint Conference on Neural Networks (IJCNN '01)*, volume 2, pages 1244–1249, Washington, DC, 2001.
6. A. Ruiz and P. E. López-de-Teruel. Nonlinear kernel-based statistical pattern analysis. *IEEE Transactions on Neural Networks*, 12(1):16–32, 2001.
7. B. Schölkopf, A. J. Smola, and K.-R. Müller. Kernel principal component analysis. In B. Schölkopf, C. J. C. Burges, and A. J. Smola, editors, *Advances in Kernel Methods: Support Vector Learning*, pages 327–352. MIT Press, Cambridge, MA, 1999.
8. A. J. Smola, O. L. Mangasarian, and B. Schölkopf. Sparse kernel feature analysis. Technical Report 99-04, University of Wisconsin, Data Mining Institute, Madison, WI, 1999.
9. B. Schölkopf and A. J. Smola. *Learning with Kernels: Support Vector Machines, Regularization, Optimization, and Beyond*. MIT Press, Cambridge, MA, 2002.
10. Z. Rong and A. I. Rudnicky. A large scale clustering scheme for kernel k-means. In *Proceedings of the Sixteenth International Conference on Pattern Recognition (ICPR 2002)*, volume 4, pages 289–292, 2002.
11. I. S. Dhillon, Y. Guan, and B. Kulis. Kernel k-means, spectral clustering and normalized cuts. In *KDD-2004: Proceedings of the Tenth ACM SIGKDD International Conference on Knowledge Discovery and Data Mining, Seattle, WA*, pages 551–556. Association for Computing Machinery, New York, 2004.

12. Z. Li, S. Tang, J. Xue, and J. Jiang. Modified FCM clustering based on kernel mapping. In J. Shen, S. Pankanti, and R. Wang, editors, *Proceedings of SPIE: Object Detection, Classification, and Tracking Technologies*, volume 4554, pages 241–245, Wuhan, China, 2001.

13. D.-Q. Zhang and S.-C. Chen. Clustering incomplete data using kernel-based fuzzy c-means algorithm. *Neural Processing Letters*, 18(3):155–162, 2003.

14. T. Graepel, M. Burger, and K. Obermayer. Self-organizing maps: Generalizations and new optimization techniques. *Neurocomputing*, 21(1–3):173–190, 1998.

15. E. Maeda and H. Murase. Multi-category classification by kernel based nonlinear subspace method. In *Proceedings of 1999 IEEE International Conference on Acoustics, Speech, and Signal Processing (ICASSP '99)*, volume 2, pages 1025–1028, 1999.

16. S. Takeuchi, T. Kitamura, S. Abe, and K. Fukui. Subspace based linear programming support vector machines. In *Proceedings of the 2009 International Joint Conference on Neural Networks (IJCNN 2009)*, pages 3067–3073, Atlanta, GA, 2009.

17. T. Kitamura, S. Abe, and K. Fukui. Subspace based least squares support vector machines for pattern classification. In *Proceedings of the 2009 International Joint Conference on Neural Networks (IJCNN 2009)*, pages 1640–1646, Atlanta, GA, 2009.

18. T. Kitamura S. Takeuchi, S. Abe, and K. Fukui. Subspace-based support vector machines. *Neural Networks*, 22(5–6):558–567, 2009.

19. C. H. Park and H. Park. Efficient nonlinear dimension reduction for clustered data using kernel functions. In *Proceedings of the Third IEEE International Conference on Data Mining (ICDM 2003)*, pages 243–250, Melbourne, FL, 2003.

20. P. Zhang, J. Peng, and C. Domeniconi. Dimensionality reduction using kernel pooled local discriminant information. In *Proceedings of the Third IEEE International Conference on Data Mining (ICDM 2003)*, pages 701–704, Melbourne, FL, 2003.

21. G. C. Cawley and N. L. C. Talbot. Efficient formation of a basis in a kernel feature space. In *Proceedings of the Tenth European Symposium on Artificial Neural Networks (ESANN 2002)*, pages 1–6, Bruges, Belgium, 2002.

22. S. Fine and K. Scheinberg. Efficient SVM training using low-rank kernel representations. *Journal of Machine Learning Research*, 2:243–264, 2001.

23. T. Evgeniou, M. Pontil, and T. Poggio. Regularization networks and support vector machines. *Advances in Computational Mathematics*, 13(1):1–50, 2000.

24. P. Zhang and J. Peng. SVM vs regularized least squares classification. In *Proceedings of the Seventeenth International Conference on Pattern Recognition (ICPR 2004)*, volume 1, pages 176–179, Cambridge, UK, 2004.

25. K. Morikawa. Pattern classification and function approximation by kernel least squares. Bachelor's thesis, Electrical and Electronics Engineering, Kobe University, Japan, 2004 (in Japanese).

26. G. H. Golub and C. F. Van Loan. *Matrix Computations, Third Edition*. The Johns Hopkins University Press, Baltimore, MD, 1996.

27. W. H. Press, S. A. Teukolsky, W. T. Vetterling, and B. P. Flannery. *Numerical Recipes in C: The Art of Scientific Computing, Second Edition*. Cambridge University Press, Cambridge, UK, 1992.

28. S. Mika, B. Schölkopf, A. Smola, K.-R. Müller, M. Scholz, and G. Rätsch. Kernel PCA and de-noising in feature spaces. In M. S. Kearns, S. A. Solla, and D. A. Cohn, editors, *Advances in Neural Information Processing Systems 11*, pages 536–542. MIT Press, Cambridge, MA, 1999.

29. B. Schölkopf, S. Mika, C. J. C. Burges, P. Knirsch, K.-R. Müller, G. Rätsch, and A. J. Smola. Input space versus feature space in kernel-based methods. *IEEE Transactions on Neural Networks*, 10(5):1000–1017, 1999.

30. T. Takahashi and T. Kurita. Robust de-noising by kernel PCA. In J. R. Dorronsoro, editor, *Artificial Neural Networks (ICANN 2002)—Proceedings of International Conference, Madrid, Spain*, pages 739–744. Springer-Verlag, Berlin, Germany, 2002.

31. J. T.-Y. Kwok and I. W.-H. Tsang. The pre-image problem in kernel methods. *IEEE Transactions on Neural Networks*, 15(6):1517–1525, 2004.

32. R. O. Duda and P. E. Hart. *Pattern Classification and Scene Analysis.* John Wiley & Sons, New York, 1973.

33. S. Mika, G. Rätsch, J. Weston, B. Schölkopf, and K.-R. Müller. Fisher discriminant analysis with kernels. In Y.-H. Hu, J. Larsen, E. Wilson, and S. Douglas, editors, *Neural Networks for Signal Processing IX—Proceedings of the 1999 IEEE Signal Processing Society Workshop,* pages 41–48, 1999.

34. G. Baudat and F. Anouar. Generalized discriminant analysis using a kernel approach. *Neural Computation,* 12(10):2385–2404, 2000.

35. H. Li, T. Jiang, and K. Zhang. Efficient and robust feature extraction by maximum margin criterion. In S. Thrun, L. K. Saul, and B. Schölkopf, editors, *Advances in Neural Information Processing Systems 16,* pages 97–104. MIT Press, Cambridge, MA, 2004.

36. J. Yang, A. F. Frangi, J.-Y. Yang, D. Zhang, and Z. Jin. KPCA plus LDA: A complete kernel fisher discriminant framework for feature extraction and recognition. *IEEE Transactions on Pattern Analysis and Machine Intelligence,* 27(2):230–244, 2005.

37. E. Pekalska and B. Haasdonk. Kernel discriminant analysis for positive definite and indefinite kernels. *IEEE Transactions on Pattern Analysis and Machine Intelligence,* 31(6):1017–1031, 2009.

# Chapter 7
# Feature Selection and Extraction

Conventional classifiers do not have a mechanism to control class boundaries. Thus if the number of features, i.e., input variables, is large compared to the number of training data, class boundaries may not overlap. In such a situation, the generalization ability of the conventional classifiers may not be good. Therefore, to improve the generalization ability, we usually generate a small set of features from the original input variables by either feature selection or feature extraction.

Because support vector machines directly determine the class boundaries by training, the generalization ability does not degrade greatly even when the number of input variables is large. Vapnik [1] even claims that feature selection or feature extraction is not necessary for support vector machines. But it is important, even using support vector machines.

In this chapter, we first survey feature selection methods using support vector machines and show how feature selection affects generalization ability of a support vector machine for some benchmark data sets. Then we discuss feature extraction by kernel principal component analysis and kernel discriminant analysis.

## 7.1 Selecting an Initial Set of Features

The most influencing factor in realizing a classifier with high generalization ability is a set of features used. But because there is no systematic way of determining an initial set of features for a given classification problem, we need to determine a set of initial features by trials and errors.

If the number of features is very large and each feature has little classification power, it is better to transform linearly or nonlinearly the set of features into a reduced set of features. In face recognition, a face image is transformed, for instance, by principal component analysis (PCA), which is

S. Abe, *Support Vector Machines for Pattern Classification*,
Advances in Pattern Recognition, DOI 10.1007/978-1-84996-098-4_7,
© Springer-Verlag London Limited 2010

a linear transformation, into a set of features that are dominant eigenvectors of the images.

If each feature in the initial set of features has a classification power, we reduce the set by feature selection or feature extraction. By feature selection we delete redundant or meaningless features so that we can realize the higher generalization performance and faster classification than by the initial set of features.

## 7.2 Procedure for Feature Selection

In feature selection we want to select the minimum subset of features, from the original set of features, that realizes the maximum generalization ability. To realize this, during the process of feature selection, the generalization ability of a subset of features needs to be estimated. This type of feature selection is called a wrapper method [2]. But because it is time-consuming to directly estimate the generalization ability, some selection criterion, which is considered to well reflect the generalization ability, is used. This method is called a filter method and various selection criteria have been developed [3, 4].

The forward or backward selection method using a selection criterion is widely used. In backward selection, we start from all the features and delete one feature at a time, which deteriorates the selection criterion the least. We delete features until the selection criterion reaches a specified value. In forward selection, we start from an empty set of features and add one feature at a time, which improves the selection criterion the most. We iterate this procedure until the selection criterion reaches a specified value. Because forward or backward selection is slow, we may add or delete more than one feature at a time based on feature ranking or we may combine backward and forward selection [5].

Because these selection methods are local optimization techniques, global optimality of feature selection is not guaranteed. Usually, backward selection is slower but is more stable in selecting optimal features than forward selection [6]. If a selection criterion is monotonic for deletion or addition of a feature, we can terminate feature selection when the selection criterion violates a predefined value [7] or we can use optimization techniques such as the branch-and-bound technique. An exception ratio defined based on the overlap of class regions approximated by hyperboxes [7] is proved to be monotonic for the deletion of features. But the exception ratio defined in the feature space is not monotonic [8].

By the introduction of support vector machines (SVMs), various selection methods suitable for support vector machines have been developed. The selection criterion for filter methods used in the literature is, except for some cases [9, 10, 8, 11, 12], the margin [13–17]. In addition, in most cases, a linear support vector machine is used. The idea of feature selection is as follows: If some elements of the coefficient vector of the hyperplane are zero, the deletion of the associated input variables does not change the optimal hyperplane for the remaining variables. But if we delete variables associated with nonzero elements, the optimal solution changes. Thus the magnitude of the margin decreases. In [18], selection of features in support vector machines with polynomial kernels is discussed, but this is for deletion of feature space variables, not input variables. In [8, 12], the objective function of kernel discriminant analysis called the KDA criterion, namely the ratio of the between-class scatter and within-class scatter, is proved to be monotonic for the deletion of features, and feature selection based on the KDA criterion was shown to be robust for benchmark data sets. Louw and Steel [11] also used the KDA criterion for feature selection.

As a wrapper method, in [19, 20], block deletion of features in backward feature selection is proposed using the generalization ability by cross-validation as the selection criterion.

In addition to filter and wrapper methods, the embedded methods combine training and feature selection; because training of support vector machines results in solving a quadratic optimization problem, feature selection can be done by modifying the objective function [21–25]. For instance, if we use an LP support vector machine with linear kernels, we can consider that the variables associated with zero coefficients of the separating hyperplane are redundant.

# 7.3 Feature Selection Using Support Vector Machines

In this section, we discuss some of the feature selection methods based on support vector machines. The methods are classified into two: backward or forward feature selection based on some selection criterion [9, 10, 13] and SVM-based feature selection, in which a feature selection criterion is added to the objective function [21] or forward feature selection is done by changing the value of the margin parameter [22, 23].

## 7.3.1 Backward or Forward Feature Selection

### 7.3.1.1 Selection Criteria

In this section, we discuss the selection criterion based on the margin.

Assume that a classification problem is linearly separable in the feature space. Then training the support vector machine with the associated kernel results in maximizing the margin $\delta$ or minimizing $\|\mathbf{w}\|$.

Now we show that the margin is nonincreasing for the deletion of the input variable so long as the classification problem is separable in the feature space for the reduced input variables, namely, the margin remains the same or decreases. First, we show that the margin is nonincreasing when the problem is linearly separable in the input space. Figure 7.1 shows the two-dimensional case where the margin decreases if $x_2$ is deleted.

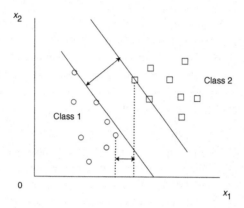

**Fig. 7.1** Decrease of the margin by deleting $x_2$

Figure 7.2 shows the two-dimensional case where the optimal separating line is parallel to $x_2$. In this case, $x_2$ does not contribute in classification and the margin remains the same even if $x_2$ is deleted.

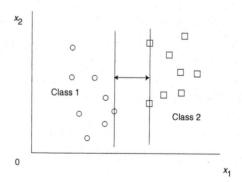

**Fig. 7.2** The margin remains the same by deleting $x_2$

In general, if the optimal separating hyperplane is not parallel to an input variable, the deletion of the variable results in the decrease in the margin.

But if the optimal hyperplane is parallel to input variables, the margin does not change for the deletion of these variables.

This is paraphrased as follows. If some elements of $\mathbf{w}$ are zero, the deletion of the associated input variables does not change the optimal hyperplane for the remaining variables. But if we delete variables associated with nonzero elements, the optimal solution changes. Thus the magnitude of the margin decreases.

We can extend this discussion to the feature space. If an input variable is deleted, some of the variables that span the feature space are deleted. Thus, the margin is nonincreasing for the deletion of input variables.

Therefore, in this situation, it is natural to delete input variables as far as possible under the constraint that the classification problem is linearly separable for the reduced input variables and under the constraint that

$$\frac{\delta - \delta'}{\delta} < \varepsilon, \tag{7.1}$$

where $\delta'$ is the margin for the reduced variables and $\varepsilon$ is a small positive value.

### 7.3.1.2 Feature Ranking

The change of $\|\mathbf{w}\|^2$ for the deletion of the $k$th input variable needs to be calculated by training the SVM with the deleted input variable. But this is time-consuming. Thus, we consider how to choose the deletion candidate from the input variables. For the linear kernel, if $w_k^2 = 0$, the optimal hyperplane is the same for the deletion of the $k$th input variable. Even if $w_k^2$ is not zero, if the value is small, the deletion of the $k$th input variable does not affect very much for the optimal hyperplane. Thus, we can choose the input variable with the minimum $w_k^2$. This is the same measure as that proposed in [13].

For the kernel other than the linear kernel, multiple variables in the feature space are deleted for the deletion of an input variable. In this case, we also choose the input variable with the minimum square sum of weights associated with the input variable. The square sum of weights in the feature space that correspond to the $k$th input variable, $\Delta^{(k)}\|\mathbf{w}\|^2$, is estimated by

$$\Delta^{(k)}\|\mathbf{w}\|^2 = \sum_{i,j \in S} \alpha_i \, y_i \alpha_j \, y_j \, (K(\mathbf{x}_i, \mathbf{x}_j) - K(\mathbf{x}_i^{(k)}, \mathbf{x}_j^{(k)})), \tag{7.2}$$

where $\mathbf{x}^{(k)}$ is the vector with the $k$th element of $\mathbf{x}$ set to zero.

If more than one $\Delta^{(k)}\|\mathbf{w}\|^2$ are zero, we can safely delete the associated variables, but this may be rare. We may choose plural variables with $\Delta^{(k)}\|\mathbf{w}\|^2$ smaller than a threshold. But how to determine the threshold is difficult. In [14], three random variables are added to a regression problem, and the average value of the three weights associated with the variables is used for the threshold.

### 7.3.1.3 Backward Feature Selection

In backward feature selection we first train the support vector machine using all the input variables. Then we delete the input variable ranked first in the feature ranking and train the support vector machine with the reduced input variables. We iterate the deletion procedure until the stopping criterion is satisfied.

The procedure of backward feature selection is as follows:

1. Train the SVM using all the input variables. Let the margin be $\delta_0$.
2. Let $\Delta^{(k)} \|\mathbf{w}\|^2$ be minimum. Then delete the $k$th input variable. (To speed up the deletion procedure, we may delete more than one variable by feature ranking.)
3. Train the SVM with the reduced input variables. If nonseparable or $(\delta_0 - \delta_c)/\delta_0 < \varepsilon$, terminate the algorithm, where $\delta_c$ is the current margin and $0 < \varepsilon < 1$. Otherwise, go to Step 2.

We can extend this method for forward selection if we calculate the margin change for the variable addition by training, which is time-consuming. But without training, estimation of the change is difficult.

## 7.3.2 Support Vector Machine-Based Feature Selection

Instead of backward or forward feature selection, feature selection can be done while training by reformulating training [21, 26, 27, 15, 18]. The idea of feature selection developed by Bradley and Mangasarian [21] is quite similar to that by Guyon et al. [13]. In the former, feature selection is done by minimizing the classification error and the number of nonzero components of $\mathbf{w}$. The problem is formulated as follows:

$$\text{minimize} \quad (1 - \lambda)\frac{1}{M}\sum_{i=1}^{M}\xi_i + \lambda\sum_{i=1}^{n} w_i^* \tag{7.3}$$

$$\text{subject to} \quad y_i\left(\mathbf{w}^\top\mathbf{x}_i + b\right) \geq 1 - \xi_i \quad \text{for } i = 1,\dots,M, \tag{7.4}$$

where $\lambda\,(1 > \lambda \geq 0)$ is a regularization parameter and

$$w_i^* = \begin{cases} 1 & \text{for } w_i \neq 0, \\ 0 & \text{otherwise.} \end{cases} \tag{7.5}$$

In (7.3), the second term minimizes the number of nonzero components of $\mathbf{w}$. The variables with zero $w_i$ are regarded as redundant. This optimization problem is written by introducing a positive vector $\mathbf{v}$,

$$\text{minimize} \quad (1 - \lambda) \frac{1}{M} \sum_{i=1}^{M} \xi_i + \lambda \sum_{i=1}^{n} v_i^* \tag{7.6}$$

$$\text{subject to} \quad y_i \left( \mathbf{w}^\top \mathbf{x}_i + b \right) \geq 1 - \xi_i \quad \text{for } i = 1, \ldots, M, \tag{7.7}$$

$$-v_i \leq w_i \leq v_i \quad \text{for } i = 1, \ldots, n, \tag{7.8}$$

where $v_i^*$ is defined as in (7.5).

Because $v_i^*$ is a step function, it is approximated by

$$1 - \exp(-\gamma v_i), \tag{7.9}$$

where $\gamma$ is a positive parameter. Then the optimization problem is solved by concave minimization technique [21].

Let the objective function be

$$Q(\mathbf{w}, \boldsymbol{\xi}) = \lambda \sum_{i=1}^{m} |w_i| + \sum_{i=1}^{M} \xi_2^2, \tag{7.10}$$

where $\lambda = 1/C$ is the regularization parameter. For $\lambda = \infty$, $w_i = 0$, which means that all the input features are not used. Brown's idea [22, 23] is to decrease the value of $\lambda$ toward zero and select the classifier with the optimal features. For $\lambda \in [0, \infty)$, the set of support vectors changes at finite points (see Section 2.6.2) and so does the set of input features. To facilitate searching the points where the set of support vectors changes, Brown proposed an iterative linear programming technique.

## 7.3.3 Feature Selection by Cross-Validation

Usually we select a feature selection criterion other than the recognition rate, because it is time-consuming to evaluate the recognition rate. But if the number of training data is small, the recognition rate can be used as a feature selection criterion.[1]

In the following we discuss backward feature selection estimating the generalization ability of the classifier by cross-validation of the training data [19].

---

[1] Professor N. Kasabov's lecture at Kobe University on June 1, 2004, showed the usefulness of this criterion.

Let the initial set of selected features be $F^m$, where $m$ is the number of input variables, and the recognition rate of the validation set by cross-validation be $R^m$.

We delete the $i$th $(i = 1, \ldots, m)$ feature temporally from $F^m$ and estimate the generalization ability by cross-validation. Let the recognition rate of the validation set be $R_i^m$. We check whether the maximum $R_i^m$ $(i \in \{1, \ldots, m\})$ is larger than or equal to $R_m$:

$$\max_{i=1,\ldots,m} R_i^m \geq R^m. \tag{7.11}$$

If (7.11) is not satisfied, we assume that the deletion of one feature results in degrading the generalization ability. Thus we cannot delete any feature from the original set of features $F_m$.

Assume that (7.11) is satisfied for $k$. Then we set

$$F^{m-1} = F^m - \{k\}, \tag{7.12}$$

namely, we assume that the set of features $F^{m-1}$ can realize the same generalization ability as $F^m$. To speed up selecting the features from $F^{m-1}$, we consider that the features that satisfy

$$R_i^m < R^m \tag{7.13}$$

are indispensable for classification and thus they cannot be deleted. Thus, we set the set of features that are candidates for deletion

$$S^{m-1} = \{i \mid R_i^m \geq R^m, \, i \neq k\}. \tag{7.14}$$

If $S^{m-1}$ is empty, we stop deleting the feature. If it is not empty, we iterate the preceding backward selection procedure.

We evaluated the method using some of the data sets listed in Table 1.3. We estimated the generalization ability by fivefold cross-validation for a given kernel changing the value of $C$. Table 7.1 shows the results. The "Deleted" column lists the features deleted according to the algorithm and the "Validation" and "Test" columns show the recognition rates of the validation sets and test data sets, respectively. If the recognition rate of the training data is not 100%, it is shown in parentheses. For the iris and numeral data sets we used a polynomial kernel with degree 2 and for the blood cell and thyroid data sets we used polynomial kernels with degree 4. For the iris and numeral data sets, the recognition rates of the test data with deleted input variables are equal to or higher than those with all the input variables. For the blood cell data, the recognition rates with deleted input variables are equal to or lower, but the differences are small.

For the thyroid data set, many redundant features are included. Based on the analysis of class regions approximated by ellipsoids, five features, i.e., the 3rd, 8th, 11th, 17th, and 21st features, were selected as important features by the forward feature selection method [6]. Thus we started from these five features. The "Deleted" column in Table 7.1 lists the remaining features. Three features, i.e., the 3rd, 8th, and 17th, were selected, and the recognition rate for the test data was higher than that for using all the features.

**Table 7.1** Feature selection by cross-validation. For the thyroid data, the "Deleted" column lists the remaining features

| Data | Deleted | $C$ | Validation (%) | Test (%) |
|------|---------|-----|----------------|----------|
| Iris | None | 5,000 | 94.67 | 93.33 |
| | 3 | 500 | 96.00 (99.00) | 96.00 (98.67) |
| | 3, 1 | 5,000 | 94.67 | 96.00 |
| | 3, 4 | 5,000 | 94.67 | 93.33 |
| Numeral | None | 1 | 99.51 (99.97) | 99.63 |
| | 4 | 1 | 99.75 | 99.63 |
| | 4, 10 | 1 | 99.75 | 99.76 |
| | 4, 10, 3 | 1 | 99.63 | 99.63 |
| | 4, 10, 3, 12 | 1 | 99.51 (99.94) | 99.76 |
| Blood cell | None | 1 | 93.77 (96.23) | 93.23 (96.51) |
| | 1 | 1 | 94.38 (96.65) | 93.06 (96.51) |
| | 1, 13 | 1 | 94.51 (96.56) | 93.03 (96.84) |
| | 1, 13, 8 | 1 | 94.35 (96.41) | 93.23 (96.67) |
| | 1, 13, 8, 10 | 1 | 94.45 (96.48) | 93.16 (96.71) |
| | 1, 13, 8, 10, 9 | 1 | 94.54 (96.67) | 92.97 (96.38) |
| | 1, 13, 8, 10, 9, 6 | 1 | 94.25 (96.00) | 92.45 (95.93) |
| Thyroid | None | $10^5$ | 97.96 | 97.93 |
| | (3, 8, 11, 17, 21) | $10^5$ | 98.44 (99.85) | 98.37 (99.81) |
| | (3, 8, 11, 21) | $10^5$ | 98.52 (99.77) | 98.45 (99.81) |
| | (3, 8, 17) | $10^4$ | 98.52 (99.76) | 98.48 (99.81) |

## 7.4 Feature Extraction

Principal component analysis is widely used for feature extraction. As a variant of PCA, kernel PCA, which was discussed in Section 6.2, has been gaining wide acceptance. In [28], computer experiments showed that the combination of KPCA and the linear support vector machine gave better generalization ability than the nonlinear support vector machine. In [29, 30], KPCA is combined with least squares, which is a variant of kernel least squares discussed in Section 6.1.

Principal component analysis does not use class information. Thus, the
first principal component is not necessarily useful for class separation. On the
other hand, linear discriminant analysis, defined for a two-class problem, finds
the axis that maximally separates training data projected on this axis into
two classes. Applications of linear discriminant analysis is limited to a case
where each class consists of one cluster and they are not heavily overlapped.
By proper selection of kernels and their parameter values, kernel discriminant
analysis solves the limitation of linear discriminant analysis. It is extended
to multiclass problems (see Section 6.5).

Kernel discriminant analysis is used as criteria for feature selection and
kernel selection as well as feature extraction.

# References

1. V. N. Vapnik. *Statistical Learning Theory.* John Wiley & Sons, New York, 1998.
2. R. Kohavi and G. H. John. Wrappers for feature subset selection. *Artificial Intelligence,* 97(1–2):273–324, 1997.
3. K. Fukunaga. *Introduction to Statistical Pattern Recognition, Second Edition.* Academic Press, San Diego, CA, 1990.
4. S. Theodoridis and K. Koutroumbas. *Pattern Recognition, Third Edition.* Academic Press, London, UK, 2006.
5. P. Somol, P. Pudil, J. Novovičová, and P. Paclík. Adaptive floating search method in feature selection. *Pattern Recognition Letters,* 20(11–13):1157–1163, 1999.
6. S. Abe. *Pattern Classification: Neuro-Fuzzy Methods and Their Comparison.* Springer-Verlag, London, UK, 2001.
7. R. Thawonmas and S. Abe. A novel approach to feature selection based on analysis of class regions. *IEEE Transactions on Systems, Man, and Cybernetics—Part B,* 27(2):196–207, 1997.
8. M. Ashihara and S. Abe. Feature selection based on kernel discriminant analysis. In S. Kollias, A. Stafylopatis, W. Duch, and E. Oja, editors, *Artificial Neural Networks (ICANN 2006)—Proceedings of the Sixteenth International Conference, Athens, Greece, Part II,* pages 282–291. Springer-Verlag, Berlin, Germany, 2006.
9. S. Mukherjee, P. Tamayo, D. Slonim, A. Verri, T. Golub, J. P. Mesirov, and T. Poggio. Support vector machine classification of microarray data. Technical Report AI Memo 1677, Massachusetts Institute of Technology, 1999.
10. T. Evgeniou, M. Pontil, C. Papageorgiou, and T. Poggio. Image representations for object detection using kernel classifiers. In *Proceedings of Asian Conference on Computer Vision (ACCV 2000),* pages 687–692, Taipei, Taiwan, 2000.
11. N. Louw and S. J. Steel. Variable selection in kernel Fisher discriminant analysis by means of recursive feature elimination. *Computational Statistics & Data Analysis,* 51(3):2043–2055, 2006.
12. T. Ishii, M. Ashihara, and S. Abe. Kernel discriminant analysis based feature selection. *Neurocomputing,* 71(13–15):2544–2552, 2008.
13. I. Guyon, J. Weston, S. Barnhill, and V. Vapnik. Gene selection for cancer classification using support vector machines. *Machine Learning,* 46(1–3):389–422, 2002.
14. J. Bi, K. P. Bennett, M. Embrechts, C. M. Breneman, and M. Song. Dimensionality reduction via sparse support vector machines. *Journal of Machine Learning Research,* 3:1229–1243, 2003.

15. S. Perkins, K. Lacker, and J. Theiler. Grafting: Fast, incremental feature selection by gradient descent in function space. *Journal of Machine Learning Research*, 3:1333–1356, 2003.

16. A. Rakotomamonjy. Variable selection using SVM-based criteria. *Journal of Machine Learning Research*, 3:1357–1370, 2003.

17. Y. Liu and Y. F. Zheng. FS_SFS: A novel feature selection method for support vector machines. *Pattern Recognition*, 39(7):1333–1345, 2006.

18. J. Weston, A. Elisseeff, B. Schölkopf, and M. Tipping. Use of the zero-norm with linear models and kernel methods. *Journal of Machine Learning Research*, 3:1439–1461, 2003.

19. S. Abe. Modified backward feature selection by cross validation. In *Proceedings of the Thirteenth European Symposium on Artificial Neural Networks (ESANN 2005)*, pages 163–168, Bruges, Belgium, 2005.

20. T. Nagatani and S. Abe. Backward variable selection of support vector regressors by block deletion. In *Proceedings of the 2007 International Joint Conference on Neural Networks (IJCNN 2007)*, pages 2117–2122, Orlando, FL, 2007.

21. P. S. Bradley and O. L. Mangasarian. Feature selection via concave minimization and support vector machines. In *Proceedings of the Fifteenth International Conference on Machine Learning (ICML '98)*, pages 82–90, Madison, WI, 1998.

22. M. Brown. Exploring the set of sparse, optimal classifiers. In *Proceedings of Artificial Neural Networks in Pattern Recognition (ANNPR 2003)*, pages 178–184, Florence, Italy, 2003.

23. M. Brown, N. P. Costen, and S. Akamatsu. Efficient calculation of the complete optimal classification set. In *Proceedings of the Seventeenth International Conference on Pattern Recognition (ICPR 2004)*, volume 2, pages 307–310, Cambridge, UK, 2004.

24. C. Gold, A. Holub, and P. Sollich. Bayesian approach to feature selection and parameter tuning for support vector machine classifiers. *Neural Networks*, 18(5–6):693–701, 2005.

25. L. Bo, L. Wang, and L. Jiao. Sparse Gaussian processes using backward elimination. In J. Wang, Z. Yi, J. M. Zurada, B.-L. Lu, and H. Yin, editors, *Advances in Neural Networks (ISNN 2006): Proceedings of Third International Symposium on Neural Networks, Chengdu, China, Part 1*, pages 1083–1088. Springer-Verlag, Berlin, Germany, 2006.

26. J. Weston, S. Mukherjee, O. Chapelle, M. Pontil, T. Poggio, and V. Vapnik. Feature selection for SVMs. In T. K. Leen, T. G. Dietterich, and V. Tresp, editors, *Advances in Neural Information Processing Systems 13*, pages 668–674. MIT Press, Cambridge, MA, 2001.

27. Y. Grandvalet and S. Canu. Adaptive scaling for feature selection in SVMs. In S. Becker, S. Thrun, and K. Obermayer, editors, *Advances in Neural Information Processing Systems 15*, pages 569–576. MIT Press, Cambridge, MA, 2003.

28. B. Schölkopf, A. J. Smola, and K.-R. Müller. Kernel principal component analysis. In B. Schölkopf, C. J. C. Burges, and A. J. Smola, editors, *Advances in Kernel Methods: Support Vector Learning*, pages 327–352. MIT Press, Cambridge, MA, 1999.

29. R. Rosipal, M. Girolami, and L. J. Trejo. Kernel PCA feature extraction of event-related potentials for human signal detection performance. In H. Malmgren, M. Borga, and L. Niklasson, editors, *Artificial Neural Networks in Medicine and Biology—Proceedings of the ANNIMAB-1 Conference, Göteborg, Sweden*, pages 321–326. Springer-Verlag, Berlin, Germany, 2000.

30. R. Rosipal, M. Girolami, L. J. Trejo, and A. Cichocki. Kernel PCA for feature extraction and de-noising in nonlinear regression. *Neural Computing & Applications*, 10(3):231–243, 2001.

# Chapter 8
# Clustering

Unlike multilayer neural networks, support vector machines can be formulated for one-class problems. This technique is called *domain description* or *one-class classification* and is applied to clustering and detection of outliers for both pattern classification and function approximation [1].

In this chapter, we first discuss domain description, in which a region for a single class is approximated by a hypersphere in the feature space. Then we discuss an extension of domain description to clustering.

## 8.1 Domain Description

In pattern classification, we consider more than one class. And, if data with only one class are available as training data, we cannot use multilayer neural networks. In such a situation, if we approximate the class region by some method and if we test whether new data are outside the region, we can detect outliers. The approximation of the class region is called the *domain description*. Tax and Duin [2–4] extended the support vector method to domain description. In the following we discuss their method.

We consider approximating the class region by the minimum-volume hypersphere, that include one-class data, with center $\mathbf{a} = (a_1, \ldots, a_m)^\top$ and radius $R$ in the input space. Let $\mathbf{x}_i \ (i = 1, \ldots, M)$ be data belonging to one class. Then the problem is as follows:

$$\text{minimize} \quad Q_\mathrm{p}(R, \mathbf{a}, \boldsymbol{\xi}) = R^2 + C \sum_{i=1}^{M} \xi_i \tag{8.1}$$

$$\text{subject to} \quad \|\mathbf{x}_i - \mathbf{a}\|^2 \le R^2 + \xi_i, \quad \xi_i \ge 0 \quad \text{for } i = 1, \ldots, M, \tag{8.2}$$

where $\boldsymbol{\xi} = (\xi_1, \ldots, \xi_M)^\top$ is the slack variable vector and $C$ determines the trade-off between the hypersphere volume and outliers. If outlier samples are

S. Abe, *Support Vector Machines for Pattern Classification*,
Advances in Pattern Recognition, DOI 10.1007/978-1-84996-098-4_8,
© Springer-Verlag London Limited 2010

available, we determine the minimum-volume hypersphere that includes one-class data and excludes outliers. Thus, the optimization problem becomes as follows:

$$\text{minimize} \quad Q_{\mathrm{p}}(R, \mathbf{a}, \boldsymbol{\xi}, \boldsymbol{\eta}) = R^2 + C_1 \sum_{i=1}^{M} \xi_i + C_1 \sum_{i=M+1}^{M'} \eta_i \tag{8.3}$$

$$\text{subject to} \quad \|\mathbf{x}_i - \mathbf{a}\|^2 \le R^2 + \xi_i, \quad \xi_i \ge 0 \quad \text{for } i = 1, \ldots, M, \tag{8.4}$$

$$\|\mathbf{x}_i - \mathbf{a}\|^2 \ge R^2 - \eta_i, \quad \eta_i \ge 0 \quad \text{for } i = M+1, \ldots, M', \tag{8.5}$$

where $\mathbf{x}_i, (i = M+1, \ldots, M')$ are outliers, $\boldsymbol{\eta} = (\eta_{M+1}, \ldots, \eta_{M'})^{\top}$ is the slack variable vector for the outliers and $C_1$ and $C_2$ determines the trade-off between the hypersphere volume and outliers. In the following we consider solving (8.1) and (8.2).

Introducing the nonnegative Lagrange multipliers $\alpha_i$ and $\gamma_i$, we obtain

$$Q_{\mathrm{d}}(R, \mathbf{a}, \boldsymbol{\alpha}, \boldsymbol{\xi}, \boldsymbol{\gamma}) = R^2 + C \sum_{i=1}^{M} \xi_i$$
$$- \sum_{i=1}^{M} \alpha_i \left( R^2 + \xi_i - \mathbf{x}_i^{\top} \mathbf{x}_i + 2\mathbf{a}^{\top} \mathbf{x}_i - \mathbf{a}^{\top} \mathbf{a} \right)$$
$$- \sum_{i=1}^{M} \gamma_i \xi_i, \tag{8.6}$$

where $\boldsymbol{\alpha} = (\alpha_1, \ldots, \alpha_M)^{\top}$ and $\boldsymbol{\gamma} = (\gamma_1, \ldots, \gamma_M)^{\top}$.

Setting the partial derivatives of $Q_{\mathrm{d}}(R, \mathbf{a}, \boldsymbol{\alpha}, \boldsymbol{\xi}, \boldsymbol{\gamma})$ with respect to $R$, $\mathbf{a}$, and $\boldsymbol{\xi}$ to zero, we obtain

$$\sum_{i}^{M} \alpha_i = 1, \tag{8.7}$$

$$\mathbf{a} = \sum_{i=1}^{M} \alpha_i \mathbf{x}_i, \tag{8.8}$$

$$C - \alpha_i - \gamma_i = 0 \quad \text{for } i = 1, \ldots, M. \tag{8.9}$$

Substituting (8.7), (8.8), and (8.9) into (8.6) gives the following maximization problem:

$$\text{maximize} \quad Q_{\mathrm{d}}(\boldsymbol{\alpha}) = \sum_{i=1}^{M} \alpha_i \mathbf{x}_i^{\top} \mathbf{x}_i - \sum_{i,j=1}^{M} \alpha_i \alpha_j \mathbf{x}_i^{\top} \mathbf{x}_j \tag{8.10}$$

$$\text{subject to} \quad \sum_{i}^{M} \alpha_i = 1, \tag{8.11}$$

$$0 \leq \alpha_i \leq C \quad \text{for } i = 1, \ldots, M. \tag{8.12}$$

The KKT conditions are as follows:

$$\alpha_i \left( R^2 + \xi_i - \mathbf{x}_i^\top \mathbf{x}_i + 2\mathbf{a}^\top \mathbf{x}_i - \mathbf{a}^\top \mathbf{a} \right) = 0 \quad \text{for } i = 1, \ldots, M, \tag{8.13}$$
$$\gamma_i \, \xi_i = 0 \quad \text{for } i = 1, \ldots, M. \tag{8.14}$$

If $0 < \alpha_i < C$, from (8.9), $\gamma_i \neq 0$. Thus $\xi_i = 0$ and

$$R = \|\mathbf{x}_i - \mathbf{a}\|, \tag{8.15}$$

where $\mathbf{a}$ is given by (8.8) and $\mathbf{x}_i$ is a support vector, namely, the unbounded support vectors form the surface of the hypersphere.

If $\alpha_i = C$, $\gamma_i = 0$. Thus if $\xi_i > 0$, the bounded support vectors are outside of the hypersphere and thus are outliers. Notice that, from (8.7), $1 \geq \alpha_i \geq 0$ is satisfied. And because at least two support vectors are necessary for defining a hypersphere, if $C \geq 1$, there are no bounded support vectors.

Thus an unknown data sample $\mathbf{x}$ is inside the hypersphere if

$$\mathbf{x}^\top \mathbf{x} - 2 \sum_{i \in S} \alpha_i \mathbf{x}^\top \mathbf{x}_i + \sum_{\substack{i \in S, \\ j \in S}} \alpha_i \, \alpha_j \, \mathbf{x}_i^\top \mathbf{x}_j \leq R^2, \tag{8.16}$$

where $S$ is the set of support vector indices.

If we want to determine the minimum volume of the hypersphere in the feature space, we change (8.2) to

$$\|\phi(\mathbf{x}_i) - \mathbf{a}\|^2 \leq R^2 + \xi_i, \quad \xi_i \geq 0 \quad \text{for } i = 1, \ldots, M, \tag{8.17}$$

where $\phi(\mathbf{x})$ is the mapping function that maps $\mathbf{x}$ into the $l$-dimensional feature space and $\mathbf{a}$ is the center of the hypersphere in the feature space.

Introducing the Lagrange multipliers $\alpha_i$ and $\gamma_i$, we obtain

$$Q_{\mathrm{d}}(R, \mathbf{a}, \boldsymbol{\alpha}, \boldsymbol{\xi}, \boldsymbol{\gamma}) = R^2 + C \sum_{i=1}^{M} \xi_i$$
$$- \sum_{i=1}^{M} \alpha_i \left( R^2 + \xi_i - K(\mathbf{x}_i, \mathbf{x}_i) + 2\mathbf{a}^\top \phi(\mathbf{x}_i) - \mathbf{a}^\top \mathbf{a} \right)$$
$$- \sum_{i=1}^{M} \gamma_i \, \xi_i, \tag{8.18}$$

where $K(\mathbf{x}_i, \mathbf{x}_i) = \phi^\top(\mathbf{x}_i) \, \phi(\mathbf{x}_i)$.

Setting the partial derivatives of $Q_{\mathrm{d}}(R, \mathbf{a}, \boldsymbol{\alpha}, \boldsymbol{\xi}, \boldsymbol{\gamma})$ with respect to $R$, $\mathbf{a}$, and $\boldsymbol{\xi}$ to zero, we obtain

$$\sum_{i}^{M} \alpha_i = 1, \tag{8.19}$$

$$\mathbf{a} = \sum_{i=1}^{M} \alpha_i\, \phi(\mathbf{x}_i), \tag{8.20}$$

$$C - \alpha_i - \gamma_i = 0 \quad \text{for } i = 1, \ldots, M. \tag{8.21}$$

Thus substituting (8.19), (8.20), and (8.21) into (8.18) gives the following maximization problem[1]:

$$\text{maximize} \quad Q_{\mathrm{d}}(\boldsymbol{\alpha}) = \sum_{i=1}^{M} \alpha_i K(\mathbf{x}_i, \mathbf{x}_i) - \sum_{i,j=1}^{M} \alpha_i\, \alpha_j K(\mathbf{x}_i, \mathbf{x}_j) \tag{8.22}$$

$$\text{subject to} \quad \sum_{i=1}^{M} \alpha_i = 1, \tag{8.23}$$

$$0 \le \alpha_i \le C \quad \text{for } i = 1, \ldots, M. \tag{8.24}$$

For the unbounded support vectors $\mathbf{x}_i$,

$$\begin{aligned} R^2 &= \|\phi(\mathbf{x}_i) - \mathbf{a}\|^2 \\ &= (\phi(\mathbf{x}_i) - \mathbf{a})^{\top}(\phi(\mathbf{x}_i) - \mathbf{a}) \\ &= K(\mathbf{x}_i, \mathbf{x}_i) - 2\sum_{j \in S} \alpha_j K(\mathbf{x}_i, \mathbf{x}_j) + \sum_{\substack{j \in S, \\ k \in S}} \alpha_j\, \alpha_k K(\mathbf{x}_j, \mathbf{x}_k), \end{aligned} \tag{8.25}$$

where $S$ is the set of support vector indices and we used (8.20). The bounded support vectors with $\xi_i > 0$ are outside of the hypersphere and thus are outliers.

An unknown data sample $\mathbf{x}$ is inside the hypersphere if

$$K(\mathbf{x}, \mathbf{x}) - 2\sum_{i \in S} \alpha_i K(\mathbf{x}, \mathbf{x}_i) + \sum_{\substack{i \in S, \\ j \in S}} \alpha_i\, \alpha_j K(\mathbf{x}_i, \mathbf{x}_j) \le R^2. \tag{8.26}$$

According to the computer experiments with polynomial kernels, an approximated region in the input space included a redundant space that did not include training data. But the use of RBF kernels showed better results [2]. To reduce the redundant space, Tax and Juszczak [6] used the kernel PCA to rescale the data in the feature space to the unit variance before one-class classification. To improve domain description capability, Tsang et al. [7] proposed using the Mahalanobis distance, instead of the Euclidian distance.

---

[1] When kernels, such as RBF kernels, that depend only on $\mathbf{x} - \mathbf{x}'$ are used, the linear term in (8.22) is constant from (8.23). Then it is shown that the problem is equivalent to maximizing the margin in separating data from the origin by the hyperplane [5, pp. 230–234].

One-class support vector machines are applied to multiclass pattern classification, namely, each class is approximated by a one-class support vector machine and a data sample is classified into a class according to the distances from class centers or the posterior probabilities calculated by them [8, 9].

*Example 8.1.* In a one-dimensional problem, assume that we have two data: $x_1 = -1$ and $x_2 = 1$. For linear kernels, the one-class classifier is given by

$$\text{maximize} \quad Q_d(\boldsymbol{\alpha}) = \alpha_1 + \alpha_2 - (\alpha_1 - \alpha_2)^2 \tag{8.27}$$

$$\text{subject to} \quad \alpha_1 + \alpha_2 = 1, \quad 0 \leq \alpha_1 \leq C, \quad 0 \leq \alpha_2 \leq C. \tag{8.28}$$

Let $C > 1$. Then the optimum solution is given by $\alpha_1 = \alpha_2 = 0.5$, and $a = 0$ and $R = 1$. Thus, the hypersphere is given by $x^2 = 1$, which is the minimum hypersphere that includes the two data in the input space.

For the polynomial kernel with degree 2, the one-class classifier is given by

$$\text{maximize} \quad Q_d(\boldsymbol{\alpha}) = 4\alpha_1 + 4\alpha_2 - 4(\alpha_1^2 + \alpha_2^2) \tag{8.29}$$

$$\text{subject to} \quad \alpha_1 + \alpha_2 = 1, \quad 0 \leq \alpha_1 \leq C, \quad 0 \leq \alpha_2 \leq C. \tag{8.30}$$

Let $C > 1$. Then the optimum solution is given by $\alpha_1 = \alpha_2 = 0.5$. The center vector is $\mathbf{a} = 0.5 \left(\phi(x_1) + \phi(x_2)\right) = (1, 0, 1)^\top$ for the coordinates of $(x^2, \sqrt{2}x, 1)^\top$, $R = \sqrt{2}$, and the hypersphere is given by $x^4 = 1$, which is equivalent to $x^2 = 1$.

Yuan and Casasent [10] proposed a support vector representation machine (SVRM) using the fact that with the RBF kernel data are on the surface of the unit hypersphere centered at the origin of the feature space because $\phi^\top(\mathbf{x})\phi(\mathbf{x}) = \exp(-\gamma\|\mathbf{x} - \mathbf{x}\|^2) = 1$. In the SVRM, we determine vector $\mathbf{h}$, in the feature space with the minimum Euclidean norm, that satisfies $\mathbf{h}^\top\phi(\mathbf{x}_i) \geq 1$ for $i = 1, \ldots, M$:

$$\text{minimize} \quad Q_p'(\mathbf{h}) = \frac{1}{2}\|\mathbf{h}\|^2 \tag{8.31}$$

$$\text{subject to} \quad \mathbf{h}^\top \phi(\mathbf{x}_i) \geq 1 \quad \text{for } i = 1, \ldots, M. \tag{8.32}$$

The dimension of the feature space associated with the RBF kernel is infinite, but for simplicity the example of an SVRM shown in Fig. 8.1 assumes a two-dimensional feature space. In the figure, let training data lie on the arc between $\phi(\mathbf{x}_1)$ and $\phi(\mathbf{x}_2)$ and $\theta$ be the angle between $\phi(\mathbf{x}_1)$ (or $\phi(\mathbf{x}_2)$) and $\mathbf{h}$. Then

$$\mathbf{h}^\top\phi(\mathbf{x}_1) = \|\mathbf{h}\|\cos\theta = 1, \quad \mathbf{h}^\top\phi(\mathbf{x}_2) = 1 \tag{8.33}$$

must be satisfied. Therefore $\mathbf{x}_1$ and $\mathbf{x}_2$ are support vectors and $\phi(\mathbf{x}_1)$ (or $\phi(\mathbf{x}_2)$) and $\mathbf{h} - \phi(\mathbf{x}_1)$ (or $\mathbf{h} - \phi(\mathbf{x}_2)$) are orthogonal.

Any data sample $\mathbf{x}$ on the arc satisfies $\mathbf{h}^\top \phi(\mathbf{x}) \geq 1$, and for any data sample $\mathbf{x}$ that is not, $\mathbf{h}^\top \phi(\mathbf{x}) < 1$.

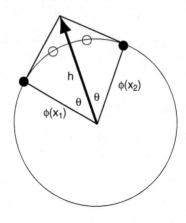

● : Support vectors

**Fig. 8.1** Concept of an SVRM

A soft-margin SVRM is given by

$$\text{minimize} \quad Q'_{\text{p}}(\mathbf{h}, \boldsymbol{\xi}) = \frac{1}{2}\|\mathbf{h}\|^2 + C\sum_{i=1}^{M}\xi_i \tag{8.34}$$

$$\text{subject to} \quad \mathbf{h}^\top \phi(\mathbf{x}_i) \geq 1 - \xi_i, \quad \xi_i \geq 0 \quad \text{for } i = 1, \ldots, M, \tag{8.35}$$

where $\xi_i\,(\geq 0)$ are slack variables associated with $\mathbf{x}_i$, $\boldsymbol{\xi} = (\xi_1, \ldots, \xi_M)^\top$, and $C$ is the margin parameter.

Introducing the nonnegative Lagrange multipliers $\alpha_i$ and $\beta_i$, we obtain

$$Q'_{\text{d}}(\mathbf{h}, \boldsymbol{\xi}, \boldsymbol{\alpha}, \boldsymbol{\beta}) = \frac{1}{2}\|\mathbf{h}\|^2 + C\sum_{i=1}^{M}\xi_i$$
$$- \sum_{i=1}^{M}\alpha_i\left(\mathbf{h}^\top\phi(\mathbf{x}_i) - 1 + \xi_i\right) - \sum_{i=1}^{M}\beta_i\,\xi_i, \tag{8.36}$$

where $\boldsymbol{\alpha} = (\alpha_1, \ldots, \alpha_M)^\top$ and $\boldsymbol{\beta} = (\beta_1, \ldots, \beta_M)^\top$.

The following conditions must be satisfied for the optimal solution:

$$\frac{\partial Q(\mathbf{h}, \boldsymbol{\xi}, \boldsymbol{\alpha}, \boldsymbol{\beta})}{\partial \mathbf{h}} = \mathbf{0}, \tag{8.37}$$

$$\frac{\partial Q(\mathbf{h}, \boldsymbol{\xi}, \boldsymbol{\alpha}, \boldsymbol{\beta})}{\partial \boldsymbol{\xi}} = \mathbf{0}. \tag{8.38}$$

Using (8.36), (8.37) and (8.38) reduce, respectively, to

$$\mathbf{h} = \sum_{i=1}^{M} \alpha_i \, \phi(\mathbf{x}_i), \tag{8.39}$$

$$\alpha_i + \beta_i = C, \quad \alpha_i \geq 0, \quad \beta_i \geq 0 \quad \text{for } i = 1, \dots, M. \tag{8.40}$$

Thus substituting (8.39) and (8.40) into (8.36), we obtain the following dual problem.

$$\text{maximize} \quad Q_{\mathrm{d}}'(\boldsymbol{\alpha}) = \sum_i^{M} \alpha_i - \frac{1}{2} \sum_{i,j=1}^{M} \alpha_i \, \alpha_j \, K(\mathbf{x}_i, \mathbf{x}_j) \tag{8.41}$$

$$\text{subject to} \quad 0 \leq \alpha_i \leq C \quad \text{for } i = 1, \dots, M, \tag{8.42}$$

where $K(\mathbf{x}_i, \mathbf{x}_j) = \exp(-\gamma \|\mathbf{x}_i - \mathbf{x}_j\|^2)$.

This problem is very similar to an L1 support vector machine. In the L1 support vector machine, deleting the equality constraint that corresponds to optimization of the bias term and setting $y_i = 1$ for $i = 1, \dots, M$, we obtain the associated SVRM.

## 8.2 Extension to Clustering

Conventional clustering methods such as $k$-means clustering algorithm and fuzzy $c$-means clustering algorithms can be extended to feature space [11, 12].

The domain description discussed in Section 8.1 defines the region of data by a hypersphere in the feature space. The hypersphere in the feature space corresponds to clustered regions in the input space. Thus domain description can be used for clustering. In the following we discuss how domain description can be extended to clustering according to Ben-Hur et al.'s method [13].

We assume that there are no outliers, namely, all the data are in or on the hypersphere. The problem is how to determine the clusters in the input space. Using Fig. 8.2, we explain the idea discussed in [13]. In the figure, two clusters are generated in the input space by approximating the region of data by a hypersphere in the feature space. The insides of the two regions correspond to the inside of the hypersphere in the feature space. In the figure, data 1, 2, and 3 are in Cluster 1 but data sample 4 is in Cluster 2. If we move along the line segment connecting data 1 and 4 from data sample 1, we go out of Cluster 1 and into Cluster 2, namely, in the feature space, we go out of and then back to the hypersphere. Therefore, if part of the line segment connecting two data is out of the hypersphere, the two data may belong to different clusters. But this is not always true, as the line segment connecting data 1 and 3 shows.

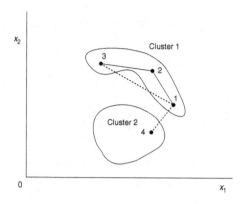

**Fig. 8.2** Cluster assignment

To avoid this, for all the data pairs we check if the associated line segments are in the hypersphere. If a line segment is included, we consider that there is a path between the two data. Then we generate sets of data that are connected by paths. Each set constitutes a cluster. In [13] the line segments are considered to be in the hypersphere if sampled data on the line segments satisfy (8.26).

In Fig. 8.2, there are paths between data 1 and 2 and between data 2 and 3. But there is no path connecting Sample 4 and the remaining data. Thus data 1 to 3 constitute a cluster, and so does Sample 4.

If cluster regions in the input space are convex, by the preceding method, all the data in a cluster are selected and no other data are selected. But if a cluster region is concave, data in a cluster may be separated into more than one set. For example, in Fig. 8.2 if part of the line segment connecting data 2 and 3 is out of the cluster region, the line segment does not form a path. Thus if there are only data 1 to 3, they are divided into two sets: $\{1, 2\}$ and $\{3\}$.

If $C \geq 1$, no outliers, namely, no bounded support vectors, are detected. Thus, if clusters are considered to be well separated and no outliers are considered to be included, we set $C = 1$. If clusters are considered to overlap or outliers are considered to be included, we set $C$ smaller than 1. From (8.7) and $0 \leq \alpha_i \leq C$,

$$|B| \leq 1/C, \tag{8.43}$$

where $|B|$ is the number of bounded support vectors.

According to this procedure for defining clusters, bounded support vectors do not belong to any cluster. But if bounded support vectors are caused by overlapping of clusters, we may include them to the nearest clusters [13].

The most crucial part of clustering is to check the paths for all pairs of data. To speed this up, in [13], only the pairs of data with one data sample being an unbounded support vector are checked. But if many support vectors are bounded, this may lead to false cluster generation. To avoid this situation, Yang et al. [14] proposed using a proximity graph, in which a node corresponds to a data sample, to model the nearness of data and to check pairs of data that are directly connected in the graph. They used the following proximity graphs:

1. A complete graph, in which all the data are connected;
2. A support vector graph, in which at least one of the data connected to a branch is an unbounded support vector;
3. The Delaunay diagram, which is the dual of the Voronoi diagram and which is composed of adjacent triangles whose edges are the data;
4. The minimum spanning tree with the minimum sum of distances; and
5. $k$ nearest neighbors, in which a data sample is connected to $k$ nearest data.

According to the experiments, Delaunay diagrams and $k$ nearest neighbors with $k$ greater than 4 showed comparable clustering performance with complete graphs with much faster clustering.

Lee and Lee [15] introduced a generalized gradient descent process for the squared radial distance $\|\phi(\mathbf{x}) - \mathbf{a}\|^2$, in which data in the same cluster converge to one of the equilibrium points for that cluster. By integrating the gradient descent process starting with training data, training data are clustered into local minima, the number of which is smaller than that of the training data. Then checking the paths for all the pairs of local minima, the training data are clustered.

Ban and Abe [16] spatially chunked the training data to speed training and to optimize RBF parameters. Chu et al. [17] speeded training, by training a one-class classifier for randomly selected data and iteratively updating the classifier, adding the data outside the hypersphere.

# References

1. J. Ma and S. Perkins. Time-series novelty detection using one-class support vector machines. In *Proceedings of International Joint Conference on Neural Networks (IJCNN 2003)*, volume 3, pages 1741–1745, Portland, OR, 2003.
2. D. M. J. Tax and R. P. W. Duin. Support vector domain description. *Pattern Recognition Letters*, 20(11–13):1191–1199, 1999.
3. D. M. J. Tax and R. P. W. Duin. Outliers and data descriptions. In *Proceedings of the Seventh Annual Conference of the Advanced School for Computing and Imaging (ASCI 2001)*, pages 234–241, Heijen, The Netherlands, 2001.
4. D. M. J. Tax and R. P. W. Duin. Support vector data description. *Machine Learning*, 54(1):45–66, 2004.
5. B. Schölkopf and A. J. Smola. *Learning with Kernels: Support Vector Machines, Regularization, Optimization, and Beyond*. MIT Press, Cambridge, MA, 2002.
6. D. M. J. Tax and P. Juszczak. Kernel whitening for one-class classification. In S.-W. Lee and A. Verri, editors, *Pattern Recognition with Support Vector Machines: Proceedings of First International Workshop, SVM 2002, Niagara Falls, Canada*, pages 40–52, Springer-Verlag, Berlin, Germany, 2002.

7. I. W. Tsang, J. T. Kwok, and S. Li. Learning the kernel in Mahalanobis one-class support vector machines. In *Proceedings of the 2006 International Joint Conference on Neural Networks (IJCNN 2006)*, pages 2148–2154, Vancouver, Canada, 2006.

8. T. Ban and S. Abe. Implementing multi-class classifiers by one-class classification methods. In *Proceedings of the 2006 International Joint Conference on Neural Networks (IJCNN 2006)*, pages 719–724, Vancouver, Canada, 2006.

9. D. Lee and J. Lee. Domain described support vector classifier for multi-classification problems. *Pattern Recognition*, 40(1):41–51, 2007.

10. C. Yuan and D. Casasent. Support vector machines for class representation and discrimination. In *Proceedings of International Joint Conference on Neural Networks (IJCNN 2003)*, volume 2, pages 1611–1616, Portland, OR, 2003.

11. M. Girolami. Mercer kernel-based clustering in feature space. *IEEE Transactions on Neural Networks*, 13(3):780–784, 2002.

12. S. Miyamoto and D. Suizu. Fuzzy c-means clustering using transformations into high dimensional spaces. In *Proceedings of the First International Conference on Fuzzy Systems and Knowledge Discovery (FSKD '02)*, volume 2, pages 656–660, Singapore, 2002.

13. A. Ben-Hur, D. Horn, H. T. Siegelmann, and V. Vapnik. Support vector clustering. *Journal of Machine Learning Research*, 2:125–137, 2001.

14. J. Yang, V. Estivill-Castro, and S. K. Chalup. Support vector clustering through proximity graph modelling. In *Proceedings of the Ninth International Conference on Neural Information Processing (ICONIP '02)*, volume 2, pages 898–903, Singapore, 2002.

15. J. Lee and D. Lee. An improved cluster labeling method for support vector clustering. *Pattern Analysis and Machine Intelligence*, 27(3):461–464, 2005.

16. T. Ban and S. Abe. Spatially chunking support vector clustering algorithm. In *Proceedings of International Joint Conference on Neural Networks (IJCNN 2004)*, volume 1, pages 413–418, Budapest, Hungary, 2004.

17. C. S. Chu, I. W. Tsang, and J. T. Kwok. Scaling up support vector data description by using core-sets. In *Proceedings of International Joint Conference on Neural Networks (IJCNN 2004)*, volume 1, pages 425–430, Budapest, Hungary, 2004.

# Chapter 9
# Maximum-Margin Multilayer Neural Networks

Three-layer (one-hidden-layer) neural networks are known to be universal approximators [1, 2], in that they can approximate any continuous functions with any accuracy. However, training a multilayer neural network by the back-propagation algorithm is slow and the generalization ability depends on the initial weights. As for improving the generalization ability, there are several approaches. One way is to add a regularization term, e.g., the square sum of the weights, in the objective function of back-propagation training. Or we may train a support vector machine with sigmoid functions as kernels. But because sigmoid functions satisfy Mercer's condition only for specific parameter values, several approaches have been proposed to overcome this problem (e.g., [3]).

Instead of using support vector machines, we may maximize margins [4–11]. For example, Jayadeva et al. [9] formulated a decision tree by linear programming and maximized margins by support vector machine-based training. Based on the network synthesis theory [12, 13], Nishikawa and Abe [10] trained a three-layer neural network layer by layer, maximizing margins.

In this chapter, based on [10], we discuss how to maximize margins of a three-layer neural network classifier that is trained layer by layer and compare the recognition performance with that of support vector machines.

## 9.1 Approach

The CARVE (constructive algorithm for real-valued examples) algorithm [12, 13] guarantees that any pattern classification problem can be synthesized in three layers if we train the hidden layer in the following way. First, we separate a part of the (or the whole) data belonging to a class from the remaining data by a hyperplane. Then we remove the separated data from the training

S. Abe, *Support Vector Machines for Pattern Classification*,
Advances in Pattern Recognition, DOI 10.1007/978-1-84996-098-4_9,
© Springer-Verlag London Limited 2010

data. We repeat this procedure until only the data belonging to one class remain.

In the following, we discuss a method for training neural network classifiers based on the CARVE algorithm. To improve generalization ability, we maximize margins of hidden-layer hyperplanes and output-layer hyperplanes. Because the training data on one side of the hidden-layer hyperplane need to belong to one class, we cannot apply the quadratic optimization technique used for training support vector machines. Therefore, we extend the heuristic training method called *direct SVMs* [14], which sequentially searches support vectors. For the output layer, because there is no such restriction, we use the quadratic optimization technique.

## 9.2 Three-Layer Neural Networks

Figure 9.1 shows a three-layer neural network with one hidden layer. In the figure, the input is fed from the left and each layer is called, from left to right, *input layer*, *hidden layer*, and *output layer*. We may have more than one hidden layer. The input layer consists of input neurons and a bias neuron whose input is constant (usually 1). The hidden layer consists of hidden neurons and a bias neuron and the output layer consists of output neurons. The number of output neurons is the number of classes for pattern classification and the number of outputs to be synthesized for function approximation. The input neurons and the hidden neurons, and the hidden neurons and the output neurons are fully connected by weights. The input and output of the $i$th neuron of the $k$th layer are denoted by $x_i(k)$ and $z_i(k)$, respectively, and the weight between the $i$th neuron of the $k$th layer and the $j$th neuron of the $(k + 1)$st layer is denoted by $w_{ji}(k)$.

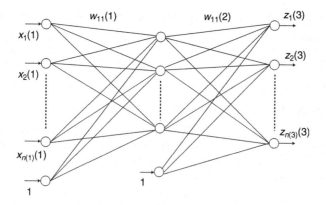

**Fig. 9.1** Structure of a three-layer neural network

Inputs to the input layer are output by the input neuron without change and the output of the bias neuron is 1, namely,

$$z_j(1) = \begin{cases} x_j(1) & \text{for } j = 1, \ldots, n(1), \\ 1 & \text{for } j = n(1) + 1, \end{cases} \tag{9.1}$$

where $n(1)$ is the number of input neurons.

Outputs of the input (hidden) neurons and the bias neuron are multiplied by the weights and their sums are input to the hidden (output) neurons as follows:

$$x_i(k+1) = \mathbf{w}_i^\top(k)\,\mathbf{z}(k) \qquad \text{for } i = 1, \ldots, n(k+1), \quad k = 1, 2, \tag{9.2}$$

where $n(k+1)$ is the number of the $(k+1)$st-layer neurons and

$$\begin{aligned} \mathbf{x}(k) &= \big(x_1(k), \ldots, x_{n(k)}(k)\big)^\top, \\ \mathbf{z}(k) &= \big(z_1(k), \ldots, z_{n(k)}(k), z_{n(k)+1}(k)\big)^\top, \\ \mathbf{w}_i(k) &= \big(w_{i1}(k), \ldots, w_{i,n(k)}(k), w_{i,n(k)+1}(k)\big)^\top \\ & \qquad \text{for } i = 1, \ldots, n(k+1). \end{aligned} \tag{9.3}$$

The output function of the hidden (output) neurons is given by the sigmoid function shown in Fig. 9.2, namely, their outputs are given by

$$z_i(k+1) = \frac{1}{1 + \exp\left(-\dfrac{x_i(k+1)}{T}\right)} \qquad \text{for } i = 1, \ldots, n(k+1), \tag{9.4}$$

where $T$ is a positive parameter that controls the slope of the sigmoid function, and usually $T = 1$.

Let the training inputs be $\mathbf{x}_1, \ldots, \mathbf{x}_M$, where $M$ is the number of training data, and the desired outputs be $\mathbf{s}_1, \ldots, \mathbf{s}_M$, respectively. Then the training of the network is to determine all the weights $\mathbf{w}_i(k)$ so that for the training input $\mathbf{x}_i$, the output becomes $\mathbf{s}_i$. Thus we want to determine the weights $\mathbf{w}_i(k)$ so that the sum-of-squares error between the target values and the network outputs

$$E = \frac{1}{2} \sum_{l=1}^{M} (\mathbf{z}_l(3) - \mathbf{s}_l)^\top (\mathbf{z}_l(3) - \mathbf{s}_l) \tag{9.5}$$

is minimized, where $\mathbf{z}_l(3)$ is the network output for $\mathbf{x}_l$.

Or for the input–output pairs $(\mathbf{x}_l, \mathbf{s}_l)$, we determine the weights $\mathbf{w}_i(k)$ so that

$$|z_{lj}(3) - s_{lj}| \le \varepsilon(3) \quad \text{for } j = 1, \ldots, n(3), \quad l = 1, \ldots, M \tag{9.6}$$

are satisfied, where $\varepsilon(3)\,(> 0)$ is the tolerance of convergence for the output neuron outputs.

When the network is used for pattern classification, the $i$th output neuron corresponds to class $i$, and for the training input $\mathbf{x}_l$ belonging to class $i$, the target values of the output neurons $j\,(j = 1, \ldots, n(3))$ are assigned as follows:

$$s_{lj} = \begin{cases} 1 & \text{for } j = i, \\ 0 & \text{for } j \neq i. \end{cases} \tag{9.7}$$

Because $E$ given by (9.5) shows how well the neural network memorizes the training data, too small a value of $E$, i.e., overfitting may result in worsening the generalization ability of the neural network. To avoid this, the validation data set is prepared as well as the training data set, and training is stopped when the value of $E$ for the validation data set starts to increase. Or, we may add a regularization term such as

$$\lambda \sum_{k=1}^{2} \sum_{i=1}^{n(k+1)} \mathbf{w}_i^\top(k)\mathbf{w}_i(k) \tag{9.8}$$

to (9.5) to prevent overfitting, where $\lambda\,(> 0)$ is the regularization constant.

In the back-propagation algorithm, for each training sample, weights in each layer are corrected by the steepest descent method so that the output errors are reduced:

$$w_{ij}^{\text{new}}(k) = w_{ij}^{\text{old}}(k) - \alpha_{\text{lr}}\frac{\partial E_l}{\partial w_{ij}^{\text{old}}(k)}, \tag{9.9}$$

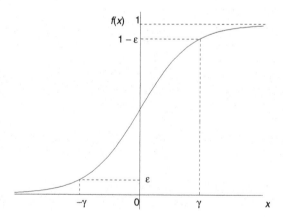

**Fig. 9.2** Sigmoid function

where $\alpha_{\mathrm{lr}}$ $(> 0)$ is the learning rate and $E_l$ is the square error for the $l$th training sample:

$$E_l = \frac{1}{2}(\mathbf{z}_l(3) - \mathbf{s}_l)^\top (\mathbf{z}_l(3) - \mathbf{s}_l). \tag{9.10}$$

Because one data sample is processed at a time, training by the back-propagation algorithm is slow for a large problem.

In the following, we discuss training of three-layer networks layer by layer, maximizing margins.

## 9.3 CARVE Algorithm

According to the CARVE algorithm [12, 13], any pattern classification problem can be synthesized in three layers. In the hidden layer, we determine the hyperplane so that the data of a single class exist on one side of the hyperplane. Then the separated data are removed from the training data. The hidden-layer training is completed when only data of a single class remain.

We explain the procedure using the example shown in Fig. 9.3. The class data shown in triangles include a data sample, which is the farthest from the center of the training data shown in the asterisk. Thus we separate the data of this class from the remaining data.

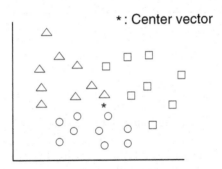

**Fig. 9.3** Sample data. Reprinted from [10, p. 323, ©IEEE 2002]

As shown in Fig. 9.4, a hyperplane is determined so that the data of this class are on one side of the hyperplane. Then the separated data in gray are removed from the training data. We call the data that are used for determining hyperplane *active data* and the deleted data *inactive data*. For the reduced training data, the class data shown in squares include a data sample that is farthest from the center. Thus we determine a hyperplane so that only the data of this class is on one side of the hyperplane, as shown in Fig. 9.5.

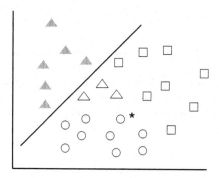

**Fig. 9.4** Create the first hyperplane. Reprinted from [10, p. 323, ©IEEE 2002]

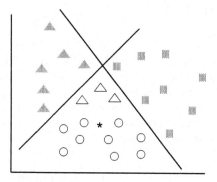

**Fig. 9.5** Create the second hyperplane. Reprinted from [10, p. 323, ©IEEE 2002]

We repeat this procedure until the remaining training data belong to a single class (see Fig. 9.6). We say that a hyperplane satisfies the CARVE condition if on one side of the hyperplane data of only one class exist.

In the output layer, for class $i$ we determine a hyperplane so that class $i$ data are separated from the remaining data. In this way we can construct a three-layer neural network for an $n$-class problem.

## 9.4 Determination of Hidden-Layer Hyperplanes

According to the CARVE algorithm, on one side of the hyperplane, data of different classes cannot coexist. We may realize this constraint, as discussed in Section 2.6.8, by training a support vector machine with different margin parameters. Because a hidden-layer hyperplane is determined in the input space, linear kernels need to be used. Therefore, the problem may be insepa-

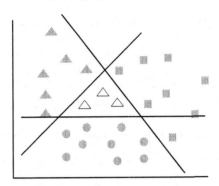

**Fig. 9.6** All active data belong to one class. Reprinted from [10, p. 323, ©IEEE 2002]

rable. However, if the problem is inseparable, there are cases where degenerate solutions are obtained. Thus, we cannot use a support vector machine.

To overcome this problem, we extend the direct SVM [14], which determines the optimal hyperplane geometrically. Let the farthest data sample from the center of the training data belong to class $i$. Initially, we consider separating class $i$ data from the remaining data and we set the target values of the class $i$ data to $+1$ and those of the remaining data to $-1$. We call the side of the hyperplane where the data with the positive targets reside the *positive side of the hyperplane.*

First, as initial support vectors we choose the data pair that has the minimum distance among the data pairs with opposite target values. The initial hyperplane is determined so that it goes through the center of the data pair and orthogonal to the line segment that connects the data pair. If there are data with negative targets on the positive side of the hyperplane, these data violate the CARVE condition. Thus, to satisfy the CARVE condition, we rotate the hyperplane until no data violate the condition. The previously violating data are added to the support vectors. In the following, we describe the procedure in more detail.

### 9.4.1 Rotation of Hyperplanes

Let the initial support vectors be $\mathbf{x}_0^+$ and $\mathbf{x}_0^-$. Then, the weight vector of the initial hyperplane is given by

$$\mathbf{w}_0 = \mathbf{x}_0^+ - \mathbf{x}_0^- \tag{9.11}$$

and the hyperplane is given by

$$\mathbf{w}_0^\top \mathbf{x} - \mathbf{w}_0^\top \mathbf{c}_0 = 0, \tag{9.12}$$

where $c_0$ is the center of the data pair $\{x_0^+, x_0^-\}$. If there are no data that violate the CARVE condition in the remaining data, we finish training. If there are data of the negative target, we add the most-violating data sample to the support vectors. Let the obtained support vector be $x_1^-$ and the center of the data pair $\{x_0^+, x_1^-\}$ be $c_1$. The hyperplane is updated so that it passes through the two points $c_0$ and $c_1$. If there are still violating data, we repeat updating.

First, we show the updating method in the two-dimensional space using Fig. 9.7, which shows the initial hyperplane.

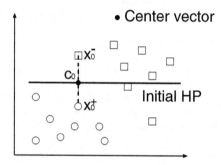

• Center vector

$\Box X_0^-$

$C_0$

$\bigcirc X_0^+$

Initial HP

**Fig. 9.7** Initial hyperplane. Reprinted from [10, p. 324, ©IEEE 2002]

Because there are data that violate the CARVE condition, we rotate the hyperplane as shown in Fig. 9.8.

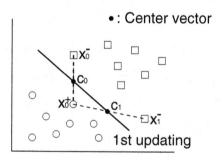

• : Center vector

$\Box X_0^-$

$C_0$

$X_0^+\bigcirc$

$C_1$

$\Box X_1^-$

1st updating

**Fig. 9.8** First updating of the hyperplane. Reprinted from [10, p. 324, ©IEEE 2002]

Define $r_1$ by $r_1 = c_1 - c_0$, where $r_1$ is a vector that the rotated hyperplane includes (see Fig. 9.9). We calculate the weight vector $w_1$ that is orthogonal to $r_1$:

$$w_1 = w_0 - w_0^\top r'_1 r'_1, \tag{9.13}$$

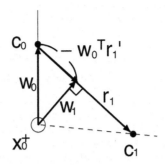

**Fig. 9.9** Detail of updating. Reprinted from [10, p. 324, ©IEEE 2002]

where $\mathbf{r}_1'$ is the normalized vector of $\mathbf{r}_1$. The resulting hyperplane satisfies the CARVE condition.

Now consider the $m$th updating in the $n$-dimensional space, which requires the hyperplane to pass through the centers of support vector pairs, $\mathbf{c}_0, \ldots, \mathbf{c}_m$. Thus the maximum number of updatings is $(n - 1)$. This is because, if we update the hyperplane $(n - 1)$ times, the hyperplane must go through $n$ centers, and no further rotation is possible. For the $m$th updating, we find the most-violating data sample with the negative target. Let the data sample be $\mathbf{x}_m^-$. Then, because $\mathbf{c}_m$ is the center of $\{\mathbf{x}_0^+, \mathbf{x}_m^-\}$, $\mathbf{r}_m$, which is included in the rotated hyperplane, is given by $\mathbf{r}_m = \mathbf{c}_m - \mathbf{c}_0$.

The hyperplane with the coefficient vector $\mathbf{w}_{m-1}$ includes the vectors $\{\mathbf{r}_1, \ldots, \mathbf{r}_{m-1}\}$. Let the vectors $\{\mathbf{p}_1, \ldots, \mathbf{p}_{m-1}\}$ be the orthogonal system of $\{\mathbf{r}_1, \ldots, \mathbf{r}_{m-1}\}$. Obviously, the hyperplane with the weight vector $\mathbf{w}_{m-1}$ includes $\{\mathbf{p}_1, \ldots, \mathbf{p}_{m-1}\}$.

We use the Gram–Schmidt orthogonalization to calculate $\mathbf{p}_m$:

$$\mathbf{p}_m = \mathbf{r}_m - \sum_{k=1}^{m-1} \mathbf{r}_m^\top \mathbf{p}_k' \mathbf{p}_k', \tag{9.14}$$

where $\mathbf{p}_k'$ is the normalized vector of $\mathbf{p}_k$. The weight vector $\mathbf{w}_{m-1}$ of the hyperplane is determined so that it is orthogonal to the orthogonal system $\{\mathbf{p}_1, \ldots, \mathbf{p}_{m-1}\}$. For the $m$th updating, it is updated to be orthogonal to $\mathbf{p}_m$, namely, $\mathbf{w}_m$ is obtained by rotating the hyperplane in the direction of $\mathbf{p}_m$ with $\mathbf{c}_0$ as the center:

$$\mathbf{w}_m = \mathbf{w}_{m-1} - \mathbf{w}_{m-1}^\top \mathbf{p}_m' \mathbf{p}_m'. \tag{9.15}$$

Then the updated hyperplane includes the vectors $\{\mathbf{r}_1, \ldots, \mathbf{r}_m\}$.

## 9.4.2 Training Algorithm

A procedure for determining hidden-layer hyperplanes is as follows:

1. Calculate the center of the active data, where initially all the training data are active.
2. Find the data sample that is the farthest from the center. Set the targets of the data that belong to the same class with the farthest data sample to 1 and those of the remaining data to $-1$.
3. Find the nearest data pair of opposite targets among all active data pairs. Let them be $(\mathbf{x}_0^+, \mathbf{x}_0^-)$. The center $\mathbf{c}_0$ and the weight $\mathbf{w}_0$ are given by

$$\mathbf{c}_0 = \frac{1}{2}(\mathbf{x}_0^+ + \mathbf{x}_0^-), \tag{9.16}$$

$$\mathbf{w}_0 = \mathbf{x}_0^+ - \mathbf{x}_0^-. \tag{9.17}$$

4. Calculate the values of the decision function:

$$D(\mathbf{x}_i) = \mathbf{w}_m^\top \mathbf{x}_i - \mathbf{w}_m^\top \mathbf{c}_0 \quad \text{for } i = 1, \ldots, M. \tag{9.18}$$

5. If all the data with negative targets satisfy

$$-D(\mathbf{x}_i^-) \geq C(2)\, D(\mathbf{x}_0^+) \quad \text{for } i \in B, \tag{9.19}$$

   the hyperplane that satisfies the CARVE condition is found; go to Step 8, where $B$ is the indices of data with negative targets and $C(2)$ plays the role of the soft margin. Otherwise, go to Step 6. Usually, $C(2)$ is set to $C(2) < 1$. If $C(2)$ is negative, the data with a negative target may exist on the positive side of the hyperplane. Thus for negative $C(2)$, the CARVE condition is violated.
6. In the $m$th updating, include the most-violating data sample $x_m^-$ in the support vectors. Then calculate $\mathbf{c}_m = (\mathbf{x}_0^+ + \mathbf{x}_m^-)/2$ and $\mathbf{r}_m = \mathbf{c}_m - \mathbf{c}_0$. From $\{\mathbf{r}_1, \ldots, \mathbf{r}_m\}$, the $m$th component of the orthogonal system, $\mathbf{p}_m$, is

$$\mathbf{p}_m = \mathbf{r}_m - \sum_{k=1}^{m-1} \mathbf{r}_m^\top \mathbf{p}_k'\, \mathbf{p}_k'. \tag{9.20}$$

   The weight vector $\mathbf{w}_m$ is written as follows:

$$\mathbf{w}_m = \mathbf{w}_{m-1} - \mathbf{w}_{m-1}^\top \mathbf{p}'_m\, \mathbf{p}'_m. \tag{9.21}$$

   The orthogonal vector $\mathbf{w}_m$ for the hyperplane that passes through $\mathbf{c}_0, \mathbf{c}_1$, $\ldots, \mathbf{c}_m$ is obtained by updating in this way.
7. If updating was repeated $n$ or $M$ times, updating is failed for the current data pair. Return to Step 4, setting the next nearest data pair as $(\mathbf{x}_0^+, \mathbf{x}_0^-)$

and recalculating the associated weight vector. Otherwise, return to Step 4 and repeat training.
8. If the active data belong to one class, terminate the algorithm. Otherwise, make the data on the positive side of the hyperplane inactive and return to Step 2.

## 9.5 Determination of Output-Layer Hyperplanes

We determine the output-layer hyperplanes using the same technique that trains support vector machines. The hyperplane is obtained by solving the following dual problem:

$$\text{maximize} \quad Q(\boldsymbol{\rho}) = \sum_{i=1}^{M} \rho_i - \frac{1}{2} \sum_{i,j=1}^{M} \rho_i \, \rho_j \, y_i \, y_j \, \mathbf{z}_i^\top \, \mathbf{z}_j \tag{9.22}$$

$$\text{subject to} \quad \sum_{i=1}^{M} y_i \, \rho_i = 0, \quad 0 \leq \rho_i \leq C(3) \quad \text{for } i = 1, \ldots, M, \tag{9.23}$$

where $\boldsymbol{\rho} = (\rho_1, \ldots, \rho_M)^\top$ and $\rho_i$ are the Lagrange multipliers, $C(3)$ is the margin parameter for the output layer, $\mathbf{z}_i$ are the output vectors of the hidden neurons, and $y_i$ are the associated labels. For training the output layer for class $i$, we set $y_i = 1$ for the class $i$ data and $-1$ for the data belonging to the remaining classes.

## 9.6 Determination of Parameter Values

Usually, the value of $T$ in the sigmoid function is set to 1. Here, we determine the value of $T$ so that the outputs of the unbounded support vectors are the same for hidden and output layers.

From (9.4), the $j$th output of the $i$th layer for the unbounded support vector is given by

$$z_j^*(i+1) = \frac{1}{1 + \exp(-(\mathbf{w}_j^\top(i)\,\mathbf{x}_j^*(i) + b_j(i))/T_j(i))}, \tag{9.24}$$

where $\mathbf{x}_j^*(i)$ and $T_j(i)$ are the support vector and the slope parameter of the $j$th hyperplane of the $i$th layer, respectively. We set the value of $\varepsilon(i+1)$ so that $0 < \varepsilon(i+1) < 0.5$. Then, setting $z_j^*(i+1) = 1 - \varepsilon(i+1)$, we get

$$T_j(i) = \frac{-\mathbf{x}_j^{*\top}(i)\,\mathbf{w}_j(i) - b_j(i)}{\log\left(\frac{\varepsilon(i+1)}{1-\varepsilon(i+1)}\right)}. \tag{9.25}$$

## 9.7 Performance Evaluation

We compared the performance of maximum-margin neural networks (MM NNs) against one-against-all fuzzy support vector machines (FSVMs) and multilayer neural networks trained by the back-propagation algorithm (BP) using the data sets listed in Table 1.3. We used a Pentium III 1 GHz personal computer.

In training FSVM and the output layer of MM NN, we used the primal–dual interior-point method combined with variable-size chunking (50 data were added). For FSVM we used polynomial kernels.

In training neural networks by the back-propagation algorithm, we set $T(2) = T(3) = 1$, the learning rate to be 1, and the maximum number of epochs to be 15,000. The number of hidden units was set to be equal to or smaller than the number generated by MM NN. The neural network was trained 10 times (3 times for hiragana data) changing the initial weights.

For MM NN, we set $C(3) = 500$, $\varepsilon(2) = 0.4$, $\varepsilon(3) = 0.2$, and we evaluated the performance changing the value of $C(2)$.

Table 9.1 lists the best performance obtained by FSVM, the neural network trained by the back-propagation algorithm (BP), and MM NN and the training time. The highest recognition rate of the test data is shown in boldface. The column "Parm" for BP shows the number of hidden neurons.

From the table, recognition performance of MM NN and FSVM is comparable, but for the thyroid data, the recognition rate of the test data for MM NN is higher than for FSVM by about 1%, although the recognition rates of the training data are almost the same. Except for the numeral data, BP showed inferior recognition performance than MM NN. Especially for the hirgana-50 data, the recognition rate of the test data was lower. This is because the number of the hiragana-50 training data per class is small and there is no overlap between classes. Thus, without the mechanism of maximizing margins, the high recognition rate was not obtained.

Training of MM NN was faster than that of BP, but except for the thyroid data slower than that of FSVM. The reason for slow training compared to that of FSVM is that by MM NN a large number of hidden neurons were generated. For example, for hiragana-13 data, 261 hidden neurons were generated.

**Table 9.1** Performance comparison of maximum-margin neural networks

| Data | Classifier | Parm | Rate (%) | Time (s) |
|------|-----------|------|----------|----------|
| | FSVM | $d3$ | 99.51 (100) | 1 |
| Numeral | BP | 18 | **99.76** (100) | 3 |
| | MM NN | $C(2) = 0.2$ | 99.63 (100) | 1 |
| | FSVM | $d3$ | 97.55 (99.26) | 64 |
| Thyroid | BP | 5 | 98.37 (99.58) | 280 |
| | MM NN | $C(2) = -0.2$ | **98.51** (99.20) | 29 |
| | FSVM | $d3$ | **93.26** (98.22) | 30 |
| Blood cell | BP | 85 | 91.61 (98.29) | 3,091 |
| | MM NN | $C(2) = -0.5$ | 93.00(98.84) | 244 |
| | FSVM | $d5$ | **99.46** (100) | 122 |
| Hiragana-50 | BP | 100 | 97.25 (98.91) | 11,262 |
| | MM NN | $C(2) = 0.6$ | 99.02 (100) | 919 |
| | FSVM | $d2$ | **100** (100) | 564 |
| Hiragana-105 | BP | 100 | 99.84 (99.99) | 28,801 |
| | MM NN | $C(2) = -0.3$ | **100** (100) | 2,711 |
| | FSVM | $d2$ | 99.51(99.92) | 292 |
| Hiragana-13 | BP | 100 | 99.17 (99.51) | 13,983 |
| | MM NN | $C(2) = -0.1$ | **99.58** (100) | 2,283 |

# References

1. K. Funahashi. On the approximate realization of continuous mappings by neural networks. *Neural Networks*, 2(3):183–192, 1989.
2. K. Hornik, M. Stinchcombe, and H. White. Multilayer feedforward networks are universal approximators. *Neural Networks*, 2(5):359–366, 1989.
3. J. A. K. Suykens and J. Vandewalle. Training multilayer perceptron classifiers based on a modified support vector method. *IEEE Transactions on Neural Networks*, 10(4):907–911, 1999.
4. Y. Freund and R. E. Schapire. Large margin classification using the perceptron algorithm. *Machine Learning*, 37(3):277–296, 1999.
5. L. Mason, P. L. Bartlett, and J. Baxter. Improved generalization through explicit optimization of margins. *Machine Learning*, 38(3):243–255, 2000.
6. K. P. Bennett, N. Cristianini, J. Shawe-Taylor, and D. Wu. Enlarging the margins in perceptron decision trees. *Machine Learning*, 41(3):295–313, 2000.
7. C. Gentile. A new approximate maximal margin classification algorithm. *Journal of Machine Learning Research*, 2:213–242, 2001.
8. Y. Li and P. M. Long. The relaxed online maximum margin algorithm. *Machine Learning*, 46(1–3):361–387, 2002.

9. Jayadeva, A. K. Deb, and S. Chandra. Binary classification by SVM based tree type neural networks. In *Proceedings of the 2002 International Joint Conference on Neural Networks (IJCNN '02)*, volume 3, pages 2773–2778, Honolulu, Hawaii, 2002.

10. T. Nishikawa and S. Abe. Maximizing margins of multilayer neural networks. In *Proceedings of the Ninth International Conference on Neural Information Processing (ICONIP '02)*, volume 1, pages 322–326, Singapore, 2002.

11. K. Crammer and Y. Singer. Ultraconservative online algorithms for multiclass problems. *Journal of Machine Learning Research*, 3:951–991, 2003.

12. S. Young and T. Downs. CARVE—A constructive algorithm for real-valued examples. *IEEE Transactions on Neural Networks*, 9(6):1180–1190, 1998.

13. S. Abe. *Pattern Classification: Neuro-Fuzzy Methods and Their Comparison*. Springer-Verlag, London, UK, 2001.

14. D. Roobaert. DirectSVM: A fast and simple support vector machine perceptron. In *Neural Networks for Signal Processing X—Proceedings of the 2000 IEEE Signal Processing Society Workshop*, volume 1, pages 356–365, 2000.

# Chapter 10
# Maximum-Margin Fuzzy Classifiers

In conventional fuzzy classifiers, fuzzy rules are defined by experts. First, we divide the ranges of input variables into several nonoverlapping intervals. And for each interval, we define a membership function, which defines the degree to which the input value belongs to the interval. Now the input space is covered with nonoverlapping hyperrectangles. We define, for each hyperrectangle, a fuzzy rule. Suppose we have two variables and each range is divided into three: small, medium, and large. An example of a fuzzy rule is as follows:

If $x_1$ is small and $x_2$ is large, then $\mathbf{x}$ belongs to Class 2.

One of the advantages of fuzzy classifiers over multilayer neural networks or support vector machines is that we can easily understand how they work. But because the fuzzy rules need to be defined by experts, development of classifiers is difficult.

To overcome this problem, many fuzzy classifiers that can be trained using numerical data have been developed [1, 2]. Trainable fuzzy classifiers are classified, from the shape of the approximated class regions, into

1. fuzzy classifiers with hyperbox regions [3, 4, 2],
2. fuzzy classifiers with polyhedral regions [5, 2, 6, 7], and
3. fuzzy classifiers with ellipsoidal regions [8, 9, 2].

For these fuzzy classifiers, fuzzy rules are defined by either preclustering or postclustering the training data. In preclustering, the training data for each class are divided into clusters in advance, and using the training data in the cluster a fuzzy rule is defined. Then we tune slopes and/or locations of the membership functions so that the recognition rate of the training data is maximized allowing the previously correctly classified data to be misclassified. In postclustering, one fuzzy rule is defined using the training data included in each class and the membership functions are tuned. Then if the recognition rate of the training data is not sufficient, we define the fuzzy rule for the misclassified training data.

S. Abe, *Support Vector Machines for Pattern Classification*,
Advances in Pattern Recognition, DOI 10.1007/978-1-84996-098-4_10,
© Springer-Verlag London Limited 2010

These fuzzy classifiers are trained so that the recognition rate of the training data is maximized. Thus, if the recognition rate reaches 100%, the training is terminated and no further tuning for the class boundaries is done. Therefore, the generalization ability of the conventional fuzzy classifiers degrades especially when the number of training data is small or the number of the input variables is large.

With the introduction of support vector machines and kernel methods, many fuzzy classifiers trained by support vector machines or maximizing margins have been proposed [7, 10–17].

In this chapter we first discuss an architecture of a kernel version of a fuzzy classifier with ellipsoidal regions and improvement of generalization ability by transductive training, in which unlabeled data are used [10, 12], and by maximizing margins [18]. Then we discuss the architecture of a fuzzy classifier with polyhedral regions and an efficient rule-generation method [7].

## 10.1 Kernel Fuzzy Classifiers with Ellipsoidal Regions

In this section, we first summarize the architecture of a fuzzy classifier with ellipsoidal regions and then discuss its kernel version [10] and improvement of generalization ability by transductive training [12].

### 10.1.1 Conventional Fuzzy Classifiers with Ellipsoidal Regions

We summarize the conventional fuzzy classifier with ellipsoidal regions, which is generated in the input space [8, 18, 2]. We call the classifier the *basic fuzzy classifier with ellipsoidal regions*. Here we consider classification of the $m$-dimensional input vector $\mathbf{x}$ into one of $n$ classes. We can define more than one fuzzy rule for each class, but to make discussions simple, we assume that we define one fuzzy rule for each class:

$$R_i: \text{if } \mathbf{x} \text{ is } \mathbf{c}_i \text{ then } \mathbf{x} \text{ belongs to class } i, \tag{10.1}$$

where $\mathbf{c}_i$ is the center vector of class $i$:

$$\mathbf{c}_i = \frac{1}{M_i} \sum_{j=1}^{M_i} \mathbf{x}_{ij}. \tag{10.2}$$

Here, $M_i$ is the number of training data for class $i$ and $\mathbf{x}_{ij}$ is the $j$th training sample for class $i$.

We define a membership function $m_i(\mathbf{x})$ $(i = 1, \dots, n)$ for input $\mathbf{x}$ by

$$m_i(\mathbf{x}) = \exp(-h_i^2(\mathbf{x})), \tag{10.3}$$

$$h_i^2(\mathbf{x}) = \frac{d_i^2(\mathbf{x})}{\alpha_i}$$

$$= \frac{1}{\alpha_i}(\mathbf{x} - \mathbf{c}_i)^\top Q_i^{-1}(\mathbf{x} - \mathbf{c}_i), \tag{10.4}$$

where $d_i(\mathbf{x})$ is the Mahalanobis distance between $\mathbf{x}$ and $\mathbf{c}_i$; $h_i(\mathbf{x})$ is the tuned distance; $\alpha_i$ is the tuning parameter for class $i$; and $Q_i$ is the covariance matrix for class $i$ in the input space.

The covariance matrix for class $i$ in the input space is given by

$$Q_i = \frac{1}{M_i} \sum_{j=1}^{M_i} (\mathbf{x}_{ij} - \mathbf{c}_i)(\mathbf{x}_{ij} - \mathbf{c}_i)^\top. \tag{10.5}$$

We calculate the membership function of input $\mathbf{x}$ for each class. If the degree of membership for class $j$ is maximum, the input is classified into class $j$. When $\alpha_i$ in (10.4) is equal to 1, this is equivalent to finding the minimum Mahalanobis distance. Function (10.3) makes the output range of (10.3) lie in [0,1], and if $m_j(\mathbf{x})$ is equal to 1, the input $\mathbf{x}$ corresponds to the center of class $j$, $\mathbf{c}_j$. We tune the membership function using $\alpha_i$ in (10.4).

When $Q_i$ is positive definite, the Mahalanobis distance given by (10.4) can be calculated using the symmetric Cholesky factorization [19]. Thus $Q_i$ can be decomposed into

$$Q_i = L_i L_i^\top, \tag{10.6}$$

where $L_i$ is the real-valued regular lower triangular matrix.

But when $Q$ is positive semidefinite, the value in the square root of diagonal element of $L_i$ is nonpositive. To avoid this, during the Cholesky factorization, if the value in the square root is smaller than $\zeta$, where $\zeta$ is a predefined threshold, the element is replaced by $\sqrt{\zeta}$ [2, 18]. This means that when the covariance matrix is positive semidefinite, the principal components in the subspace that the training data do not span are taken into consideration to calculate the Mahalanobis distance. Thus by this method, the generalization ability does not decrease even when the training data are degenerate, namely, the training data do not span the input space.

## 10.1.2 Extension to a Feature Space

In a kernel fuzzy classifier with ellipsoidal regions [10], the input space is mapped into the feature space by a nonlinear mapping function. Because we map the input space into the feature space, we assume that each class consists of one cluster and we define a fuzzy rule for each class.

We define the following fuzzy rule for class $i$[1]:

$$R_i : \text{if } \phi(\mathbf{x}) \text{ is } \mathbf{c}_i \text{ then } \mathbf{x} \text{ belongs to class } i, \qquad (10.7)$$

where $\phi(\mathbf{x})$ is a mapping function that maps the input space into the $l$-dimensional feature space, $\mathbf{c}_i$ is the center of class $i$ in the feature space and is calculated by the training data included in class $i$:

$$\mathbf{c}_i = \frac{1}{|X_i|} \sum_{\mathbf{x} \in X_i} \phi(\mathbf{x}), \qquad (10.8)$$

where $X_i$ is the set of training data included in class $i$ and $|X_i|$ is the number of data included in $X_i$.

For center $\mathbf{c}_i$, we define the membership function $m_{\phi_i}(\mathbf{x})$ that defines the degree to which $\mathbf{x}$ belongs to $\mathbf{c}_i$:

$$m_{\phi_i}(\mathbf{x}) = \exp(-h_{\phi_i}^2(\mathbf{x})), \qquad (10.9)$$

$$h_{\phi_i}^2(\mathbf{x}) = \frac{d_{\phi_i}^2(\mathbf{x})}{\alpha_i}, \qquad (10.10)$$

$$d_{\phi_i}^2(\mathbf{x}) = (\phi(\mathbf{x}) - \mathbf{c}_i)^\top Q_{\phi_i}^+ (\phi(\mathbf{x}) - \mathbf{c}_i), \qquad (10.11)$$

where $h_{\phi_i}(\mathbf{x})$ is a tuned distance, $d_{\phi_i}(\mathbf{x})$ is a kernel Mahalanobis distance between $\phi(\mathbf{x})$ and $\mathbf{c}_i$, $\alpha_i$ is a tuning parameter for class $i$, $Q_{\phi_i}$ is the $l \times l$ covariance matrix for class $i$. And $Q_{\phi_i}^+$ denotes the pseudo-inverse of the covariance matrix $Q_{\phi_i}$. Here we calculate the covariance matrix $Q_{\phi_i}$ using the data belonging to class $i$ as follows:

$$Q_{\phi_i} = \frac{1}{|X_i|} \sum_{\mathbf{x} \in X_i} (\phi(\mathbf{x}) - \mathbf{c}_i)(\phi(\mathbf{x}) - \mathbf{c}_i)^\top. \qquad (10.12)$$

In calculating the kernel Mahalanobis distance, we use either of the methods discussed in Section 6.3.

For an input vector $\mathbf{x}$ we calculate the degrees of membership for all the classes. If $m_{\phi_k}(\mathbf{x})$ is the maximum, we classify the input vector into class $k$.

## 10.1.3 Transductive Training

### 10.1.3.1 Concept

We discussed the SVD-based Mahalanobis distance in Section 6.3.1 and the KPCA-based Mahalanobis distance in Section 6.3.2. But with those methods,

---

[1] If we use the empirical feature space instead of the feature space, discussions will be as simple as in the input space.

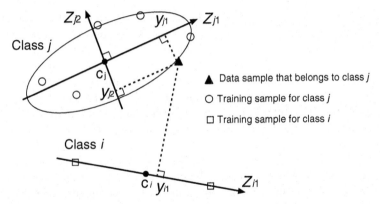

**Fig. 10.1** The training data of class $i$ are degenerate. Because the second eigenvalue of class $i$ is zero, the kernel Mahalanobis distance for class $i$, $d_{\phi_i}(\mathbf{x})$, cannot be calculated correctly. Reprinted from [12, p. 201] with permission from Elsevier

when the training data are degenerate, i.e., the space spanned by the mapped training data is a proper subspace in the feature space, the generalization ability of the fuzzy classifier is decreased [10]. It is because the principal components are zero in the subspace complementary to that spanned by the mapped training data.

We now explain the reason using the example shown in Fig. 10.1. In the figure, training data for class $j$ are in the two-dimensional space, but those of class $i$ are in one dimension.

Assume that we have a data sample $\mathbf{x}$ belonging to class $j$. This data sample is in the subspace spanned by the mapped training data belonging to class $j$, but not in the subspace spanned by class $i$. Then the kernel Mahalanobis distance between $\mathbf{x}$ and $\mathbf{c}_j$ is correctly calculated by

$$d_{\phi_i}^2(\mathbf{x}) = \frac{y_{j1}^2}{\lambda_{j1}} + \frac{y_{j2}^2}{\lambda_{j2}}. \tag{10.13}$$

But for class $i$, because the second eigenvalue is zero due to the degeneracy of the training data, $d_{\phi_i}^2(\mathbf{x})$ is erroneously given by

$$d_{\phi_i}^2(\mathbf{x}) = \frac{y_{i1}^2}{\lambda_{i1}}. \tag{10.14}$$

Because the kernel Mahalanobis distance for class $i$ does not change even if $\mathbf{x}$ moves in the direction orthogonal to the eigenvector $\mathbf{z}_{i1}$, classification by the Mahalanobis distance becomes erroneous.[2]

---

[2] This problem will be alleviated if we use the empirical feature space with the mapping function common to all classes.

Similar to (6.47), we can overcome this problem, adding the vector that is orthogonal to the existing eigenvectors for the covariance matrix with a small positive eigenvalue. The next problem is how to select the appropriate vectors for addition. If the linear kernel is used and the training data are degenerate, i.e., the training data do not span the input space, we can add the basis vectors in the input space that are in the complementary subspace of the training data. But if the nonlinear kernel is used, the mapped training data may not span the feature space even though the training data are not degenerate.

To solve this problem, we use transductive training of the classifier. In training, we add the basis vectors of the input space as the unlabeled training data, and if the mappings of these data are in the complementary subspace associated with the covariance matrix, we generate associated eigenvectors with small positive eigenvalues, namely, first we map the basis vectors into the feature space and then we perform the Cholesky factorization to judge whether each mapped basis vector is included in the subspace associated with the covariance matrix. If it is not included in the subspace, we calculate the vector that is orthogonal to the subspace and modify the covariance matrix.

By this method, however, the complementary subspace may remain. Thus, in classification, whenever an unknown data sample is given, we judge whether the mapped data sample is included in the subspace associated with the covariance matrix by the Cholesky factorization. If it is not included, we calculate the vector orthogonal to the previously selected vectors by the Gram–Schmidt orthogonalization and modify the covariance matrix. We can speed up factorization, storing the previously factorized matrices and factorizing the row and column associated with only the unknown data sample.

### 10.1.3.2 Transductive Training Using Basis Vectors

In this section, we discuss how to improve approximation of the feature space using the basis vectors $\{e_1, \ldots, e_m\}$ in the input space in addition to the vector of the mapped training data $\phi(X_i)$. Because the class labels of $\{e_1, \ldots, e_m\}$ are not known, these data should not affect the principal components associated with $\phi(X_i)$. Therefore, first we calculate the eigenvalues $\{\lambda_1, \ldots, \lambda_{M_i'}\}$ and eigenvectors $\{z_1, \ldots, z_{M_i'}\}$ of $Q_{\phi_i}$ and add eigenvectors with small positive eigenvalues that are orthogonal to $\{z_1, \ldots, z_{M_i'}\}$ using some of the $\{\phi(e_1), \ldots, \phi(e_m)\}$.

Now we will explain the procedure in more detail. In a similar way as discussed in Section 6.1, we can select linearly independent vectors by the Cholesky factorization of the kernel matrix for $\{x_1, \ldots, x_{M_i}\}$ and $\{e_1, \ldots, e_m\}$:

$$K(X_i', X_i'^{\top}) = \phi(X_i') \, \phi^{\top}(X_i'), \tag{10.15}$$

where

$$\phi(X_i') = \big( \phi(\mathbf{x}_1) \cdots \phi(\mathbf{x}_{M_i}) \; \phi(\mathbf{e}_1) \cdots \phi(\mathbf{e}_m) \big)^\top . \tag{10.16}$$

Placing $\{\phi(\mathbf{e}_1), \ldots, \phi(\mathbf{e}_m)\}$ after $\phi(\mathbf{x}_{M_i})$, we can select linearly independent vectors from $\{\mathbf{e}_1, \ldots, \mathbf{e}_m\}$ that are not included in the subspace spanned by $\phi(X_i)$. Assume that $m'$ basis vectors, $\phi(\mathbf{e}_r)\,(r = 1, \ldots, m')$, are selected as the linearly independent vectors in addition to $M_i'$ linearly independent training vectors. From the $M_i'$ linearly independent training data, the orthogonal system $\{\mathbf{z}_1, \ldots, \mathbf{z}_{M_i'}\}$ is calculated by the method discussed in Section 6.3.2.

Next, using $\phi(\mathbf{e}_r)$ we generate the $(M_i' + 1)$st extra eigenvector $\mathbf{z}_{M_i'+1}$ by the Gram–Schmidt orthogonalization:

$$\mathbf{z}_{M_i'+1} = \frac{\mathbf{p}_r}{\|\mathbf{p}_r\|}, \tag{10.17}$$

where $\|\mathbf{p}_r\|$ is the norm of $\mathbf{p}_r$:

$$\mathbf{p}_r = \phi(\mathbf{e}_r) - \sum_{i=1}^{M_i'} (\phi^\top(\mathbf{e}_r)\,\mathbf{z}_i)\,\mathbf{z}_i$$

$$= \phi(\mathbf{e}_r) - \sum_{i=1}^{M_i'} \omega_i\,\mathbf{z}_i. \tag{10.18}$$

Here, $\omega_i = \phi^\top(\mathbf{e}_r)\,\mathbf{z}_i$.

By substituting (6.26) into (10.18), $\mathbf{z}_{M_i'+1}$ is expressed by the linear sum of $\{\phi(\mathbf{x}_1), \ldots, \phi(\mathbf{x}_{M_i'}), \phi(\mathbf{e}_r)\}$:

$$\mathbf{z}_{M_i'+1} = \big( \phi(\mathbf{x}_1), \ldots, \phi(\mathbf{x}_{M_i'}), \phi(\mathbf{e}_r) \big) \begin{pmatrix} \rho_{M_i'+1,1} \\ \vdots \\ \rho_{M_i'+1,M_i'} \\ \frac{1}{\|\mathbf{p}_r\|} \end{pmatrix}, \tag{10.19}$$

where

$$\rho_{M_i'+1,j} = -\frac{1}{\|\mathbf{p}_r\|} \sum_{i=1}^{M_i'} \omega_i\,\rho_{ij} \quad \text{for} \quad j = 1, \ldots, M_i'. \tag{10.20}$$

Using (10.19), we can calculate the $(M_i' + 1)$st principal component as follows:

$$y_{M_i'+1} = \mathbf{z}_{M_i'+1}^\top (\phi(\mathbf{x}) - \mathbf{c}_i)$$

$$= \sum_{i=1}^{M_i'} \rho_{M_i'+1,i}\,K(\mathbf{x}, \mathbf{x}_i) + \frac{1}{\|\mathbf{p}_r\|}K(\mathbf{e}_r, \mathbf{x}) - \mathbf{z}_{M_i'+1}^\top \mathbf{c}_i. \tag{10.21}$$

We iterate this procedure until the rest of the principal components $y_i$ ($i = M_i' + 1, \ldots, M_i' + m'$) are calculated.

We add these new principal components to the right-hand side of (6.51), and finally the Mahalanobis distance becomes

$$d_{\phi_i}^2(\mathbf{x}) = \frac{y_1^2}{\lambda_1} + \cdots + \frac{y_{M_i'}^2}{\lambda_{M_i'}} + \frac{y_{M_i'+1}^2}{\varepsilon} + \cdots + \frac{y_{M_i'+m'}^2}{\varepsilon}, \qquad (10.22)$$

where for the added eigenvectors the associated eigenvalues are assumed to be small and we set a small value to $\varepsilon$. We call (10.22) the *KPCA-based Mahalanobis distance with Gram–Schmidt orthogonalization.*

### 10.1.3.3 Transductive Training Using Unknown Data

In the previous section, we discussed approximation of a subspace using the mapped basis vectors $\phi(\mathbf{e}_1), \ldots, \phi(\mathbf{e}_m)$. With this method, the input space is spanned if linear kernels are used, but for nonlinear kernels the whole feature space cannot be covered. Thus approximation with the basis vectors may not be enough to prevent generalization ability from decreasing.

To overcome this problem, we need to make unknown data lie in the subspace spanned by the eigenvectors associated with the covariance matrix. To do so, when an unknown data sample $\mathbf{t}$ is given, we judge whether this sample is included in the space spanned by $\phi(X_i')$. If the sample is not in the space, we generate the extra eigenvector in a similar way as discussed previously, namely, for the kernel matrix

$$K(X_i'', X_i''^{\top}) = \phi(X_i'')\phi^{\top}(X_i''), \qquad (10.23)$$

where

$$\phi(X_i'') = (\phi(X_i'), \ \phi(\mathbf{t}))^{\top}, \qquad (10.24)$$

we perform the Cholesky factorization. But because the factorization of $K(X_i', X_i'^{\top})$ can be done off-line, we only have to calculate the $(M_i+m+1)$st row and column of $K(X_i'', X_i''^{\top})$ (see (6.11) and (6.12)). If the value in the square root in (6.12) is larger than $\eta$ ($> 0$) (see (6.13)), the unknown sample $\mathbf{t}$ is not included in the space spanned by $\phi(X_i')$. Thus we generate the extra eigenvector to span the subspace.

This method is not time-consuming because we do not need to calculate the whole elements. When we are not given enough training data, this online approximation may be effective.

## 10.1.4 Maximizing Margins

### 10.1.4.1 Concept

In the fuzzy classifier with ellipsoidal regions, if there are overlaps between classes, the overlaps are resolved by tuning the membership functions, i.e., by tuning $\alpha_i$ one at a time. When $\alpha_i$ is increased, the slope of $m_{\phi_i}(\mathbf{x})$ is decreased and the degree of membership is increased. Then misclassified data may be correctly classified and correctly classified data may be misclassified. Based on this, we calculate the net increase of the correctly classified data. Likewise, by increasing the slope, we calculate the net increase of the correctly classified data. Then allowing new misclassification, we tune the slope so that the recognition rate is maximized. In this way we tune fuzzy rules successively until the recognition rate of the training data is not improved [2, pp. 121–129].

But if there is no overlap, the membership functions are not tuned. When the number of training data is small, usually the overlaps are scarce. Thus, the generalization ability is degraded. To tune membership functions even when the overlaps are scarce, we use the idea used in training support vector machines. In training support vector machines for a two-class problem, the separating margin between the two classes is maximized to improve the generalization ability. In the kernel fuzzy classifier with ellipsoidal regions, we tune the slopes of the membership functions so that the slope margins are maximized.

Initially, we set the values of $\alpha_i$ to be 1, namely, the kernel fuzzy classifier with ellipsoidal regions is equivalent to the classifier based on the kernel Mahalanobis distance. Then we tune $\alpha_i$ so that the slope margins are maximized. Here we maximize the margins without causing new misclassification. When the recognition rate of the training data is not 100% after tuning, we tune $\alpha_i$ so that the recognition rate is maximized as discussed in [2]. Here we discuss how to maximize slope margins.

Unlike the support vector machines, tuning of $\alpha_i$ is not restricted to two classes, but for ease of illustration we explain the concept of tuning using two classes. In Fig. 10.2 the filled square and circle show the training data, belonging to classes $i$ and $j$, that are nearest to classes $j$ and $i$, respectively. The class boundary of the two classes is somewhere between the two curves shown in the figure. We assume that the generalization ability is maximized when it is in the middle of the two.

In Fig. 10.3, if the data sample belongs to class $i$, it is correctly classified because the degree of membership for class $i$ is larger. This sample remains correctly classified until the degree of membership for class $i$ is decreased, as shown in the dotted curve. Similarly, in Fig. 10.4, if the sample belongs to class $j$, it is correctly classified because the degree of membership for class $j$ is larger. This sample remains correctly classified until the degree of membership for class $i$ is increased as shown in the dotted curve. Thus, for each $\alpha_i$, there is an interval of $\alpha_i$ that makes correctly classified data remain correctly classified. Therefore, if we change the value of $\alpha_i$ so that it is in the middle of the interval, the slope margins are maximized.

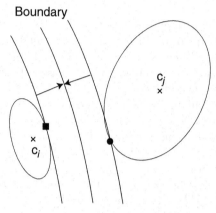

**Fig. 10.2** Concept of maximizing margins. Reprinted from [18, p. 209, ©IEEE 2001]

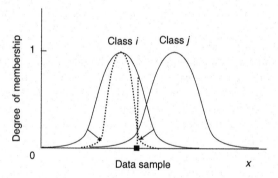

**Fig. 10.3** Upper bound of $\alpha_i$ that does not cause misclassification. Reprinted from [18, p. 209, ©IEEE 2001]

In the following we discuss how to tune $\alpha_i$.

### 10.1.4.2 Upper and Lower Bounds of $\alpha_i$

Let $X$ be the set of training data that are correctly classified for the initial $\alpha_i$. Let $\mathbf{x}\,(\in X)$ belong to class $i$.

If $m_{\phi_i}(\mathbf{x})$ is the largest, there is a lower bound of $\alpha_i$, $L_i(\mathbf{x})$, to keep $\mathbf{x}$ correctly classified:

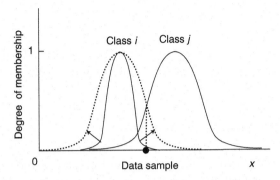

**Fig. 10.4** Lower bound of $\alpha_i$ that does not cause misclassification. Reprinted from [18, p. 209, ©IEEE 2001]

$$L_i(\mathbf{x}) = \frac{d_{\phi_i}^2(\mathbf{x})}{\min\limits_{j \neq i} h_{\phi_j}^2(\mathbf{x})}. \tag{10.25}$$

Then the lower bound $L_i(1)$ that does not cause new misclassification is given by

$$L_i(1) = \max_{\mathbf{x} \in X} L_i(\mathbf{x}). \tag{10.26}$$

Similarly, for $\mathbf{x}\,(\in\,X)$ belonging to a class other than class $i$, we can calculate the upper bound of $\alpha_i$. Let $m_{\phi_j}(\mathbf{x})\,(j \neq i)$ be the largest. Then the upper bound $U_i(\mathbf{x})$ of $\alpha_i$ that does not cause misclassification of $\mathbf{x}$ is given by

$$U_i(\mathbf{x}) = \frac{d_{\phi_i}^2(\mathbf{x})}{\min\limits_{j} h_{\phi_j}^2(\mathbf{x})}. \tag{10.27}$$

The upper bound $U_i(1)$ that does not make new misclassification is given by

$$U_i(1) = \min_{\mathbf{x} \in X} U_i(\mathbf{x}). \tag{10.28}$$

In [2, p. 122], $L_i(l)$ and $U_i(l)$ are defined as the lower and upper bounds in which $l - 1$ correctly classified data are misclassified, respectively. Thus, $L_i(1)$ and $U_i(1)$ are the special cases of $L_i(l)$ and $U_i(l)$.

### 10.1.4.3 Tuning Procedure

The correctly classified data remain correctly classified even when $\alpha_i$ is set to some value in the interval $(L_i(1), U_i(1))$. The tuning procedure of $\alpha_i$ becomes as follows. For $\alpha_i$ we calculate $L_i(1)$ and $U_i(1)$ and set the value of $\alpha_i$ with the middle point of $L_i(1)$ and $U_i(1)$:

$$\alpha_i = \frac{1}{2}(L_i(1) + U_i(1)). \tag{10.29}$$

We successively tune one $\alpha_i$ after another. Tuning results depend on the order of tuning $\alpha_i$, but usually tuning from $\alpha_1$ to $\alpha_n$ leads to a good result.

## 10.1.5 Performance Evaluation

We evaluated the methods using two groups of classification problems: (1) multiclass problems listed in Table 1.3 and (2) two-class problems in Table 1.1. In the first group, each problem consists of one training data set and one test data set, and in the second group, each problem consists of 100 (or 20) training data sets and associated test data sets. We used a Pentium III 1 GHz personal computer to evaluate the methods.

### 10.1.5.1 Multiclass Data Sets

For the basic fuzzy classifier with ellipsoidal regions, we need to set the value of $\zeta$, which detects the singularity of the covariance matrix and the value of $l_M$, in which $l_M - 1$ is the maximum number of misclassification allowed for tuning. We selected the value of $\zeta$ so that the recognition rate of the test data was maximized. And we set $l_M = 10$, which usually gives good generalization ability.

We compared performance of the four kinds of the kernel Mahalanobis distance:

1. The singular value decomposition-based kernel Mahalanobis distance discussed in Section 6.3.1 (denoted hereafter as "SVD")
2. The KPCA-based Mahalanobis distance discussed in Section 6.3.2 in which the kernel matrix is used to select the linearly independent vectors ("KPCA I")
3. The KPCA-based Mahalanobis distance given by (10.22) by transductive training using the basis vectors ("KPCA II")
4. The KPCA-based Mahalanobis distance by transductive training using the basis vectors and test data ("KPCA III")

Because SVD-based training took a long time, we determined the value of $\sigma$ in (6.47) so that the recognition rates of the test data were the highest. Table 10.1 shows the difference between the conventional method, in which the subspace associated with the small eigenvalues is neglected, and the improved method (6.47) to perform singular value decomposition for the numeral data. "Initial" and "Final" denote the initial recognition rates of the test (training) data with the tuning parameters $\alpha_i = 1$ and after tuning, respectively. "Num." denotes the number of selected diagonal elements. From the table, the improved method of singular value decomposition is effective to prevent generalization ability from decreasing, especially for initial recognition rates for linear kernels. Thus in the following experiments, we use this method to calculate the pseudo-inverse when the training data are degenerate (numeral, thyroid, and hiragana-50 data sets).

**Table 10.1** Comparison of singular value decomposition. Reprinted from [12, p. 209] with permission from Elsevier

| Type | Kernel | Initial (%) | Final (%) | Num. |
|------|--------|-------------|-----------|------|
| Conventional | Linear | 88.78 (90.00) | 98.78 (98.40) | 100 |
| $(\sigma = 10^{-8})$ | $\gamma 0.01$ | 98.90 (99.75) | 99.02 (100) | 206 |
| Improved | Linear | 98.54 (97.28) | 98.78 (98.52) | 810 |
| $(\sigma = 10^{-8})$ | $\gamma 0.01$ | 99.39 (99.88) | 99.27 (100) | 810 |

To calculate a KPCA-based Mahalanobis distance, we need to set two parameters, $\varepsilon$, which determine the minimum value of the eigenvalues, and $\eta$, which determines how strictly we select the linearly independent vectors. From our computer experiments, we know that the generalization ability is more sensitive to the value of $\varepsilon$ than that of $\eta$. In addition, the best value of $\varepsilon$ depends on the training data sets and kernel functions.

Thus we determined the value of $\eta$ by fivefold cross-validation for the blood cell data and made the value of $\eta$ common to all training data sets and kernel functions. Then we performed fivefold cross-validation to determine the value of $\varepsilon$ for each training data set and kernel function.

For linear kernels, polynomial kernels with $d = [2, 3]$ or RBF kernels with $\gamma = [0.001, 0.01, 0.1, 1, 10]$, we performed fivefold cross-validation for $\varepsilon = [10^{-8}, 10^{-7}, \ldots, 10^{-1}]$. We iterated cross-validation five times to decrease the influence of random selection and determined the value of $\varepsilon$ with the highest average recognition rate. For KPCA I, II, and III, we determined the parameters for KPCA II and used the values for KPCA I and III.

Table 10.2 shows the recognition rates of the test data. In the "Type" column, (1), (2), and (3) denote that linear, polynomial, and RBF kernels were used, respectively. For each data set the highest recognition rate is shown in boldface. The table also includes the recognition rates for the pairwise L1 fuzzy support vector machine evaluated in Section 3.2.3 and "Basic" denotes the conventional fuzzy classifier with ellipsoidal regions.

If the training data set is not degenerate and KPCA I (1) selects linearly independent vectors that span the input space, the basic fuzzy classifier, SVD (1), and KPCA I (1) will give the same recognition rate. Except for the numeral, thyroid, and hiragana-50 data sets, the training data sets are nondegenerate. Thus, for these sets, the recognition rates are almost the same.

For the degenerate training data sets, the recognition rates of SVD (1) and KPCA I (1) are inferior to that of the basic fuzzy classifier. This is especially evident for the thyroid data set. Using nonlinear kernels, the recognition rates were improved but still lower for the thyroid data set. In KPCA II (1), because the basis vectors in the input space were added, the degeneracy was resolved and the recognition rates are comparable with those of the basic fuzzy classifier. For the linear kernels there were no online additions of the basis vectors. Thus KPCA II (1) and KPCA III (1) give the same results. For nonlinear kernels KPCA II shows better performance than SVD. But there is not much difference between KPCA II and KPCA III. This means that the training data and the basis vectors in the input space are sufficient to represent the space spanned by the test data.

Performance of the SVM and that of the KPCA-based methods are comparable.

Table 10.3 shows the recognition rates of the test data without tuning the tuning parameters. This means that the classifier classifies a data sample according to the (kernel) Mahalanobis distance. The recognition rates that are higher than those by tuning are shown in boldface.

From Tables 10.2 and 10.3, in most cases by tuning the tuning parameters, the recognition rates of the test data are improved. The effect of tuning is especially evident for blood cell and thyroid data sets. But for the iris data set, the recognition rates were not improved by tuning for all the cases tried.

Each classifier has a different number of parameters to optimize. Thus, it is difficult to compare the training time of each classifier fairly. Therefore,

**Table 10.2** Recognition rates of the test data in percentage. Reprinted from [12, p. 212] with permission from Elsevier

| Type | | Blood | Numeral | Iris | Thyroid | H-50 | H-105 | H-13 |
|---|---|---|---|---|---|---|---|---|
| SVM | | 92.03 | **99.63** | **97.33** | **97.61** | **99.11** | 99.95 | 99.74 |
| Basic | | 91.32 | 99.27 | **97.33** | 95.62 | 98.85 | 100 | 99.34 |
| SVD | (1) | 91.32 | 98.78 | **97.33** | 89.53 | 96.38 | 99.99 | 99.25 |
| | (2) | 92.35 | 97.68 | 94.67 | 90.58 | 95.99 | 99.90 | 98.96 |
| | (3) | 92.45 | 99.27 | 96.00 | 94.40 | 93.51 | 100 | 99.74 |
| KPCA I (1) | | 91.23 | 99.02 | 96.00 | 87.25 | 97.53 | 100 | 99.43 |
| | (2) | 92.87 | 99.27 | **97.33** | 90.14 | 97.25 | 99.99 | 99.86 |
| | (3) | 91.35 | 99.39 | 94.67 | 95.01 | 97.57 | 100 | 99.86 |
| KPCA II (1) | | 91.23 | 99.51 | 96.00 | 94.75 | 98.31 | 100 | 99.43 |
| | (2) | 93.16 | 99.51 | **97.33** | 96.82 | 97.61 | 100 | 99.86 |
| | (3) | 91.32 | 99.39 | 94.67 | 95.01 | 98.52 | 100 | **99.88** |
| KPCA III (1) | | 91.23 | 99.51 | 96.00 | 94.75 | 98.31 | 100 | 99.43 |
| | (2) | **93.23** | 99.51 | **97.33** | 96.76 | 97.55 | 100 | 99.86 |
| | (3) | 91.03 | 99.39 | 94.67 | 95.71 | 98.52 | 100 | **99.88** |

**Table 10.3** Recognition rates for test data without tuning by the Mahalanobis distance in percentage. Reprinted from [12, p. 212] with permission from Elsevier

| Type | Blood | Numeral | Iris | Thyroid | H-50 | H-105 | H-13 |
|------|-------|---------|------|---------|------|-------|------|
| Basic | 87.45 | 99.63 | **98.67** | 86.41 | 98.79 | 100 | 98.36 |
| SVD (1) | 87.45 | 98.54 | **98.67** | 74.77 | 80.89 | 99.98 | 98.36 |
| (2) | **92.42** | 95.98 | **97.33** | 86.03 | 67.53 | 97.75 | **99.63** |
| (3) | 91.58 | **99.39** | **98.67** | 83.46 | 82.43 | 99.99 | **99.86** |
| KPCA I (1) | 88.65 | 96.34 | **98.67** | 72.40 | 94.84 | 100 | 99.15 |
| (2) | 92.29 | 97.93 | **98.67** | 83.72 | 85.36 | 98.90 | **99.90** |
| (3) | 89.52 | 97.20 | **98.67** | 80.25 | 81.13 | 99.96 | 99.84 |
| KPCA II (1) | 88.65 | 99.27 | **98.67** | 84.92 | 97.85 | 100 | 99.15 |
| (2) | 92.26 | 99.15 | **98.67** | 95.60 | **97.96** | 99.89 | **99.90** |
| (3) | 89.48 | 99.39 | **98.67** | 87.95 | 98.50 | 100 | 99.86 |
| KPCA III (1) | 88.65 | 99.27 | **98.67** | 84.92 | 97.85 | 100 | 99.15 |
| (2) | 92.16 | 99.27 | **98.67** | 95.68 | **98.42** | 100 | **99.89** |
| (3) | 89.58 | 99.39 | **98.67** | 90.32 | 98.48 | 100 | 99.87 |

here we only compare the training time of fuzzy classifiers for the parameters determined by model selection. Table 10.4 lists the training time of each method for the given conditions. In the table, 0 means that the calculation time is shorter than 0.5 s. Because training of the classifiers for the iris data set was very short, we do not include the result in the table.

We do not include the training time of KPCA III because all the test data were added at once. For KPCA III, we measured the time to process one unknown data sample for classification by the fast online method, but it was too short to be measured correctly.

Training of the basic fuzzy classifier is the fastest and training of KPCA I and II is the second fastest. Training of SVD is the slowest due to singular value decomposition. From this table the effectiveness of the KPCA-based methods is evident.

## 10.1.5.2 Two-Class Data Sets

We compared recognition rates of the SVM, the conventional fuzzy classifier with ellipsoidal regions, KPCA I, and KPCA II using the two-class benchmark data sets. As a reference we used recognition performance of the SVM listed in the home page [20]. Because the number of inputs for the splice data sets is 60, we normalized the input.

Throughout the experiment, we set $\eta = 10^{-5}$ for linearly independent data selection. As for the determination of the values of $\gamma$ and $\varepsilon$, first we fixed $\varepsilon$ to $10^{-7}$ and performed fivefold cross-validation of the first five training data sets for $\gamma = [0.001, 0.01, 0.1, 1, 10]$ and selected the median of the best value

**Table 10.4** Training time comparison in seconds. Reprinted from [12, p. 214] with permission from Elsevier

| Type       | Blood | Numeral | Thyroid | H-50 | H-105 | H-13 |
|------------|-------|---------|---------|------|-------|------|
| Basic      | 0     | 0       | 0       | 2    | 8     | 1    |
| SVD   (1)  | 228   | 2       | 96,660  | 98   | 906   | 991  |
|       (2)  | 218   | 2       | 96,120  | 101  | 933   | 885  |
|       (3)  | 222   | 2       | 94,440  | 100  | 841   | 983  |
| KPCA I (1) | 2     | 0       | 31      | 7    | 52    | 5    |
|        (2) | 10    | 1       | 194     | 33   | 239   | 31   |
|        (3) | 5     | 0       | 364     | 21   | 40    | 21   |
| KPCA II (1)| 2     | 0       | 31      | 8    | 69    | 6    |
|         (2)| 12    | 1       | 201     | 50   | 395   | 35   |
|         (3)| 6     | 0       | 384     | 33   | 72    | 23   |

for each data set. Then fixing the value of $\gamma$ with the median we performed fivefold cross-validation for $\varepsilon = [10^{-8}, 10^{-7}, \ldots, 10^{-1}, 0.5]$. We also determine the value of $\zeta$ by fivefold cross-validation. For the ringnorm and titanic data sets we performed cross-validation, including $d = [2, 3]$.

Table 10.5 lists the mean classification errors and their standard deviations. For each row in the results the smallest mean classification error among the four methods is shown in boldface. Comparing basic and KPCA I, except for the breast cancer and flare-solar data sets, KPCA I is better than or comparable to the conventional fuzzy classifier with ellipsoidal regions (basic). KPCA II is better than the conventional fuzzy classifier with ellipsoidal regions and better than or comparable to KPCA I. Thus, the generalization improvement of KPCA II over the conventional fuzzy classifier with ellipsoidal regions is clear. As for KPCA II and the SVM, except for the ringnorm and splice data sets, they are comparable. To improve the recognition performance for the ringnorm and splice data sets we need to do more extensive parameter survey. We tested the performance of KPCA III for these data sets, but the generalization performance was not improved. This may mean that the transductive training using the basis vectors was enough for these data sets.

## 10.2 Fuzzy Classifiers with Polyhedral Regions

The generalization ability of a classifier depends on how well we approximate the input region for each class. Approximation by polyhedrons is one way to improve the approximation accuracy [5, 21, 6, 7]. In [2, 6], starting from an initial convex hull for a class, the convex hull is expanded for the training data that are not in the convex hull. By this method, however, the number of generated facets explodes as the numbers of input variables and training data increase. To overcome this problem, in this section we discuss a different

**Table 10.5** Comparison among the four methods. Reprinted from [12, p. 215] with permission from Elsevier

| Data | SVM | Basic | KPCA I | KPCA II |
|------|------|------|--------|---------|
| Banana | 11.5±0.7 | 35.8±4.2 | **10.9±0.6** | **10.9±0.6** |
| Breast cancer | **26.0±4.7** | 28.9±5.3 | 33.5±4.9 | 26.5±4.4 |
| Diabetes | **23.5±1.7** | 25.8±2.2 | 25.3±2.0 | 25.3±2.0 |
| Flare solar | **32.4±1.8** | 34.6±1.8 | 47.5±2.0 | 34.4±2.3 |
| German | **23.6±2.1** | 27.3±2.6 | 25.2±2.5 | 25.2±2.4 |
| Heart | **16.0±3.3** | 20.2±3.7 | 16.5±3.6 | 16.5±3.6 |
| Image | 3.0±0.6 | 11.4±1.2 | 3.1±0.9 | **2.9±0.7** |
| Ringnorm | **1.7±0.1** | 27.6±1.7 | 3.6±0.4 | 3.2±0.3 |
| Splice | **10.9±0.7** | 15.9±1.1 | 15.2±1.0 | 15.2±1.0 |
| Thyroid | **4.8±2.2** | 8.9±2.8 | 4.9±2.4 | 5.0±2.2 |
| Titanic | **22.4±1.0** | 23.0±1.1 | 23.0±1.3 | 22.5±1.2 |
| Twonorm | 3.0±0.2 | 3.6±0.4 | **2.6±0.2** | **2.6±0.3** |
| Waveform | **9.9±0.4** | 19.5±1.9 | 12.0±0.9 | 11.9±0.9 |

approach. We start from a hyperbox that includes all the class data and cut the hyperbox in the region where class data overlap [7]. We evaluate the performance of the method for some benchmark data sets.

## 10.2.1 Training Methods

### 10.2.1.1 Concept

In the following we discuss how to approximate a region for class $i$ by a convex hull:

1. As shown in Fig. 10.5, we first approximate the class region by a hyperbox calculating the minimum and maximum values of the training inputs belonging to class $i$.
2. We sequentially read data belonging to classes other than class $i$. Let $\mathbf{q}$ be the current data sample. If $\mathbf{q}$ is in the convex hull (initially the hyperbox), we cut it so that class $i$ data are on the facet or inside the convex hull and that the facet is far from $\mathbf{q}$. This is the same idea as support vector machines that maximize margins. In Fig. 10.5, let $\mathbf{c}_i$ be the center of class $i$. To make class $i$ data in the convex hull or on the facet, we project class $i$ data in the direction orthogonal to $\mathbf{q} - \mathbf{c}_i$. Then, let $\mathbf{p}'$ be the data sample nearest to $\mathbf{q}$ on the line that includes $\mathbf{q} - \mathbf{c}_i$. If we cut the convex hull by the facet that goes through $\mathbf{p}'$ and that is orthogonal to $\mathbf{q} - \mathbf{c}_i$, the distance between $\mathbf{q}$ and the facet is not maximized. Thus as shown in Fig. 10.6, we cut the convex hull by the facet that goes through $\mathbf{p}'$ and that is orthogonal to $\mathbf{q} - \mathbf{p}'$.

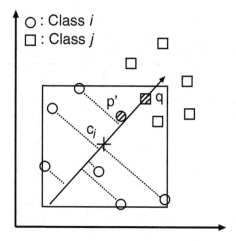

**Fig. 10.5** Extracting a boundary data sample. Reprinted from [7, p. 674]

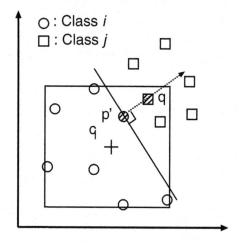

**Fig. 10.6** Class $i$ data exist on the negative side of the hyperplane. Reprinted from [7, p. 674]

Then, as shown in Figs. 10.7 and 10.8, if data sample $q$ and class $i$ data are outside of facet h1, we rotate the orthogonal vector of the facet from $q - c_i$ to $q - p'$ until the facet touches the class $i$ data (the facet h2 in each figure). By this method, however, two convex hulls may overlap as shown in Fig. 10.8, but we do not consider this case here.

According to this procedure, if the data of other classes are in the convex hull, the convex hull is cut by a facet. Then in the case shown in Fig. 10.9, the class $i$ convex hull is not cut by a facet. In this case, although the data in the shaded region are nearer to class $j$ than to class $i$, they are classified

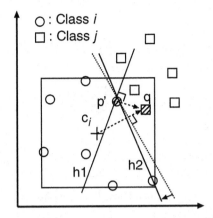

**Fig. 10.7** Class $i$ data exist on both sides of the hyperplane. Reprinted from [7, p. 674]

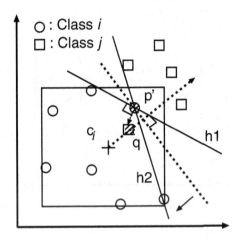

**Fig. 10.8** Class $i$ data exist on the positive side of the hyperplane. Reprinted from [7, p. 674]

into class $i$. To avoid this we use data that are near the convex hull (initially the hyperbox), namely, as shown in Fig. 10.10, we expand the convex hull and use the data in the expanded convex hull.

In the following, we discuss the detailed procedure of convex hull generation, definition of membership functions, and tuning of membership functions.

### 10.2.1.2 Generation of Hyperboxes

Let the set of $m$-dimensional data for class $i$ be $X_i$. Then for each input variable we calculate the minimum and maximum values

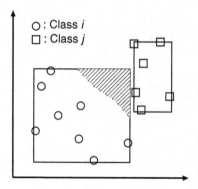

**Fig. 10.9** Degradation of generalization ability. Reprinted from [7, p. 674]

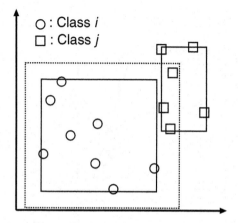

**Fig. 10.10** Generalization improvement by expansion of a polyhedron. Reprinted from [7, p. 675]

$$V_{ij} = \max_{\mathbf{p} \in X_i} p_j \quad \text{for } j = 1, \ldots, m, \tag{10.30}$$

$$v_{ij} = \min_{\mathbf{p} \in X_i} p_j \quad \text{for } j = 1, \ldots, m, \tag{10.31}$$

and we generate the hyperbox.

### 10.2.1.3 Generation of Polyhedral Regions

In the following we discuss in detail how to cut the hyperbox for class $i$ by facets:

1. We define the center of class $i$ by the average of class $i$ data:

$$c_i = \frac{1}{|X_i|} \sum_{p \in X_i} p. \tag{10.32}$$

2. As shown in Fig. 10.11, we expand the convex hull by $s\,d$ where $d$ is the distance from the center to a facet and $s$ is an expansion parameter. We call this the *expanded convex hull*. Then we read a data sample not belonging to class $i$, $q$ ($q \in X_k, k \neq i$), sequentially and check whether it is in the expanded convex hull. The expansion parameter $s$ specifies how much the convex hull is expanded, and for $s = 0$, we use only the data in the convex hull. As $s$ increases, the number of facets increases.

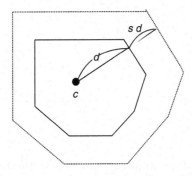

**Fig. 10.11** An expanded polyhedron. Reprinted from [7, p. 675]

We first check whether the data sample is in the expanded hyperbox. If it is, we further check whether it is in the convex hull, namely, if for $q_j$ ($j = 1, \ldots, m$)

$$v_{ij} - d'_j\,s < q_j < V_{ij} + d''_j\,s \tag{10.33}$$

is satisfied, $q$ is in the hyperbox, where $d'_j, d''_j$ are the one-dimensional distances given by

$$d'_j = c_j - v_{ij}, \tag{10.34}$$
$$d''_j = V_{ij} - c_j, \tag{10.35}$$

respectively.

If $q$ is in the hyperbox, we further check whether it is in the convex hull. Let the facets that form the convex hull for class $i$ be $F_{ij}(j = 1, \ldots, f)$, where $f$ is the number of facets. Let $F_{ij}$ go through $p^j$ and its outer orthogonal vector be $n_{ij}$. As shown in Fig. 10.12, we project $q - p^j$ and $p^j - c_i$ in the direction of $n_{ij}$. If

$$n_{ij}^\top (q - p^j) \leq s\,n_{ij}^\top(p^j - c_i) \quad \text{for } j = 1, \ldots, f$$

is satisfied, **q** is in the convex hull. If the point is in the convex hull, go
to Step 3. Otherwise, read the next data sample. If all the data have been
read, we finish training.

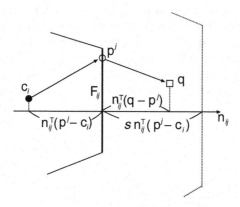

**Fig. 10.12** Check whether a data sample is in the expanded polyhedron. Reprinted from
[7, p. 675]

3. We find the class $i$ data sample $\mathbf{p}'$, which is farthest in the direction from
   the class $i$ center $\mathbf{c}_i$ to the data sample $\mathbf{q}$:

$$\mathbf{p}' = \arg\max_{\mathbf{p}\in X_i}(\mathbf{q} - \mathbf{c}_i)^\top(\mathbf{p} - \mathbf{c}_i). \tag{10.36}$$

   Go to Step 4.
4. We generate a facet that goes through $\mathbf{p}'$, which includes class $i$ data on
   the facet or in the convex hull, and from which $\mathbf{q}$ is the farthest in the
   outer direction of the facet. Here, the orthogonal vector of the facet, which
   goes through $\mathbf{p}'$ and in which $\mathbf{q}$ is the farthest in the outer direction of the
   facet, is $\mathbf{q} - \mathbf{p}'$.

   If we generate a facet with $\mathbf{q} - \mathbf{p}'$ as the orthogonal vector, there may
   be cases where class $i$ data are outside of the facet. To avoid this, we check
   if the data are outside of the facet with $\mathbf{q} - \mathbf{p}'$ as the orthogonal vector.
   If there are no data (see Fig. 10.6), we generate the facet and go back to
   Step 2. If there are (see Fig. 10.7), go to Step 5. Here, for $\mathbf{p} \in X_i$, if

$$(\mathbf{q} - \mathbf{p}')^\top(\mathbf{p}' - \mathbf{p}) \le 0 \tag{10.37}$$

is satisfied, class $i$ data sample $\mathbf{p}$ is within the facet, and if

$$(\mathbf{q} - \mathbf{p}')^\top(\mathbf{p}' - \mathbf{p}) > 0 \tag{10.38}$$

is satisfied, it is outside of the facet.

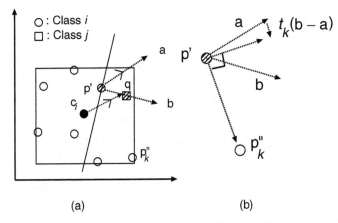

(a)                              (b)

**Fig. 10.13** Rotation of a hyperplane. Reprinted from [7, p. 676]

5. We rotate the orthogonal vector of the facet from $\mathbf{q} - \mathbf{c}_i$ to $\mathbf{q} - \mathbf{p}'$ and stop rotation at the point where class $i$ data are included on the facet. As shown in Fig. 10.13a, for the orthogonal vector $\mathbf{q} - \mathbf{p}'$, let there be $o$ data outside of the facet and let them be $\mathbf{p}''_k (k = 1, 2, \ldots, o)$ and $\mathbf{a}$ and $\mathbf{b}$ be

$$\mathbf{a} = \mathbf{q} - \mathbf{c}_i, \tag{10.39}$$
$$\mathbf{b} = \mathbf{q} - \mathbf{p}'. \tag{10.40}$$

As shown in Fig. 10.13b, let $\mathbf{p}''_k$ be on the facet and $\mathbf{p}' - \mathbf{p}''_k$ be the orthogonal vector. We define the parameter $t_k$ for rotation by

$$(\mathbf{a} + t_k(\mathbf{b} - \mathbf{a}))^\top (\mathbf{p}' - \mathbf{p}''_k) = 0. \tag{10.41}$$

We denote the minimum $t_k$ by $t$:

$$t = \min_{k=1,\ldots,o} t_k. \tag{10.42}$$

To determine the facet with a minimum rotation, we determine the orthogonal vector $\mathbf{n}$ by

$$\mathbf{n} = \mathbf{a} + t(\mathbf{b} - \mathbf{a}). \tag{10.43}$$

Next, we generate the facet that goes through $\mathbf{p}'$ with the orthogonal vector $\mathbf{n}$ and return to Step 2.

## 10.2.1.4 Membership Functions

We define a membership function in which the degree of membership is 1 at the center of the convex hull and it decreases as the location moves away from the center.

We define a one-dimensional membership function, $m_i(\mathbf{p})$, of data sample $\mathbf{p}$ for the facet $F_i$ by

$$m_i(\mathbf{p}) = \exp\left(-h_i(\mathbf{p})\right) = \exp\left(-\frac{d_i(\mathbf{p})}{\alpha_c}\right), \qquad (10.44)$$

where $h_i(\mathbf{p})$ is a tuned distance, $\alpha_c$ is a tuning parameter for class $c$, and $d_i(\mathbf{p})$ is given by

$$d_i(\mathbf{p}) = \begin{cases} \dfrac{\mathbf{a}_i^\top (\mathbf{p} - \mathbf{c})}{|\mathbf{a}_i|w_i} & \mathbf{a}_i^\top (\mathbf{p} - \mathbf{c}) \geq 0, \\ 0 & \mathbf{a}_i^\top (\mathbf{p} - \mathbf{c}) < 0, \end{cases} \qquad (10.45)$$

where $\mathbf{a}_i$ is the outer orthogonal vector for the facet and $w_i$ is the distance from the center to the facet $F_i$.

Then using (10.44), the membership function of $\mathbf{p}$ for the convex hull, $m(\mathbf{p})$, is given by

$$m(\mathbf{p}) = \min_i m_i(\mathbf{p}) = \exp\left(-\max_i h_i(\mathbf{p})\right). \qquad (10.46)$$

## 10.2.1.5 Tuning of Membership Functions

We can improve the recognition performance by tuning membership functions. In the following, we explain the idea [2].

Now suppose we tune the tuning parameter $\alpha_i$. Up to some value we can increase or decrease $\alpha_i$ without making the correctly classified data sample $\mathbf{x}$ be misclassified. Thus we can calculate the upper bound $U_i(\mathbf{x})$ or lower bound $L_i(\mathbf{x})$ of $\alpha_i$ that causes no misclassification of $\mathbf{x}$. Now let $U_i(1)$ and $L_i(1)$ denote the upper and lower bounds that do not make the correctly classified data be misclassified, respectively. Likewise, $U_i(l)$ and $L_i(l)$ denote the upper and lower bounds in which $l - 1$ correctly classified data are misclassified, respectively.

Similarly, if we increase or decrease $\alpha_i$, misclassified data may be correctly classified. Let $\beta_i(l)$ denote the upper bound of $\alpha_i$ that is smaller than $U_i(l)$ and that resolves misclassification. Let $\gamma_i(l)$ denote the lower bound of $\alpha_i$ that is larger than $L_i(l)$ and that resolves misclassification.

Then the next task is to find which interval among $(L_i(l), \gamma_i(l))$ and $(\beta_i(l), U_i(l))$ $(l = 1, \ldots)$ gives the maximum recognition rate. To limit the search space, we introduce the maximum $l$, i.e., $l_M$. Let $(L_i(l), \gamma_i(l))$ be the interval that gives the maximum recognition rate of the training data among $(L_i(k), \gamma_i(k))$ and $(\beta_i(k), U_i(k))$ for $k = 1, \ldots, l_M$. Then even if we set any value in the interval to $\alpha_i$, the recognition rate of the training data does not change but the recognition rate of the test data may change. To control the generalization ability, we set $\alpha_i$ as follows:

$$\alpha_i = \beta_i(l) + \delta(U_i(l) - \beta_i(l)) \tag{10.47}$$

for $(\beta_i(l), U_i(l))$, where $\delta$ satisfies $0 < \delta < 1$ and

$$\alpha_i = \gamma_i(l) - \delta(\gamma_i(l) - L_i(l)) \tag{10.48}$$

for $(L_i(l), \gamma_i(l))$.

In the following performance evaluation, we set $l_M = \infty$. The value of $\delta$ does not affect the recognition rate of the test data very much, and we set $\delta = 0.1$.

## 10.2.2 Performance Evaluation

Using the numeral data, thyroid data, blood cell data, hiragana-13 data, hiragana-50 data, and hiragan-105 data listed in Table 1.3, we compared the performance of the fuzzy classifier with polyhedral regions (FCPR) discussed in this section with that of the conventional fuzzy classifier with polyhedral regions based on the dynamic convex hull generation method (C-FCPR) [6]. The performance of the FCPR was measured with $s = 0$ and the best recognition rate of the test data obtained by changing $s$. We used a Pentium III (1 GHz) personal computer.

If the number of inputs is large, the C-FCPR takes a long time generating polyhedral regions. Thus, in this computer experiment, we bounded the number of facets to be generated by 5,000.

Table 10.6 shows the results for the FCPR and C-FCPR. For comparison, we also include the results of the fuzzy classifier with ellipsoidal regions (FCER) and the one-against-all support vector machine (SVM). In the table, the "Parm" column shows the parameter value and, for example, $s5$ means that 5 is set to $s$ and $d2$ and $\gamma0.5$ mean that polynomial kernels with degree 2 and RBF kernels with $\gamma = 0.5$ are used for the support vector machine, respectively. "Initial" means the recognition rate before tuning membership functions and "Final" means the recognition rates after tuning membership functions. If the recognition rates of the training data were not 100%, they are shown in parentheses.

The training time of the FCPR with $s = 0$ is much shorter than that of C-FCPR and the recognition rates of the test data are better. For the FCPR with nonzero $s$, training was slowed, but the recognition rates of the test data were improved and comparable with those of the fuzzy classifier with ellipsoidal regions and the one-against-all support vector machine.

In evaluating the thyroid data, for the conventional FCPR, the number of variables was decreased from 21 to 5 so that the convex hulls could be generated. For the FCPR, it was not necessary to reduce the number.

**Table 10.6** Comparison of recognition rates

| Data | Method | Parm | Initial (%) | Final (%) | Time (s) |
|------|--------|------|-------------|-----------|----------|
| | FCPR | $s5$ | 99.63 | 99.63 | 0.10 |
| | FCPR | $s0$ | 99.51 | 99.51 | 0.10 |
| Numeral | C-FCPR | | 99.39 | 99.39 | 251 |
| | FCER | | 99.63 (99.63) | 99.39 (99.88) | 0.14 |
| | SVM | $d2$ | – | 99.88 | 1.93 |
| | FCPR | $s0.004$ | 99.21 (99.92) | 99.18 (99.97) | 6 |
| | FCPR | $s0$ | 99.18 (99.92) | 99.18 (99.97) | 5 |
| Thyroid | C-FCPR | | 98.16 (98.25) | 98.22 (99.81) | 72 |
| | FCER | | 86.41 (86.77) | 97.29 (99.02) | 9 |
| | SVM | $d2$ | – | 98.02 (99.34) | 7 |
| | FCPR | $s2$ | 88.29 (94.51) | 90.23 (96.45) | 101 |
| | FCPR | $s0$ | 88.81 (94.12) | 89.29 (95.83) | 15 |
| Blood cell | C-FCPR | | 82.13 (89.09) | 86.39 (93.51) | 1,514 |
| | FCER | | 87.45 (92.64) | 92.03 (96.29) | 3 |
| | SVM | $\gamma0.5$ | – | 92.87 (99.23) | 10 |
| | FCPR | $s5$ | 98.56 (99.51) | 98.92 (99.84) | 4,380 |
| | FCPR | $s0$ | 97.34 (98.97) | 97.57 (99.31) | 17 |
| Hiragana-13 | C-FCPR | | 94.78 (99.45) | 97.74 (99.94) | 8,161 |
| | FCER | | 99.41 (99.84) | 99.59 (99.96) | 9 |
| | SVM | $\gamma1$ | – | 99.77 | 131 |
| | FCPR | $s2$ | 96.07 (99.87) | 96.23 (99.98) | 840 |
| Hiragana-50 | FCPR | $s0$ | 90.69 (99.50) | 90.78 (99.89) | 11 |
| | FCER | | 98.83 (99.87) | 98.85 | 107 |
| | SVM | $d2$ | – | 98.89 | 238 |
| | FCPR | $s0.6$ | 99.90 | 99.90 | 203 |
| Hiragana-105 | FCPR | $s0$ | 97.82 | 97.82 | 25 |
| | FCER | | 99.94 (99.92) | 100 | 512 |
| | SVM | $d2$ | – | 100 | 836 |

Figure 10.14 shows the recognition rates of the hiragana-50 data for the change of $s$. For $0 \leq s < 2$, the recognition rate of the test data increases, but for $s$ larger than 2, the recognition rate decreases.

For all the data sets used in the study, training by the FCPR with $s = 0$ was very fast but the recognition rates of the test data were lower for some data sets than those by other methods. Thus we need to set positive $s$, which

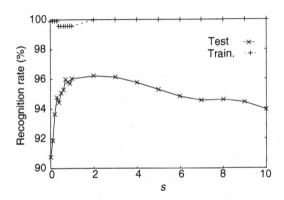

**Fig. 10.14** Recognition rates of the hiragana-50 data for the different values of $s$. Reprinted from [7, p. 679]

controls the expansion of convex hulls. From the simulation results (only the result of the hiragana-50 data was included in this book), the optimal value of $s$ was around the point where the recognition rate of the training data reached a peak value.

# References

1. L. I. Kuncheva. *Fuzzy Classifier Design*. Physica-Verlag, Heidelberg, 2000.
2. S. Abe. *Pattern Classification: Neuro-Fuzzy Methods and Their Comparison.* Springer-Verlag, London, 2001.
3. P. K. Simpson. Fuzzy min-max neural networks—Part 1: Classification. *IEEE Transactions on Neural Networks*, 3(5):776–786, 1992.
4. S. Abe and M.-S. Lan. A method for fuzzy rules extraction directly from numerical data and its application to pattern classification. *IEEE Transactions on Fuzzy Systems*, 3(1):18–28, 1995.
5. V. Uebele, S. Abe, and M.-S. Lan. A neural network–based fuzzy classifier. *IEEE Transactions on Systems, Man, and Cybernetics*, 25(2):353–361, 1995.
6. T. Shimozaki, T. Takigawa, and S. Abe. A fuzzy classifier with polyhedral regions. *Transactions of the Institute of Systems, Control and Information Engineers*, 14(7):365–372, 2001 (in Japanese).
7. T. Takigawa and S. Abe. High speed training of a fuzzy classifier with polyhedral regions. *Transactions of the Institute of Systems, Control and Information Engineers*, 15(12):673–680, 2002 (in Japanese).
8. S. Abe and R. Thawonmas. A fuzzy classifier with ellipsoidal regions. *IEEE Transactions on Fuzzy Systems*, 5(3):358–368, 1997.
9. S. Abe. Dynamic cluster generation for a fuzzy classifier with ellipsoidal regions. *IEEE Transactions on Systems, Man, and Cybernetics—Part B*, 28(6):869–876, 1998.
10. K. Kaieda and S. Abe. A kernel fuzzy classifier with ellipsoidal regions. In *Proceedings of International Joint Conference on Neural Networks (IJCNN 2003)*, volume 3, pages 2043–2048, Portland, OR, 2003.
11. Y. Chen and J. Z. Wang. Support vector learning for fuzzy rule-based classification systems. *IEEE Transactions on Fuzzy Systems*, 11(6):716–728, 2003.

12. K. Kaieda and S. Abe. KPCA-based training of a kernel fuzzy classifier with ellipsoidal regions. *International Journal of Approximate Reasoning*, 37(3):189–217, 2004.

13. J.-H. Chiang and P.-Y. Hao. Support vector learning mechanism for fuzzy rule-based modeling: A new approach. *IEEE Transactions on Fuzzy Systems*, 12(1):1–12, 2004.

14. S.-M. Zhou and J. Q. Gan. Constructing L2-SVM-based fuzzy classifiers in high-dimensional space with automatic model selection and fuzzy rule ranking. *IEEE Transactions on Fuzzy Systems*, 15(3):398–409, 2007.

15. R. Hosokawa and S. Abe. Fuzzy classifiers based on kernel discriminant analysis. In J. Marques de Sá, L. A. Alexandre, W. Duch, and D. Mandic, editors, *Artificial Neural Networks (ICANN 2007)—Proceedings of the Seventeenth International Conference, Porto, Portugal, Part II*, pages 180–189. Springer-Verlag, Berlin, Germany, 2007.

16. K. Morikawa and S. Abe. Improved parameter tuning algorithms for fuzzy classifiers. In M. Köppen, N. Kasabov, and G. Coghill, editors, *Advances in Neuro-Information Processing: Proceedings of the 15th International Conference (ICONIP 2008), Auckland, New Zealand*, pages 937–944. Springer-Verlag, Heidelberg, Germany, 2009.

17. K. Morikawa, S. Ozawa, and S. Abe. Tuning membership functions of kernel fuzzy classifiers by maximizing margins. *Memetic Computing*, 1(3):221–228, 2009.

18. S. Abe and K. Sakaguchi. Generalization improvement of a fuzzy classifier with ellipsoidal regions. In *Proceedings of the Tenth IEEE International Conference on Fuzzy Systems*, volume 1, pages 207–210, Melbourne, Australia, 2001.

19. G. H. Golub and C. F. Van Loan. *Matrix Computations, Third Edition*. The Johns Hopkins University Press, Baltimore, MD, 1996.

20. Intelligent Data Analysis Group. http://ida.first.fraunhofer.de/projects/bench/benchmarks.htm.

21. S. Abe. *Neural Networks and Fuzzy Systems: Theory and Applications*. Kluwer Academic Publishers, Norwell, MA, 1997.

# Chapter 11
# Function Approximation

Support vector regressors, which are extensions of support vector machines, have shown good generalization ability for various function approximation and time series prediction problems (e.g., [1, 2]). In this chapter, first we discuss definitions of L1 and L2 support vector regressors and their training methods. Then, we discuss extensions of various support vector machines for pattern classification to function approximation: LP support vector regressors, $\nu$-support vector regressors, and LS support vector regressors. Finally, we discuss variable selection methods.

## 11.1 Optimal Hyperplanes

In function approximation, we determine the input–output relation using the input–output pairs $(\mathbf{x}_i, y_i)$ $(i = 1, \ldots, M)$, where $\mathbf{x}_i$ is the $i$th $m$-dimensional input vector, $y_i$ is the $i$th scalar output,[1] and $M$ is the number of training data.

In support vector regression, we map the input space into the high-dimensional feature space, and in the feature space we determine the optimal hyperplane given by

$$f(\mathbf{x}) = \mathbf{w}^\top \phi(\mathbf{x}) + b, \tag{11.1}$$

where $\mathbf{w}$ is the $l$-dimensional weight vector, $\phi(\mathbf{x})$ is the mapping function that maps $\mathbf{x}$ into the $l$-dimensional feature space, and $b$ is the bias term.

In linear regression usually the square error function shown in Fig. 11.1a is used, where $r$ is the residual and $r = y - f(\mathbf{x})$. Using this error function, how-

---

[1] There are several extensions to a multidimensional output [3–6]. Pérez-Cruz et al. [3] constrained data in the hyperspherical insensitive zone with radius $\varepsilon$. Brudnak [6] proposed vector-valued support vector regression (VV-SVR), in which the $\varepsilon$-insensitive zone is applied to the $p$-norm of the error vector $\mathbf{e}$, $\|\mathbf{e}\|_p = (\sum_{i=1}^n |e_i|^p)^{1/p}$, where $n$ is the number of output variables.

S. Abe, *Support Vector Machines for Pattern Classification*,
Advances in Pattern Recognition, DOI 10.1007/978-1-84996-098-4_11,
© Springer-Verlag London Limited 2010

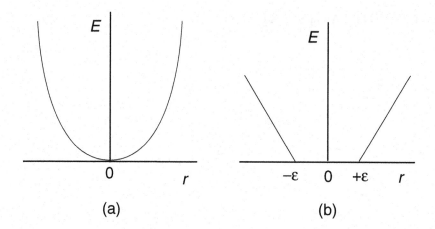

**Fig. 11.1** Error functions: (a) a square function; (b) a piecewise linear function

ever, the large residuals caused by outliers worsen the estimation accuracy significantly. To avoid this, in support vector regressors, the piecewise linear function, shown in Fig. 11.1b, which is originally used for robust function approximation, is used:

$$E(r) = \begin{cases} 0 & \text{for } |r| \le \varepsilon, \\ |r| - \varepsilon & \text{otherwise,} \end{cases} \tag{11.2}$$

where $\varepsilon$ is a small positive parameter.

Now define the residual of output $y$ and the estimate $f(\mathbf{x})$ by

$$D(\mathbf{x}, y) = y - f(\mathbf{x}). \tag{11.3}$$

According to (11.2), the ideal estimation is realized when all the absolute residuals are within $\varepsilon$, namely

$$-\varepsilon \le D(\mathbf{x}, y) \le +\varepsilon, \tag{11.4}$$

which is rewritten as follows:

$$|D(\mathbf{x}, y)| \le \varepsilon. \tag{11.5}$$

Figure 11.2 illustrates this; in the original input–output space, if all the training data are within the zone with radius $\varepsilon$ shown in Fig. 11.2a, the ideal estimation is realized. We call this $\varepsilon$-*insensitive zone*. In the feature-input–output space, this tube corresponds to the straight tube shown in Fig. 11.2b.

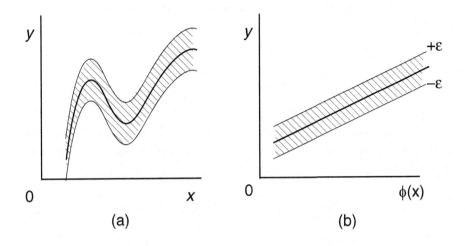

(a)                                    (b)

**Fig. 11.2** Insensitive zone: **(a)** in the original input space, and **(b)** in the feature space

There are infinite possibilities of solutions that satisfy (11.5). Here, we consider obtaining a solution with the maximum generalization ability. Assuming that all the training data satisfy (11.5), the data sample that satisfies $D(\mathbf{x}, y) = \pm \varepsilon$ is the farthest from the hyperplane. We call the associated distance the *margin*. Maximizing the margin leads to maximizing the possibility that the unknown data get into this tube. Thus, it leads to maximizing the generalization ability.

The distance from the hyperplane $D(\mathbf{x}, y) = 0$ to a data sample $(\mathbf{x}, y)$ is given by $|D(\mathbf{x}, y)|/\|\mathbf{w}^*\|$, where $\mathbf{w}^*$ is given by

$$\mathbf{w}^* = (1, -\mathbf{w}^\top)^\top. \tag{11.6}$$

Assuming that the maximum distance of data from the hyperplane is $\delta$, all the training data satisfy the following:

$$\frac{|D(\mathbf{x}, y)|}{\|\mathbf{w}^*\|} \leq \delta, \tag{11.7}$$

namely,

$$|D(\mathbf{x}, y)| \leq \delta \|\mathbf{w}^*\|. \tag{11.8}$$

From (11.5) and (11.8), the data that are farthest from the hyperplane satisfy $|D(\mathbf{x}, y)| = \varepsilon$. Thus

$$\delta \|\mathbf{w}^*\| = \varepsilon. \tag{11.9}$$

Therefore, to maximize the margin $\delta$, we need to minimize $\|\mathbf{w}^*\|$. Because $\|\mathbf{w}^*\|^2 = \|\mathbf{w}\|^2 + 1$ holds, minimizing $\|\mathbf{w}\|$ leads to maximizing the margin.

Now the regression problem is expressed by the following optimization problem:

$$\text{minimize} \quad \frac{1}{2}\|\mathbf{w}\|^2 \tag{11.10}$$

$$\text{subject to} \quad y_i - \mathbf{w}^\top \phi(\mathbf{x}_i) - b \le \varepsilon \quad \text{for } i = 1,\dots,M, \tag{11.11}$$

$$\mathbf{w}^\top \phi(\mathbf{x}_i) + b - y_i \le \varepsilon \quad \text{for } i = 1,\dots,M. \tag{11.12}$$

In the formulation, we assumed that all the training data are within the tube with radius $\varepsilon$. To allow the data that are outside of the tube to exist, we introduce the nonnegative slack variables $\xi_i$ and $\xi_i^*$ as shown in Fig. 11.3, where

$$\xi_i = \begin{cases} 0 & \text{for } D(\mathbf{x}_i, y_i) - \varepsilon \le 0, \\ D(\mathbf{x}_i, y_i) - \varepsilon & \text{otherwise,} \end{cases} \tag{11.13}$$

$$\xi_i^* = \begin{cases} 0 & \text{for } \varepsilon + D(\mathbf{x}_i, y_i) \ge 0, \\ -\varepsilon - D(\mathbf{x}_i, y_i) & \text{otherwise.} \end{cases} \tag{11.14}$$

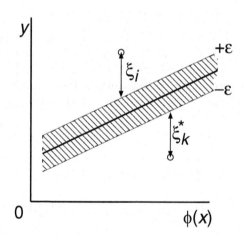

**Fig. 11.3** Slack variables

Now the regression problem is solved by

$$\text{minimize} \quad Q(\mathbf{w}, b, \boldsymbol{\xi}, \boldsymbol{\xi}^*) = \frac{1}{2}\|\mathbf{w}\|^2 + \frac{C}{p}\sum_{i=1}^{M}(\xi_i^p + \xi_i^{*p}) \tag{11.15}$$

$$\text{subject to} \quad y_i - \mathbf{w}^\top \phi(\mathbf{x}_i) - b \le \varepsilon + \xi_i \quad \text{for } i = 1,\dots,M, \tag{11.16}$$

$$\mathbf{w}^\top \phi(\mathbf{x}_i) + b - y_i \le \varepsilon + \xi_i^* \quad \text{for } i = 1,\dots,M, \tag{11.17}$$

$$\xi_i \ge 0, \quad \xi_i^* \ge 0 \quad \text{for } i = 1,\dots,M, \tag{11.18}$$

where $C$ is the margin parameter that determines the trade-off between the magnitude of the margin and the estimation error of the training data and $p$ is either 1 or 2. If $p = 1$, we call the support vector regressor the *L1 soft-margin support vector regressor (L1 SVR)* and $p = 2$, the *L2 soft-margin support vector regressor (L2 SVR)*.

The optimization problem given by (11.15), (11.16), (11.17), and (11.18) is solved by the quadratic programming technique, by converting (11.15), (11.16), (11.17), and (11.18) into the dual problem.

## 11.2 L1 Soft-Margin Support Vector Regressors

In this section we derive the dual problem of the L1 soft-margin support vector regressors. Introducing the Lagrange multipliers $\alpha_i, \alpha_i^*, \eta_i,$ and $\eta_i^*$ ($\geq 0$), we convert the original constrained problem into an unconstrained one:

$$
\begin{aligned}
&Q(\mathbf{w}, b, \boldsymbol{\xi}, \boldsymbol{\xi}^*, \boldsymbol{\alpha}, \boldsymbol{\alpha}^*, \boldsymbol{\eta}, \boldsymbol{\eta}^*) \\
&= \frac{1}{2}\|\mathbf{w}\|^2 + C \sum_{i=1}^{M}(\xi_i + \xi_i^*) - \sum_{i=1}^{M} \alpha_i \left(\varepsilon + \xi_i - y_i + \mathbf{w}^\top \boldsymbol{\phi}(\mathbf{x}_i) + b\right) \\
&\quad - \sum_{i=1}^{M} \alpha_i^* \left(\varepsilon + \xi_i^* + y_i - \mathbf{w}^\top \boldsymbol{\phi}(\mathbf{x}_i) - b\right) - \sum_{i=1}^{M}(\eta_i \xi_i + \eta_i^* \xi_i^*). \quad (11.19)
\end{aligned}
$$

This function has the saddle point that corresponds to the optimal solution for the original problem. At the optimal solution, the partial derivatives of $Q$ with respect to $\mathbf{w}, b, \boldsymbol{\xi},$ and $\boldsymbol{\xi}^*$ vanish, namely,

$$
\frac{\partial Q(\mathbf{w}, b, \boldsymbol{\xi}, \boldsymbol{\xi}^*, \boldsymbol{\alpha}, \boldsymbol{\alpha}^*, \boldsymbol{\eta}, \boldsymbol{\eta}^*)}{\partial \mathbf{w}} = \mathbf{w} - \sum_{i=1}^{M}(\alpha_i - \alpha_i^*)\boldsymbol{\phi}(\mathbf{x}_i) = \mathbf{0}, \quad (11.20)
$$

$$
\frac{\partial Q(\mathbf{w}, b, \boldsymbol{\xi}, \boldsymbol{\xi}^*, \boldsymbol{\alpha}, \boldsymbol{\alpha}^*, \boldsymbol{\eta}, \boldsymbol{\eta}^*)}{\partial b} = \sum_{i=1}^{M}(\alpha_i^* - \alpha_i) = 0, \quad (11.21)
$$

$$
\frac{\partial Q(\mathbf{w}, b, \boldsymbol{\xi}, \boldsymbol{\xi}^*, \boldsymbol{\alpha}, \boldsymbol{\alpha}^*, \boldsymbol{\eta}, \boldsymbol{\eta}^*)}{\partial \xi_i} = C - \alpha_i - \eta_i = 0 \quad \text{for } i = 1, \dots, M, \quad (11.22)
$$

$$
\frac{\partial Q(\mathbf{w}, b, \boldsymbol{\xi}, \boldsymbol{\xi}^*, \boldsymbol{\alpha}, \boldsymbol{\alpha}^*, \boldsymbol{\eta}, \boldsymbol{\eta}^*)}{\partial \xi_i^*} = C - \alpha_i^* - \eta_i^* = 0 \quad \text{for } i = 1, \dots, M. \quad (11.23)
$$

From (11.20),

$$
\mathbf{w} = \sum_{i=1}^{M}(\alpha_i - \alpha_i^*)\boldsymbol{\phi}(\mathbf{x}_i). \quad (11.24)
$$

Therefore, using $\alpha_i$ and $\alpha_i^*$, $f(\mathbf{x})$ is expressed by

$$f(\mathbf{x}) = \sum_{i=1}^{M} (\alpha_i - \alpha_i^*) \, \boldsymbol{\phi}^\top(\mathbf{x}_i) \, \boldsymbol{\phi}(\mathbf{x}) + b. \tag{11.25}$$

Substituting (11.20), (11.21), (11.22), and (11.23) into (11.19), we obtain the following dual problem:

$$\text{maximize} \quad Q(\boldsymbol{\alpha}, \boldsymbol{\alpha}^*) = -\frac{1}{2} \sum_{i,j=1}^{M} (\alpha_i - \alpha_i^*)(\alpha_j - \alpha_j^*) \, \boldsymbol{\phi}^\top(\mathbf{x}_i) \, \boldsymbol{\phi}(\mathbf{x}_j)$$

$$-\varepsilon \sum_{i=1}^{M} (\alpha_i + \alpha_i^*) + \sum_{i=1}^{M} y_i \, (\alpha_i - \alpha_i^*) \tag{11.26}$$

$$\text{subject to} \quad \sum_{i=1}^{M} (\alpha_i - \alpha_i^*) = 0, \tag{11.27}$$

$$0 \le \alpha_i \le C, \quad 0 \le \alpha_i^* \le C \quad \text{for } i = 1, \dots, M. \tag{11.28}$$

The optimal solution must satisfy the following KKT complementarity conditions:

$$\alpha_i \left( \varepsilon + \xi_i - y_i + \mathbf{w}^\top \boldsymbol{\phi}(\mathbf{x}_i) + b \right) = 0 \quad \text{for } i = 1, \dots, M, \tag{11.29}$$
$$\alpha_i^* \left( \varepsilon + \xi_i^* + y_i - \mathbf{w}^\top \boldsymbol{\phi}(\mathbf{x}_i) - b \right) = 0 \quad \text{for } i = 1, \dots, M, \tag{11.30}$$
$$\eta_i \, \xi_i = (C - \alpha_i) \, \xi_i = 0 \quad \text{for } i = 1, \dots, M, \tag{11.31}$$
$$\eta_i^* \, \xi_i^* = (C - \alpha_i^*) \, \xi_i^* = 0 \quad \text{for } i = 1, \dots, M. \tag{11.32}$$

From (11.31), when $0 < \alpha_i < C$, $\xi_i = 0$ holds. Likewise, from (11.32), when $0 < \alpha_i^* < C$, $\xi_i^* = 0$ holds. When either of them is satisfied, in (11.29) or (11.30) the equation in the parentheses vanishes, namely, either of the following equation holds:

$$\varepsilon - y_i + \mathbf{w}^\top \boldsymbol{\phi}(\mathbf{x}_i) + b = 0 \quad \text{for } 0 < \alpha_i < C, \tag{11.33}$$

$$\varepsilon + y_i - \mathbf{w}^\top \boldsymbol{\phi}(\mathbf{x}_i) - b = 0 \quad \text{for } 0 < \alpha_i^* < C. \tag{11.34}$$

This means that for the data sample with the residual $y - f(\mathbf{x}) = +\varepsilon$, $\alpha_i$ satisfies $0 < \alpha_i < C$, and the data sample with the residual $y - f(\mathbf{x}) = -\varepsilon$, $0 < \alpha_i^* < C$. Thus, $b$ satisfies

$$b = y_i - \mathbf{w}^\top \boldsymbol{\phi}(\mathbf{x}_i) - \varepsilon \quad \text{for } 0 < \alpha_i < C, \tag{11.35}$$

$$b = y_i - \mathbf{w}^\top \boldsymbol{\phi}(\mathbf{x}_i) + \varepsilon \quad \text{for } 0 < \alpha_i^* < C. \tag{11.36}$$

In calculating $b$, to avoid calculation errors, we average $b$'s that satisfy (11.35) and (11.36).

From (11.31) and (11.32), if either $\xi_i$ or $\xi_i^*$ is not zero, namely, the data sample is outside of the tube of radius $\varepsilon$, $\alpha_i$ equals $C$ when the data sample is above the tube and $\alpha_i^*$ equals $C$ when the data sample is under the tube.

From (11.29) and (11.30), when data satisfy $|y - f(\mathbf{x})| < \varepsilon$, both $\alpha_i$ and $\alpha_i^*$ are zero, and these data do not contribute in constructing the function given by (11.25). Conversely, for the data that satisfy $|y - f(\mathbf{x})| \geq \varepsilon$, $\alpha_i$ and $\alpha_i^*$ are not zero, and they contribute in constructing the function. The training data $\mathbf{x}_i$ with $0 < \alpha_i \leq C$ or $0 < \alpha_i^* \leq C$ are called *support vectors*, especially those with $0 < \alpha_i < C$ or $0 < \alpha_i^* < C$, *unbounded support vectors*, and those with $\alpha_i = C$ or $\alpha_i^* = C$, *bounded support vectors*.

Define

$$K(\mathbf{x}, \mathbf{x}') = \boldsymbol{\phi}^\top(\mathbf{x})\, \boldsymbol{\phi}(\mathbf{x}). \tag{11.37}$$

Then $K(\mathbf{x}, \mathbf{x}')$ satisfies Mercer's condition and is called a *kernel*. Using kernels, we need not treat the high-dimensional feature space explicitly.

## 11.3 L2 Soft-Margin Support Vector Regressors

In this section we derive the dual problem of the L2 soft-margin support vector regressors. Introducing the Lagrange multipliers $\alpha_i$ and $\alpha_i^*$, we convert the original constrained problem given by (11.15), (11.16), (11.17), and (11.18) into an unconstrained one:

$$
\begin{aligned}
&Q(\mathbf{w}, b, \boldsymbol{\xi}, \boldsymbol{\xi}^*, \boldsymbol{\alpha}, \boldsymbol{\alpha}^*) \\
&= \frac{1}{2}\|\mathbf{w}\|^2 + \frac{C}{2}\sum_{i=1}^M (\xi_i^2 + \xi_i^{*2}) - \sum_{i=1}^M \alpha_i \left(\varepsilon + \xi_i - y_i + \mathbf{w}^\top \boldsymbol{\phi}(\mathbf{x}_i) + b\right) \\
&\quad - \sum_{i=1}^M \alpha_i^* \left(\varepsilon + \xi_i^* + y_i - \mathbf{w}^\top \boldsymbol{\phi}(\mathbf{x}_i) - b\right). 
\end{aligned}
\tag{11.38}
$$

Here, unlike the L1 support vector regressor, we do not need to introduce the Lagrange multipliers associated with $\xi_i$ and $\xi_i^*$.

This function has the saddle point that corresponds to the optimal solution for the original problem. For the optimal solution, the partial derivatives of $Q$ with respect to $\mathbf{w}, b, \boldsymbol{\xi}$, and $\boldsymbol{\xi}^*$ vanish, namely,

$$\frac{\partial Q(\mathbf{w}, b, \boldsymbol{\xi}, \boldsymbol{\xi}^*, \boldsymbol{\alpha}, \boldsymbol{\alpha}^*)}{\partial \mathbf{w}} = \mathbf{w} - \sum_{i=1}^M (\alpha_i - \alpha_i^*)\, \boldsymbol{\phi}(\mathbf{x}_i) = \mathbf{0}, \tag{11.39}$$

$$\frac{\partial Q(\mathbf{w}, b, \boldsymbol{\xi}, \boldsymbol{\xi}^*, \boldsymbol{\alpha}, \boldsymbol{\alpha}^*)}{\partial b} = \sum_{i=1}^M (\alpha_i^* - \alpha_i) = 0, \tag{11.40}$$

$$\frac{\partial Q(\mathbf{w}, b, \boldsymbol{\xi}, \boldsymbol{\xi}^*, \boldsymbol{\alpha}, \boldsymbol{\alpha}^*)}{\partial \xi_i} = C\,\xi_i - \alpha_i = 0 \quad \text{for } i = 1, \ldots, M, \tag{11.41}$$

$$\frac{\partial Q(\mathbf{w}, b, \boldsymbol{\xi}, \boldsymbol{\xi}^*, \boldsymbol{\alpha}, \boldsymbol{\alpha}^*)}{\partial \xi_i^*} = C\,\xi_i^* - \alpha_i^* = 0 \quad \text{for } i = 1, \ldots, M. \quad (11.42)$$

From (11.39),

$$\mathbf{w} = \sum_{i=1}^M (\alpha_i - \alpha_i^*)\,\phi(\mathbf{x}_i). \quad (11.43)$$

Therefore, using $\alpha_i$ and $\alpha_i^*$, $f(\mathbf{x})$ is expressed by

$$f(\mathbf{x}) = \sum_{i=1}^M (\alpha_i - \alpha_i^*)\,K(\mathbf{x}_i, \mathbf{x}) + b, \quad (11.44)$$

where $K(\mathbf{x}_i, \mathbf{x}) = \phi^\top(\mathbf{x}_i)\,\phi(\mathbf{x})$.

Substituting (11.39), (11.40), (11.41), and (11.42) into (11.38), we obtain the following dual problem:

$$\text{maximize} \quad Q(\boldsymbol{\alpha}, \boldsymbol{\alpha}^*) = -\frac{1}{2} \sum_{i,j=1}^M (\alpha_i - \alpha_i^*)(\alpha_j - \alpha_j^*)\left(K(\mathbf{x}_i, \mathbf{x}_j) + \frac{\delta_{ij}}{C}\right)$$

$$-\varepsilon \sum_{i=1}^M (\alpha_i + \alpha_i^*) + \sum_{i=1}^M y_i\,(\alpha_i - \alpha_i^*) \quad (11.45)$$

$$\text{subject to} \quad \sum_{i=1}^M (\alpha_i - \alpha_i^*) = 0, \quad (11.46)$$

$$\alpha_i \geq 0, \quad \alpha_i^* \geq 0 \quad \text{for } i = 1, \ldots, M, \quad (11.47)$$

where $\delta_{ij}$ is Kronecker's delta function.

The optimal solution must satisfy the following KKT complementarity conditions:

$$\alpha_i\left(\varepsilon + \xi_i - y_i + \mathbf{w}^\top \phi(\mathbf{x}_i) + b\right) = 0 \quad \text{for } i = 1, \ldots, M, \quad (11.48)$$

$$\alpha_i^*\left(\varepsilon + \xi_i^* + y_i - \mathbf{w}^\top \phi(\mathbf{x}_i) - b\right) = 0 \quad \text{for } i = 1, \ldots, M, \quad (11.49)$$

$$C\,\xi_i = \alpha_i, \quad C\,\xi_i^* = \alpha_i^* \quad \text{for } i = 1, \ldots, M. \quad (11.50)$$

Thus, $\alpha_i = 0$, $\alpha_i^* = 0$ or

$$b = y_i - \mathbf{w}^\top \phi(\mathbf{x}_i) - \varepsilon - \frac{\alpha_i}{C} \quad \text{for } \alpha_i > 0, \quad (11.51)$$

$$b = y_i - \mathbf{w}^\top \phi(\mathbf{x}_i) + \varepsilon + \frac{\alpha_i^*}{C} \quad \text{for } \alpha_i^* > 0. \quad (11.52)$$

From (11.48) and (11.49), when data satisfy $|y - f(\mathbf{x})| < \varepsilon$, both $\alpha_i$ and $\alpha_i^*$ are zero, and these data do not contribute in constructing the function given by (11.44). The data $\mathbf{x}_i$ with nonzero $\alpha_i$ are called *support vectors*.

Comparing L1 and L2 support vector regressors, the latter does not have bounded support vectors and the Hessian matrix is positive definite.

# 11.4 Model Selection

Model selection, i.e., selection of optimal parameter values, e.g., $\varepsilon$, $C$, and $\gamma$ values for RBF kernels, is a difficult task for support vector regressors. Cross-validation is one of the most reliable model selection methods. However, compared to support vector machines we need to determine one more parameter value in addition to two parameter values.

To speedup model selection, several methods have been proposed [7–11]. Chang and Lin [9] derived the leave-one-out error bounds for support vector regressors and compared them with those for classification. Nakano's group [7, 10] proposed determining these parameters by alternating training of a support vector regressor (determination of $\alpha_i$ and $\alpha_i^*$) and optimizing the parameters by steepest descent for the cross-validation data. This method works for L2 support vector regressors but not for L1 support vector regressors due to the nonsmooth output for the parameter $\varepsilon$.

Smets et al. [11] evaluated the several performance measures for model selection and showed that $k$-fold cross-validation, leave-one-out, and span bound worked equally well but the selected values were different according to the methods.

# 11.5 Training Methods

## 11.5.1 Overview

Support vector regressors have all the advantages and disadvantages that support vector machines have. Because a support vector regressor is expressed by a quadratic optimization problem, the solution is globally optimal. However, because we usually use nonlinear kernels, we need to solve the dual optimization problem whose number of variables is twice the number of training data. Therefore, if the number of training data is very large, training becomes difficult. To solve this problem, as discussed in Section 5.2, we can use the decomposition technique. Then, in selecting a working set, should we select $\alpha_i$ and $\alpha_i^*$ simultaneously? Liao, Lin, and Lin [12] showed that there is not much difference in convergence in selecting both or either.

The decomposition technique can be classified into the fixed-size chunking and variable-size chunking. In fixed-size chunking, if the support vector candidates exceed the working set size, some of them with nonzero $\alpha_i$ or $\alpha_i^*$ are move out of the working set. Thus, by this method we can train a large-size problem but because nonzero $\alpha_i$ or $\alpha_i^*$ are in the fixed set, many iterations are necessary for convergence.

The most well-known fixed-size chunking method is the SMO (sequential minimal optimization) extended to function approximation [13]. Mattera et al. [14] redefined variables $\alpha_i$ and $\alpha_i^*$ in (11.26), (11.27), and (11.28) by $u_i = \alpha_i^* - \alpha_i$ and $|u_i| = \alpha_i^* + \alpha_i$ and proposed solving the quadratic programming problem by optimizing two variables $u_i$ and $u_j$ at a time. In each training cycle, all the $M(M-1)/2$ pairs of variables are selected and optimized, and when all the changes in the variables are within the specified limit, training is terminated. According to the computer experiment for the Lorenz chaotic process, the method was faster for a small value of $C$ ($C = 0.1$) than the QP solver. But it slowed down for large $C$ ($C = 10, 100$). Based on this formulation, Flake and Lawrence [15] derived a more efficient training method and Takahashi et al. [16, 17] proved convergence of the SMO.

Vogt [18] extended SMO for function approximation when the bias term is not present. Veropoulos [19, pp. 104–106] extended the kernel Adatron to function approximation. Kecman et al. [20] proved that these algorithms are equivalent.

In variable-size chunking, $\alpha_i$ and $\alpha_i^*$ that are not included in the working set are zero. Therefore, usually the number of iterations of the variable-size chunking is smaller than that of the fixed-size chunking. Training methods using variable-size chunking are sometimes called *active set training* because the working set include variables associated with the active constraints, where equalities are satisfied. The training methods using variable-size chunking developed for pattern classification can be extended to function approximation [21, 22].

There are many other ideas to speed up training. De Freitas et al. [23] used the Kalman filtering technique for sequential training of support vector regressors. Anguita et al. [24] proposed speeding up training and reduce memory storage when the training inputs are positioned on a grid using the fact that the matrix associated with the kernel is expressed by the Toeplitz block matrix [25]. They showed the effectiveness of the method for an image interpolation problem.

In the following we discuss two methods of training support vector regressors: Newton's methods with fixed-size chunking [26, 27], which is an extension of SMO for support vector regressors and active set training, which is an extension of the training method discussed in Section 5.8.

## 11.5.2 Newton's Methods

In this section, we discuss Newton's methods with fixed-size chunking to training support vector regressors. The working set size is increased from two and the variables in the working set are optimized by Newton's method. Calculations of corrections of variables in the working set include inversion of the associated Hessian matrix. But because the Hessian matrix for the L1 support vector machine is not guaranteed to be positive definite, we calculate the corrections only for the linearly independent variables in the working set.

We change the notations: $\alpha_i^* = \alpha_{M+i}$ and $\xi_i^* = \xi_{M+i}$. Then the optimization problem for the L1 support vector regressor given by (11.26), (11.27), and (11.28) becomes as follows:

$$\text{maximize} \quad Q(\boldsymbol{\alpha}) = -\frac{1}{2} \sum_{i,j=1}^{M} (\alpha_i - \alpha_{M+i})(\alpha_j - \alpha_{M+j}) K(\mathbf{x}_i, \mathbf{x}_j)$$

$$+ \sum_{i=1}^{M} y_i (\alpha_i - \alpha_{M+i}) - \varepsilon \sum_{i=1}^{M} (\alpha_i + \alpha_{M+i}) \quad (11.53)$$

$$\text{subject to} \quad \sum_{i=1}^{M} (\alpha_i - \alpha_{M+i}) = 0, \quad (11.54)$$

$$0 \le \alpha_i \le C \quad \text{for } i = 1, \ldots, M. \quad (11.55)$$

The optimization problem for the L2 support vector regressor given by (11.45) and (11.47) becomes as follows:

$$\text{maximize} \quad Q(\boldsymbol{\alpha}) = -\frac{1}{2} \sum_{i,j=1}^{M} (\alpha_i - \alpha_{M+i})(\alpha_j - \alpha_{M+j}) \left( K(\mathbf{x}_i, \mathbf{x}_j) + \frac{\delta_{ij}}{C} \right)$$

$$+ \sum_{i=1}^{M} y_i (\alpha_i - \alpha_{M+i}) - \varepsilon \sum_{i=1}^{M} (\alpha_i + \alpha_{M+i}) \quad (11.56)$$

$$\text{subject to} \quad \sum_{i=1}^{M} (\alpha_i - \alpha_{M+i}) = 0, \quad (11.57)$$

$$\alpha_i \ge 0 \quad \text{for } i = 1, \ldots, M. \quad (11.58)$$

SMO solves two-variable subproblems without using a QP solver. In the following we solve subproblems with more than two variables by Newton's method.

Now we define a candidate set $V$ as the index set of variables that are candidates of support vectors and a working set $W$ as the index set of variables in $V$.

The rough flow of the training procedure of the support vector regressor is as follows:

1. Add all the indexes that are associated with the training data to $V$.
2. Select indices from $V$ randomly and set them to $W$. Selected indices are removed from $V$.
3. Calculate corrections of the variables in the working set by Newton's method so that the objective function is maximized.
4. If a convergence condition is satisfied, finish training. Otherwise, if $V$ is empty, add new candidates that violate the KKT complementarity conditions. Return to Step 2.

### 11.5.2.1 Subproblem Optimization

In this subsection we explain Step 3 in more detail.

Let $\alpha_W$ be vectors whose elements are $\alpha_i$ $(i \in W)$. From (11.54), $\alpha_s \in \alpha_W$ is expressed as follows:

$$\alpha_s = -\sum_{i \neq s, i=1}^{M} \alpha_i + \sum_{i=M+1}^{2M} \alpha_i \quad \text{if } s \leq M, \tag{11.59}$$

$$\alpha_s = \sum_{i=1}^{M} \alpha_i - \sum_{i \neq s, i=M+1}^{2M} \alpha_i \quad \text{if } s > M. \tag{11.60}$$

Substituting (11.59) or (11.60) into the objective function, we eliminate constraint (11.54) from the dual problem. Here $W'$ is defined as the set in which $s$ is removed from $W$, namely $W' = W - \{s\}$.

Because the objective function is quadratic, the change of the objective function, $\Delta Q(\alpha_{W'})$, for the change of variables, $\Delta \alpha_{W'}$, is given by

$$\Delta Q(\alpha_{W'}) = \frac{\partial Q(\alpha_{W'})}{\partial \alpha_{W'}} \Delta \alpha_{W'} + \frac{1}{2} \Delta \alpha_{W'}^\top \frac{\partial^2 Q(\alpha_{W'})}{\partial \alpha_{W'}^2} \Delta \alpha_{W'}. \tag{11.61}$$

If $\partial^2 Q(\alpha_{W'})/\partial \alpha_{W'}^2$ is positive definite, we can calculate corrections by the following formula so that $\Delta Q(\alpha_{W'})$ is maximized:

$$\Delta \alpha_{W'} = -\left(\frac{\partial^2 Q(\alpha_{W'})}{\partial \alpha_{W'}^2}\right)^{-1} \frac{\partial Q(\alpha_{W'})}{\partial \alpha_{W'}}. \tag{11.62}$$

For the L1 support vector regressor, from (11.53), (11.59), and (11.60), the element of $\partial Q(\alpha_{W'})/\partial \alpha_{W'}$ is

$$\frac{\partial Q(\alpha_{W'})}{\partial \alpha_i} = p_i \{y_{i^*} - y_{s^*} - \varepsilon (p_i + q)$$

$$- \sum_{j=1}^{M} (\alpha_j - \alpha_{M+j}) (K_{ij} - K_{sj})\} \quad \text{for } i, s \in \{1, \ldots, 2M\}, \tag{11.63}$$

where $K_{ij} = K(\mathbf{x}_{i^*}, \mathbf{x}_{j^*})$ and $i^*$ $(s^*)$, $p_i$, and $q$ are defined as

$$i^* = \begin{cases} i & \text{for } i \leq M, \\ i - M & \text{for } i > M, \end{cases} \tag{11.64}$$

$$p_i = \begin{cases} +1 & \text{for } i \leq M, \\ -1 & \text{for } i > M, \end{cases} \tag{11.65}$$

$$q = \begin{cases} -1 & \text{for } s \leq M, \\ +1 & \text{for } s > M. \end{cases} \tag{11.66}$$

The element of $\partial^2 Q(\boldsymbol{\alpha}_{W'})/\partial \boldsymbol{\alpha}_{W'}^2$ is

$$\frac{\partial^2 Q(\boldsymbol{\alpha}_{W'})}{\partial \alpha_i \partial \alpha_j} = -p_i\, p_j\, (K_{ij} + K_{ss} - K_{is} - K_{js})$$
$$\text{for } i, j, s \in \{1, \dots, 2M\}. \tag{11.67}$$

For the L2 support vector regressor, the element of $\partial Q(\boldsymbol{\alpha}_{W'})/\partial \boldsymbol{\alpha}_{W'}$ is

$$\frac{\partial Q(\boldsymbol{\alpha}_{W'})}{\partial \alpha_i} = p_i\, \{y_{i^*} - y_{s^*} - \varepsilon\,(p_i + q) - \sum_{j=1}^{M} (\alpha_j - \alpha_{M+j})\,(K_{ij} - K_{sj})$$

$$-\frac{1}{C}\,(\alpha_{i^*} - \alpha_{M+i^*} - \alpha_{s^*} + \alpha_{M+s^*})\} \quad \text{for } i, s \in \{1, \dots, 2M\} \tag{11.68}$$

and the element of $\partial^2 Q(\boldsymbol{\alpha}_{W'})/\partial \boldsymbol{\alpha}_{W'}^2$ is

$$\frac{\partial^2 Q(\boldsymbol{\alpha}_{W'})}{\partial \alpha_i \partial \alpha_j} = -p_i\, p_j\, (K_{ij} + K_{ss} - K_{is} - K_{js}) - \frac{1}{C}\,(\delta_{ij} + p_i\, p_j)$$
$$\text{for } i, j, s \in \{1, \dots, 2M\}. \tag{11.69}$$

The solution given by (11.62) is obtained by solving a set of simultaneous equations.

To solve (11.62), we decompose $\partial^2 Q(\boldsymbol{\alpha}_{W'})/\partial \boldsymbol{\alpha}_{W'}^2$ into the upper and lower triangular matrices by the Cholesky factorization. If the Hessian matrix is not positive definite, the Cholesky factorization stops because the input of the square root becomes nonpositive. When this happens, we solve (11.62) only for the variables that are decomposed so far. And we delete the associated variable and the variables that are not decomposed from the working set. Alternatively, we may add a small positive value to the diagonal element and resume the Cholesky factorization. For the L2 support vector regressor, because the Hessian matrix is positive definite, the procedure is not necessary.

Now we can calculate the correction of $\alpha_s$. In the case of $s \leq M$,

$$\Delta \alpha_s = - \sum_{i \in W', i \leq M} \Delta \alpha_i + \sum_{i \in W', i > M} \Delta \alpha_i, \tag{11.70}$$

or in the case of $s > M$,

$$\Delta\alpha_s = \sum_{i \in W', i \leq M} \Delta\alpha_i - \sum_{i \in W', i > M} \Delta\alpha_i. \tag{11.71}$$

Now confine our consideration to the L1 support vector regressor. Although the solution calculated by the described procedure satisfies (11.54), it may not satisfy (11.55), namely, when there are variables that cannot be corrected,

$$\Delta\alpha_i < 0 \quad \text{when} \quad \alpha_i = 0,$$
$$\Delta\alpha_i > 0 \quad \text{when} \quad \alpha_i = C,$$

we remove these variables from the working set and again solve (11.62) for the reduced working set. This does not require much time because recalculation of the kernels for the reduced working set is not necessary by caching them.

Suppose the solution that can make some corrections for all the variables in the working set is obtained. Then the corrections are adjusted so that all the updated variables go into the range $[0, C]$. Let $\Delta\alpha_i'$ be the allowable corrections if each variable is corrected separately. Then

$$\Delta\alpha_i' = \begin{cases} C - \alpha_i^{\text{old}} & \text{if} \quad \alpha_i^{\text{old}} + \Delta\alpha_i > C, \\ -\alpha_i^{\text{old}} & \text{if} \quad \alpha_i^{\text{old}} + \Delta\alpha_i < 0, \\ \Delta\alpha_i & \text{otherwise.} \end{cases} \tag{11.72}$$

Using $\Delta\alpha_i'$ we calculate the minimum ratio of corrections:

$$r = \min_{i \in W} \frac{\Delta\alpha_i'}{\Delta\alpha_i}. \tag{11.73}$$

Then the variables are updated by

$$\alpha_i^{\text{new}} = \alpha_i^{\text{old}} + r\Delta\alpha_i. \tag{11.74}$$

Clearly the updated variables satisfy (11.55).

For the L2 support vector regressor, we can calculate corrections without considering the upper bound $C$ for $\alpha_i$.

### 11.5.2.2 Convergence Check

After the update of the variables in the working set, we check whether training should be finished, namely, in Step 4, when $V$ becomes empty, we check if there are variables that violate the KKT complementarity conditions and terminate training if there are no violating variables. But by this method the training may be slow in some cases.

To accelerate training, if an increase of the objective function becomes very small, we consider the solution sufficiently near the optimal solution. Thus we finish training when the following inequality is satisfied for $N$ consecutive iterations:

$$\frac{Q_n - Q_{n-1}}{Q_{n-1}} < \eta, \tag{11.75}$$

where $\eta$ is a small positive parameter.

### 11.5.2.3 Candidate Set Selection

If all the variables satisfy KKT complementarity conditions the optimal solution is obtained. Thus to speed up training we need to select violating variables as members of the candidate set.

Inexact KKT Conditions

The KKT conditions for the L1 support vector regressor are as follows:

$$\begin{cases} \alpha_i = C, \alpha_{M+i} = 0 & \text{if} \quad y_i - f(\mathbf{x}_i) > \varepsilon, \\ 0 < \alpha_i < C, \alpha_{M+i} = 0 & \text{if} \quad y_i - f(\mathbf{x}_i) = \varepsilon, \\ \alpha_i = 0, \alpha_{M+i} = 0 & \text{if} \quad |y_i - f(\mathbf{x}_i)| \le \varepsilon, \\ \alpha_i = 0, 0 < \alpha_{M+i} < C & \text{if} \quad y_i - f(\mathbf{x}_i) = -\varepsilon, \\ \alpha_i = 0, \alpha_{M+i} = C & \text{if} \quad y_i - f(\mathbf{x}_i) < -\varepsilon. \end{cases} \tag{11.76}$$

And the KKT conditions for the L2 support vector regressor are as follows:

$$\begin{cases} \alpha_i > 0, \alpha_{M+i} = 0 & \text{if} \quad y_i - f(\mathbf{x}_i) = \varepsilon + \dfrac{\alpha_i}{C}, \\ \alpha_i = 0, \alpha_{M+i} = 0 & \text{if} \quad |y_i - f(\mathbf{x}_i)| \le \varepsilon, \\ \alpha_i = 0, \alpha_{M+i} > 0 & \text{if} \quad y_i - f(\mathbf{x}_i) = -\varepsilon - \dfrac{\alpha_{M+i}}{C}. \end{cases} \tag{11.77}$$

In checking the KKT conditions, we need to calculate the value of $f(\mathbf{x}_i)$. But because the bias term $b$ is not included in the dual problem, during training the value is inexact.

For the L1 support vector regressor, variables $\alpha_i$ have the upper and lower bounds, and for the L2 support vector regressor, variables have the lower bound. If $\alpha_i$ is bounded, the possibility that the variable is modified in the next iteration is small. Thus, we choose the unbounded variables with high priority.

Exact KKT Conditions

The KKT conditions discussed in the previous section are inexact in that $b$ is estimated during training. Keerthi et al. [28, 29] proposed exact KKT conditions and showed by computer experiments that selecting the violating variables led to training speedup. In the following, we first discuss their method for the L1 support vector regressor.

We define $F_i$ by

$$F_i = y_i - \sum_{j=1}^{M} (\alpha_j - \alpha_{M+j}) K_{ij}, \tag{11.78}$$

where $K_{ij} = K(\mathbf{x}_i, \mathbf{x}_j)$.

We can classify KKT conditions into the following five cases:

$$\text{Case 1.} \quad 0 < \alpha_i < C$$
$$F_i - b = \varepsilon, \tag{11.79}$$
$$\text{Case 2.} \quad 0 < \alpha_{M+i} < C$$
$$F_i - b = -\varepsilon, \tag{11.80}$$
$$\text{Case 3.} \quad \alpha_i = \alpha_{M+i} = 0$$
$$-\varepsilon \le F_i - b \le \varepsilon, \tag{11.81}$$
$$\text{Case 4.} \quad \alpha_{M+i} = C$$
$$F_i - b \le -\varepsilon, \tag{11.82}$$
$$\text{Case 5.} \quad \alpha_i = C$$
$$F_i - b \ge \varepsilon. \tag{11.83}$$

Then we define $\tilde{F}_i, \bar{F}_i$ as follows:

$$\tilde{F}_i = \begin{cases} F_i - \varepsilon & \text{if} \quad 0 < \alpha_i < C \quad \text{or} \quad \alpha_i = \alpha_{M+i} = 0, \\ F_i + \varepsilon & \text{if} \quad 0 < \alpha_{M+i} < C \quad \text{or} \quad \alpha_{M+i} = C, \end{cases} \tag{11.84}$$

$$\bar{F}_i = \begin{cases} F_i - \varepsilon & \text{if} \quad 0 < \alpha_i < C \quad \text{or} \quad \alpha_i = C, \\ F_i + \varepsilon & \text{if} \quad 0 < \alpha_{M+i} < C \quad \text{or} \quad \alpha_i = \alpha_{M+i} = 0. \end{cases} \tag{11.85}$$

Then the KKT conditions are simplified as follows:

$$\bar{F}_i \ge b \ge \tilde{F}_i \quad \text{for } i = 1, \ldots, M. \tag{11.86}$$

We must notice that according to the values of $\alpha_i$ and $\alpha_{M+i}$, $\bar{F}_i$ or $\tilde{F}_i$ is not defined. For instance, if $\alpha_i = C$, $\tilde{F}_i$ is not defined.

To detect the violating variables, we define $b_{\text{low}}, b_{\text{up}}$ as follows:

$$b_{\text{low}} = \max_i \tilde{F}_i,$$
$$b_{\text{up}} = \min_i \bar{F}_i. \tag{11.87}$$

Then if the KKT conditions are not satisfied, $b_{\mathrm{up}} < b_{\mathrm{low}}$ and the data sample $i$ that satisfies

$$b_{\mathrm{up}} < \tilde{F}_i - \tau \quad \text{or} \quad b_{\mathrm{low}} > \bar{F}_i + \tau \quad \text{for } i \in \{1, \ldots, M\} \tag{11.88}$$

violates the KKT conditions, where $\tau$ is a positive parameter to loosen the KKT conditions. By this, without calculating $b$, we can detect the variables that violate the KKT conditions. As training proceeds, $b_{\mathrm{up}}$ and $b_{\mathrm{low}}$ approach each other and at the optimal solution, $b_{\mathrm{up}}$ and $b_{\mathrm{low}}$ have the same value if the solution is unique. If not, $b_{\mathrm{up}} > b_{\mathrm{low}}$.

For the L2 support vector regressor, we define $\tilde{F}_i$ and $\bar{F}_i$ as follows:

$$\tilde{F}_i = \begin{cases} F_i - \varepsilon & \text{if} \quad \alpha_i = \alpha_{M+i} = 0, \\ F_i - \varepsilon - \dfrac{\alpha_i}{C} & \text{if} \quad \alpha_i > 0, \alpha_{M+i} = 0, \\ F_i + \varepsilon + \dfrac{\alpha_i}{C} & \text{if} \quad \alpha_i = 0, \alpha_{M+i} > 0, \end{cases} \tag{11.89}$$

$$\bar{F}_i = \begin{cases} F_i + \varepsilon & \text{if} \quad \alpha_i = \alpha_{M+i} = 0, \\ F_i - \varepsilon - \dfrac{\alpha_i}{C} & \text{if} \quad \alpha_i > 0, \alpha_{M+i} = 0, \\ F_i + \varepsilon + \dfrac{\alpha_{M+i}}{C} & \text{if} \quad \alpha_i = 0, \alpha_{M+i} > 0. \end{cases} \tag{11.90}$$

The remaining procedure is the same as that of the L1 support vector regressor.

For selection of variables for the working set, it may be desirable to select the most violating variables. In the following, we discuss one such method.

### 11.5.2.4 Selection of Violating Variables

The degree of violation is larger as $\tilde{F}_i$ becomes larger and $\bar{F}_i$ smaller. Thus a procedure of candidate set selection for Newton's method is as follows:

1. Sort $\bar{F}_i$ in increasing order and $\tilde{F}_k$ in decreasing order and set $i = 1, k = 1$.
2. Compare the value of $\tilde{F}_i$ with $b_{\mathrm{up}}$ and if the KKT conditions are violated, add $i$ to the candidate set $V$ and increment $i$ by 1.
3. Compare the value of $\bar{F}_k$ with $b_{\mathrm{low}}$ and if the KKT conditions are violated, add $i$ to the candidate set $V$ and increment $k$ by 1.
4. Iterate Steps 2 and 3, so that the violating data for $\tilde{F}_i$ and $\bar{F}_k$ are selected alternately until there are no violating data for $\tilde{F}_i$ and $\bar{F}_k$.
5. Move indices in $V$ to the working set $W$ in decreasing order of violation.

In function approximation, for $\mathbf{x}_i$, there are two variables $\alpha_i$ and $\alpha_{M+i}$. In Newton's method, we select either variable according to the following conditions. If one is zero and the other is nonzero, we select the nonzero variable.

If both are zero, we evaluate the error and if $y_i - f(\mathbf{x}_i) > 0$, we select $\alpha_i$, and otherwise, $\alpha_{M+i}$. To estimate $f(\mathbf{x}_i)$, we use $b = (b_{\mathrm{up}} + b_{\mathrm{low}})/2$.

### 11.5.2.5 Performance Evaluation

Evaluation Conditions

We evaluated the performance of L1 and L2 support vector regressors (SVRs) using the noisy data for a water purification plant and the noiseless Mackey–Glass data listed in Table 1.4. For the Mackey–Glass data, we measured the performance by NRMSE. Unless stated otherwise, we used $\varepsilon = 0.01$ and $C = 10,000$. In addition to the training data, we artificially generated the outliers and evaluated the robustness of the estimation.

For the water purification plant data, because the number of nonstationary data is too small, we used only the stationary data for most of the evaluation. We evaluated the performance by the average and the maximum estimation errors for the training and test data. We used $\varepsilon = 1$ and $C = 1,000$ unless stated otherwise.

We used linear kernels, polynomial kernels with $d = 3$, and RBF kernels with $\gamma = 10$.

We set $\eta = 10^{-10}$ in (11.75) and if (11.75) is satisfied for consecutive 10 times, we stopped training. We used an AthlonMP2000+ personal computer under Linux.

We evaluated

- the effect of the working set size on training time using Newton's method (NM),
- performance difference of L1 SVRs and L2 SVRs,
- convergence difference by inexact and exact KKT conditions,
- performance difference among several estimation methods, and
- robustness of SVRs to outliers.

In the following L1 NM and L2 NM mean that the L1 and L2 SVRs are trained by Newton's method, respectively.

Effect of Working Set Size on Speedup

We evaluated how an increase of the working set size accelerates training by Newton's method. In the following we show the results using the L1 SVR when the inexact KKT conditions were used for selecting violating variables. The tendency was the same when the L2 SVR or the exact KKT conditions were used.

Table 11.1 shows the results of the water purification data with polynomial kernels. In the table, "Size," "SVs, " and "Time" denote the working set size, the number of support vectors, and the training time. From the table, the training time was shortened by increasing the working set size, and the 50 times speedup against the working set size of 2 was obtained for the working set size of 50. The maximum errors and the numbers of support vectors are slightly different for different working set sizes because the stopping conditions are not so strict.

**Table 11.1** Effect of working set size for the water purification data with polynomial kernels

| Size | Time | Training data | | Test data | | SVs |
|------|------|---------------|---------------|---------------|---------------|-----|
| | (s) | Ave.err (mg/l) | Max.err (mg/l) | Ave.err (mg/l) | Max.err (mg/l) | |
| 2 | 151 | 0.73 | 4.30 | 1.03 | 11.55 | 98 |
| 3 | 112 | 0.73 | 4.31 | 1.03 | 11.56 | 97 |
| 5 | 59 | 0.73 | 4.27 | 1.03 | 11.51 | 97 |
| 10 | 22 | 0.73 | 4.34 | 1.03 | 11.57 | 98 |
| 20 | 7.3 | 0.73 | 4.37 | 1.03 | 11.59 | 98 |
| 30 | 3.8 | 0.73 | 4.33 | 1.03 | 11.55 | 98 |
| 40 | 3.4 | 0.73 | 4.44 | 1.03 | 11.65 | 95 |
| 50 | 3.0 | 0.73 | 4.31 | 1.03 | 11.53 | 96 |

Table 11.2 shows the training speedup for the Mackey–Glass data with RBF kernels when the working set size was increased from 2 to 50. By increasing the working set size, training was sped up and 8.3 times speedup against the working set size of 2 was obtained for the working set size of 30. When the size was larger than 30, training slowed down. This is because the calculation of the inverse matrix becomes longer as the size increases. Thus there is an optimal size for training speedup.

**Table 11.2** Effect of working set size for the Mackey–Glass data with RBF kernels

| Size | Time (s) | Training data (NRMSE) | Test data (NRMSE) | SVs |
|------|----------|----------------------|-------------------|-----|
| 2 | 58 | 0.028 | 0.027 | 55 |
| 3 | 48 | 0.028 | 0.027 | 57 |
| 5 | 29 | 0.028 | 0.027 | 52 |
| 10 | 16 | 0.028 | 0.027 | 52 |
| 20 | 8.8 | 0.028 | 0.027 | 54 |
| 30 | 7.0 | 0.028 | 0.027 | 52 |
| 40 | 7.7 | 0.028 | 0.027 | 50 |
| 50 | 8.7 | 0.029 | 0.028 | 49 |

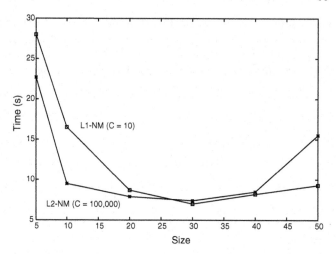

**Fig. 11.4** Training time comparison of L1 and L2 support vector regressors with RBF kernels trained by Newton's method for the Mackey–Glass data. Reprinted from [27, p. 2069]

## Comparison of L1 and L2 Support Vector Regressors

Here, we compare performance of L1 SVRs and L2 SVRs using the Mackey–Glass data. Because the fitting capabilities of L1 SVRs and L2 SVRs are different for the same value of $C$, we set the values of $C$ so that the L1 SVR and the L2 SVR showed the similar NRMSEs for the training data.

Figure 11.4 shows the training time of the L1 NM and L2 NM with RBF kernels when the working set size was changed. We set $C = 10$ for the L1 NM and $C = 100,000$ for the L2 NM. When the size is small, the L2 NM is faster. But for sizes between 20 and 40, where training was the fastest, the difference is small. The reasons may be as follows: Because of RBF kernels the variables did not reach the upper bounds for the L1 SVR and the Hessian matrix for the L1 SVR with RBF kernels was positive definite. Thus, the advantages of the L2 SVR did not appear.

Figure 11.5 shows the training time comparison of the L1 NM ($C = 1,000$) and the L2 NM ($C = 100,000$) when the working set size was changed. For all the sizes, training by the L2 NM is faster. For sizes larger than 10, training by the L1 NM was slowed but that by the L2 NM was sped up.

Table 11.3 shows the training time of the L1 NM ($C = 1,000$) and the L2 NM ($C = 10$) with linear kernels. The numeral in parentheses shows the working set size. For sizes 5 and 30, training by the L2 NM is faster. Training by L1 NM with size 30 is slower than with size 5. This is because for linear kernels the number of linearly independent variables is the number of input variables plus 1. For the Mackey–Glass data, there are four input variables.

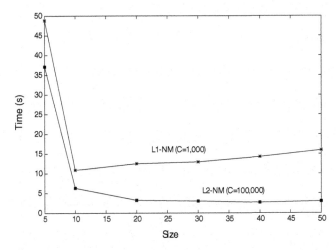

**Fig. 11.5** Training time comparison of L1 and L2 support vector regressors for the Mackey–Glass data with polynomial kernels. Reprinted from [27, p. 2069]

Thus, if we set the working set size of more than five, unnecessary calculations increase. But for the L2 NM, because the Hessian matrix is always positive definite, this sort of thing did not happen.

**Table 11.3** Training time comparison of L1 and L2 support vector regressors with linear kernels trained by Newton's method

| Method | Time (s) | Training data (NRMSE) | Test data (NRMSE) | SVs |
|---|---|---|---|---|
| L1 NM (5) | 7.4 | 0.46 | 0.46 | 446 |
| L1 NM (30) | 16.9 | 0.46 | 0.46 | 446 |
| L2 NM (5) | 1.0 | 0.43 | 0.43 | 475 |
| L2 NM (30) | 0.8 | 0.43 | 0.43 | 475 |

## Comparison of Exact and Inexact KKT Conditions

In selecting violating variables we use either the inexact or the exact KKT conditions. Here, we compare training time of Newton's method by these conditions. We set $\tau = 0.001$ for the exact KKT conditions and used RBF kernels.

Figure 11.6 shows the training time comparison by the L1 NM for the water purification data. In general, training by the exact KKT conditions is faster.

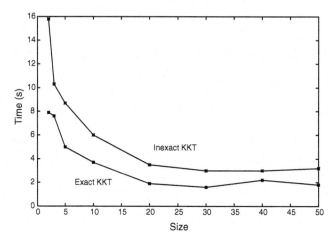

**Fig. 11.6** Training time comparison by the L1 support vector regressor trained by Newton's method for the water purification data. Reprinted from [27, p. 2070]

Figure 11.7 shows the training time comparison by the L2 NM for the
water purification data. In general, training by the inexact KKT conditions
is faster, but the difference is small. But for sizes 2 and 4, training by the
exact KKT conditions is faster.

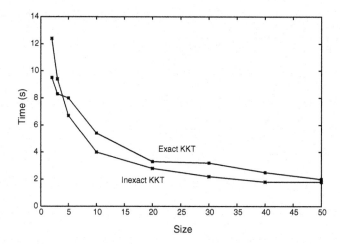

**Fig. 11.7** Training time comparison by the L2 support vector regressor trained by Newton's method for water purification data

Figure 11.8 shows training time comparison by the L1 NM for the Mackey–
Glass data. With sizes 2–10, training by the exact KKT conditions is faster,

but with sizes larger than or equal to 30, training by the inexact KKT conditions is faster.

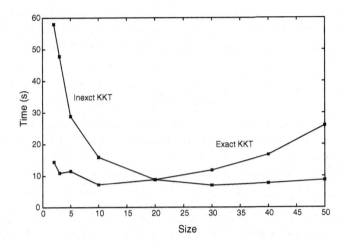

**Fig. 11.8** Training time comparison of the L1 support vector regressor trained by Newton's method for the Mackey–Glass data. Reprinted from [27, p. 2070]

Figure 11.9 shows training time comparison of the L2 support vector regressor trained by Newton's method for the Mackey–Glass data. Except for the sizes 2 and 20, training by the inexact KKT conditions is faster.

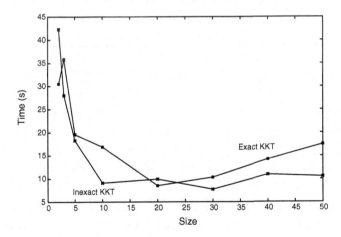

**Fig. 11.9** Training time comparison of the L2 support vector regressor trained by Newton's method for the Mackey–Glass data

Using the inexact KKT conditions for the Mackey–Glass data, as the working set size became larger, longer training was needed. For the exact KKT conditions, at the initial stage of training, most of the data are detected as violating data. Thus for the Mackey–Glass data, in which the ratio of support vectors for the training data is low, redundant calculation was increased. Thus training was considered slower than by the inexact KKT conditions.

## Comparison with Other Training Methods

Using polynomial and RBF kernels, we compared training times of Newton's method and the primal–dual interior-point method (PDIP) [30] with and without chunking.

In the following tables a numeral in parentheses shows the working set size. In chunking, we changed the number of added data from 10 to 100 with the increment of 10 and listed the fastest result.

The value of $C$ for the L2 SVR was set so that the estimation performance was comparable to that of the L1 SVR. For the water purification data, because there was not much difference, we set $C = 1,000$.

Table 11.4 lists the training time comparison for the Mackey–Glass data. For polynomial kernels, training by Newton's method is faster than by PDIP with and without chunking. But for RBF kernels, PDIP with chunking is the fastest. This is because the number of support vectors is small and the size of the problem of each iteration was very small.

**Table 11.4** Training time comparison for the Mackey–Glass data

| Kernel | Size | Time (s) | Training data (NRMSE) | Test data (NRMSE) | SVs |
|--------|------|----------|----------------------|-------------------|-----|
| Poly | L1 NM (10) | 11.9 | 0.067 | 0.066 | 277 |
|        | L2 NM (30) | 2.7 | 0.067 | 0.066 | 282 |
|        | PDIP (–) | 84.0 | 0.067 | 0.066 | 283 |
|        | PDIP (70) | 50.6 | 0.067 | 0.066 | 269 |
| RBF | L1 NM (30) | 7.0 | 0.028 | 0.027 | 48 |
|        | L2 NM (20) | 6.6 | 0.028 | 0.027 | 55 |
|        | PDIP (–) | 79.9 | 0.028 | 0.027 | 54 |
|        | PDIP (50) | 0.9 | 0.028 | 0.027 | 54 |

Table 11.5 shows training time comparison for the water purification data with linear kernels. Training by L2 NM (30) was the fastest, and the PDIP with chunking was the second–fastest.

According to the results, when the number of support vectors is small, the primal–dual interior-point method with chunking was the fastest, but for the data with many support vectors, L2 NM was the fastest.

**Table 11.5** Training comparison for the water purification data with linear kernels

| Size | Time | Training data | | Test data | | SVs |
|---|---|---|---|---|---|---|
| | (s) | Ave. err (mg/l) | Max.err (mg/l) | Ave. err (mg/l) | Max. err (mg/l) | |
| L1 NM (10) | 2.5 | 1.14 | 18.36 | 1.18 | 5.77 | 113 |
| L2 NM (30) | 0.7 | 1.26 | 14.43 | 1.26 | 7.64 | 124 |
| PDIP (–) | 10.8 | 1.13 | 18.60 | 1.16 | 5.96 | 115 |
| PDIP (30) | 2.6 | 1.13 | 18.60 | 1.16 | 5.96 | 115 |

## Performance Comparison with Other Approximation Methods

*Stationary Data*: Table 11.6 shows the best estimation performance for different methods, namely, the three-layer neural network (NN), the fuzzy function approximator with inhibition (FAMI), the fuzzy function approximator with center-of-gravity defuzzification (FACG), and the fuzzy function approximator with a linear combination output (FALC) [31]. The support vector regressor showed the minimum estimation errors for the training data and the second minimum average estimation error for the test data following the neural network. But the maximum estimation error for the test data was the maximum.

With noisy data, usually overfitting to the training data causes low generalization ability. But for the support vector regressor, although the estimation error for the training data is very small, the estimation error for the test data is comparable to those of the other methods.

**Table 11.6** Performance comparison for the stationary data

| Method | Training data | | Test data | |
|---|---|---|---|---|
| | Ave. err (mg/l) | Max. err (mg/l) | Ave. err (mg/l) | Max. err (mg/l) |
| L1 SVR | 0.69 | 1.04 | 1.03 | 6.99 |
| NN | 0.84 | 4.75 | 0.99 | 6.95 |
| FAMI | 1.07 | 4.75 | 1.18 | 5.57 |
| FACG | 0.91 | 5.06 | 1.05 | 5.33 |
| FALC | 1.09 | 4.34 | 1.16 | 5.22 |

*Nonstationary Data*: Because the nonstationary training data included only 45 data and they were noisy, overfitting occurred easily for conventional methods. We evaluated whether this is the case for the support vector regressor.

The best estimation performance was achieved for the polynomial kernel with $C = 10$, $d = 4$, and $\varepsilon = 0.5$. Table 11.7 lists the best estimation performance for different methods. The average estimation error for the training data by the support vector regressor was the smallest, but the maximum estimation error was the largest. The average estimation error of the test data by the support vector regressor was the second smallest following FAMI but the maximum estimation error was the largest.

**Table 11.7** Performance comparison for the nonstationary data

| Method | Training data | | Test data | |
|--------|---------------|--|-----------|--|
| | Ave. err (mg/l) | Max. err (mg/l) | Ave. err (mg/l) | Max. err (mg/l) |
| L1 SVR | 0.95 | 9.95 | 1.62 | 7.39 |
| NN | 1.59 | 6.83 | 1.74 | 6.78 |
| FAMI | 1.56 | 7.20 | 1.46 | 4.97 |
| FACG | 1.91 | 6.30 | 1.95 | 7.18 |
| FALC | 1.63 | 5.79 | 1.92 | 6.30 |

*Mackey–Glass Data*: The best estimation performance was obtained when RBF kernels with $\gamma = 10$ and $\varepsilon = 0.001$ were used. Table 11.8 lists the best estimation results for different methods. In the table, the support vector regressor showed the smallest NRMSE.

**Table 11.8** Performance comparison for the Mackey–Glass data

| Approximator | Test data (NRMSE) |
|--------------|-------------------|
| L1 SVR | 0.003 |
| NN [32] | 0.02 |
| ANFIS [33] | 0.007 |
| Cluster estimation based [34] | 0.014 |
| FAMI [31] | 0.092 |
| FACG [31] | 0.005 |
| FALC [31] | 0.006 |

## Robustness for Outliers

The support vector regressor has a mechanism to suppress outliers. To evaluate this, we randomly selected five data in the Mackey–Glass time series data and multiplied them by 2. Therefore, the five input–output pairs that are generated from the time series data included outliers, namely, 25 outliers were included in the training data. We used the RBF kernels with $\gamma = 10$ and $\varepsilon = 0.005$ that showed best performance without outliers. We evaluated

**Table 11.9** Performance of the Mackey–Glass data with outliers

| C | Training data (NRMSE) | Test data (NRMSE) | SVs | Time (s) |
|---|---|---|---|---|
| 100,000 | 0.335 (0.019) | 0.033 | 244 | 34822 |
| 10,000 | 0.342 (0.019) | 0.023 | 182 | 6348 |
| 1,000 | 0.345 (0.018) | 0.019 | 146 | 2174 |
| 100 | 0.346 (0.017) | 0.018 | 117 | 700 |
| 10 | 0.348 (0.017) | 0.017 | 95 | 414 |
| 1 | 0.349 (0.018) | 0.018 | 114 | 251 |
| 0.1 | 0.363 (0.031) | 0.030 | 181 | 129 |

**Table 11.10** Performance of the Mackey–Glass data without outliers

| C | Training data (NRMSE) | Test data (NRMSE) | SVs | Time (s) |
|---|---|---|---|---|
| 100,000 | 0.014 | 0.014 | 80 | 414 |
| 10,000 | 0.014 | 0.014 | 80 | 393 |
| 1,000 | 0.014 | 0.014 | 80 | 375 |
| 100 | 0.014 | 0.014 | 80 | 381 |
| 10 | 0.015 | 0.014 | 82 | 381 |
| 1 | 0.017 | 0.016 | 101 | 270 |
| 0.1 | 0.026 | 0.026 | 188 | 155 |

the performance changing $C$. The larger the magnitude of $C$, the larger the effect of outliers to the estimation.

Table 11.9 shows the estimation performance when outliers were included. The numerals in the parentheses are the NRMSEs excluding the outliers. Table 11.10 shows the estimation performance when outliers were not included. When the outliers were included, the best estimation was achieved for $C = 10$. The NRMSE for the test data decreased as the value of $C$ changed from 100,000 to 10. This means that the effect of outliers was effectively suppressed. But when the value of $C$ was further decreased to 0.1, the NRMSE increased. In this case the number of support vectors increased. Thus too small a value of $C$ excluded not only outliers but also normal data. Therefore, to effectively eliminate outliers, we need to set a proper value to $C$.

The NRMSE for the training data is large when the outliers were included to calculate the NRMSE, but when they were excluded, the NRMSEs were almost the same as those without outliers. For example, for $C = 10$, the difference was only 0.002. Thus the support vector regressor can be trained irrespective of inclusion of outliers.

Summary

By computer experiments, we evaluated support vector regressors (SVRs) for Mackey–Glass and water purification data. The results are summarized as follows:

1. Training by Newton's method (NM), which optimizes plural data at a time, was sped up by increasing the working set size from 2. There was an optimal working set size, namely, if a working set size is too large, the training is slowed down.
2. The difference between the L1 NM and the L2 NM was small for kernels with a high-dimensional feature space such as RBF kernels. But for kernels with a low-dimensional feature space such as linear kernels, training by L2 NM was much faster. In such a situation, the advantages of positive definiteness of the Hessian matrix and the unbounded variables became evident.
3. The exact KKT conditions did not always lead to faster convergence than the inexact KKT conditions. For the water purification data and for the Mackey–Glass data with a small working set size, the exact KKT conditions performed better for the L1 NM. But for the Mackey–Glass data, as we increased the working set size, the inexact KKT conditions performed better. This is because, in the initial stage of training, exact KKT conditions detect almost all variables as violating variables, and this leads to useless calculations.
4. For problems with a small number of support vectors, training by the primal–dual interior-point method with chunking was faster than that by the NM. But for problems with a relatively large number of support vectors, the NM was faster.
5. There was not much difference in performance among different approximation methods.
6. By proper selection of the value of $C$, SVRs could suppress the effect of outliers.

## 11.5.3 Active Set Training

In this section, we reformulate support vector regressors so that the number of variables is equal to the number of training data. Then we discuss active set training of L2 support vector regressors, which is an extension of active set training for support vector machines discussed in Section 5.8.

### 11.5.3.1 Reformulation of Support Vector Regressors

The number of variables of the support vector regressor in the dual form is twice the number of the training data. But because nonnegative $\alpha_i$ and $\alpha_i^*$ appear only in the forms of $\alpha_i - \alpha_i^*$ and $\alpha_i + \alpha_i^*$ and both $\alpha_i$ and $\alpha_i^*$ are not positive at the same time, we can reduce the number of variables to half by replacing $\alpha_i - \alpha_i^*$ with real-valued $\alpha_i$ and $\alpha_i + \alpha_i^*$ with $|\alpha_i|$ [14]. Then L1 support vector regressor given by (11.26), (11.27), and (11.28) are rewritten as follows:

$$\text{maximize} \quad Q(\boldsymbol{\alpha}) = -\frac{1}{2} \sum_{i,j=1}^{M} \alpha_i \, \alpha_j \, K(\mathbf{x}_i, \mathbf{x}_j) - \varepsilon \sum_{i=1}^{M} |\alpha_i|$$

$$+ \sum_{i=1}^{M} y_i \, \alpha_i \tag{11.91}$$

$$\text{subject to} \quad \sum_{i=1}^{M} \alpha_i = 0, \tag{11.92}$$

$$C \geq |\alpha_i| \quad \text{for } i = 1, \ldots, M. \tag{11.93}$$

The KKT complementarity conditions become

$$\alpha_i \left( \varepsilon + \xi_i - y_i + \mathbf{w}^\top \boldsymbol{\phi}(\mathbf{x}_i) + b \right) = 0 \quad \text{for } \alpha_i \geq 0, \quad i = 1, \ldots, M, \tag{11.94}$$

$$\alpha_i \left( \varepsilon + \xi_i + y_i - \mathbf{w}^\top \boldsymbol{\phi}(\mathbf{x}_i) - b \right) = 0 \quad \text{for } \alpha_i < 0, \quad i = 1, \ldots, M, \tag{11.95}$$

$$\eta_i \, \xi_i = (C - |\alpha_i|) \, \xi_i = 0 \quad \text{for } i = 1, \ldots, M. \tag{11.96}$$

Thus $b$ is obtained by

$$b = \begin{cases} y_i - \mathbf{w}^\top \boldsymbol{\phi}(\mathbf{x}_i) - \varepsilon & \text{for } C > \alpha_i > 0 \\ y_i - \mathbf{w}^\top \boldsymbol{\phi}(\mathbf{x}_i) + \varepsilon & \text{for } -C < \alpha_i < 0 \end{cases} \quad i \in \{1, \ldots, M\}. \tag{11.97}$$

Similarly, replacing $\alpha_i - \alpha_i^*$ with $\alpha_i$ and $\alpha_i + \alpha_i^*$ with $|\alpha_i|$ in (11.45) to (11.47), we obtain the following dual problem for the L2 support vector regressor:

$$\text{maximize} \quad Q(\boldsymbol{\alpha}) = -\frac{1}{2} \sum_{i,j=1}^{M} \alpha_i \, \alpha_j \left( K(\mathbf{x}_i, \mathbf{x}_j) + \frac{\delta_{ij}}{C} \right)$$

$$-\varepsilon \sum_{i=1}^{M} |\alpha_i| + \sum_{i=1}^{M} y_i \, \alpha_i \tag{11.98}$$

$$\text{subject to} \quad \sum_{i=1}^{M} \alpha_i = 0, \tag{11.99}$$

where $\delta_{ij}$ is Kronecker's delta function.
The KKT complementarity conditions become

$$\alpha_i \left(\varepsilon + \xi_i - y_i + \mathbf{w}^\top \phi(\mathbf{x}_i) + b\right) = 0 \quad \text{for } \alpha_i \geq 0, \quad i = 1, \ldots, M, \quad (11.100)$$

$$\alpha_i \left(\varepsilon + \xi_i + y_i - \mathbf{w}^\top \phi(\mathbf{x}_i) - b\right) = 0 \quad \text{for } \alpha_i < 0 \quad i = 1, \ldots, M, \quad (11.101)$$

$$C\,\xi_i = |\alpha_i| \quad \text{for } i = 1, \ldots, M. \quad (11.102)$$

Therefore, $b$ is obtained by

$$b = \begin{cases} y_i - \mathbf{w}^\top \phi(\mathbf{x}_i) - \varepsilon - \dfrac{\alpha_i}{C} & \text{for } \alpha_i > 0 \\[2mm] y_i - \mathbf{w}^\top \phi(\mathbf{x}_i) + \varepsilon - \dfrac{\alpha_i}{C} & \text{for } \alpha_i < 0 \end{cases} \quad i \in \{1, \ldots, M\}. \quad (11.103)$$

By the reformulation, the number of variables reduced from $2M$ to $M$ and the obtained support vector regressors are very similar to support vector machines. If the value of $\varepsilon$ is very small almost all training data become support vectors. In such a case the L2 support vector regressor behaves very similar to the LS support vector regressors, which will be discussed in Section 11.6.3. And for $\varepsilon = 0$ the L2 support vector regressor is equivalent to the LS support vector regressor.

*Example 11.1.* Consider the regression problem with one input with two training data: $(x_1, y_1) = (0, 0)$ and $(x_2, y_2) = (1, 1)$. The linear solution that includes the two data is $y = x$. Then the L1 support vector regressor with linear kernels becomes

$$\text{maximize} \quad Q(\alpha_2) = -\frac{1}{2}\alpha_2^2 + (1 - 2\varepsilon)\,\alpha_2$$

$$= -\frac{1}{2}(\alpha_2 - 1 + 2\varepsilon)^2 + \frac{1}{2}(1 - 2\varepsilon)^2 \quad (11.104)$$

$$\text{subject to} \quad C \geq \alpha_2 \geq 0. \quad (11.105)$$

Here, we used that $K(0,0) = K(0,1) = 0$, $K(1,1) = 1$, $\alpha_1 = -\alpha_2$, and $w = \alpha_2 > 0$. For $C \geq 1 - 2\varepsilon$, the solution of the above optimization problem is

$$\alpha_2 = -\alpha_1 = 1 - 2\varepsilon, \quad (11.106)$$

$$w = 1 - 2\varepsilon, \quad (11.107)$$

$$b = \varepsilon. \quad (11.108)$$

Therefore, the approximation function is given by

$$y = (1 - 2\varepsilon)x + \varepsilon. \quad (11.109)$$

Thus, so long as $C \geq 1 - 2\varepsilon$ is satisfied, the solution does not change for the change of the value of $C$ (see Fig. 11.10).

Now the L2 support vector regressor with linear kernels becomes

$$\text{maximize} \quad Q(\alpha_2) = -\frac{1}{2}\left(1 + \frac{2}{C}\right)\alpha_2^2 + (1 - 2\varepsilon)\,\alpha_2$$

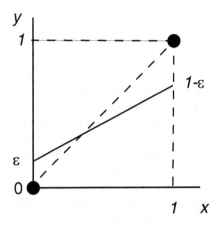

**Fig. 11.10** Regression for a one-dimensional input

$$= -\frac{C+2}{2\,C}\left(\alpha_2 - \frac{C(1-2\varepsilon)}{C+2}\right)^2 + \frac{C\,(1-2\,\varepsilon)^2}{2\,(C+2)} \quad (11.110)$$

subject to $\quad \alpha_2 \geq 0.$ $\hspace{6cm}$ (11.111)

The solution of the above optimization problem is given by

$$\alpha_2 = -\alpha_1 = \frac{C\,(1-2\,\varepsilon)}{C+2}, \quad (11.112)$$

$$w = \frac{C\,(1-2\,\varepsilon)}{C+2}, \quad (11.113)$$

$$b = \frac{1+\varepsilon\,C}{C+2}. \quad (11.114)$$

Therefore, the approximation function is given by

$$y = \frac{C\,(1-2\,\varepsilon)}{C+2}x + \frac{1+\varepsilon\,C}{C+2}. \quad (11.115)$$

Unlike the solution of the L1 support vector regressor, the solution given by (11.115) changes as the value of $C$ changes and it converge to (11.109) as the value of $C$ approaches infinity.

## 11.5.3.2 Training Methods

The reformulated L1 and L2 support vector regressors can be solved either by fixed-size chunking or by variable-size chunking so long as the solution is constrained to be feasible. And active set training for L2 support vector machines discussed in Section 5.8 can be extended to L2 support vector regressors. In the following we discuss this training method.

We solve the equality constraint (11.99) for one variable and substitute it into (11.98). Then the optimization problem is reduced to the maximization problem without constraints. We divide the variables into the working set and the fixed set and solve the subproblem for the working set fixing the variables in the fixed set. In the next iteration process, we delete the variables that are within the $\varepsilon$-tube from the working set and add, from the fixed set, the variables that do not satisfy the KKT conditions and iterate optimizing the subproblem until the same solution is obtained. We discuss the method more in detail.

Consider solving (11.98) and (11.99) for the index set $S$. Solving the equality constraint in (11.99) for $\alpha_s$ $(s \in S)$, we obtain

$$\alpha_s = - \sum_{\substack{i \neq s, \\ i \in S}} \alpha_i. \tag{11.116}$$

Substituting (11.116) into (11.98), we obtain the following optimization problem

$$\text{maximize} \quad Q(\boldsymbol{\alpha}_S) = \mathbf{c}_S^\top \boldsymbol{\alpha}_S' - \frac{1}{2} {\boldsymbol{\alpha}_S'}^\top K_S \, \boldsymbol{\alpha}_S' \tag{11.117}$$

where $\boldsymbol{\alpha}_S = \{\alpha_i | i \in S\}$, $\boldsymbol{\alpha}_S' = \{\alpha_i | i \neq s, i \in S\}$, $\mathbf{c}_S$ is the $(|S| - 1)$-dimensional vector, $K_S$ is the $(|S|-1) \times (|S|-1)$ positive definite matrix, and

$$c_{S_i} = \begin{cases} y_i - y_s & \text{for } D(\mathbf{x}_i, y_i) \geq 0, \ D(\mathbf{x}_s, y_s) \geq 0, \\ y_i - y_s - 2\varepsilon & \text{for } D(\mathbf{x}_i, y_i) \geq 0, \ D(\mathbf{x}_s, y_s) < 0, \\ y_i - y_s + 2\varepsilon & \text{for } D(\mathbf{x}_i, y_i) < 0, \ D(\mathbf{x}_s, y_s) < 0, \end{cases}$$
$$\qquad\qquad\qquad\qquad i \neq s, \ \ i \in S \tag{11.118}$$

$$K_{S_{ij}} = K(\mathbf{x}_i, \mathbf{x}_j) - K(\mathbf{x}_i, \mathbf{x}_s) - K(\mathbf{x}_s, \mathbf{x}_j)$$
$$+ K(\mathbf{x}_s, \mathbf{x}_s) + \frac{1 + \delta_{ij}}{C} \quad \text{for } i, j \neq s, \ \ i, j \in S, \tag{11.119}$$

where $c_{S_i}$ is the $i$th element of $\mathbf{c}_{S_i}$.

If $\varepsilon = 0$, in (11.118), $c_{S_i} = y_i - y_s$ irrespective of the signs of $D(\mathbf{x}_i, y_i)$ and $D(\mathbf{x}_s, y_s)$. Thus, similar to LS support vector machines, we can solve (11.117) by a single matrix inversion.

Initially, for positive $\varepsilon$, we set some indices to $S$ and set $\alpha_i = 0$ $(i \in S)$ and $b = 0$. Therefore, $f(\mathbf{x}_i) = 0$. Thus, in (11.118), $c_{S_i}$ is set according to the signs of $D(\mathbf{x}_i, y_i)$ and $D(\mathbf{x}_s, y_s)$. We solve (11.117) by

$$\alpha'_S = K_S^{-1} c_S. \tag{11.120}$$

We calculate $b$ using (11.103). Because of $\varepsilon$, the $b$ values calculated by different $\alpha_i$ in $S$ may be different. Thus we calculate the average of $b$'s. If training data associated with the variables in $S$ are within the $\varepsilon$ tube, we delete these variables and add the indices of the variables to $S$ that violate KKT conditions. If the working sets and $c_{S_i}$ are the same for the consecutive two iterations, the solution is obtained and we stop training. This stopping condition is different from that for the support vector machine discussed in Section 5.8 because of the absolute value of $\alpha_i$ in the objective function.

The procedure for training the L2 support vector regressor is as follows.

1. Set the indices associated with $h$ training data to set $S$ and go to Step 2.
2. Calculate $\alpha'_S$ using (11.120) and using (11.116) obtain $\alpha_s$. Calculate $b$'s for $i \in S$ using (11.103) and calculate the average of $b$'s.
3. Delete from $S$ the indices of $\mathbf{x}_i$ that satisfy $|D(\mathbf{x}_i, y_i)| \leq \varepsilon$. And add to $S$ the indices associated with at most $h$ most violating data, namely $\mathbf{x}_i$ that satisfy $|D(\mathbf{x}_i, y_i)| > \varepsilon$ from the largest $|D(\mathbf{x}_i, y_i)|$ in order. If the solution is obtained, stop training. Otherwise, go to Step 2.

For a large value of $\varepsilon$, there may be cases where a proper working set is not obtained and thus the solution does not converge.

### 11.5.3.3 Performance Comparison

We compared the training time of the LS SVR, L1 SVR, and L2 SVR using the data sets listed in Table 1.4. For the data sets that are not divided into training and test data sets, we randomly divided into two sets with almost the equal sizes.

We used the Cholesky factorization in training the LS SVR, the primal–dual interior-point method for the L1 SVR, and the active set training for the L2 SVR. We used the RBF kernel with the $\gamma$ value selected from $\{0.1, 0.5, 1, 5, 10, 15, 20\}$. The value of the margin parameter was selected from $\{1, 10, 50, 100, 500, 1,000, 2,000, 3,000, 5,000, 7,000, 10,000, 100,000\}$ and the $\varepsilon$ value for L1 and L2 SVRs from $\{0.001, 0.01, 0.05, 0.1, 0.5, 1\}$ and determined the parameter values by fivefold cross-validation. For the parameter values of the LS SVR and L1 SVR, we used the grid search. But for the L2 SVR, for a large value of $\varepsilon$, active set training sometimes failed to converge. Therefore, we first carried out cross-validation with $\varepsilon = 0.001$ and selected the $\gamma$ value. Then fixing the $\gamma$ value we carried out cross-validation changing $C$ and $\varepsilon$ values. Table 11.11 lists the determined parameter values. The parameter values for the LS SVR and L2 SVR are very similar. Especially, the $\gamma$ values are the same because we started cross-validation of the L2 SVR with $\varepsilon = 0.001$, which is very close to the LS SVR.

**Table 11.11** Parameter values for LS, L1, and L2 support vector regressors

| Data | LS | | L1 | | | L2 | | |
|---|---|---|---|---|---|---|---|---|
| | $\gamma$ | $C$ | $\gamma$ | $C$ | $\varepsilon$ | $\gamma$ | $C$ | $\varepsilon$ |
| Abalone | 5 | 50 | 1 | 500 | 0.01 | 5 | 50 | 0.001 |
| Boston 5 | 15 | 50 | 20 | 1 | 0.001 | 10 | 50 | 0.001 |
| Boston 14 | 0.5 | 1,000 | 0.1 | 2,000 | 0.05 | 0.5 | 1,000 | 0.1 |
| Orange juice | 0.1 | $10^5$ | 0.1 | $10^4$ | 0.01 | 0.1 | $10^5$ | 0.1 |
| Water purification | 20 | 10 | 0.01 | 2,000 | 0.01 | 20 | 50 | 0.001 |

Table 11.12 shows the numbers of support vectors and the approximation errors. The smallest number of support vectors and the smallest approximation error are shown in boldface. For the LS SVR, the number of support vectors is the number of training data. From the table, the solutions for the L1 and L2 SVRs are not so sparse. This is because the data outside of the $\varepsilon$ tube are support vectors and all the data sets used for evaluation include noise. The approximation errors for the three regressors are almost identical.

**Table 11.12** Comparison of the batch training method and the interior-point method with the decomposition technique using L1 and L2 support vector regressors

| Data | SVs | | | Error | | |
|---|---|---|---|---|---|---|
| | LS | L1 | L2 | LS | L1 | L2 |
| Abalone | 2,068 | **1,791** | 2,068 | 1.51 | **148** | 1.51 |
| Boston 5 | 245 | 239 | **237** | **0.0264** | 0.0270 | 0.0270 |
| Boston 14 | 245 | **115** | 235 | **2.19** | **2.19** | **2.19** |
| Orange juice | 150 | **136** | 146 | 4.78 | **4.68** | 4.79 |
| Water purification | 241 | **205** | 241 | 0.976 | **0.958** | 0.976 |

Table 11.13 lists the training time in seconds, measured by a personal computer (3 GHz, 2 GB memory, Windows XP operating system). In the table "L1 PDIP" denotes that the L1 SVR was trained by the primal–dual interior-point method and "L2 NM" denotes that the L2 SVR was trained by Newton's method. For active set training, we set $h = 500$. Therefore, except for the abalone problem, initially all the training data were used for training. For all the cases, active set training for the L2 SVR is the fastest.

To check how the support vectors are decreased for the increase of the $\varepsilon$ value, fixing the $\gamma$ value of the L2 SVR as in Table 11.11, we increased the value of $\varepsilon$ within the range in which the solution converged. The $C$ value was determined by cross-validation. Table 11.14 shows the results. For the Boston 14 problem the solution for the $\varepsilon = 0.5$ did not converge. Thus, the result is the same as in Table 11.12. The reduction of support vectors was

**Table 11.13** Training time comparison (s)

| Data | L1 PDIP | L2 NM | Active |
|------|---------|-------|--------|
| Abalone | 446 | 198 | **58** |
| Boston 5 | 1.0 | 4.5 | **0.09** |
| Boston 14 | 1.6 | 1.4 | **0.09** |
| Orange juice | 0.83 | 36 | **0.16** |
| Water purification | 0.80 | 0.23 | **0.13** |

not so significant and the average approximation error did not deteriorate so much compared to that in Table 11.12. For the water purification problem the average approximation error was slightly smaller.

**Table 11.14** Performance for the increase of the $\varepsilon$ value for the L2 support vector regressor

| Data | $C$ | $\varepsilon$ | SVs | Error |
|------|-----|---------------|-----|-------|
| Abalone | 1 | 1 | 1,248 | 1.63 |
| Boston 5 | 50 | 0.01 | 181 | 0.0281 |
| Orange juice | $10^5$ | 1 | 123 | 4.86 |
| Water purification | 50 | 0.1 | 226 | 0.973 |

# 11.6 Variants of Support Vector Regressors

Various variants of support vector regressors have been developed, e.g., linear programming support vector regressors, $\nu$-support vector regressors, least-squares support vector regressors, and sparse support vector regressors [35–40].

In addition, enhancement or addition of functions such as improvement of generalization ability by multiple kernels and outlier detection is considered. To enhance approximation ability, mixtures of different kernels are used [41–43]. To enhance interpolation and extrapolation abilities, Smits and Jordaan [43] proposed mixing global kernels such as polynomials and local kernels such as RBF kernels by $\rho K_{poly} + (1 - \rho)K_{RBF}$, where $0 \geq \rho \geq 1$.

Inspired by multiresolution signal analysis, such as wavelet analysis, Shao and Cherkassky [41] proposed multiresolution support vector machines. Instead of using one kernel, two kernels with different resolutions are combined. For multiresolution signal analysis with orthogonal basis functions, components of different resolutions can be computed separately. But because the kernels in support vector regressors are not orthogonal, the support vector training is reformulated to determine the Lagrange multipliers for two types of kernels simultaneously.

Jordaan and Smits [44] detected, as an outlier, the data sample that frequently becomes a bounded support vector whose average slack variable value is the largest for several trained support vector regressors with different values of $\varepsilon$.

In the following, we discuss linear programming support vector regressors, $\nu$-support vector regressors, and least-squares support vector regressors.

### 11.6.1 Linear Programming Support Vector Regressors

Similar to LP support vector machines, LP support vector regressors can be defined. Assume that for $m$-dimensional input vector $\mathbf{x}$ and the associated scalar output $y$, we have $M$ training input–output pairs $(\mathbf{x}_i, y_i)$ $(i = 1, \ldots, M)$. First, we define the approximation function in the dual form as follows:

$$f(\mathbf{x}) = \sum_{i=1}^{M} \alpha_i K(\mathbf{x}, \mathbf{x}_i) + b, \qquad (11.121)$$

where $\alpha_i$ take on real values, $K(\mathbf{x}, \mathbf{x}')$ is a kernel, and $b$ is the bias term. Then the LP support vector regressor is given by

$$\text{minimize} \quad Q(\boldsymbol{\alpha}, b, \boldsymbol{\xi}) = \sum_{i=1}^{M} \left( |\alpha_i| + C \left( \xi_i + \xi_i^* \right) \right) \qquad (11.122)$$

$$\text{subject to} \quad -\varepsilon - \xi_j^* \leq \sum_{i=1}^{M} \alpha_i K(\mathbf{x}_j, \mathbf{x}_i) + b - y_j \leq \varepsilon + \xi_j$$

$$\text{for } j = 1, \ldots, M, \qquad (11.123)$$

where $\varepsilon$ is a small positive parameter and $\xi_i^*$ and $\xi_i$ are nonnegative slack variables.

Letting $\alpha_i = \alpha_i^+ - \alpha_i^-$ and $b = b^+ - b^-$, where $\alpha_i^+$, $\alpha_i^-$, $b^+$, and $b^-$ are nonnegative, we can solve (11.122) and (11.123) for $\boldsymbol{\alpha}$, $b$, and $\boldsymbol{\xi}$ by linear programming. For large problems, we use decomposition techniques [45–48] (see Section 5.9).

There are variants of LP support vector machines [49, 50, 47]. Kecman, et al. [49, 50] formulated the linear programming support vector regressors differently, namely, minimize $\sum_{i=1}^{m} |w_i|$ subject to $|\boldsymbol{\phi}^\top \mathbf{w} - y| \leq \varepsilon$. When RBF kernels are used, elements of $\boldsymbol{\phi}$ are given by $\exp(-\gamma \|\mathbf{x} - \mathbf{x}'\|^2)$, where $\mathbf{x}'$ is a training sample. The advantage of this formulation is that, unlike support vector regressors, we can place multiple basis functions (i.e., basis functions with different values of $\gamma$) for each training sample.

## 11.6.2 $\nu$-Support Vector Regressors

Usually it is difficult to set the optimal value of $\varepsilon$ in the support vector regressor. One approach to overcome this problem is to estimate the value assuming that it is in proportion to the standard deviation of the noise [51, 52]. Another approach is to modify the model so that it can be optimized during training [53].

Schölkopf et al. [53] proposed $\nu$-support vector regressors, in which the accuracy of regression is controlled by parameter $\nu$ as follows:

$$\text{minimize} \quad Q(\mathbf{w}, b, \boldsymbol{\xi}, \boldsymbol{\xi}^*, \varepsilon) = \frac{1}{2}\|\mathbf{w}\|^2 + C\left(\nu\varepsilon + \frac{1}{M}\sum_{i=1}^{M}(\xi_i + \xi_i^*)\right) \quad (11.124)$$

$$\text{subject to} \quad y_i - \mathbf{w}^\top\phi(\mathbf{x}_i) - b \leq \varepsilon + \xi_i \quad \text{for } i = 1,\ldots,M,$$

$$\mathbf{w}^\top\phi(\mathbf{x}_i) + b - y_i \leq \varepsilon + \xi_i^* \quad \text{for } i = 1,\ldots,M, \quad (11.125)$$

$$\xi_i \geq 0, \quad \xi_i^* \geq 0 \quad \text{for } i = 1,\ldots,M,$$

where $\nu$ is a positive parameter.

Introducing the Lagrange multipliers $\alpha_i, \alpha_i^*, \eta_i, \eta_i^*$, and $\beta \, (\geq 0)$, we convert the original constrained problem into an unconstrained one:

$$Q(\mathbf{w}, b, \beta, \boldsymbol{\xi}, \boldsymbol{\xi}^*, \varepsilon, \boldsymbol{\alpha}, \boldsymbol{\alpha}^*, \boldsymbol{\eta}, \boldsymbol{\eta}^*)$$

$$= \frac{1}{2}\|\mathbf{w}\|^2 + C\nu\varepsilon + \frac{C}{M}\sum_{i=1}^{M}(\xi_i + \xi_i^*) - \sum_{i=1}^{M}\alpha_i\left(\varepsilon + \xi_i - y_i + \mathbf{w}^\top\phi(\mathbf{x}_i) + b\right)$$

$$- \sum_{i=1}^{M}\alpha_i^*\left(\varepsilon + \xi_i^* + y_i - \mathbf{w}^\top\phi(\mathbf{x}_i) - b\right) - \beta\varepsilon - \sum_{i=1}^{M}(\eta_i\xi_i + \eta_i^*\xi_i^*). \quad (11.126)$$

Setting the derivatives of (11.126) with respect to the primal variables to zero, we obtain

$$\mathbf{w} = \sum_{i=1}^{M}(\alpha_i - \alpha_i^*)\,\phi(\mathbf{x}_i), \quad (11.127)$$

$$\sum_{i=1}^{M}(\alpha_i - \alpha_i^*) = 0, \quad (11.128)$$

$$\alpha_i + \eta_i = \frac{C}{M} \quad \text{for } i = 1,\ldots,M, \quad (11.129)$$

$$\alpha_i^* + \eta_i^* = \frac{C}{M} \quad \text{for } i = 1,\ldots,M, \quad (11.130)$$

$$C\nu - \sum_{i=1}^{M}(\alpha_i + \alpha_i^*) - \beta = 0. \quad (11.131)$$

Then the dual problem is given by

$$\text{maximize} \quad Q(\alpha, \alpha^*) = -\frac{1}{2} \sum_{i,j=1}^{M} (\alpha_i - \alpha_i^*)(\alpha_j - \alpha_j^*) K(\mathbf{x}_i, \mathbf{x}_j)$$

$$+ \sum_{i=1}^{M} y_i (\alpha_i - \alpha_i^*) \tag{11.132}$$

$$\text{subject to} \quad \sum_{i=1}^{M} (\alpha_i - \alpha_i^*) = 0, \tag{11.133}$$

$$0 \le \alpha_i \le \frac{C}{M}, \quad 0 \le \alpha_i^* \le \frac{C}{M} \quad \text{for } i = 1, \dots, M, \tag{11.134}$$

$$\sum_{i=1}^{M} (\alpha_i + \alpha_i^*) \le C\nu, \tag{11.135}$$

where $K(\mathbf{x}_i, \mathbf{x}_j) = \boldsymbol{\phi}^\top(\mathbf{x}_i)\,\boldsymbol{\phi}(\mathbf{x}_j)$.

From (11.134) and $\alpha_i \alpha_i^* = 0$, the left-hand side of (11.135) is bounded by $C$. Thus, the solution with $\nu > 1$ is the same as that with $\nu = 1$. Assume that the obtained $\varepsilon$ is not zero ($\varepsilon > 0$). Then the following relations hold:

$$\frac{\text{Number of errors}}{M} \le \nu \le \frac{\text{Number of support vectors}}{M}, \tag{11.136}$$

where the number of errors is the number of data that are outside of the $\varepsilon$ tube. This can be proved as follows. For the training data outside of the tube, either $\alpha_i = C/M$ or $\alpha_i^* = C/M$. Thus from (11.134) and (11.135), the training data outside of the tube is at most $\nu M$. Therefore, the first inequality holds. From the KKT conditions, $\varepsilon > 0$ implies $\beta = 0$. Thus from (11.131), the equality holds in (11.135). Therefore, from (11.134) and (11.135), the number of support vectors is at least $\nu M$. Thus the second inequality holds.

### 11.6.3 Least-Squares Support Vector Regressors

Similar to the discussions for least-squares (LS) support vector machines for pattern classification, Suykens [35] proposed least-squares support vector regressors, in which the inequality constraints in the L2 support vector regressors are converted into equality constraints.

Using the $M$ training data pairs $(\mathbf{x}_i, y_i)$ $(i = 1, \dots, M)$, we consider determining the following scalar function:

$$f(\mathbf{x}) = \mathbf{w}^\top \boldsymbol{\phi}(\mathbf{x}) + b, \tag{11.137}$$

where $\mathbf{w}$ is the $l$-dimensional vector, $b$ is the bias term, and $\boldsymbol{\phi}(\mathbf{x})$ is the mapping function that maps the $m$-dimensional vector $\mathbf{x}$ into the $l$-dimensional feature space.

The LS support vector regressor is trained by

$$\text{minimize} \quad Q(\mathbf{w}, b, \boldsymbol{\xi}) = \frac{1}{2}\mathbf{w}^\top \mathbf{w} + \frac{C}{2}\sum_{i=1}^{M}\xi_i^2 \qquad (11.138)$$

$$\text{subject to} \quad y_i = \mathbf{w}^\top \boldsymbol{\phi}(\mathbf{x}_i) + b + \xi_i \quad \text{for } i = 1,\ldots,M, \qquad (11.139)$$

where $\xi_i$ are the slack variables for $\mathbf{x}_i$, $\boldsymbol{\xi} = (\xi_1,\ldots,\xi_M)^\top$, and $C$ is the margin parameter.

Introducing the Lagrange multipliers $\alpha_i$ into (11.138) and (11.139), we obtain the unconstrained objective function:

$$Q(\mathbf{w}, b, \boldsymbol{\alpha}, \boldsymbol{\xi}) = \frac{1}{2}\mathbf{w}^\top \mathbf{w} + \frac{C}{2}\sum_{i=1}^{M}\xi_i^2$$

$$- \sum_{i=1}^{M}\alpha_i\left(\mathbf{w}^\top\boldsymbol{\phi}(\mathbf{x}_i) + b + \xi_i - y_i\right), \qquad (11.140)$$

where $\boldsymbol{\alpha} = (\alpha_1,\ldots,\alpha_M)^\top$.

Taking the partial derivatives of (11.140) with respect to $\mathbf{w}, b, \boldsymbol{\alpha}$, and $\boldsymbol{\xi}$, respectively, and equating them to zero, we obtain the optimal conditions as follows:

$$\mathbf{w} = \sum_{i=1}^{M}\alpha_i\,\boldsymbol{\phi}(\mathbf{x}_i), \qquad (11.141)$$

$$\sum_{i=1}^{M}\alpha_i = 0, \qquad (11.142)$$

$$\mathbf{w}^\top\boldsymbol{\phi}(\mathbf{x}_i) + b + \xi_i - y_i = 0, \qquad (11.143)$$

$$\alpha_i = C\,\xi_i \quad \text{for } i = 1,\ldots,M. \qquad (11.144)$$

Substituting (11.141) and (11.144) into (11.143) and expressing it and (11.142) in matrix form, we obtain

$$\begin{pmatrix} \Omega & \mathbf{1} \\ \mathbf{1}^\top & 0 \end{pmatrix}\begin{pmatrix} \boldsymbol{\alpha} \\ b \end{pmatrix} = \begin{pmatrix} \mathbf{y} \\ 0 \end{pmatrix}, \qquad (11.145)$$

where $\mathbf{1}$ is the $M$-dimensional vector and

$$\{\Omega_{ij}\} = \boldsymbol{\phi}^\top(\mathbf{x}_i)\,\boldsymbol{\phi}(\mathbf{x}_j) + \frac{\delta_{ij}}{C}, \qquad (11.146)$$

$$\delta_{ij} = \begin{cases} 1 & i = j, \\ 0 & i \neq j, \end{cases} \tag{11.147}$$

$$\mathbf{y} = (y_1, \dots, y_M), \tag{11.148}$$

$$\mathbf{1} = (1, \dots, 1)^\top. \tag{11.149}$$

Like the support vector machine, selecting $K(\mathbf{x}, \mathbf{x}') = \boldsymbol{\phi}^\top(\mathbf{x})\,\boldsymbol{\phi}(\mathbf{x}')$ that satisfies Mercer's condition, we can avoid the explicit treatment of the feature space. The resulting approximation function becomes

$$f(\mathbf{x}) = \sum_{i=1}^{M} \alpha_i\, K(\mathbf{x}, \mathbf{x}_i) + b. \tag{11.150}$$

The advantages of the LS support vector regressors over regular support vector regressors are as follows.

1. Training can be done by solving the set of simultaneous linear equations (11.145) for $\boldsymbol{\alpha}$ and $b$, not by solving the quadratic programming problems.
2. Least-squares support vector regressors are exactly the same with two-class LS support vector machines discussed in Section 4.1.1. Thus unlike regular support vector regressors, complexity of model selection is the same with that for two-class LS support vector machines.

But the disadvantage of LS support vector regressors is that unlike regular support vector regressors, the solution is not sparse. There are several approaches to solve this problem [35–39]. The ideas to realize sparse LS support vector regressors are essentially the same with those for sparse support vector machines discussed in Section 4.3.1. The major approaches are summarized as follows:

1. Pruning support vectors with small absolute values of $\alpha_i$ [35, 36]. First, we solve (11.145) using all the training data. Then we sort $\alpha_i$ according to their absolute values and delete a portion of the training data set (say 5% of the set) starting from the data with the minimum absolute value in order. Then we solve (11.145) using the reduced training data set and iterate this procedure while the user-defined performance index is not degraded.
2. Training LS support vector machines in the reduced empirical feature space [39]. We train the LS support vector machine in the reduced empirical feature space, which is obtained by reducing the dimension of the empirical feature space.
3. Selecting basis vectors by forward selection [38, 40]. We select the training data that map the empirical feature space and that reduce the approximation most by forward selection.

Another disadvantage is that unlike L1 or L2 support vector regressors, LS support vector regressors are not robust to outliers because of the square error function. To avoid this, the weighted LS support vector regressor, in which small weights are assigned to the data considered to be outliers, is used and two-stage training method is proposed. At the first stage an LS support vector regressor is trained and in the second stage the weights in the weighted LS support vector regressor are iteratively updated using a robust error function until the solution of the weighted LS support vector regressor converges [54].

## 11.7 Variable Selection

### 11.7.1 Overview

Variable selection is one of the effective ways in reducing computational complexity and improving the generalization ability of the regressor. The goal of variable selection is to obtain the smallest set of variables that realizes the generalization ability comparable with the original set of variables or to get the set of variables that maximizes the generalization ability. This method is called a wrapper method [55]. But because it will be time consuming to use the generalization ability as the selection criterion, we usually use another selection criterion with less computational burden. This method is called a filter method [56–58]. Although the computational cost of the filter methods may be small, it will take a risk of selecting a subset of input variables that may deteriorate the generalization ability of the regressor.

Recently, along with the improvement of computational power, the use of wrapper methods becomes possible for large-size problems. Wrapper methods provide good generalization ability but spend much computational cost. To alleviate this, a combination of both methods [59, 60] and selecting variables during training called imbedded methods [61–63] are considered.

In general, the generalization ability by wrapper methods is higher than that by filter methods. But usually, wrapper methods are not efficient [64]. In addition, from the standpoint of the computational cost, it is hard to test the performance of all the subsets of input variables. Thus, we generally perform backward selection or forward selection for the filter method or the wrapper method. There is also a combination of forward selection with backward selection [65, 66].

To speed up wrapper methods, some variable selection methods using LP support vector regressors with linear kernels [67–69] are proposed. After training an LP support vector regressor, input variables are ranked according to the absolute values of the weights and the input variables with small absolute values are deleted. But because nonlinear function approximation problems are approximated by the LP support vector regressors with linear kernels, the nonlinear relations may be overlooked. To overcome this problem, in [67], the data set is divided into 20 subsets and for each subset the weight vector is obtained by training a linear LP support vector regressor. Then the variables with large absolute weight values that often appear among 20 subsets are selected for nonlinear function approximation.

Backward variable selection by block deletion [70, 71] can select useful input variables faster than the conventional backward variable selection. This method uses as a selection criterion the generalization ability estimated by cross-validation. To speed up variable selection, it deletes multiple candidate variables. In the following we discuss variable selection by block deletion.

## 11.7.2 Variable Selection by Block Deletion

How and when to stop variable selection is one of the important problems in variable selection. To obtain a set of variables whose generalization ability is comparable with or better than that of the initial set of variables, as the selection criterion we use the approximation error of the validation data set in cross-validation.

Our purpose is to obtain the smallest set of variables with generalization ability comparable to the initial set of variables. To do this, before variable selection, we set the threshold value for the selection criterion evaluating the approximation error using all the variables. Let the threshold be $T$. Then $T$ is determined by

$$T = E^m, \qquad (11.151)$$

where $m$ is the number of initial input variables and $E^m$ is the approximation error of the validation set by cross-validation. We fix $T$ to $E^m$ throughout variable selection and delete variables so long as the approximation error of the current set of variables is below the threshold.

Now we explain the idea of block deletion of variables. First, we calculate the approximation error $E^m$ from the initial set of input variables $I^m = \{1, \ldots, m\}$ and set the threshold value of the stopping condition $T = E^m$.

Assume that by block deletion, we have reduced the set of features from $I^m$ to $I^j$. We delete the $i$th variable in $I^j$ temporarily from $I^j$ and calculate $E_i^j$, where $E_i^j$ is the approximation error when we delete the $i$th variable from $I^j$. Then we consider the input variables that satisfy

$$E_i^j \leq T, \qquad (11.152)$$

as candidates of deletion and generate the set of input variables that are candidates for deletion by

$$S^j = \{i \mid E_i^j \leq T, i \in I^j\}. \qquad (11.153)$$

We rank the candidates in the ascending order of $E_i^j$ and delete all the candidates from $I^j$ temporarily. We compare the error $E^{j'}$ with the threshold $T$, where $j'$ is the number of input variables after the deletion. If

$$E^{j'} \leq T, \tag{11.154}$$

block deletion has succeeded and we delete the candidate variables permanently from $I^j$. If block deletion has failed, we backtrack and delete half of the variables previously deleted. We iterate the procedure until block deletion succeeds. When the accuracy is not improved by deleting any input variable, we finish variable selection.

The algorithm of block deletion is as follows.

1. Calculate $E^m$ for $I^m$. Set $T = E^m$, $j = m$, and go to Step 2.
2. Delete temporarily the $i$th input variable in $I^j$ and calculate $E_i^j$. Iterate the above procedure for all $i$ in $I^j$.
3. Calculate $S^j$. If $S^j$ is empty, stop variable selection. If only one input variable is included in $S^j$, set $I^{j-1} = I^j - S^j$, $j \leftarrow j - 1$ and go to Step 2. If $S^j$ has more than one input variable, generate $V^j$ and go to Step 4.
4. Delete all the variables in $V^j$ from $I^j$: $I^{j'} = I^j - V^j$, where $j' = j - |V^j|$ and $|V^j|$ denotes the number of elements in $V^j$. Then, calculate $E^{j'}$ and if (11.154) is not satisfied, go to Step 5. Otherwise, update $j$ with $j'$ and go to Step 2.
5. Let $V'^j$ include the upper half elements of $V^j$. Set $I^{j'} = I^j - \{V'^j\}$, where $\{V'^j\}$ is the set that includes all the variables in $V'^j$ and $j' = j - |\{V'^j\}|$. Then, if (11.154) is satisfied, delete input variables in $V'^j$ and go to Step 2 updating $j$ with $j'$. Otherwise, update $V^j$ with $V'^j$ and iterate Step 5 so long as (11.154) is not satisfied.

## 11.7.3 Performance Evaluation

We evaluated the effectiveness of block deletion using the data sets for regression listed in Table 1.4. In our experiments, we combined the training data set with the test data set and randomly divided the set into two. In doing so, we made 20 pairs of training and test data sets.

We used the LS support vector regressor and determined the values of the kernel parameter $\gamma$, and the margin parameter $C$ by fivefold cross-validation. We changed $\gamma = \{0.1, 0.5, 1.0, 5.0, 10, 15, 20, 50, 100\}$ and $C = \{1, 10, 100, 1,000, 5,000, 10^4, 10^5\}$.

We use an Athlon 64 XII 4800+ personal computer (2GB memory, Linux operating system) in measuring variable selection time.

We evaluated the average approximation errors, the average number of selected variables, and the average training time using the multiple data sets. Table 11.15 shows the result. The "Vali." and "Test" columns list the average approximation errors of the validation and test data sets, respectively. And

"Initial" and "Selected" mean that the results are obtained by using the initial set of variables and the selected set of variables, respectively. The column "Num." lists the number of input variables selected by variable selection; the column "Time" lists the computation time.

By the Welch $t$ test with a 5% significance level, we tested the statistical differences of the approximation errors between the initial set of variables and the selected set of variables for the validation data sets and the test data sets. If there is significant difference, we mark the asterisk to the better result. For example, the asterisk in the "Selected" "Vali." column of the Triazines data set means that the approximation error of the validation sets after variable selection is significantly smaller than that before variable selection.

**Table 11.15** Evaluation results of block deletion

| Data | Initial | | Selected | | | |
| | Vali. | Test | Vali | Test | Num. | Time (s) |
| --- | --- | --- | --- | --- | --- | --- |
| Boston 14 | 2.34±0.23 | 2.36±0.16* | 2.08±0.73 | 2.55±0.19 | 7.5±2.6 | 21±2.6 |
| Boston 5 | 0.030±0.002 | 0.029±0.002 | 0.026±0.002* | 0.026±0.002* | 3.1±0.22 | 19±3.4 |
| Water P. | 0.975±0.064 | 0.971±0.039 | 0.958±0.066 | 0.992±0.040 | 4.7±1.6 | 13.5±4.4 |
| Pyrimid. | 0.029±0.014 | 0.030±0.007 | 0.013±0.012* | 0.020±0.009* | 1.1±0.22 | 0.75±0.54 |
| Triazines | 0.006±0.003 | 0.004±0.003 | 0.004±0.002* | 0.004±0.002 | 2.0±0.77 | 14.4±4.39 |
| Phenetyl. | 0.186±0.051 | 0.237±0.139 | 0.138±0.067* | 0.263±0.080 | 3.0±1.07 | 12.6±3.61 |
| O. Juice | 4.45±0.52 | 6.96±2.07 | 4.14±0.47 | 7.05±1.87 | 6.0±2.0 | 2,337±2,631 |

From the table for the seven regression problems, variable selection by block deletion can delete variables without deteriorating the approximation accuracy. Especially for the pyrimidines, triazines, and phenetylamines problems one to three input variables are enough to get a comparable accuracy.

Except for the orange juice problem, the variable selection time did not take long. For the orange juice problem, the standard deviations of the computation time are considerably large. The reason is as follows. For some data sets only several input variables were deleted at a time. This led to slowing down variable selection more than ten times compared to that for the other data sets.

# References

1. S. Mukherjee, E. Osuna, and F. Girosi. Nonlinear prediction of chaotic time series using support vector machines. In *Neural Networks for Signal Processing VII—Proceedings of the 1997 IEEE Signal Processing Society Workshop*, pages 511–520, 1997.
2. K.-R. Müller, A. J. Smola, G. Rätsch, B. Schölkopf, J. Kohlmorgen, and V. Vapnik. Predicting time series with support vector machines. In W. Gerstner, A. Germond,

M. Hasler, and J.-D. Nicoud, editors, *Artificial Neural Networks (ICANN '97)—Proceedings of the Seventh International Conference, Lausanne, Switzerland*, pages 999–1004. Springer-Verlag, Berlin, Germany, 1997.

3. F. Pérez-Cruz, G. Camps-Valls, E. Soria-Olivas, J. J. Pérez-Ruixo, A. R. Figueiras-Vidal, and A. Artés-Rodríguez. Multi-dimensional function approximation and regression estimation. In J. R. Dorronsoro, editor, *Artificial Neural Networks (ICANN 2002)—Proceedings of International Conference, Madrid, Spain*, pages 757–762. Springer-Verlag, Berlin, Germany, 2002.

4. C. A. Micchelli and M. Pontil. On learning vector-valued functions. *Neural Computation*, 17(1):177–204, 2005.

5. M. Lázaro, I. Santamaría, F. Pérez-Cruz, and A. Artés-Rodriguez. Support vector regression for the simultaneous learning of a multivariate function and its derivatives. *Neurocomputing*, 69(1–3):42–61, 2005.

6. M. Brudnak. Vector-valued support vector regression. In *Proceedings of the 2006 International Joint Conference on Neural Networks (IJCNN 2006)*, pages 2871–2878, Vancouver, Canada, 2006.

7. K. Ito and R. Nakano. Optimizing support vector regression hyperparameters based on cross-validation. In *Proceedings of International Joint Conference on Neural Networks (IJCNN 2003)*, volume 3, pages 2077–2082, Portland, OR, 2003.

8. V. Cherkassky and Y. Ma. Practical selection of SVM parameters and noise estimation for SVM regression. *Neural Networks*, 17(1):113–126, 2004.

9. M.-W. Chang and C.-J. Lin. Leave-one-out bounds for support vector regression model selection. *Neural Computation*, 17(5):1188–1222, 2005.

10. M. Karasuyama, D. Kitakoshi, and R. Nakano. Revised optimizer of SVR hyperparameters minimizing cross-validation error. In *Proceedings of the 2006 International Joint Conference on Neural Networks (IJCNN 2006)*, pages 711–718, Vancouver, Canada, 2006.

11. K. Smets, B. Verdonk, and E. M. Jordaan. Evaluation of performance measures for SVR hyperparameter selection. In *Proceedings of the 2007 International Joint Conference on Neural Networks (IJCNN 2007)*, pages 637–642, Orlando, FL, 2007.

12. S.-P. Liao, H.-T. Lin, and C.-J. Lin. A note on the decomposition methods for support vector regression. In *Proceedings of International Joint Conference on Neural Networks (IJCNN '01)*, volume 2, pages 1474–1479, Washington, DC, 2001.

13. R. Collobert and S. Bengio. SVMTorch: Support vector machines for large-scale regression problems. *Journal of Machine Learning Research*, 1:143–160, 2001.

14. D. Mattera, F. Palmieri, and S. Haykin. An explicit algorithm for training support vector machines. *IEEE Signal Processing Letters*, 6(9):243–245, 1999.

15. G. W. Flake and S. Lawrence. Efficient SVM regression training with SMO. *Machine Learning*, 46(1–3):271–290, 2002.

16. J. Guo, N. Takahashi, and T. Nishi. Convergence proof of a sequential minimal optimization algorithm for support vector regression. In *Proceedings of the 2006 International Joint Conference on Neural Networks (IJCNN 2006)*, pages 747–754, Vancouver, Canada, 2006.

17. N. Takahashi, J. Guo, and T. Nishi. Global convergence of SMO algorithm for support vector regression. *IEEE Transactions on Neural Networks*, 19(6):971–982, 2008.

18. M. Vogt. SMO algorithms for support vector machines without bias term. Technical report, Institute of Automatic Control, TU Darmstadt, Germany, 2002.

19. K. Veropoulos. Machine learning approaches to medical decision making. PhD thesis, Department of Computer Science, University of Bristol, Bristol, UK, 2001.

20. V. Kecman, M. Vogt, and T. M. Huang. On the equality of kernel AdaTron and Sequential Minimal Optimization in classification and regression tasks and alike algorithms for kernel machines. In *Proceedings of the Eleventh European Symposium on Artificial Neural Networks (ESANN 2003)*, pages 215–222, Bruges, Belgium, 2003.

21. J. Ma, J. Theiler, and S. Perkins. Accurate on-line support vector regression. *Neural Computation*, 15(11):2683–2703, 2003.

22. D. R. Musicant and A. Feinberg. Active set support vector regression. *IEEE Transactions on Neural Networks*, 15(2):268–275, 2004.

23. N. de Freitas, M. Milo, P. Clarkson, M. Niranjan, and A. Gee. Sequential support vector machines. In *Neural Networks for Signal Processing IX—Proceedings of the 1999 IEEE Signal Processing Society Workshop*, pages 31–40, 1999.

24. D. Anguita, A. Boni, and S. Pace. Fast training of support vector machines for regression. In *Proceedings of the IEEE-INNS-ENNS International Joint Conference on Neural Networks (IJCNN 2000)*, volume 5, pages 210–214, Como, Italy, 2000.

25. G. H. Golub and C. F. Van Loan. *Matrix Computations, Third Edition*. The Johns Hopkins University Press, Baltimore, MD, 1996.

26. Y. Hirokawa and S. Abe. Training of support vector regressors based on the steepest ascent method. In *Proceedings of the Ninth International Conference on Neural Information Processing (ICONIP '02)*, volume 2, pages 552–555, Singapore, 2002.

27. Y. Hirokawa and S. Abe. Steepest ascent training of support vector regressors. *IEEJ Transactions of Electronics, Information and Systems*, 124(10):2064–2071, 2004 (in Japanese).

28. S. S. Keerthi and E. G. Gilbert. Convergence of a generalized SMO algorithm for SVM classifier design. *Machine Learning*, 46(1–3):351–360, 2002.

29. S. S. Keerthi, S. K. Shevade, C. Bhattacharyya, and K. R. K. Murthy. Improvements to Platt's SMO algorithm for SVM classifier design. *Neural Computation*, 13(3):637–649, 2001.

30. R. J. Vanderbei. LOQO: An interior point code for quadratic programming. Technical Report SOR-94-15, Princeton University, 1998.

31. S. Abe. *Pattern Classification: Neuro-Fuzzy Methods and Their Comparison*. Springer-Verlag, London, UK, 2001.

32. R. S. Crowder. Predicting the Mackey-Glass time series with cascade-correlation learning. In D. S. Touretzky, J. L. Elman, T. J. Sejnowski, and G. E. Hinton, editors, *Connectionist Models: Proceedings of the 1990 Summer School*, pages 117–123, Morgan Kaufmann, San Mateo, CA, 1991.

33. J.-S. R. Jang. ANFIS: Adaptive-network-based fuzzy inference system. *IEEE Transactions on Systems, Man, and Cybernetics*, 23(3):665–685, 1993.

34. S. L. Chiu. Fuzzy model identification based on cluster estimation. *Journal of Intelligent and Fuzzy Systems*, 2(3):267–278, 1994.

35. J. A. K. Suykens. Least squares support vector machines for classification and nonlinear modelling. *Neural Network World*, 10(1–2):29–47, 2000.

36. J. A. K. Suykens, L. Lukas, and J. Vandewalle. Sparse least squares support vector machine classifiers. In *Proceedings of the Eighth European Symposium on Artificial Neural Networks (ESANN 2000)*, pages 37–42, Bruges, Belgium, 2000.

37. G. C. Cawley and N. L. C. Talbot. Fast exact leave-one-out cross-validation of sparse least-squares support vector machines. *Neural Networks*, 17(10):1467–1475, 2004.

38. X. X. Wang, S. Chen, D. Lowe, and C. J. Harris. Sparse support vector regression based on orthogonal forward selection for the generalised kernel model. *Neurocomputing*, 70(1–3):462–474, 2006.

39. S. Abe and K. Onishi. Sparse least squares support vector regressors trained in the reduced empirical feature space. In J. Marques de Sá, L. A. Alexandre, W. Duch, and D. Mandic, editors, *Artificial Neural Networks (ICANN 2007)—Proceedings of the Seventeenth International Conference, Porto, Portugal, Part II*, pages 527–536. Springer-Verlag, Berlin, Germany, 2007.

40. S. Muraoka and S. Abe. Sparse support vector regressors based on forward basis selection. In *Proceedings of the 2009 International Joint Conference on Neural Networks (IJCNN 2009)*, pages 2183–2187, Atlanta, GA, 2009.

41. X. Shao and V. Cherkassky. Multi-resolution support vector machine. In *Proceedings of International Joint Conference on Neural Networks (IJCNN '99)*, volume 2, pages 1065–1070, Washington, DC, 1999.

42. C. Ap. M. Lima, A. L. V. Coelho, and F. J. Von Zuben. Ensembles of support vector machines for regression problems. In *Proceedings of the 2002 International Joint Conference on Neural Networks (IJCNN '02)*, volume 3, pages 2381–2386, Honolulu, Hawaii, 2002.

43. G. F. Smits and E. M. Jordaan. Improved SVM regression using mixtures of kernels. In *Proceedings of the 2002 International Joint Conference on Neural Networks (IJCNN '02)*, volume 3, pages 2785–2790, Honolulu, Hawaii, 2002.

44. E. M. Jordaan and G. F. Smits. Robust outlier detection using SVM for regression. In *Proceedings of International Joint Conference on Neural Networks (IJCNN 2004)*, volume 3, pages 2017–2022, Budapest, Hungary, 2004.

45. P. S. Bradley and O. L. Mangasarian. Massive data discrimination via linear support vector machines. *Optimization Methods and Software*, 13(1):1–10, 2000.

46. T. B. Trafalis and H. Ince. Benders decomposition technique for support vector regression. In *Proceedings of the 2002 International Joint Conference on Neural Networks (IJCNN '02)*, volume 3, pages 2767–2772, Honolulu, Hawaii, 2002.

47. O. L. Mangasarian and D. R. Musicant. Large scale kernel regression via linear programming. *Machine Learning*, 46(1–3):255–269, 2002.

48. Y. Torii and S. Abe. Decomposition techniques for training linear programming support vector machines. *Neurocomputing*, 72(4–6):973–984, 2009.

49. V. Kecman and I. Hadzic. Support vectors selection by linear programming. In *Proceedings of the IEEE-INNS-ENNS International Joint Conference on Neural Networks (IJCNN 2000)*, volume 5, pages 193–198, Como, Italy, 2000.

50. V. Kecman, T. Arthanari, and I. Hadzic. LP and QP based learning from empirical data. In *Proceedings of International Joint Conference on Neural Networks (IJCNN '01)*, volume 4, pages 2451–2455, Washington, DC, 2001.

51. A. J. Smola, N. Murata, B. Schölkopf, and K.-R. Müller. Asymptotically optimal choice of ε-loss for support vector machines. In *Proceedings of the Eighth International Conference on Artificial Neural Networks (ICANN '98)*, volume 1, pages 105–110, Skövde, Sweden, 1998.

52. J.-T. Jeng and C.-C. Chuang. A novel approach for the hyperparameters of support vector regression. In *Proceedings of the 2002 International Joint Conference on Neural Networks (IJCNN '02)*, volume 1, pages 642–647, Honolulu, Hawaii, 2002.

53. B. Schölkopf, P. Bartlett, A. Smola, and R. Williamson. Support vector regression with automatic accuracy control. In *Proceedings of the Eighth International Conference on Artificial Neural Networks (ICANN '98)*, volume 2, pages 111–116, Skövde, Sweden, 1998.

54. K. De Brabanter, K. Pelckmans, J. De Brabanter, M. Debruyne, J. A. K. Suykens, M. Hubert, and B. De Moor. Robustness of kernel based regression: A comparison of iterative weighting schmes. In C. Alippi, M. Polycarpou, C. Panayiotou, and G. Ellinas, editors, *Artificial Neural Networks (ICANN 2009)—Proceedings of the Nineteenth International Conference, Limassol, Cyprus, Part I*, pages 100–110. Springer-Verlag, Berlin, Germany, 2009.

55. R. Kohavi and G. H. John. Wrappers for feature subset selection. *Artificial Intelligence*, 97(1, 2):273–324, 1997.

56. I. Guyon and A. Elisseeff. An introduction to variable and feature selection. *Journal of Machine Learning Research*, 3:1157–1182, 2003.

57. V. Sindhwani, S. Rakshit, D. Deodhare, D. Erdogmus, J. C. Principe, and P. Niyogi. Feature selection in MLPs and SVMs based on maximum output information. *IEEE Transactions on Neural Networks*, 15(4):937–948, 2004.

58. L. J. Herrera, H. Pomares, I. Rojas, M. Verleysen, and A. Guilén. Effective input variable selection for function approximation. In S. Kollias, A. Stafylopatis, W. Duch, and E. Oja, editors, *Artificial Neural Networks (ICANN 2006)—Proceedings of the Sixteenth International Conference, Athens, Greece, Part I*, pages 41–50. Springer-Verlag, Berlin, Germany, 2006.

59. I. Guyon, J. Weston, S. Barnhill, and V. Vapnik. Gene selection for cancer classification using support vector machines. *Machine Learning*, 46(1–3):389–422, 2002.

60. Y. Liu and Y. F. Zheng. FS_SFS: A novel feature selection method for support vector machines. *Pattern Recognition*, 39(7):1333–1345, 2006.

61. R. Tibshirani. Regression shrinkage and selection via the lasso. *Journal of the Royal Statistical Society, Series B*, 58(1):267–288, 1996.

62. C. Gold, A. Holub, and P. Sollich. Bayesian approach to feature selection and parameter tuning for support vector machine classifiers. *Neural Networks*, 18(5–6):693–701, 2005.

63. L. Bo, L. Wang, and L. Jiao. Sparse Gaussian processes using backward elimination. In J. Wang, Z. Yi, J. M. Zurada, B.-L. Lu, and H. Yin, editors, *Advances in Neural Networks (ISNN 2006): Proceedings of Third International Symposium on Neural Networks, Chengdu, China, Part 1*, pages 1083–1088. Springer-Verlag, Berlin, Germany, 2006.

64. A. Rakotomamonjy. Variable selection using SVM-based criteria. *Journal of Machine Learning Research*, 3:1357–1370, 2003.

65. S. D. Stearns. On selecting features for pattern classifiers. In *Proceedings of the Third International Conference on Pattern Recognition*, pages 71–75, 1976.

66. P. Pudil, J. Novovičová, and J. Kittler. Floating search methods in feature selection. *Pattern Recognition Letters*, 15(11):1119–1125, 1994.

67. J. Bi, K. P. Bennett, M. Embrechts, C. M. Breneman, and M. Song. Dimensionality reduction via sparse support vector machines. *Journal of Machine Learning Research*, 3:1229–1243, 2003.

68. G. M. Fung and O. L. Mangasarian. A feature selection Newton method for support vector machine classification. *Computational Optimization and Applications*, 28(2):185–202, 2004.

69. E. Pranckeviciene, R. Somorjai, and M. N. Tran. Feature/model selection by the linear programming SVM combined with state-of-art classifiers: what can we learn about the data. In *Proceedings of the International Joint Conference on Neural Networks (IJCNN 2007)*, volume 6, pages 1627–1632, Orlando, FL, 2007.

70. S. Abe. Modified backward feature selection by cross validation. In *Proceedings of the Thirteenth European Symposium on Artificial Neural Networks (ESANN 2005)*, pages 163–168, Bruges, Belgium, 2005.

71. T. Nagatani and S. Abe. Backward variable selection of support vector regressors by block deletion. In *Proceedings of the 2007 International Joint Conference on Neural Networks (IJCNN 2007)*, pages 2117–2122, Orlando, FL, 2007.

# Appendix A
# Conventional Classifiers

In this appendix, we briefly discuss two conventional classifiers: Bayesian classifiers and nearest-neighbor classifiers. As for multiplayer neural networks, see Chapter 9.

## A.1 Bayesian Classifiers

Bayesian classifiers are based on probability theory and give the theoretical basis for pattern classification.

Let $\omega$ be a random variable and take one of $n$ states $\omega_1, \ldots, \omega_n$, where $\omega_i$ indicates class $i$, and an $m$-dimensional feature vector $\mathbf{x}$ be a random variable vector. We assume that we know the a priori probabilities $P(\omega_i)$ and conditional densities $p(\mathbf{x} \mid \omega_i)$. Then when $\mathbf{x}$ is observed, the a posteriori probability of $\omega_i$, $P(\omega_i \mid \mathbf{x})$, is calculated by the Bayes' rule:

$$P(\omega_i \mid \mathbf{x}) = \frac{p(\mathbf{x} \mid \omega_i)\, P(\omega_i)}{p(\mathbf{x})}, \qquad (A.1)$$

where

$$p(\mathbf{x}) = \sum_{i=1}^{n} p(\mathbf{x} \mid \omega_i)\, P(\omega_i). \qquad (A.2)$$

Assume that the cost $c_{ij}$ is given when $\mathbf{x}$ is classified into class $i$ although it is class $j$. Then the expected conditional cost in classifying $\mathbf{x}$ into class $i$, $C(\omega_i \mid \mathbf{x})$, is given by

$$C(\omega_i \mid \mathbf{x}) = \sum_{j=1}^{n} c_{ij}\, P(\omega_j \mid \mathbf{x}). \qquad (A.3)$$

The conditional cost is minimized when $\mathbf{x}$ is classified into the class

S. Abe, *Support Vector Machines for Pattern Classification*,
Advances in Pattern Recognition, DOI 10.1007/978-1-84996-098-4,
© Springer-Verlag London Limited 2010

$$\arg \min_{i=1,\ldots,n} C(\omega_i \mid \mathbf{x}). \tag{A.4}$$

This is called *Bayes' decision rule*.

In diagnosis problems, usually there are normal and abnormal classes. Misclassification of normal data into the abnormal class is less fatal than misclassification of abnormal data into the normal class. In such a situation, we set a smaller cost to the former than the latter.

If we want to minimize the average probability of misclassification, we set the cost as follows:

$$c_{ij} = \begin{cases} 0 & \text{for} \quad i = j, \\ 1 & \text{for} \quad i \neq j, \end{cases} \quad i, j = 1, \ldots, n. \tag{A.5}$$

Then, from (A.1) and (A.2) the conditional cost given by (A.3) becomes

$$C(\omega_i \mid \mathbf{x}) = \sum_{\substack{j \neq i, \\ j = 1}}^{n} P(\omega_j \mid \mathbf{x})$$

$$= 1 - P(\omega_i \mid \mathbf{x}). \tag{A.6}$$

Therefore, the Bayes decision rule given by (A.4) becomes

$$\arg \max_{i=1,\ldots,n} P(\omega_i \mid \mathbf{x})$$

$$= \arg \max_{i=1,\ldots,n} p(\mathbf{x} \mid \omega_i) P(\omega_i). \tag{A.7}$$

Now, we assume that the conditional densities $p(\mathbf{x} \mid \omega_i)$ are normal:

$$p(\mathbf{x} \mid \omega_i) = \frac{1}{\sqrt{(2\pi)^n \det(Q_i)}} \exp\left(-\frac{(\mathbf{x} - \mathbf{c}_i)^\top Q_i^{-1} (\mathbf{x} - \mathbf{c}_i)}{2}\right), \tag{A.8}$$

where $\mathbf{c}_i$ is the mean vector and $Q_i$ is the covariance matrix of the normal distribution for class $i$. If the a priori probabilities $P(\omega_i)$ are the same for $i = 1, \ldots, n$, $\mathbf{x}$ is classified into class $i$ with the maximum $p(\mathbf{x} \mid \omega_i)$.

## A.2 Nearest-Neighbor Classifiers

Nearest-neighbor classifiers use all the training data as templates for classification. In the simplest form, for a given input vector, the nearest-neighbor classifier searches the nearest template and classifies the input vector into the class to which the template belongs. In the complex form the classifier treats $k$ nearest neighbors. For a given input vector, the $k$ nearest templates are searched and the input vector is classified into the class with the maximum number of templates. The classifier architecture is simple, but as the number of training data becomes larger, the classification time becomes longer. Therefore many methods for speeding up classification are studied [1, pp. 181–191], [2, pp. 191–201]. One uses the branch-and-bound method [3, pp. 360–362] and another edits the training data, i.e., selects or replaces the data with the suitable templates. It is proved theoretically that as the number of templates becomes larger, the expected error rate of the nearest-neighbor classifier is bounded by twice that of the Bayesian classifier [4, pp. 159–175].

Usually the Euclidean distance is used to measure the distance between two data $\mathbf{x}$ and $\mathbf{y}$:

$$d(\mathbf{x}, \mathbf{y}) = \sqrt{\sum_{i=1}^{m} (x_i - y_i)^2}, \tag{A.9}$$

but other distances, such as the Manhattan distance

$$d(\mathbf{x}, \mathbf{y}) = \sum_{i=1}^{m} |x_i - y_i|, \tag{A.10}$$

are used. It is clear from the architecture that the recognition rate of the training data for the one-nearest-neighbor classifier is 100%. But for the $k$-nearest-neighbor classifier with $k > 1$, the recognition rate of the training data is not always 100%.

Because the distances such as the Euclidean and Manhattan distances are not invariant in scaling, classification performance varies according to the scaling of input ranges.

# References

1. J. C. Bezdek, J. Keller R. Krisnapuram, and N. R. Pal. *Fuzzy Models and Algorithms for Pattern Recognition and Image Processing*. Kluwer Academic Publishers, Norwell, MA, 1999.
2. B. D. Ripley. *Pattern Recognition and Neural Networks*. Cambridge University Press, Cambridge, 1996.
3. K. Fukunaga. *Introduction to Statistical Pattern Recognition, Second Edition*. Academic Press, San Diego, 1990.
4. E. Gose, R. Johnsonbaugh, and S. Jost. *Pattern Recognition and Image Analysis*. Prentice Hall, Upper Saddle River, NJ, 1996.

# Appendix B
# Matrices

In this appendix, we summarize some basic properties of matrices, least-squares methods and singular value decomposition, and covariance matrices.

## B.1 Matrix Properties

In this section, we summarize the matrix properties used in this book. For more detailed explanation, see, e.g., [1].

Vectors $\mathbf{x}_1, \ldots, \mathbf{x}_n$ are *linearly independent* if

$$a_1 \mathbf{x}_1 + \cdots + a_n \mathbf{x}_n = 0 \tag{B.1}$$

holds only when $a_1 = \cdots = a_n = 0$. Otherwise, $\mathbf{x}_1, \ldots, \mathbf{x}_n$ are *linearly dependent*.

Let $A$ be an $m \times m$ matrix:

$$A = \begin{pmatrix} a_{11} & \cdots & a_{1m} \\ \vdots & \ddots & \vdots \\ a_{m1} & \cdots & a_{mm} \end{pmatrix}. \tag{B.2}$$

Then the *transpose* of $A$, denoted by $A^\top$, is

$$A^\top = \begin{pmatrix} a_{11} & \cdots & a_{m1} \\ \vdots & \ddots & \vdots \\ a_{1m} & \cdots & a_{mm} \end{pmatrix}. \tag{B.3}$$

If $A$ satisfies $A = A^\top$, $A$ is a *symmetric matrix*. If $A$ satisfies $A^\top A = A A^\top = I$, $A$ is an *orthogonal matrix*, where $I$ is the $m \times m$ *unit matrix*.

$$
I = \begin{pmatrix} 1\,0\,\cdots\,0 \\ 0\,1\,\cdots\,0 \\ \vdots\,\vdots\,\cdots\,\vdots \\ 0\,0\,\cdots\,1 \end{pmatrix}. \tag{B.4}
$$

If $m \times m$ matrices $A$ and $B$ satisfy $AB = I$, $B$ is called the *inverse* of $A$ and is denoted by $A^{-1}$. If $A$ has the inverse, $A$ is *regular* (or *nonsingular*). Otherwise, $A$ is singular.

**Lemma B.1.** *Let matrices $A$, $B$, $C$, and $D$ be $n \times n$, $n \times m$, $m \times n$, and $m \times m$ matrices, respectively, and $A$, $D$, and $D + CA^{-1}B$ be nonsingular. Then the following relation holds:*

$$
\left(A + BD^{-1}C\right)^{-1} = A^{-1} - A^{-1}B\left(D + CA^{-1}B\right)^{-1}CA^{-1}. \tag{B.5}
$$

This is called the *matrix inversion lemma*.

Let $m = 1$, $B = \mathbf{b}$, $C = \mathbf{c}^{\top}$, and $D = 1$. Then (B.5) reduces to

$$
\left(A + \mathbf{b}\mathbf{c}^{\top}\right)^{-1} = A^{-1} - \frac{A^{-1}\mathbf{b}\mathbf{c}^{\top}A^{-1}}{1 + \mathbf{c}^{\top}A^{-1}\mathbf{b}}. \tag{B.6}
$$

Using (B.6), LOO error rate estimation of linear equation-based machines such as LS support vector machines can be sped up [2–5].

The *determinant* of an $m \times m$ matrix $A = \{a_{ij}\}$, $\det(A)$, is defined recursively by

$$
\det(A) = \sum_{i=1}^{m} (-1)^{i+1} a_{1i} \det(A_{1i}), \tag{B.7}
$$

where $A_{1i}$ is the $(m-1) \times (m-1)$ matrix obtained by deleting the first row and the $i$th column from $A$. When $m = 1$, $\det(A) = a_{11}$.

If, for the $m \times m$ matrix $A$, a nonzero $m$-dimensional vector $\mathbf{x}$ exists for a constant $\lambda$:

$$
A\mathbf{x} = \lambda\mathbf{x}, \tag{B.8}
$$

$\lambda$ is called an *eigenvalue* and $\mathbf{x}$ an *eigenvector*. Rearranging (B.8) gives

$$
(A - \lambda I)\mathbf{x} = 0. \tag{B.9}
$$

Thus, (B.9) has nonzero $\mathbf{x}$, when

$$
\det(A - \lambda I) = 0, \tag{B.10}
$$

which is called the *characteristic equation*.

**Theorem B.1.** *All the eigenvalues of a real symmetric matrix are real.*

**Theorem B.2.** *Eigenvectors associated with different eigenvalues for a real symmetric matrix are orthogonal.*

For an $m$-dimensional vector $\mathbf{x}$ and an $m \times m$ symmetric matrix $A$, $Q = \mathbf{x}^\top A \mathbf{x}$ is called the *quadratic form*. If for any nonzero $\mathbf{x}$, $Q = \mathbf{x}^\top A \mathbf{x} \geq 0$, $Q$ is *positive semidefinite*. Matrix $Q$ is *positive definite* if the strict inequality holds. Let $L$ be an $m \times m$ orthogonal matrix. By $\mathbf{y} = L\mathbf{x}$, $\mathbf{x}$ is transformed into $\mathbf{y}$. This is the transformation from one orthonormal base into another orthonormal basis. The quadratic form $Q$ is

$$
\begin{aligned}
Q &= \mathbf{x}^\top A \mathbf{x} \\
&= \mathbf{y}^\top L A L^\top \mathbf{y}.
\end{aligned}
\tag{B.11}
$$

**Theorem B.3.** *The characteristic equations for $A$ and $L A L^\top$ are the same.*

**Theorem B.4.** *An $m \times m$ real symmetric matrix $A$ is diagonalized by $L$:*

$$
L A L^\top = \begin{pmatrix} \lambda_1 & 0 & \cdots & 0 \\ 0 & \lambda_2 & \cdots & 0 \\ \vdots & \vdots & \cdots & \vdots \\ 0 & 0 & \cdots & \lambda_m \end{pmatrix},
\tag{B.12}
$$

*where $\lambda_1, \ldots, \lambda_m$ are the eigenvalues of $A$ and the $i$th row of $L$ is the eigenvector associated with $\lambda_i$.*

If all the eigenvalues of $A$ are positive, $A$ is *positive definite*. If all the eigenvalues are nonnegative, $A$ is *positive semidefinite*.

# B.2 Least-Squares Methods and Singular Value Decomposition

Assume that we have $M$ input–output pairs $\{(\mathbf{a}_1', b_1), \ldots, (\mathbf{a}_M', b_M)\}$ in the $(n-1)$-dimensional input space $\mathbf{x}'$ and one-dimensional output space $y$. Now using the least-squares method, we determine the linear relation of the input–output pairs:

$$
y = \mathbf{p}^\top \mathbf{x}' + q,
\tag{B.13}
$$

where $\mathbf{p}$ is the $(n-1)$-dimensional vector, $q$ is a scalar constant, and $M \geq n$. Rewriting (B.13), we get

$$
(\mathbf{x}'^\top, 1) \begin{pmatrix} \mathbf{p} \\ q \end{pmatrix} = y.
\tag{B.14}
$$

Substituting $\mathbf{a}_i'$ and $b_i$ into $\mathbf{x}'$ and $y$ in (B.14), respectively, and replacing $(\mathbf{p}^\top, q)^\top$ with the $n$-dimensional parameter vector $\mathbf{x}$, we obtain

$$
\mathbf{a}_i^\top \mathbf{x} = b_i \quad \text{for } i = 1, \ldots, M,
\tag{B.15}
$$

where $\mathbf{a}_i = (\mathbf{a}_i'^\top, 1)^\top$.

We determine the parameter vector $\mathbf{x}$ so that the sum-of-squares error

$$E = (A\mathbf{x} - \mathbf{b})^{\top}(A\mathbf{x} - \mathbf{b}) \tag{B.16}$$

is minimized, where $A$ is the $M \times n$ matrix and $\mathbf{b}$ is the $M$-dimensional vector:

$$A = \begin{pmatrix} \mathbf{a}_1^{\top} \\ \mathbf{a}_2^{\top} \\ \vdots \\ \mathbf{a}_M^{\top} \end{pmatrix}, \quad \mathbf{b} = \begin{pmatrix} b_1 \\ b_2 \\ \vdots \\ b_M \end{pmatrix}. \tag{B.17}$$

Here, if the rank of $A$ is smaller than $n$, there is no unique solution. In that situation, we determine $\mathbf{x}$ so that the Euclidean norm of $\mathbf{x}$ is minimized.

Matrix $A$ is decomposed into singular values [1]:

$$A = USV^{\top}, \tag{B.18}$$

where $U$ and $V$ are $M \times M$ and $n \times n$ orthogonal matrices, respectively, and $S$ is an $M \times n$ diagonal matrix given by

$$S = \begin{pmatrix} \sigma_1 & & 0 \\ & \ddots & \\ 0 & & \sigma_n \\ \hline & 0_{M-n,n} & \end{pmatrix}. \tag{B.19}$$

Here, $\sigma_i$ are singular values and $\sigma_1 \geq \sigma_2 \geq \cdots \geq \sigma_n \geq 0$, and $0_{M-n,n}$ is the $(M - n) \times n$ zero matrix.

It is known that the columns of $U$ and $V$ are the eigenvectors of $AA^{\top}$ and $A^{\top}A$, respectively, and that the singular values correspond to the square roots of the eigenvalues of $AA^{\top}$, which are the same as those of $A^{\top}A$ [6, pp. 434–435]. Thus when $A$ is a symmetric square matrix, $U = V$ and $A = USU^{\top}$. This is similar to the diagonalization of the square matrix given by Theorem B.4. The difference is that the singular values $A$ are the absolute values of the eigenvalues of $A$. Thus, if $A$ is a positive (semi)definite matrix, both decompositions are the same.

Rewriting (B.16), we get [1, p. 256]

$$\begin{aligned} E &= (A\mathbf{x} - \mathbf{b})^{\top}(A\mathbf{x} - \mathbf{b}) \\ &= (USV^{\top}\mathbf{x} - UU^{\top}\mathbf{b})^{\top}(A\mathbf{x} - \mathbf{b}) \\ &= (SV^{\top}\mathbf{x} - U^{\top}\mathbf{b})^{\top}(SV^{\top}\mathbf{x} - U^{\top}\mathbf{b}) \\ &= \sum_{i=1}^{n}(\sigma_i \mathbf{v}_i^{\top}\mathbf{x} - \mathbf{u}_i^{\top}\mathbf{b})^2 + \sum_{i=n+1}^{M}(\mathbf{u}_i^{\top}\mathbf{b})^2, \end{aligned} \tag{B.20}$$

where $U = (\mathbf{u}_1, \ldots, \mathbf{u}_M)$ and $V = (\mathbf{v}_1, \ldots, \mathbf{v}_n)$. Assuming that the rank of $A$ is $r\,(\leq n)$, (B.20) is minimized when

$$\sigma_i \, \mathbf{v}_i^\top \mathbf{x} = \mathbf{u}_i^\top \mathbf{b} \quad \text{for } i = 1, \ldots, r, \tag{B.21}$$

$$\mathbf{v}_i^\top \mathbf{x} = 0 \quad \text{for } i = r+1, \ldots, n. \tag{B.22}$$

Equation (B.22) is imposed to obtain the minimum Euclidean norm solution. From (B.21) and (B.22), we obtain

$$\mathbf{x} = V\,S^+U^\top \mathbf{b} = A^+\mathbf{b}, \tag{B.23}$$

where $S^+$ is the $n \times M$ diagonal matrix given by

$$S^+ = \left( \begin{array}{ccc|c} \dfrac{1}{\sigma_1} & & 0 & \\ & \ddots & & 0 \\ & & \dfrac{1}{\sigma_r} & \\ 0 & & 0 & \end{array} \right). \tag{B.24}$$

We call $A^+$ the pseudo-inverse of $A$. We must bear in mind that in calculating the pseudo-inverse, we replace the reciprocal of 0 with 0, not with infinity. This ensures the minimum norm solution.

From (B.18) and (B.23),

$$A^+A = V\,S^+U^\top\,U\,S\,V^\top$$
$$= V\,S^+\,S\,V^\top$$
$$= V \begin{pmatrix} I_r & 0_{r,n-r} \\ 0_{n-r,r} & 0_{n-r} \end{pmatrix} V^\top$$
$$= \begin{pmatrix} I_r & 0_{r,n-r} \\ 0_{n-r,r} & 0_{n-r} \end{pmatrix}, \tag{B.25}$$
$$A\,A^+ = U\,S\,S^+\,U^\top$$
$$= \begin{pmatrix} I_r & 0_{r,M-r} \\ 0_{M-r,r} & 0_{M-r} \end{pmatrix}, \tag{B.26}$$

where $I_r$ is the $r \times r$ unit matrix, $0_i$ is the $i \times i$ zero matrix, and $0_{i,j}$ is the $i \times j$ zero matrix. Therefore, if $A$ is a square matrix with rank $n$, $A^+ A = A A^+ = I$, namely, the pseudo-inverse of $A$ coincides with the inverse of $A$, $A^{-1}$. If $M > n$ and the rank of $A$ is $n$, $A^+ A = I$ but $A A^+ \neq I$. In this case $A^+$ is given by

$$A^+ = (A^\top A)^{-1} A^\top. \tag{B.27}$$

This is obtained by taking the derivative of (B.16) with respect to $\mathbf{x}$ and equating the result to zero.

When $M > n$ and the rank of $A$ is smaller than $n$, $A^+ A \neq I$ and $A A^+ \neq I$.

Even when $A^\top A$ is nonsingular, it is recommended to calculate the pseudo-inverse by singular value decomposition, not using (B.27). Because if $A^\top A$ is near singular, $(A^\top A)^{-1} A^\top$ is vulnerable to the small singular values [7, pp. 59–70].

## B.3 Covariance Matrices

Let $\mathbf{x}_1, \ldots, \mathbf{x}_M$ be $M$ samples of the $m$-dimensional random variable $X$. Then the sample covariance matrix of $X$ is given by

$$Q = \frac{1}{M} \sum_{i=1}^{M} (\mathbf{x}_i - \mathbf{c})(\mathbf{x}_i - \mathbf{c})^\top, \tag{B.28}$$

where $\mathbf{c}$ is the mean vector:

$$\mathbf{c} = \frac{1}{M} \sum_{i=1}^{M} \mathbf{x}_i. \tag{B.29}$$

To get the unbiased estimate of the covariance matrix, we replace $1/M$ with $1/(M-1)$ in (B.28), but in this book we use (B.28) as the sample covariance matrix.

Let

$$\mathbf{y}_i = \mathbf{x}_i - \mathbf{c}. \tag{B.30}$$

Then (B.28) becomes

$$Q = \frac{1}{M} \sum_{i=1}^{M} \mathbf{y}_i \mathbf{y}_i^\top. \tag{B.31}$$

From (B.29) and (B.30), $\mathbf{y}_1, \ldots, \mathbf{y}_M$ are linearly dependent.

According to the definition, the covariance matrix $Q$ is symmetric. Matrix $Q$ is positive (semi)definite, as the following theorem shows.

**Theorem B.5.** *The covariance matrix $Q$ given by (B.31) is positive definite if $\mathbf{y}_1, \ldots, \mathbf{y}_M$ have at least $m$ linearly independent data. Matrix $Q$ is positive semidefinite, if any $m$ data from $\mathbf{y}_1, \ldots, \mathbf{y}_M$ are linearly dependent.*

*Proof.* Let $\mathbf{z}$ be an $m$-dimensional nonzero vector. From (B.31),

$$\mathbf{z}^\top Q \mathbf{z} = \mathbf{z}^\top \left( \frac{1}{M} \sum_{i=1}^{M} \mathbf{y}_i \mathbf{y}_i^\top \right) \mathbf{z}$$

$$= \frac{1}{M} \sum_{i=1}^{M} \left( \mathbf{z}^\top \mathbf{y}_i \right) \left( \mathbf{z}^\top \mathbf{y}_i \right)^\top$$

$$= \frac{1}{M} \sum_{i=1}^{M} \left( \mathbf{z}^\top \mathbf{y}_i \right)^2 \geq 0. \tag{B.32}$$

Thus $Q$ is positive semidefinite. If there are $m$ linearly independent data in $\{\mathbf{y}_1, \ldots, \mathbf{y}_M\}$, they span the $m$-dimensional space. Because any $\mathbf{z}$ is expressed by a linear combination of these data, the strict inequality holds for (B.32).

Because $\mathbf{y}_1, \ldots, \mathbf{y}_M$ are linearly dependent, at least $m + 1$ samples are necessary so that $Q$ becomes positive definite. ∎

Assuming that $Q$ is positive definite, the following theorem holds.

**Theorem B.6.** *If $Q$ is positive definite, the mean square weighted distance for $\{\mathbf{y}_1, \ldots, \mathbf{y}_M\}$ is $m$:*

$$\frac{1}{M} \sum_{i=1}^{M} \mathbf{y}_i^\top Q^{-1} \mathbf{y}_i = m. \tag{B.33}$$

*Proof.* Let $P$ be the orthogonal matrix that diagonalizes $Q$, namely,

$$P Q P^\top = \mathrm{diag}(\lambda_1, \ldots, \lambda_m), \tag{B.34}$$

where diag denotes the diagonal matrix and $\lambda_1, \ldots, \lambda_m$ are the eigenvalues of $Q$. From (B.34),

$$Q = P^\top \mathrm{diag}(\lambda_1, \ldots, \lambda_m) P, \tag{B.35}$$

$$Q^{-1} = P^\top \mathrm{diag}(\lambda_1^{-1}, \ldots, \lambda_m^{-1}) P. \tag{B.36}$$

Let

$$\tilde{\mathbf{y}}_i = P \mathbf{y}_i. \tag{B.37}$$

Then from (B.31) and (B.37), (B.34) becomes

$$\frac{1}{M} \sum_{i=1}^{M} \tilde{\mathbf{y}}_i \tilde{\mathbf{y}}_i^\top = \mathrm{diag}(\lambda_1, \ldots, \lambda_m). \tag{B.38}$$

Thus for the diagonal elements of (B.38),

$$\frac{1}{M} \sum_{i=1}^{M} \tilde{y}_{ik}^2 = \lambda_k \quad \text{for } k = 1, \ldots, m, \tag{B.39}$$

where $\tilde{y}_{ik}$ is the $k$th element of $\tilde{\mathbf{y}}_i$. From (B.36) and (B.37), the left-hand side of (B.33) becomes

$$\frac{1}{M} \sum_{i=1}^{M} \mathbf{y}_i^\top Q^{-1} \mathbf{y}_i = \frac{1}{M} \sum_{i=1}^{M} \tilde{\mathbf{y}}_i^\top \mathrm{diag}(\lambda_1^{-1}, \ldots, \lambda_m^{-1}) \tilde{\mathbf{y}}_i$$

$$= \frac{1}{M} \sum_{i=1}^{M} \sum_{k=1}^{m} \lambda_k^{-1} \tilde{y}_{ik}^2. \qquad (B.40)$$

Thus from (B.39) and (B.40), the theorem holds. ∎

# References

1. G. H. Golub and C. F. Van Loan. *Matrix Computations, Third Edition*. The Johns Hopkins University Press, Baltimore, MD, 1996.
2. M. Brown, N. P. Costen, and S. Akamatsu. Efficient calculation of the complete optimal classification set. In *Proceedings of the Seventeenth International Conference on Pattern Recognition (ICPR 2004)*, volume 2, pages 307–310, Cambridge, UK, 2004.
3. G. C. Cawley and N. L. C. Talbot. Efficient model selection for kernel logistic regression. In *Proceedings of the Seventeenth International Conference on Pattern Recognition (ICPR 2004)*, volume 2, pages 439–442, Cambridge, UK, 2004.
4. K. Saadi, N. L. C. Talbot, and G. C. Cawley. Optimally regularised kernel Fisher discriminant analysis. In *Proceedings of the Seventeenth International Conference on Pattern Recognition (ICPR 2004)*, volume 2, pages 427–430, Cambridge, UK, 2004.
5. Z. Ying and K. C. Keong. Fast leave-one-out evaluation and improvement on inference for LS-SVMs. In *Proceedings of the Seventeenth International Conference on Pattern Recognition (ICPR 2004)*, volume 3, pages 494–497, Cambridge, UK, 2004.
6. V. Cherkassky and F. Mulier. *Learning from Data: Concepts, Theory, and Methods*. John Wiley & Sons, New York, 1998.
7. W. H. Press, S. A. Teukolsky, W. T. Vetterling, and B. P. Flannery. *Numerical Recipes in C: The Art of Scientific Computing, Second Edition*. Cambridge University Press, Cambridge, 1992.

# Appendix C
# Quadratic Programming

Quadratic programming is the basis of support vector machines. Here we summarize some of the basic properties of quadratic programming.

## C.1 Optimality Conditions

Consider the following optimization problem:

$$\text{minimize} \quad f(\mathbf{x}) = \frac{1}{2}\mathbf{x}^\top Q\,\mathbf{x} + \mathbf{c}^\top \mathbf{x} \tag{C.1}$$

$$\text{subject to} \quad g_i(\mathbf{x}) = \mathbf{a}_i^\top \mathbf{x} + b_i \geq 0 \quad \text{for } i = 1,\dots,k, \tag{C.2}$$

$$h_i(\mathbf{x}) = \mathbf{d}_i^\top \mathbf{x} + e_i = 0 \quad \text{for } i = 1,\dots,o, \tag{C.3}$$

where $\mathbf{x}$, $\mathbf{c}$, $\mathbf{a}_i$, and $\mathbf{d}_i$ are $m$-dimensional vectors; $Q$ is an $m \times m$ positive semidefinite matrix; and $b_i$ and $e_i$ are scalar constants. This problem is called the *quadratic programming problem*. Because of the linear equality and inequality constraints, $\mathbf{x}$ is in a closed convex domain.

We introduce the Lagrange multipliers $\alpha_i$ and $\beta_i$:

$$L(\mathbf{x}, \boldsymbol{\alpha}, \boldsymbol{\beta}) = f(\mathbf{x}) - \sum_{i=1}^{k} \alpha_i\, g_i(\mathbf{x}) + \sum_{i=1}^{o} \beta_i\, h_i(\mathbf{x}), \tag{C.4}$$

where $\boldsymbol{\alpha} = (\alpha_1,\dots,\alpha_k)^\top$, $\alpha_i \geq 0$ for $i = 1,\dots,k$, and $\boldsymbol{\beta} = (\beta_1,\dots,\beta_o)^\top$. Then the following theorem holds.

**Theorem C.1.** *The optimal solution* $(\mathbf{x}^*, \boldsymbol{\alpha}^*, \boldsymbol{\beta}^*)$ *exists if and only if the following conditions are satisfied:*

$$\frac{\partial L(\mathbf{x}^*, \boldsymbol{\alpha}^*, \boldsymbol{\beta}^*)}{\partial \mathbf{x}} = \mathbf{0}, \tag{C.5}$$

$$\alpha_i^*\, g_i(\mathbf{x}^*) = 0 \quad \text{for } i = 1,\dots,k, \tag{C.6}$$

$$\alpha_i^* \geq 0 \quad for \ i = 1, \ldots, k, \tag{C.7}$$

$$h_i(\mathbf{x}^*) = 0 \quad for \ i = 1, \ldots, o. \tag{C.8}$$

These conditions are called the *Karush–Kuhn–Tucker (KKT) conditions* and the conditions given by (C.6) are called the *Karush–Kuhn–Tucker complementarity conditions*. If there is no confusion, the KKT complementarity conditions are abbreviated as the *KKT conditions*.

The KKT complementarity condition means that if $\alpha_i^* > 0$, $g_i(\mathbf{x}^*) = 0$ (it is called *active*); and if $\alpha_i^* = 0$, $g_i(\mathbf{x}^*) \geq 0$ (it is called *inactive*).

## C.2 Properties of Solutions

The optimal solution can be interpreted geometrically. From (C.4),

$$\frac{\partial L(\mathbf{x}^*, \boldsymbol{\alpha}^*, \boldsymbol{\beta}^*)}{\partial \mathbf{x}} = \frac{\partial f(\mathbf{x}^*)}{\partial \mathbf{x}} - \sum_{i=1}^{k} \alpha_i \frac{\partial g_i(\mathbf{x}^*)}{\partial \mathbf{x}} + \sum_{i=1}^{o} \beta_i \frac{\partial h_i(\mathbf{x}^*)}{\partial \mathbf{x}} = \mathbf{0}. \tag{C.9}$$

If some inequality constraints are inactive (i.e., the associated Lagrange multipliers are zero) for the optimal solution, we can discard the associated terms from (C.9). If they are active, they can be treated as the equality constraints. Thus, without loss of generality, we can assume that the inequality constraints are all inactive. Then the optimal solution must satisfy

$$-\frac{\partial f(\mathbf{x}^*)}{\partial \mathbf{x}} = \sum_{i=1}^{o} \beta_i^* \frac{\partial h_i(\mathbf{x}^*)}{\partial \mathbf{x}}. \tag{C.10}$$

The negative gradient of $f(\mathbf{x})$, $-\partial f(\mathbf{x})/\partial \mathbf{x}$, points the direction in which $f(\mathbf{x})$ decreases the most. And at the optimal solution the negative gradient must be perpendicular to the equality constraint $h_i(\mathbf{x}) = 0$ or parallel to $\partial h_i(\mathbf{x}^*)/\partial \mathbf{x}$. Therefore, the negative gradient must be in the subspace spanned by $\partial g_i(\mathbf{x}^*)/\partial \mathbf{x}$ ($i = 1, \ldots, o$), which is equivalent to (C.10).

If $Q$ is positive definite, the solution is unique. And if $Q$ is positive semidefinite, the solution may not be unique. But if $\mathbf{x}_o$ and $\mathbf{x}_o'$ are solutions, $\lambda \mathbf{x}_o + (1 - \lambda)\mathbf{x}_o'$, where $1 \geq \lambda \geq 0$, is also a solution.

For the optimal solution $(\mathbf{x}^*, \boldsymbol{\alpha}^*, \boldsymbol{\beta}^*)$, the following relation holds:

$$L(\mathbf{x}, \boldsymbol{\alpha}^*, \boldsymbol{\beta}^*) \geq L(\mathbf{x}^*, \boldsymbol{\alpha}^*, \boldsymbol{\beta}^*) \geq L(\mathbf{x}^*, \boldsymbol{\alpha}, \boldsymbol{\beta}), \tag{C.11}$$

namely, $L(\mathbf{x}, \boldsymbol{\alpha}, \boldsymbol{\beta})$ is minimized with respect to $\mathbf{x}$ and maximized with respect to $\boldsymbol{\alpha}$ and $\boldsymbol{\beta}$. Thus the optimal point is a saddle point.

*Example C.1.* Consider the following problem:

$$\text{minimize} \quad f(\mathbf{x}) = \frac{1}{2}\mathbf{x}^\top Q\,\mathbf{x} \tag{C.12}$$

$$\text{subject to} \quad 2 \geq x_1 + x_2 \geq 1, \tag{C.13}$$

where $\mathbf{x} = (x_1\,x_2)^\top$ and $Q$ is positive definite:

$$Q = \begin{pmatrix} 1 & \frac{1}{2} \\ \frac{1}{2} & 1 \end{pmatrix}. \tag{C.14}$$

Because

$$L(\mathbf{x}, \boldsymbol{\alpha}) = \frac{1}{2}\left(x_1^2 + x_1 x_2 + x_2^2\right) - \alpha_1\left(x_1 + x_2 - 1\right) - \alpha_2\left(2 - x_1 - x_2\right), \tag{C.15}$$

the KKT conditions are given by

$$\frac{\partial L(\mathbf{x}, \boldsymbol{\alpha})}{\partial x_1} = x_1 + \frac{1}{2}x_2 - \alpha_1 + \alpha_2 = 0, \tag{C.16}$$

$$\frac{\partial L(\mathbf{x}, \boldsymbol{\alpha})}{\partial x_2} = \frac{1}{2}x_1 + x_2 - \alpha_1 + \alpha_2 = 0, \tag{C.17}$$

$$\alpha_1\left(x_1 + x_2 - 1\right) = 0, \quad \alpha_2\left(2 - x_1 - x_2\right) = 0, \tag{C.18}$$

$$\alpha_1 \geq 0, \quad \alpha_2 \geq 0. \tag{C.19}$$

Subtracting (C.17) from (C.16), we obtain $x_1 = x_2$. Therefore, from $f(\mathbf{x}) = 3\,x_1^2/2$ and the KKT conditions, the optimal solution satisfies

$$x_1 = x_2 = \frac{1}{2}, \quad \alpha_1 = \frac{3}{2}, \quad \alpha_2 = 0. \tag{C.20}$$

Thus the solution is unique (see Fig. C.1).

Let $Q$ be positive semidefinite:

$$Q = \begin{pmatrix} 1 & 1 \\ 1 & 1 \end{pmatrix}. \tag{C.21}$$

Because

$$L(\mathbf{x}, \boldsymbol{\alpha}) = \frac{1}{2}\left(x_1 + x_2\right)^2 - \alpha_1\left(x_1 + x_2 - 1\right) - \alpha_2\left(2 - x_1 - x_2\right), \tag{C.22}$$

the KKT conditions are given by

$$\frac{\partial L(\mathbf{x}, \boldsymbol{\alpha})}{\partial x_1} = \frac{\partial L(\mathbf{x}, \boldsymbol{\alpha})}{\partial x_2} = x_1 + x_2 - \alpha_1 + \alpha_2 = 0, \tag{C.23}$$

$$\alpha_1\left(x_1 + x_2 - 1\right) = 0, \quad \alpha_2\left(2 - x_1 - x_2\right) = 0, \tag{C.24}$$

$$\alpha_1 \geq 0, \quad \alpha_2 \geq 0. \tag{C.25}$$

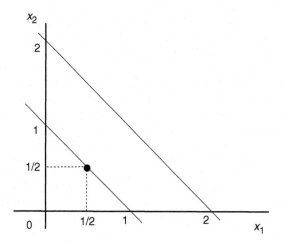

**Fig. C.1** Unique solution with a positive definite matrix

So long as $x_1 + x_2$ is constant, the value of the objective function does not change. Thus the optimal solution satisfies $x_1 + x_2 = 1$. Therefore, from (C.23) and (C.24), the optimal solution satisfies

$$x_1 + x_2 = 1, \quad \alpha_1 = 1, \quad \alpha_2 = 0. \tag{C.26}$$

Thus the solution is nonunique (see Fig. C.2).

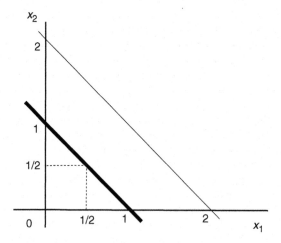

**Fig. C.2** Nonunique solution with a positive semidefinite matrix

# Appendix D
# Positive Semidefinite Kernels
# and Reproducing Kernel Hilbert Space

Support vector machines are based on the theory of reproducing kernel Hilbert space. Here, we summarize some of the properties of positive semidefinite kernels and reproducing kernel Hilbert space based on [1].

## D.1 Positive Semidefinite Kernels

**Definition D.1.** Let $K(\mathbf{x}, \mathbf{x}')$ be a real-valued symmetric function with $\mathbf{x}$ and $\mathbf{x}'$ being $m$-dimensional vectors. For any set of data $\{\mathbf{x}_1, \ldots, \mathbf{x}_M\}$ and $\mathbf{h}_M = (h_1, \ldots, h_M)^\top$ with $M$ being any natural number, if

$$\mathbf{h}_M^\top K_M \mathbf{h}_M \geq 0 \tag{D.1}$$

is satisfied (i.e., $K_M$ is a positive semidefinite matrix), we call $K(\mathbf{x}, \mathbf{x}')$ a *positive semidefinite kernel*, where

$$K_M = \begin{pmatrix} K(\mathbf{x}_1, \mathbf{x}_1) & \cdots & K(\mathbf{x}_1, \mathbf{x}_M) \\ \vdots & \ddots & \vdots \\ K(\mathbf{x}_M, \mathbf{x}_1) & \cdots & K(\mathbf{x}_M, \mathbf{x}_M) \end{pmatrix}. \tag{D.2}$$

If (D.1) is satisfied under the constraint

$$\sum_{i=1}^{M} h_i = 0, \tag{D.3}$$

$K(\mathbf{x}, \mathbf{x}')$ is called a *conditionally positive semidefinite kernel*.

From the definition it is obvious that if $K(\mathbf{x}, \mathbf{x}')$ is positive semidefinite, it is also conditionally positive semidefinite.

In the following we discuss several properties of (conditionally) positive semidefinite kernels that are useful for constructing positive semidefinite kernels.

**Theorem D.1.** *If*

$$K(\mathbf{x}, \mathbf{x}') = a, \tag{D.4}$$

*where* $a > 0$, $K(\mathbf{x}, \mathbf{x}')$ *is positive semidefinite.*

*Proof.* Because for any natural number $M$,

$$K_M = (\sqrt{a}, \ldots, \sqrt{a})^\top (\sqrt{a}, \ldots, \sqrt{a}), \tag{D.5}$$

$K(\mathbf{x}, \mathbf{x}')$ is positive semidefinite. ∎

**Theorem D.2.** *If* $K_1(\mathbf{x}, \mathbf{x}')$ *and* $K_2(\mathbf{x}, \mathbf{x}')$ *are positive semidefinite kernels,*

$$K(\mathbf{x}, \mathbf{x}') = a_1 K_1(\mathbf{x}, \mathbf{x}') + a_2 K_2(\mathbf{x}, \mathbf{x}') \tag{D.6}$$

*is also positive semidefinite, where* $a_1$ *and* $a_2$ *are positive.*

*Proof.* Because for any $M$, $h_i$, and $\mathbf{x}_i$

$$\mathbf{h}_M^\top (a_1 K_{1M} + a_2 K_{2M}) \mathbf{h}_M = a_1 \mathbf{h}_M^\top K_{1M} \mathbf{h}_M + a_2 \mathbf{h}_M^\top K_{2M} \mathbf{h}_M \geq 0, \tag{D.7}$$

$K(\mathbf{x}, \mathbf{x}')$ is positive semidefinite. ∎

**Theorem D.3.** *If* $K(\mathbf{x}, \mathbf{x}') = f(\mathbf{x}) f(\mathbf{x}')$, *where* $f(\mathbf{x})$ *is an arbitrary scalar function,* $K(\mathbf{x}, \mathbf{x}')$ *is positive semidefinite.*

*Proof.* Because for any $M$, $h_i$, and $\mathbf{x}_i$

$$\sum_{i,j=1}^{M} h_i h_j f(\mathbf{x}_i) f(\mathbf{x}_j) = \left( \sum_{i=1}^{M} h_i f(\mathbf{x}_i) \right)^2 \geq 0, \tag{D.8}$$

$K(\mathbf{x}, \mathbf{x}')$ is positive semidefinite. ∎

**Theorem D.4.** *If* $K_1(\mathbf{x}, \mathbf{x}')$ *and* $K_2(\mathbf{x}, \mathbf{x}')$ *are positive semidefinite,*

$$K(\mathbf{x}, \mathbf{x}') = K_1(\mathbf{x}, \mathbf{x}') K_2(\mathbf{x}, \mathbf{x}') \tag{D.9}$$

*is also positive semidefinite.*

*Proof.* It is sufficient to show that if $M \times M$ matrices $A = \{a_{ij}\}$ and $B = \{b_{ij}\}$ are positive semidefinite, $\{a_{ij} b_{ij}\}$ is also positive semidefinite.

Because $A$ is positive semidefinite, $A$ is expressed by $A = F^\top F$, where $F$ is an $M \times M$ matrix. Then $a_{ij} = \mathbf{f}_i^\top \mathbf{f}_j$, where $\mathbf{f}_j$ is the $j$th column vector of $F$. Thus for arbitrary $h_1, \ldots, h_M$,

$$\sum_{i,j=1}^{M} h_i h_j \mathbf{f}_i^\top \mathbf{f}_j b_{ij} = \sum_{i,j=1}^{M} (h_i \mathbf{f}_i)^\top (h_j \mathbf{f}_j) b_{ij} \geq 0. \blacksquare \qquad (\text{D.10})$$

*Example D.1.* The linear kernel $K(\mathbf{x}, \mathbf{x}') = \mathbf{x}^\top \mathbf{x}$ is positive semidefinite because $K_M = (\mathbf{x}_1, \ldots, \mathbf{x}_M)^\top (\mathbf{x}_1, \ldots, \mathbf{x}_M)$. Thus, from Theorems D.1 to D.4 the polynomial kernel given by $K(\mathbf{x}, \mathbf{x}') = (1 + \mathbf{x}^\top \mathbf{x}')^d$ is positive semidefinite.

**Corollary D.1.** *If $K(\mathbf{x}, \mathbf{x}')$ and $K'(\mathbf{y}, \mathbf{y}')$ are positive semidefinite kernels, where $\mathbf{x}$ and $\mathbf{y}$ may be of different dimensions, $K(\mathbf{x}, \mathbf{x}') K'(\mathbf{y}, \mathbf{y}')$ is also a positive semidefinite kernel.*

**Corollary D.2.** *Let $K(\mathbf{x}, \mathbf{x}')$ be positive semidefinite and satisfy*

$$|K(\mathbf{x}, \mathbf{x}')| \leq \rho, \qquad (\text{D.11})$$

*where $\rho > 0$. Then if*

$$f(y) = \sum_{i=1}^{\infty} a_i y^i \qquad (\text{D.12})$$

*converges for $|y| \leq \rho$, where $a_i \geq 0$ for all integers $i$, the composed kernel $f(K(\mathbf{x}, \mathbf{x}'))$ is also positive semidefinite.* $\blacksquare$

*Proof.* From Theorem D.4, $K^i(\mathbf{x}, \mathbf{x}')$ is positive semidefinite. Then from Theorem D.4,

$$\sum_{i=0}^{N} a_i K^i(\mathbf{x}, \mathbf{x}') \qquad (\text{D.13})$$

is positive semidefinite for all integers $N$. Therefore, so is $f(K(\mathbf{x}, \mathbf{x}'))$. $\blacksquare$

From Corollary D.2, especially for positive semidefinite kernel $K(\mathbf{x}, \mathbf{x}')$, $\exp(K(\mathbf{x}, \mathbf{x}'))$ is also positive semidefinite.

In the following we clarify the relations between positive and conditionally positive semidefinite kernels.

**Lemma D.1.** *Let*

$$K(\mathbf{x}, \mathbf{x}') = L(\mathbf{x}, \mathbf{x}') + L(\mathbf{x}_0, \mathbf{x}_0) - L(\mathbf{x}, \mathbf{x}_0) - L(\mathbf{x}', \mathbf{x}_0). \qquad (\text{D.14})$$

*Then $K(\mathbf{x}, \mathbf{x}')$ is positive semidefinite, if and only if $L(\mathbf{x}, \mathbf{x}')$ is conditionally positive semidefinite.*

*Proof.* For $\{\mathbf{x}_1, \ldots, \mathbf{x}_M\}$ and $\mathbf{h}_M = (h_1, \ldots, h_M)^\top$ with

$$\sum_{i=1}^{M} h_i = 0, \qquad (\text{D.15})$$

we have

$$\mathbf{h}_M^\top K_M \mathbf{h}_M = \mathbf{h}_M^\top L_M \mathbf{h}_M. \tag{D.16}$$

Thus, if $K(\mathbf{x}, \mathbf{x}')$ is positive semidefinite, $L(\mathbf{x}, \mathbf{x}')$ is conditionally positive semidefinite.

On the other hand, suppose that $L(\mathbf{x}, \mathbf{x}')$ is conditionally positive semidefinite. Then for $\{\mathbf{x}_1, \dots, \mathbf{x}_M\}$ and $\mathbf{h}_M = (h_1, \dots, h_M)^\top$ with

$$h_0 = -\sum_{i=1}^M h_i, \tag{D.17}$$

we have

$$
\begin{aligned}
0 \le & \sum_{i,j=0}^M h_i\, h_j\, L(\mathbf{x}_i, \mathbf{x}_j) \\
= & \sum_{i,j=1}^M h_i\, h_j\, L(\mathbf{x}_i, \mathbf{x}_j) + \sum_{i=1}^M h_i\, h_0\, L(\mathbf{x}_i, \mathbf{x}_0) + \sum_{j=1}^M h_0\, h_j\, L(\mathbf{x}_0, \mathbf{x}_j) \\
& + h_0^2\, L(\mathbf{x}_0, \mathbf{x}_0) \\
= & \sum_{i,j=1}^M h_i\, h_j\, K(\mathbf{x}_i, \mathbf{x}_j). 
\end{aligned} \tag{D.18}
$$

Therefore, $K(\mathbf{x}, \mathbf{x}')$ is positive semidefinite. ∎

**Theorem D.5.** *Kernel $L(\mathbf{x}, \mathbf{x}')$ is conditionally positive semidefinite if and only if $\exp(\gamma\, L(\mathbf{x}, \mathbf{x}'))$ is positive semidefinite for any positive $\gamma$.*

*Proof.* If $\exp(\gamma\, L(\mathbf{x}, \mathbf{x}'))$ is positive semidefinite, $\exp(\gamma\, L(\mathbf{x}, \mathbf{x}')) - 1$ is conditionally positive semidefinite. So is the limit

$$L(\mathbf{x}, \mathbf{x}') = \lim_{\gamma \to +0} \frac{\exp(\gamma\, L(\mathbf{x}, \mathbf{x}')) - 1}{\gamma}. \tag{D.19}$$

Now let $L(\mathbf{x}, \mathbf{x}')$ be conditionally positive semidefinite and choose some $\mathbf{x}_0$ and $K(\mathbf{x}, \mathbf{x}')$ as in Lemma D.1. Then for positive $\gamma$

$$\gamma L(\mathbf{x}, \mathbf{x}') = \gamma K(\mathbf{x}, \mathbf{x}') - \gamma L(\mathbf{x}_0, \mathbf{x}_0) + \gamma L(\mathbf{x}, \mathbf{x}_0) + \gamma L(\mathbf{x}', \mathbf{x}_0). \tag{D.20}$$

Thus,

$$
\begin{aligned}
\exp(\gamma L(\mathbf{x}, \mathbf{x}')) = & \exp(\gamma K(\mathbf{x}, \mathbf{x}')) \exp(-\gamma L(\mathbf{x}_0, \mathbf{x}_0)) \\
& \times \exp(\gamma L(\mathbf{x}, \mathbf{x}_0)) \exp(\gamma L(\mathbf{x}', \mathbf{x}_0)).
\end{aligned} \tag{D.21}
$$

From Theorems D.3 and D.4 and Corollary D.2, $\exp(\gamma L(\mathbf{x}, \mathbf{x}'))$ is positive semidefinite. ∎

*Example D.2.* Kernel $K(\mathbf{x}, \mathbf{x}') = -\|\mathbf{x} - \mathbf{x}'\|^2$ is conditionally positive semidefinite because for $\sum_i^M h_i = 0$,

$$\mathbf{h}_M^\top K_M \mathbf{h}_M = -\sum_{i=1}^M h_i h_j \|\mathbf{x}_i - \mathbf{x}_j\|^2$$

$$= -\sum_{i,j=1}^M h_i h_j (\mathbf{x}_i^\top \mathbf{x}_i - 2\mathbf{x}_i^\top \mathbf{x}_j + \mathbf{x}_j^\top \mathbf{x}_j)$$

$$= 2\left(\sum_{i=1}^M h_i \mathbf{x}_i\right)^\top \left(\sum_{i=1}^M h_i \mathbf{x}_i\right) \geq 0. \tag{D.22}$$

Thus, $\exp(-\gamma \|\mathbf{x} - \mathbf{x}'\|^2)$ is positive semidefinite.

## D.2 Reproducing Kernel Hilbert Space

Because a positive semidefinite kernel has the associated feature space called the *reproducing kernel Hilbert space (RKHS)*, support vector machines can determine the optimal hyperplane in that space using the kernel trick. In this section, we discuss reproducing kernel Hilbert spaces for positive and conditionally positive semidefinite kernels.

For the positive semidefinite kernels, the following theorem holds.

**Theorem D.6.** *Let $X$ be the input space and $K(\mathbf{x}, \mathbf{x}') (\mathbf{x}, \mathbf{x}' \in X)$ be a positive semidefinite kernel. Let $K_0$ be the space spanned by the functions $\{K_\mathbf{x} \mid \mathbf{x} \in X\}$ where*

$$K_\mathbf{x}(\mathbf{x}') = K(\mathbf{x}, \mathbf{x}'). \tag{D.23}$$

*Then there exist a Hilbert space $H$, which is a complete space of $K_0$, and the mapping from $X$ to $H$ such that*

$$K(\mathbf{x}, \mathbf{x}') = \langle K_\mathbf{x}, K_{\mathbf{x}'} \rangle. \tag{D.24}$$

*Here, instead of $\mathbf{x}^\top \mathbf{x}'$, we use $\langle \mathbf{x}, \mathbf{x}' \rangle$ to denote the dot product.*

*Proof.* Let $K_\mathbf{x}(\mathbf{x}') = K(\mathbf{x}, \mathbf{x}')$ and $K_0$ be a linear subspace generated by the functions $\{K_\mathbf{x} \mid \mathbf{x} \in X\}$. Then for $f, g \in K_0$ expressed by

$$f = \sum_{\mathbf{x}_i \in X} c_i K_{\mathbf{x}_i}, \tag{D.25}$$

$$g = \sum_{\mathbf{x}_j' \in X} d_j K_{\mathbf{x}_j'}, \tag{D.26}$$

we define the dot product as follows:

$$\langle f, g \rangle = \sum_{\mathbf{x}_j' \in X} d_j f(\mathbf{x}_j')$$

$$= \sum_{\mathbf{x}_i, \mathbf{x}'_j \in X} c_i\, d_j\, K(\mathbf{x}_i, \mathbf{x}'_j)$$

$$= \sum_{\mathbf{x}_i \in X} c_i\, g(\mathbf{x}_i). \tag{D.27}$$

Now we show that (D.27) satisfies the properties of the dot product. Clearly, (D.27) is symmetric and linear. Also, according to the assumption of $K(\mathbf{x}, \mathbf{x}')$ being positive semidefinite,

$$\langle f, f \rangle = \sum_{\mathbf{x}_i, \mathbf{x}_j \in X} c_i\, c_j\, K(\mathbf{x}_i, \mathbf{x}_j) \geq 0 \tag{D.28}$$

is satisfied. Here, the strict equality holds if and only if $f$ is identically zero. Thus, (D.27) is the dot product. Hence, $K_0$ is a pre-Hilbert space and its completion $H$ is a Hilbert space, which is called *RKHS associated with $K_{\mathbf{x}}$*.

From (D.27) the following reproducing property is readily obtained:

$$\langle f, K_{\mathbf{x}} \rangle = f(\mathbf{x}). \tag{D.29}$$

In particular,

$$\langle K_{\mathbf{x}}, K_{\mathbf{x}'} \rangle = K(\mathbf{x}, \mathbf{x}'). \ \blacksquare \tag{D.30}$$

For a conditionally positive semidefinite kernel, for $f \in K_0$ the following theorem holds.

**Theorem D.7.** *Let $K(\mathbf{x}, \mathbf{x}')\,(\mathbf{x}, \mathbf{x}' \in X)$ be a conditionally positive semidefinite kernel. Then there exist a Hilbert space $H$ and a mapping $L_{\mathbf{x}}$ from $X$ to $H$ such that*

$$K(\mathbf{x}, \mathbf{x}') - \frac{1}{2}K(\mathbf{x}, \mathbf{x}) - \frac{1}{2}K(\mathbf{x}', \mathbf{x}') = -\|L_{\mathbf{x}} - L_{\mathbf{x}'}\|^2. \tag{D.31}$$

*Proof.* For $\mathbf{x}_0$ we define

$$L(\mathbf{x}, \mathbf{x}') = \frac{1}{2}\left(K(\mathbf{x}, \mathbf{x}') + K(\mathbf{x}_0, \mathbf{x}_0) - K(\mathbf{x}, \mathbf{x}_0) - K(\mathbf{x}', \mathbf{x}_0)\right), \tag{D.32}$$

which is a positive semidefinite kernel from Lemma D.1. Let $H$ be the associated RKHS for $L(\mathbf{x}, \mathbf{x}')$ and $L_{\mathbf{x}}(\mathbf{x}') = L(\mathbf{x}, \mathbf{x}')$. Then

$$\|L_{\mathbf{x}} - L_{\mathbf{x}'}\|^2 = L(\mathbf{x}, \mathbf{x}) + L(\mathbf{x}', \mathbf{x}') - 2L(\mathbf{x}, \mathbf{x}')$$

$$= -K(\mathbf{x}, \mathbf{x}') + \frac{1}{2}K(\mathbf{x}, \mathbf{x}) + \frac{1}{2}K(\mathbf{x}', \mathbf{x}'). \tag{D.33}$$

Thus the theorem holds. $\blacksquare$

# References

1. C. Berg, J. P. R. Christensen, and P. Ressel. *Harmonic Analysis on Semigroups: Theory of Positive Definite and Related Functions.* Springer-Verlag, New York, 1984.

# Index